*Fritz Aldinger and
Volker A. Weberruß*

**Advanced Ceramics and
Future Materials**

Related Titles

Riedel, R., Chen, I-W. (eds.)

Ceramics Science and Technology

Volume 2: Properties

2010
ISBN: 978-3-527-31156-9

Riedel, R., Chen, I-W. (eds.)

Ceramics Science and Technology

Volume 1: Structures

2008
ISBN: 978-3-527-31155-2

Krenkel, W. (ed.)

Ceramic Matrix Composites

Fiber Reinforced Ceramics and their Applications

2008
ISBN: 978-3-527-31361-7

Wachtman, J. B., Cannon, W. R., Matthewson, M. J.

Mechanical Properties of Ceramics

2009
ISBN: 978-0-471-73581-6

Boch, P., Niepce, J.-C. (eds.)

Ceramic Materials

Processes, Properties, and Applications

2007
ISBN: 978-1-905209-23-1

Fritz Aldinger and Volker A. Weberruß

Advanced Ceramics and Future Materials

An Introduction to Structures, Properties, Technologies, Methods

WILEY-VCH Verlag GmbH & Co. KGaA

The Authors

Prof. Dr. Fritz Aldinger
Max-Planck-Institut für Metallforschung
and
Universität Stuttgart
Institut für Nichtmetallische-Anorganische
Materialien
Heisenbergstr. 3
70569 Stuttgart
Germany

Dr. Volker A. Weberruß
Max-Planck-Institut für Metallforschung
Pulvermetallurgisches Laboratorium (PML)
Heisenbergstr. 1
70569 Stuttgart
and
V.A.W. Scientific Consultation
Im Lehenbach 18
73650 Winterbach
Germany

Cover

The cover image shows a brake disc with titanium bell.
The image was kindly supplied by
SGL Brakes GmbH
Werner-von-Siemens-Str. 18
86405 Meitingen, Germany,

All books published by **Wiley-VCH** are carefully produced. Nevertheless, authors, editors, and publisher do not warrant the information contained in these books, including this book, to be free of errors. Readers are advised to keep in mind that statements, data, illustrations, procedural details or other items may inadvertently be inaccurate.

Library of Congress Card No.:
applied for

British Library Cataloguing-in-Publication Data
A catalogue record for this book is available from the British Library.

Bibliographic information published by the Deutsche Nationalbibliothek
The Deutsche Nationalbibliothek lists this publication in the Deutsche Nationalbibliografie; detailed bibliographic data are available on the Internet at http://dnb.d-nb.de.

© 2010 WILEY-VCH GmbH & Co. KGaA, Weinheim

All rights reserved (including those of translation into other languages). No part of this book may be reproduced in any form - by photoprinting, microfilm, or any other means - nor transmitted or translated into a machine language without written permission from the publishers. Registered names, trademarks, etc. used in this book, even when not specifically marked as such, are not to be considered unprotected by law.

Printing and Binding   betz-druck GmbH, Darmstadt
Cover Design   Schulz Grafik-Design, Fußgönheim

Printed in the Federal Republic of Germany
Printed on acid-free paper

ISBN: 978-3-527-32157-5

Preface

Chemical and physical basics paired with engineering "black box" concepts as well as practical and theoretical experience in conventional and advanced ceramics paired with unorthodox ideas for future materials are the humus of the book in hand. On the one hand, it is a guideline for students, scientists, and engineers, and on the other hand, it presents ideas for visionaries. It is the alliance of a view into the past, the present, and the future. Not more, not less. May the readers have exactly the same pleasure with the book in hand as we have had with its creation.

The book is based upon lectures/papers/posters that are held/published/presented at/from the Department Aldinger of the Max-Planck-Institut für Metallforschung (Germany) and the Chair Aldinger on Nichtmetallische-Anorganische Materialien of the Universität Stuttgart (Germany) during the past decades – joint institutions which became also known as "Pulvermetallurgisches Laboratorium (PML)". A selection of these works is accumulated in [76]–[189]. Some of the knowledge of [76]–[189] is used as "case studies" in the book. We truly regret that it was not possible to include all works that were done at the "Pulvermetallurgisches Laboratorium (PML)" during that time. Nevertheless, we hope that most of the readers consider the selection of works presented in this book satisfactory. Furthermore, we hope that most of the readers consider our decision to discuss ceramic materials within an extended frame of reference, which also incorporates metallic materials and organic materials, satisfactory. Moreover, we hope that at least some of our readers agree with us that it is a good time to carry out first steps towards future materials paving the way to artifical gravitation technologies and energy production technologies that go beyond chemical and nuclear reactions – exactly this is what we want to do in the final chapter of the book.

We wish to thank Dorothee Klink for the proofreading, Bernd Heinze for supplying us with nice photographs, CeramTec [71] for the abandonment of product images, microstructure images, and pictures of technical facilities, and SGL Brakes GmbH [72] for the release of images of ceramic (carbon) brake discs. Furthermore, we would like to thank Joachim Bill and Jerzy Golczewski for their valuable help. Moreover, the coauthor would like to thank Arndt Simon and Jürgen Köhler for the stimulating time at the Department Simon of the Max-Planck-Institut für Festkörperforschung.

Fritz Aldinger und Volker A. Weberruß,
Stuttgart, January 2010.

Prof. Dr. Fritz Aldinger

Max-Planck-Institut für Metallforschung
and
Universität Stuttgart, Institut für Nichtmetallische-Anorganische Materialien
Heisenbergstr. 3, D-70569 Stuttgart

Dr. rer. nat. Volker A. Weberruß

Max-Planck-Institut für Metallforschung
Pulvermetallurgisches Laboratorium (PML)
Heisenbergstr. 3, D-70569 Stuttgart
and
V.A.W. scientific consultation
Im Lehenbach 18, D-73650 Winterbach

Max-Planck-Campus, Stuttgart (Germany)

Max-Planck-Institut für Metallforschung,
Pulvermetallurgisches Laboratorium (PML)

Silicon carbide materials.

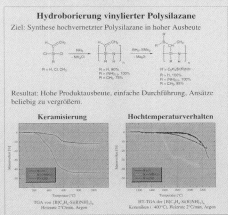

Precursor-derived materials.

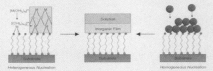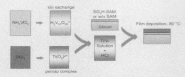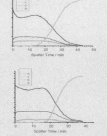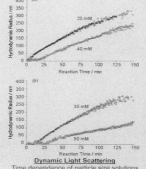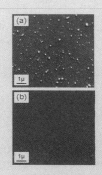

Bio-inspired materials.

Contents

Preface V

Defining Ceramics — XIV

1. **Advanced Ceramics** 1

Understanding Ceramics — 14

2. **Structures** 15
   2.1 *Chemical Structures* 16
      2.1.1 Ionic Bond 22
      2.1.2 Covalent Bond 29
      2.1.3 Van der Waals Bond 47
      2.1.4 Metallic Bond 49
   2.2 *Crystal Structures* 59
      2.2.1 Lattice Types 64
      2.2.2 Crystal Types 74
      2.2.3 Defect Types 111
   2.3 *Microstructures* 143
      2.3.1 Microstructure Features 143
      2.3.2 Microstructure Design 150
      2.3.3 Microstructure Visualization 157

Observing Ceramics — 170

3. **Properties** 171
   3.1 *Chemical Properties* 171
      3.1.1 Phase Equilibria 171
      3.1.2 Oxygen Environments 199
      3.1.3 Technical Environments 199
      3.1.4 Biological Environments 199
   3.2 *Physical Properties* 210
      3.2.1 Thermal Properties 210
      3.2.2 Electromagnetic Properties 218
      3.2.3 Optical Properties 236

	3.3	*Mechanical Properties*	238
		3.3.1 Elastic Properties	238
		3.3.2 Plastic Properties	248
		3.3.3 Strength and Fracture	260
	3.4	*Special Issues*	272
		3.4.1 Superplasticity	272
		3.4.2 Superconductivity	272
		3.4.3 Bond Sensitivity	326

Manufacturing Ceramics — 334

4.	**Technologies**	335
	4.1 *Powder-Based Technologies*	341
	4.1.1 Basic Procedures	341
	4.1.2 Advanced Procedures	372
	4.2 *Powder-Free Technologies*	380
	4.2.1 Chemical Procedures	380
	4.2.2 Vapour Deposition Procedures	398
	4.2.3 Bio-Inspired Mineralisation Procedures	408

Pointing into the Future of Materials — 428

5.	**Future Materials**	429
	5.1 *Advanced Conceptions*	430
	5.1.1 Advanced Modeling of Particles/Spins	430
	5.1.2 Advanced Modeling of Substances/Materials .	445
	5.1.3 Advanced Modeling of Threshold Ranges	459
	5.2 *Advanced Applications*	488
	5.2.1 Artifical Gravitation Technologies	489
	5.2.2 Energy Production Technologies	489

Bibliography 491

Index 501

Defining Ceramics

1. Advanced Ceramics

What would the ancient China have been without china? Imperfect! What would be the life of today without dental prostheses? Painful! What would be the space shuttle without thermal tiles? Disastrous! What would be the daily communication without mobile phones? Unimaginable! Is it thus carrying things too far that ceramic materials are objects of technologies of the past, the present, and the future? No way! But what are ceramic materials actually? Which structure do they have? How are their properties related to their structure? How are these manufactured? How are these modeled? Going somewhat beyond the conventional textbooks and the conventional monographs on modern ceramics, in the chapters that follow, an elaborate picture of modern ceramics is developed.

Going beyond the hellenic word *keramos* ("fired soil"), on the one hand, *ceramics* is defined as a name for products made out of non-metallic inorganic substances, and on the other hand, *ceramics* is defined as the art and science of making materials and products of non-metallic inorganic substances [28]. Naturally, following the standards of modern physics and modern chemistry, we want to associate the notion *non-metallic* with non-metallic energy band structures and their dependence on further constraints such as temperature and pressure, and we want to associate the notion *inorganic* with substances not showing the chemical structures of hydrocarbons. Certainly, we want to appreciate *conventional ceramics*, but we want to focus on *advanced ceramics*. We here compare with Figure 1.1, which shows the classifications, the definitions, and the terminology used in this book, and we here compare with Figures 1.2–1.13, which show some selected starting materials as well as some selected products of various branches of the ceramic industry.

We should already here appreciate that advanced ceramics meets the highest demands of present technologies. For instance, many ceramic materials are extremly resistant against abrasion! For instance, many ceramic materials are extremly resistant against heat! For instance, depending on further constraints such as temperature and pressure, ceramic materials can be insulators, ceramic materials can be conductors, and ceramic materials can be semi-conductors. Dear reader, ceramic materials even can be ferroelectrics without which the performance of present communication technology would be more than poor. Dear reader, the hip joint endoprostheses that are shown in Figure 1.10 are implants meanwhile doing their job in hundreds of millions of human bodies worldwide! Isn't this a feat?

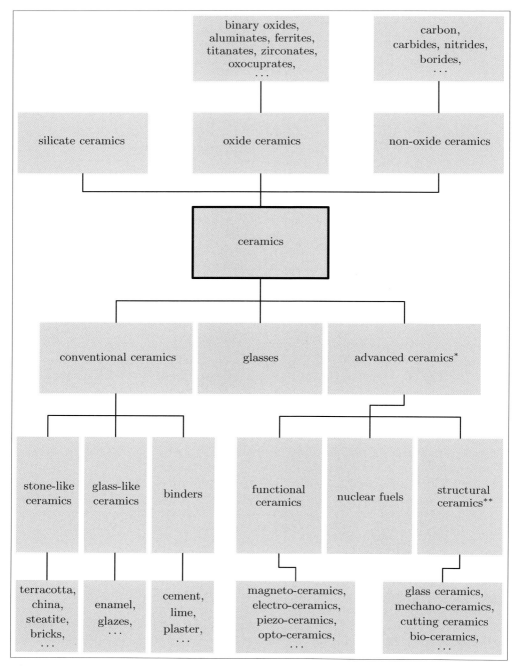

Figure 1.1. Classifications, definitions, and terminology. * frequently noted as "fine ceramics" (Japan) or "Hochleistungskeramik" (Germany). ** frequently noted as "engineering ceramics" (USA, GB) or "Ingenieurkeramik" (Germany).

Figure 1.2. Mineral stones and powdered artifical non-metallic inorganic materials.

4   1. Advanced Ceramics

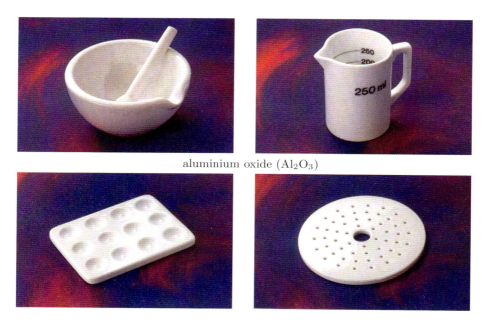

aluminium oxide (Al_2O_3)

Figure 1.3. Laboratory ware.

images kindly provided by CeramTec

magnesium silicate
($Mg_3Si_4O_{10}$ hydroxide: steatite)

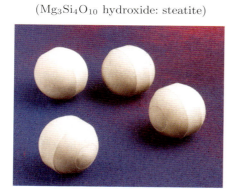

aluminium oxide (Al_2O_3)

Figure 1.4. Grinding balls.

Tribofil® (CeramTec)

Figure 1.5. Textile machinery components. Pigtails (top) and oiler guides (bottom).

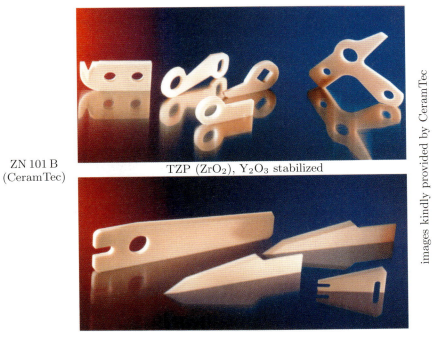

ZN 101 B (CeramTec) TZP (ZrO$_2$), Y$_2$O$_3$ stabilized

Figure 1.6. Ceramic cutters. Scissors (top) and knives (bottom).

Figure 1.7. Components for industrial machinery and industrial facilities. SSiC: sintered SiC. SiSiC: Si-infiltrated SiC.

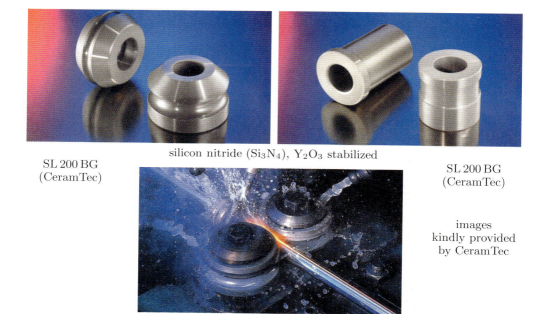

Figure 1.8. Welding rollers (top) and their application (bottom). The lower picture shows a facility for pipe welding.

1. Advanced Ceramics   7

SL 200 BG
(CeramTec)

liquid-phase sintered
silicon nitride (Si_3N_4)

images kindly provided by CeramTec

Figure 1.9. Centering pins for welding processes (top) and their application (bottom). The lower picture shows a facility for metal active gas welding.

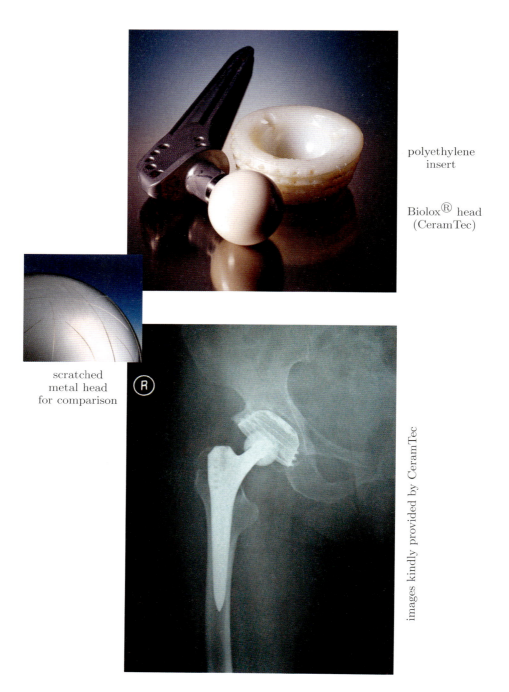

Figure 1.10. Hip joint endoprosthesis (top) and its application (bottom).

image kindly provided
by FHG/IWM [191]

Figure 1.11. Ball (roller) bearings.

brake disc with stainless steel bell

images kindly provided by SGL Brakes GmbH

brake disc with titanium bell

Figure 1.12. Ceramic (carbon) brake discs.

taken from the VAWsc picture archive

Figure 1.13. PC motherboard with chip carriers.

Going beyond advanced ceramics, in the final Chapter 5, first steps towards new fields of materials research are carried out. The following first introductory remarks already here should be useful. On the one hand, a branch of research is started up dealing with the question "How must materials be structured and how must their technical environment be structured that these can be used as the material basis for an oscillatory system generating a measurable oscillatory gravitational field, so to speak, as the material basis for a gravitational laser?" On the other hand, a branch of research is started up dealing with the question "How must materials be structured and how must their technical environment be structured that these – fundamentally going beyond nuclear techniques and matter–antimatter destruction – can be disaggregated into radiation, which in turn can be used as starting point for new types of energy plants, new types of spacecraft propulsions, and new types of matter conversion techniques?" For this purpose, a self-consistent network of model conceptions is launched, which some colleagues may consider as a direct extension of the Ginzburg–Landau theory of superconductivity [20] or as a semi-classical approach to quantum systems comparable to the Wunner–Main approach to quantum systems [34, 35, 64], however, which could turn out to be the true nonlinear extension of conventional quantum mechanics [63], on all accounts, which works without any known restrictions, and which firstly can supply us with much more information about the principles that govern the constitution, the stabilities, and the instabilities of materials than the methods of thermodynamics and the methods of quantum mechanics can do, and which secondly can supply us with paths to these new fields of materials research. Certainly, it goes far beyond the scope of this book to develop these new fields of research in great detail already here. However, the first steps towards this goal indeed are taken here.

Why do we think that a book about advanced ceramics is the adequate place for such first steps? Well, on the one hand, in the framework of advanced fuels such as nuclear fuel rods, ceramic materials play an important role so that we expect that ceramic materials will also play an important role in our more advanced context. Well, on the other hand, wanting to gain an access to artifical gravitation, it should not be the worst choice to depart from materials with a relatively high mass density and a relatively low free charge carrier density, promising us high gravitational radiation and low electromagnetic radiation, as it is the case for a lot of ceramic materials. Anyway, departing from advanced ceramics of the past and the present, let us also point into the future of materials.

Understanding Ceramics

2. Structures

Analyzing the structural landscapes of the forms of appearance of matter, one discovers that each system mathematically and technically can be decomposed in subsystems, which in turn also can be decomposed in subsystems etc., defining inner hierarchies of the forms of appearance of matter. This is so in the domain of quantum objects and in the domain of cosmic objects. This is also so in the domain of the animated nature and in the domain of the inanimate nature. And no doubt this is also so in the domain of materials. Focussing on ceramic materials, let us here study the inner composition of ceramics, namely their *chemical bond*, *crystal structure*, and *microstructure*.

Ranging from hot plasmas via fuel cells and liquid crystals to cold superconductors, we nowadays know a wealth of forms of appearance of matter. Fundamental notions important in this context are the notions *gaseous*, *liquid*, and *solid*. It is well-known that in the gas phase, a substance takes the total space available. It is also well-known that in the liquid phase, the substance maintains the volume, but the shape of the volume depends on the shape of the space available, and the atoms or molecules are surrounded by other atoms or molecules, which are connected by van der Waals interactions or bridge (H, O) interactions, leading to short-range order. And it is also well-known that in the solid phases, the substance maintains the volume as well as the shape of the volume, and we speak of *cristalline solids* if long-range order is detectable, but we speak of *amorphous solids* if only short-range order is detectable. Connecting these states of matter – in the present context termed *phases* –, there are phase transitions depending on further quantities such as *temperature* and *pressure*.

Equations of state such as the *ideal gas law*, which combines the state quantities pressure, volume, and temperature, and the *van der Waals equation*, which includes correction terms that measure the forces between the particles and the real volume of the particles, make possible quick access to the behavior of gases. Equations of motion such as the *Navier–Stockes equations*, which especially includes viscosity properties, and the *Fokker–Planck equation*, which especially includes diffusion properties, make possible quick access to the static/dynamic behavior of liquids. Equations of motion such as the *Maxwell equations*, which relate currents to electromagnetic fields, and the *London equations*, which relate currents to electromagnetic fields in the special case of superconducting materials, make possible quick access to the static/dynamic behavior of solids. However, let us here emphasize that in some special cases the methods of quantum mechanics are the better choice. Therefore, in the sections that follow, let us consider the methods of quantum mechanics, too.

2.1 Chemical Structures

Space and time and properties such as color and flavor or mass and charge are mutually conditional, i.e. space and time cannot exist without matter and *vice versa*. Physical properties such as the named ones adjust oneself to each other, and this is ascertainable by interaction principles and interaction terms. Certainly, exclusively dealing with practical materials such as advanced ceramics, we are neither interested in the strong interaction and the weak interaction, which dominate elementary processes, nor in the gravitational interaction, which dominates cosmic processes. But we are interested in the electromagnetic interaction in all its varieties, in particular, in the context of chemical bonds between atoms. The explanations that follow may supply the reader with an insight into the notion *chemical bond*.

The mainspring for the evolution of solids is the pursuit of relatively low energy. Following this driving force, chemical bonds come into being. The remarks that follow might be helpful for the understanding of chemical bonds. Gases, liquids, and solids show a broad spectrum of electron states generating the chemical bonds. The notions *ionic (heteropolar) bond*, *covalent (homopolar) bond*, *van der Waals (molecular) bond*, and *metallic bond* are of particular importance. Dealing with polymers and glasses, also the notion *bridge bond* is of particular importance. We here note that if electrons are transferred from one atom or molecule to another one, the bond state is termed *ionic*. A crystal that is based upon this type of bond state is termed *ionic*, too. We here also note that if electrons of a pair of atoms/molecules are not transferred from one partner to the other, but are shared *pari passu*, the bond state is termed *covalent*. A crystal that is based upon this type of bond state is termed *covalent*, too. We here already note that the *van der Waals bond inter alia* is based upon electric fluctuations around the point of electric neutrality, inducing dipole–dipole interactions within gases, liquids, and solids. We here already note that the *metallic bond* is based upon electrons in a special energy domain spanned by *energy bands*, leading to attractive forces between the components of metallic solids. We here note, too, that the term *bridge bond* describes the fact that atoms or molecules *inter alia* are linked via hydrogen or oxygen atoms. We compare with Figure 2.1, which explains these notions by example.

Most of the chemical compounds are known to reside somewhere in between these notions, i.e. their bond state is a mixture of bond states reviewed above. We compare with Figure 2.2, which presents a schematic representation of this fact for the four main bond types. (i) Idealized representatives of the four main bond types residing at the corners of the bond tetrahedron are the examples that are shown in Figure 2.1, namely diamond (C), CsF, Ar, and Na. Elements and compounds revealing two, three, or all four bond types are consequently located at the edges, at the faces, or in the volume of the bond tetrahedron. (ii) The examples along the edge between the notions *ionic bond* and *covalent bond* reflect the Pauling system describing the ionic character of inorganic compounds by means of the *electronegativity difference* $\Delta\chi$ of the bond partners, i.e. by using the *electronegativity* χ measuring the ability of atoms/molecules to attract electrons, we also compare with Figure 2.2. (iii) Consistent with our definition of ceramics, in this figure, we only show solid phases of the chemical compounds.

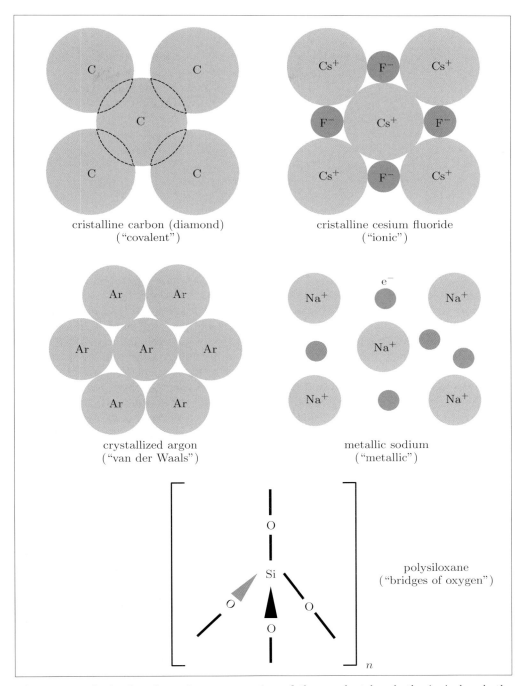

Figure 2.1. A simple schematic representation of the covalent bond, the ionic bond, the van der Waals bond, the metallic bond, and the bridge bond for special examples.

It is especially worth knowing that oxidic ceramic materials in most cases are ionic with covalent contributions and in some cases are covalent with ionic contributions. ZrO_2 and SiO_2 are examples of the first case, whereas B_4C is an example of the second case. It is worth knowing that very much stronger covalent contributions are observable for the nitrides and the carbides of the chemical elements of the main groups III–V of the periodic table of the elements. Si_3N_4 is an example of the first case, whereas SiC is an example of the second case. It is worth knowing that chemical elements which are neighbors in the periodic table of the elements lead to compounds with small value for $\Delta\chi$. Naturally, a bigger electronegativity difference $\Delta\chi$ leads to a bigger ionic character. Naturally, B and C, which are direct neighbors in the periodic table of the elements, lead to a compound with small value for $\Delta\chi$ (i.e. with a more covalent character), but Mg and O, which are not direct neighbors in the periodic table of the elements, lead to a compound with a big value for $\Delta\chi$ (i.e. with a more ionic character). It is worth knowing that SiC, besides its covalent and ionic character and consistent with its electric conductivity, also shows a metallic character. SiC thus is an example for compounds located on the face between the covalent, the ionic, and the metallic bond, whereas layered silicates are examples for compounds located in the volume between the covalent, the ionic, the metallic, and the van der Waals bond.

It is especially worth knowing that the formal classification "ceramic materials" and "metallic materials" of Figure 2.2 reflects our definition of "ceramics". In particular, gallium arsenide (GaAs), according to its energy band structure, under normal pressure is a semi-conductor without any metallic and semi-metallic character, according to our definition of "ceramics", then representing a ceramic material, and under high pressure is a semi-metal, according to our definition of "ceramics", then not representing a ceramic material. In particular, argon (Ar) shows solid non-metallic phases, according to our definition of "ceramics", then representing a ceramic material. We here firstly annotate that the rectangles point out materials that according to newer *and* older definitions including our definition, which is based upon the notion "band structure", are "metallic". We here secondly annotate that the circles point out materials that according to newer *and* older definitions including our definition are "ceramic". We here thirdly annotate that the triangles point out materials that according to our definition also have to be included when we speak of "ceramic materials".

Important coordination geometries (cg) and important coordination numbers (cn) are presented in Figure 2.3. Certainly, in crystals of compounds in the first instance the coordination numbers 4 and 6 are observable. However, in special cases also the coordination numbers 2, 3, and 8, and in very special cases also 12, are observable. Covalent crystals, according to the main group position and the therewith associated electronic structure of the constituting elements, exhibit the coordination number 4. Ionic crystals, according to the main group position and the therewith associated relative size and electrical charge of the constituting elements, however, exhibit the coordination number 6. In the case of crystals with more or less ionic character, which starting from the end of these introductory explanations are also called "ionic crystals", further coordination numbers have to be taken into consideration depending on the relative size and electrical charge of the constituting elements.

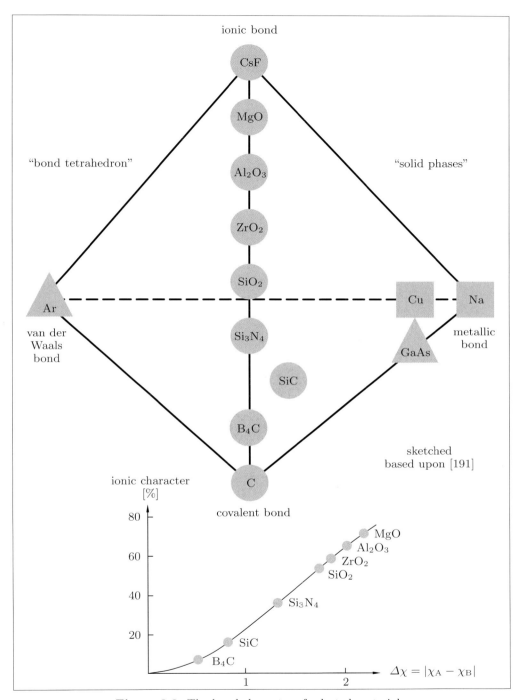

Figure 2.2. The bond character of selected materials.

We here take notice that such notions are simplifications of the reality, so to speak, such notions are "black boxes" for students, engineers, and material scientists. We here take notice that the methods of quantum mechanics absolutely do not need such simplifications of the reality. Above all, these methods neither do need the notion of a covalent bond nor do need the notion of an ionic bond. In lieu thereof, these methods introduce discrete, in special cases also continuous sets of energy states associated with wavefunctions, together fixing the wave properties and the particle properties of the electrons and the nuclei. In particular, the symmetry properties of such a wavefunction define the coordination geometry of the system. In particular, electron contributions to such a wavefunction can be localized around a nucleus or can be localized between two nuclei, fixing the location of the electrons associated with the electron contributions, in this manner, replacing the simplified notions of electrons that are totally shifted ("ionic bond") and electrons that are shared *pari passu* ("covalent bond").

Henceforth, we want to apply the term "ionic" also in the case of crystals with more or less ionic character. In doing so, the following statements can be made. (i) Modeling ionic crystals, a cation–anion radius ratio model is applied, and this means to describe anions (cations) as spheres with well-defined radii of ions, depending on the cation–anion radius ratio, forming a linear, triangular, tetrahedral, octahedral, cubic etc. arrangement of anions (cations) the void of which is spaciously filled with a central cation (anion). (ii) Modeling covalent crystals, a simple sphere model is applied, and this means to describe atoms/molecules as spheres with well-defined radii, forming a linear, triangular, tetrahedral, octahedral, cubic etc. arrangement of atoms/molecules the void of which is spaciously filled with a central atom/molecule. (iii) It is important to know that if we are in the domain of bonds with dominant covalent contributions, a hybrid orbital model is applied, too, and this means to construct bonding orbitals as superpositions of atomic/molecular orbitals, depending on the bond orientations, also forming a linear, triangular, tetrahedral, octahedral, cubic etc. arrangement of atoms/molecules the void of which is spaciously filled with a central atom/molecule. (iv) It is important to know that in the domain of bonds with dominant ionic contributions the notion of charged particles attracting or repelling each other as well as the notion of *undirected bonds* fixed by the required space is valuable and that in the domain of bonds with dominant covalent contributions the notion of *directed bonds* fixed by the directions of the axes of the "electron clouds" is valuable.

We here take notice that structures of complex ionic crystals obey *Pauling's rules* [42]. (1) Around each cation, a coordination polyhedron of anions is formed. The sum of radii of ions determines the ion distance, i.e. the distance between cations and anions. The radius ratio determines the coordination number. (2) The charge of an anion is approximately or exactly equal to the negative sum of the electrostatic bond strengths of the cations that surround the anion. (3) The stability of an ionic structure decreases if edges and particularly faces are shared by two anion polyhedra. For cations with high charge and low coordination number, this effect is particularly large. (4) In a crystal with different kinds of cations, those of high valency and small coordination number tend not to share polyhedrons with one another. (5) The number of different kinds of constituents in a crystal tends to be small.

cation–anion radius ratio model
$r_c/r_a|_{stable} > 0$

a hybrid orbital model is also known

cn = 2
cg = linear

cation–anion radius ratio model
$r_c/r_a|_{stable} \geq 0.155$

sp² hybrid orbital model
(for example, graphite)

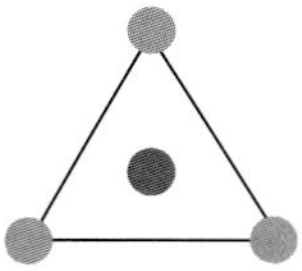

cn = 3
cg = triangular

cation–anion radius ratio model
$r_c/r_a|_{stable} \geq 0.225$

sp³ hybrid orbital model
(for example, diamond)

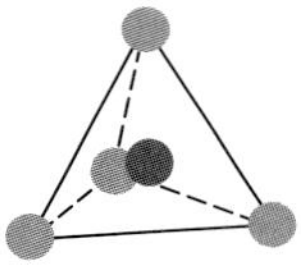

cn = 4
cg = tetrahedron

cation–anion radius ratio model
$r_c/r_a|_{stable} \geq 0.414$

d²sp³ hybrid orbital model
(for example, SF$_6$)

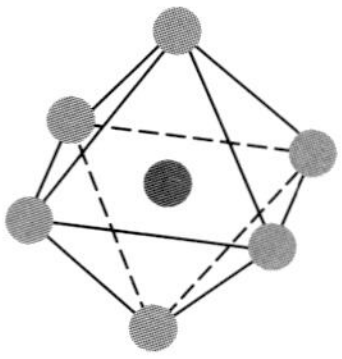

cn = 6
cg = octahedron

cation–anion radius ratio model
$r_c/r_a|_{stable} \geq 0.732$

a hybrid orbital model is also known

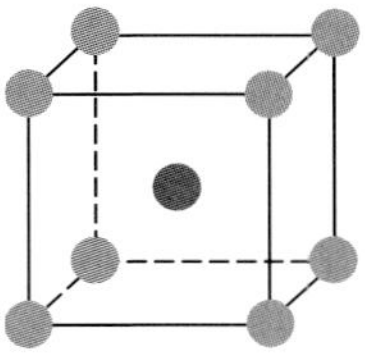

cn = 8
cg = cube

Figure 2.3. Important coordination geometries (cg) and coordination numbers (cn).

2.1.1 Ionic Bond

Making use of Coulomb's law and resorting to the notion of "point masses" with the property "point charges", a simplified model for the computation of the bond energy of bond partners showing distinctive ionic characteristics is easily derived. Moreover, a simplified model for the computation of the lattice energy (lattice enthalpy) of a crystal showing distinctive ionic characteristics is easily derived in almost the same manner by taking all attractive and all repulsive contributions of the crystal into account.

Resorting to the notion of "point masses" with the property "point charges", the Coulomb energy E_C of two ions with charge Ze and $Z'e$, respectively, is given by (2.1) of Box 2.1, with $\boldsymbol{F}_C$ being the Coulomb force, r being the distance of the two centers of the two ions, and ϵ_0 being the permittivity of the vacuum, also called absolute permittivity or dielectric constant. Focussing on two oppositely charged ions, i.e. $ZZ' = -|ZZ'|$, the Coulomb energy E_C passes into the energy of attraction E_a that is given by (2.2). Focussing on two oppositely charged ions, i.e. $ZZ' = -|ZZ'|$, the bond energy E_b that is given by (2.4) results, where the energy of repulsion E_r, which in most cases can be modeled as (2.3), originates from the atomic cores of the two ions, i.e. the nuclei and those electrons of complete shells which cannot be excited optically or chemically evoke an energy of repulsion E_r which in most cases can be modeled as (2.3). In Figure 2.4, this is illustrated for potassium chloride (KCl), where $Z = -Z'$ and $|ZZ'| = Z^2 = 1$. As it is outlined in Figure 2.4, the bond energy of two partners of an ionic bond corresponds to the point of equilibrium, which is nothing else but the minimum of the resulting energy graph. As it is outlined in Figure 2.4, the distance of equilibrium r_0 for KCl is $r_0 = 2.79\,\text{Å} = 0.279\,\text{nm}$ and the bond energy E_b for KCl is $E_b = -4.4\,\text{eV}$, assuming a space lattice characterized by the coordination number 6. We here note in passing that the decay constant ρ follows from elastic constants, that the formation of an ionic bond always is an exothermic process, and that the exothermic property is directly reflected by the negative sign of the bond energy E_b.

The reader should take notice that the considerations above certainly are valid for isolated K$^+$ and Cl$^-$ ions, but the values $r_0 = 0.279\,\text{nm}$ and $E_b = -4.4\,\text{eV}$ are valid for K$^+$ and Cl$^-$ ions of an ionic crystal arranged in an octahedral coordination geometry associated with the coordination number 6, and such a surrounding has an influence on ionic parameters such as the ion radii and the bond energies. We compare with Figure 2.5, where this and other ionic surroundings are illustrated by means of the cation–anion radius ratio model. We note that in this cation–anion radius ratio model the ions of an ionic crystal are described as spheres imposing non-directional forces on other ions, depending on the cation–anion radius ratio, leading to a specific dense arrangement of ligands. We also note that the stability of an ionic crystal increases when the volume of the central ions increases so that the ligands are kept at distance to each other. We further note that the stability of an ionic crystal decreases when the volume of the central ions decreases. However, in the latter case, following the principle of minimization of energy and thus the principle of balance of forces, the central ions are moved out of the center. For example, this is the origin of the ferroelectric domains of titanates, we compare with Figure 3.34.

> **Box 2.1** (Important formulae: bond energy, lattice energy (lattice enthalpy)).
>
> Bond energy of a pair of ions:
>
> $$E_\text{C} = \int_\infty^r \boldsymbol{F}_\text{C}\,\mathrm{d}\boldsymbol{s} = -\frac{1}{4\pi\epsilon_0}\int_\infty^r \frac{ZZ'e^2}{r'^2}\mathrm{d}r' = \frac{1}{4\pi\epsilon_0}\frac{ZZ'e^2}{r}\,, \tag{2.1}$$
>
> $$E_\text{a} = -\frac{1}{4\pi\epsilon_0}\frac{|ZZ'|e^2}{r}\,, \tag{2.2}$$
>
> $$E_\text{r} \propto +\exp(-r/\rho) \text{ or } E_\text{r} \propto +\frac{1}{r^n}\quad (n \approx 9)\,, \tag{2.3}$$
>
> $$E_\text{b} = E_\text{a}|_{r_0} + E_\text{r}|_{r_0}\,, \tag{2.4}$$
>
> $$E_\text{a}|_{r_0} = -\frac{1}{4\pi\epsilon_0}\frac{|ZZ'|e^2}{r_0}\,,\ E_\text{r}|_{r_0} \propto +\exp(-r_0/\rho) \text{ or } E_\text{r}|_{r_0} \propto +\frac{1}{r_0^n}\quad (n\approx 9)\,. \tag{2.5}$$
>
> Lattice energy (lattice enthalpy) of an ionic crystal:
>
> $$E_\text{iil} = +\sum_{i=1}^N \frac{1}{4\pi\epsilon_0}\frac{ZZ_i e^2}{r_0}\frac{n_i}{c_i} + n_1\,E_\text{r}|_{r_0}\,, \tag{2.6}$$
>
> $$E_\text{iil} = -\sum_{i=1}^N \frac{1}{4\pi\epsilon_0}\frac{|ZZ'|e^2}{r_0}\frac{n_i}{c_i}z_i + n_1\,E_\text{r}|_{r_0}\,, \tag{2.7}$$
>
> $$E_\text{iil} = \alpha\,E_\text{a}|_{r_0} + n_1\,E_\text{r}|_{r_0}\,,\quad \alpha = \sum_{i=1}^N \frac{n_i}{c_i}z_i\,, \tag{2.8}$$
>
> $$E_\text{l} = \nu N_\text{A}\left[\alpha\,E_\text{a}|_{r_0} + n_1\,E_\text{r}|_{r_0}\right]\,,\quad \alpha = \sum_{i=1}^{N\to\infty}\frac{n_i}{c_i}z_i\,. \tag{2.9}$$
>
> Madelung constant of NaCl:
>
> $$E_\text{l,NaCl} = \nu N_\text{A}\alpha_\text{NaCl}\,E_\text{a,NaCl}|_{r_0} + \nu N_\text{A}n_1\,E_\text{r,NaCl}|_{r_0}\,, \tag{2.10}$$
>
> $$E_\text{a,NaCl}|_{r_0} = -\frac{1}{4\pi\epsilon_0}\frac{e^2}{r_0}\,, \tag{2.11}$$
>
> $$n_1\,E_\text{r,NaCl}|_{r_0} \propto 6\exp(-r_0/\rho) \text{ or } n_1\,E_\text{r,NaCl}|_{r_0} \propto 6\frac{1}{r_0^n}\quad (n\approx 9)\,,$$
>
> $$\alpha_\text{NaCl} = \sum_{i=1}^{N\to\infty}\frac{n_i}{c_i}(-1)^{i+1} = 6 - \frac{12}{\sqrt{2}} + \frac{8}{\sqrt{3}} - \frac{6}{\sqrt{4}} + \frac{24}{\sqrt{5}} - \cdots \approx 1.748\,. \tag{2.12}$$

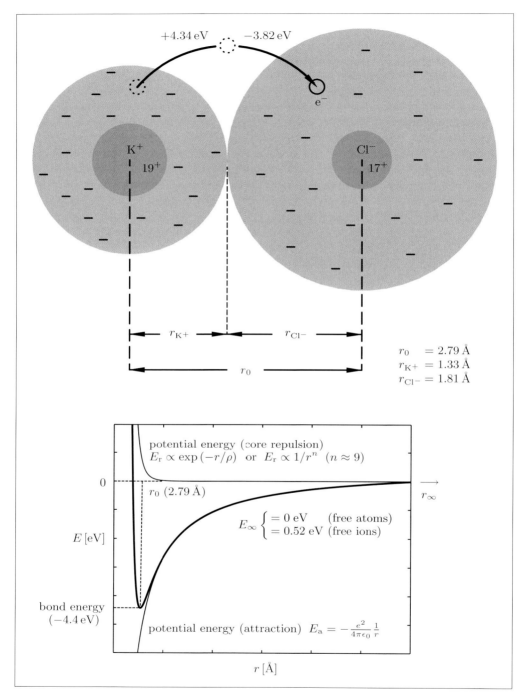

Figure 2.4. Ionic bond. Coulomb model of potassium chloride (KCl).

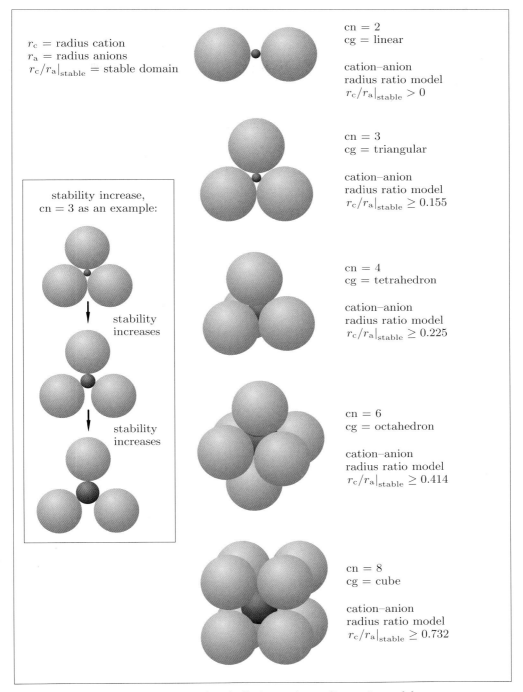

Figure 2.5. Ionic bond. Cation–anion radius ratio model.

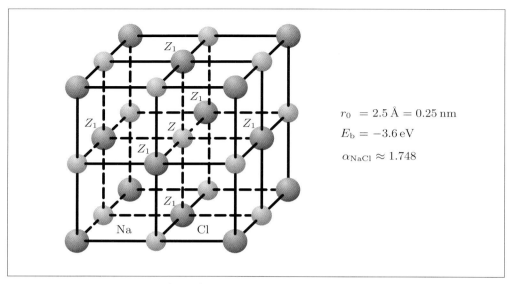

Figure 2.6. Sodium chloride (NaCl). Individual ion (Ze) and first coordination sphere ($Z_1 e$).

$$r_0 = 2.5\,\text{Å} = 0.25\,\text{nm}$$
$$E_b = -3.6\,\text{eV}$$
$$\alpha_{NaCl} \approx 1.748$$

The reader surely knows that the first derivative of the resulting potential energy defines the resulting force F that acts between the two bond partners, while the second derivative of the resulting potential energy defines the bond stiffness S. For covalent crystals, we typically observe $S \approx 20\text{–}200\,\text{N/m}$. For ionic crystals, we typically observe $S \approx 15\text{–}210\,\text{N/m}$. For van der Waals crystals, we typically observe $S \approx 0.2\text{–}5\,\text{N/m}$.

Table 2.1. The Madelung constant α of ionic crystals.

crystal type	compound type	α
α–ZnS (zinc blende)	AX	1.638
β–ZnS (wurtzite)	AX	1.641
cesium chloride	AX	1.763
sodium chloride (rock salt)	AX	1.748
calcium fluoride	AX$_2$	2.52
titanium oxide (rutile)	AX$_2$	4.816
α–Al$_2$O$_3$ (corundum)	A$_2$X$_3$	4.17

Table 2.2. Lattice energy (lattice enthalpy). 25°C. On the influence of the radius.

crystal type	compound	cation radius (alkali metal) [pm]	$|E_1|$ [kJ/mole]
lithium fluoride	LiF (Li$^+$, F$^-$)	74	1039
sodium fluoride	NaF (Na$^+$, F$^-$)	102	920
potassium fluoride	KF (K$^+$, F$^-$)	138	816
rubidium fluoride	RbF (Rb$^+$, F$^-$)	149	780
cesium fluoride	CsF (Cs$^+$, F$^-$)	170	749

Considering wide-stretched, three-dimensional assemblies of bond partners, we speak of *lattice energy (lattice enthalpy)*. Following the above lines, we construct a simplified model for the computation of the lattice energy (lattice enthalpy) as follows. In a first step, we construct the energy of equilibrium of a central individual ion of the crystal under the influence of the remaining ions of the crystal according to (2.6). The first part describes the attraction and repulsion energies caused by the remaining ions, where $i = 1 \ldots N$ counts the coordination spheres around the individual ion that complete the crystal, and where n_i, Z_i, and c_i describe that the ith coordination sphere is occupied with n_i ions of the same charge $Z_i e$ and is placed at intervals of $r_0 c_i$ around the individual ion, we also compare with Figure 2.6. The second part describes the repulsion effects caused by the cores of the direct neigbours of the individual ion, i.e. by the n_1 ions of the first coordination sphere, we also compare with Figure 2.6. In a second step, playing back this situation to the above situation, we additionally set $ZZ_i = -|ZZ'|z_i$, where the first contribution $(-|ZZ'|)$ implements the above basis, i.e. oppositely charged two ions, and the second contribution (z_i) manages changes of the charge quantum number from coordination sphere to coordination sphere including the $\pm$ sign such that the resulting equation (2.7) and the original equation (2.6) are completely identical. In a third step, introducing the *Madelung constant* α, we are led to (2.8). In a fourth step, multiplying with the Avogadro constant N_A and the amount of substance (mole number) ν, and setting $N \to \infty$, we are led to (2.9), which enables us to calculate the lattice energy (lattice enthalpy) of wide-stretched, three-dimensional assemblies of bond partners by using basic lattice and substance parameters.

Table 2.3. Lattice energy (lattice enthalpy). 25°C. On the influence of the charge.

crystal type	compound	$|E_1|$ [kJ/mole]
sodium chloride	NaCl (Na$^+$, Cl$^-$)	780
sodium sulfide	Na$_2$S (Na$^+$, S^{2-})	2207
magnesium chloride	MgCl$_2$ (Mg^{2+}, Cl$^-$)	2502
magnesium sulfide	MgS (Mg^{2+}, S^{2-})	3360
aluminum oxide	Al$_2$O$_3$ (Al^{3+}, O^{2-})	15157

Table 2.4. Ionic radii (ir, $[10^{-10}$ m]) and atomic radii (ar, $[10^{-10}$ m]).

element	ar	ir	element	ar	ir
Al^{+3} (aluminum)	1.43	0.50	N^{-3} (nitrogen)	0.70	1.71
B^{+3} (boron)	0.88	0.20	K^{+1} (potassium)	2.02	1.33
C^{+4} (carbon)	0.77	0.16	Na^{+1} (sodium)	1.86	0.95
Cl^{-1} (chlorine)	0.99	1.81	O^{-2} (oxygen)	0.66	1.45
Mg^{+2} (magnesium)	1.60	0.65	Si^{+4} (silicon)	1.17	0.39

Focussing on sodium chloride (NaCl), i.e. $r_0 = 2.5\,\text{Å} = 0.25\,\text{nm}$, $E_b = -3.6\,\text{eV}$, $Z = -Z'$, and $|ZZ'| = Z^2 = 1$ so that $z_i = (-1)^{i+1}$, we immediately arrive at the lattice energy (lattice enthalpy) (2.10) with the Madelung constant (2.12), where the $n_i = \{6, 12, 8, 6, 24 \text{ etc.}\}$ are the ion numbers of the coordination spheres, and where the $r_0 c_i = \{r_0, r_0\sqrt{2}, r_0\sqrt{3}, r_0\sqrt{4}, r_0\sqrt{5} \text{ etc.}\}$ are the distances of equilibrium of the coordination spheres. We compare with Figure 2.6, where the first coordination sphere (Cl^- ions) with respect to an individual ion (Na^+ ion) is shown, and with Table 2.1, which collects Madelung constants. As it becomes clear from this example, including the bond energies of neighbors, the Madelung constant describes the bond strength in the crystal relative to the bond strength of a molecule.

Besides the type of chemical bond, the most important physical parameters for the formation of a space lattice are the lattice energy (lattice enthaply) and the ratios of the ionic radii. We particularly compare with Table 2.2 and Table 2.3, which collect the lattice energy (lattice enthalpy) for selected ionic crystals. We immediately realize that larger ions call for smaller absolute values of the lattice enthalpies. We note that this is caused by the smaller attractive forces. We immediately realize that larger charges call for larger absolute values of the lattice enthalpies. We note that this is caused by the larger attractive forces. Furthermore, Table 2.4 tells us that if the ionic charge is positive, i.e. electrons are released, the ionic radius becomes smaller in comparison to the atomic radius, and if the ionic charge is negative, i.e. electrons are affiliated, the ionic radius becomes bigger in comparison to the atomic radius, in both cases, reflecting the alteration of the Coulomb attraction/repulsion. Moreover, Table 2.5 tells us that the ionic radius increases with increasing coordination number (cn), reflecting the available space, we compare with Figure 2.5.

Table 2.5. Ionic radii ([nm]) as functions of the coordination number (cn).

element	cn 4	cn 6	cn 8	cn 12	element	cn 4	cn 6	cn 8	cn 12
Al^{3+}	0.039	0.054	-	-	O^{2-}	0.138	0.140	0.142	-
Ba^{2+}	-	0.135	0.142	0.161	Pb^{2+}	0.098	0.119	0.129	0.149
C^{4+}	0.015	0.016	-	-	Si^{4+}	0.026	0.040	-	-
Ca^{2+}	-	0.100	0.112	0.134	Sr^{2+}	-	0.118	0.126	0.144

2.1.2 Covalent Bond

On the one hand, matter shows solid properties, if microscopic systems are studied, due to the orders of magnitude of microscopic systems, better termed *particle properties*. On the other hand, matter shows field properties, if microscopic systems are studied, due to the shapes of microscopic fields, better termed *wave properties*. It is a significant feature of microscopic systems that wave and particle properties occur coexistently and inseparably. For example, let us again point at the fact that covalent bond forces are directional forces, whereas ionic bond forces are non-directional forces, and let us firstly annotate that the reason for it is that in the first case the wave properties of electrons dominante, evoking oriented bonds, reflecting that "standing waves" are only possible for special numbers and special values which are called quantum numbers and eigenvalues respectively, whereas in the second case the particle properties of electrons dominante, evoking non-oriented bonds, reflecting the properties of Coulomb forces, in the first case leading to a directed arrangement of ligands and in the second case leading to a dense arrangement of ligands, and let us secondly annotate that in both cases the complementary (particle or wave) properties are needed in order to build up an exact quantum-mechanical picture of both types of chemical bonds, by the way, which is the reason why we also use the term "wave–particle system".

The reader should know that the covalent bond between two atoms is immediately characterized as states of matter where two cores are connected by *electron pairs* that are shared *pari passu*. As it is outlined in Figure 2.7 and Figure 2.8 for one electron pair, the entire description of such a state of matter needs the inclusion of a *wavefunction* ψ that records wave-related aspects of the electron pairs including wave-related aspects of the electron pair spins, in a common manner of speaking, defining a *molecular orbital*. In particular, we read off from these figures that oppositely oriented spins lead to a bonding molecular orbital, namely to a state of matter with a relatively low energy, and that equally oriented spins lead to an antibonding molecular orbital, namely to a state of matter with a relatively high energy. In particular, we read off from these figures that the covalent bond state is consistent with Coulomb's law, namely the covalent bond state involves a negatively charged area between the cores so that positively charged cores are finally attracted by a negatively charged area in between. We note that in the context of the covalent bond, the Coulomb interaction is not the decisive aspect. We note that in the context of the covalent bond, the evolution of an "overlapping" electron state that is covered by an "overlapping" wavefunction is the decisive aspect. Of course, in the context of the ionic bond, the bond electrons more or less are shifted from an atom to another atom, leading to a more or less "localized" electron state that is covered by a more or less "localized" wavefunction, in last consequence, evoking a situation where the Coulomb interaction becomes the decisive aspect. We compare with Figure 2.7, top, and Figure 2.7, bottom, which illustrates the difference between the covalent bond and the ionic bond. We appreciate that the symbol ψ indicates the existence of a wavefunction, that the symbol $\uparrow$ indicates the existence of an electron spin, and that oppositely oriented spins of an electron pair are formally outlined by the symbol pair $\uparrow\downarrow$ ("up–down").

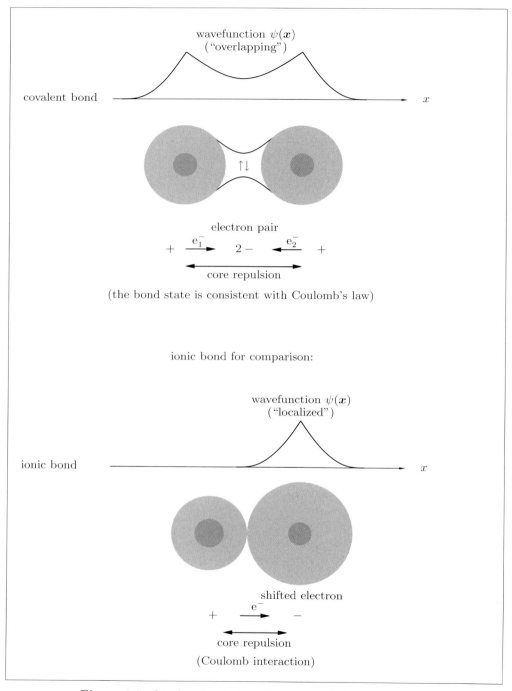

Figure 2.7. Covalent bond. Wavefunction and charge structure.

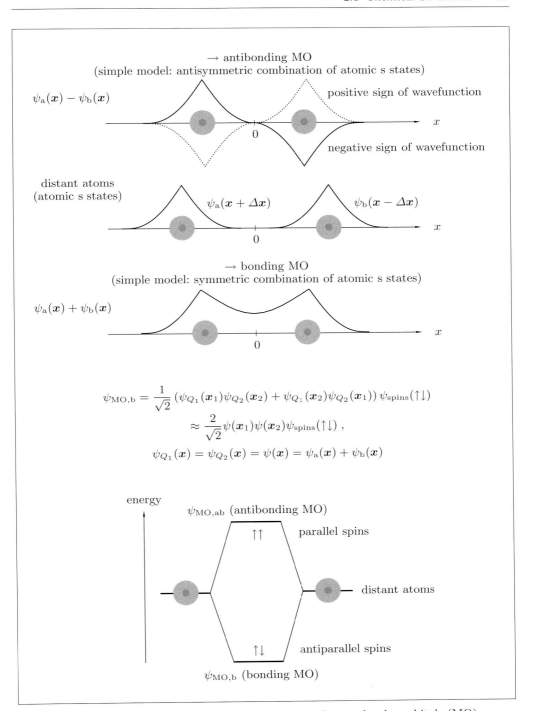

Figure 2.8. Covalent bond. Bonding and antibonding molecular orbitals (MO).

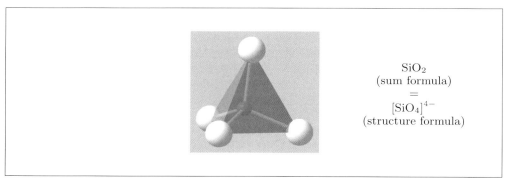

Figure 2.9. An example of a tetrahedral assembly with dominant covalent contributions.

Certainly, for students, engineers, and material scientists in many cases it will do to know that ionic bonds are non-directional bonds and thus are associated with relatively high packing densities, while covalent bonds are directional bonds and thus are associated with relatively low packing densities. Certainly, for students, engineers, and material scientists in many cases it will do to know that ionic bonds are evoked by more or less shifted electrons, while covalent bonds are evoked by electron pairs shared *pari passu*. Of course, the reader should know that covalent bonds such as the graphite bonds and the diamond bonds in some approximation can be modeled by the hybrid orbital technique. Of course, the reader should also know that bonds that show stronger covalent contributions than ionic contributions, for example, the SiO_2 bonds showing the tetrahedral geometry, we here compare with Figure 2.9, or the SF_6 bonds showing the octahedral geometry, we here compare with Figure 2.10, in some approximation can be modeled by the hybrid orbital technique. However, these "black boxes" for students, engineers, and material scientists in special cases surely will not do. Moreover, the material science of the near future surely will make advanced demands on us. Therefore, in addition to the advanced ideas of quantum mechanics already introduced above, in the segments that follow, let us present a little bit more of these trailblazing ideas.

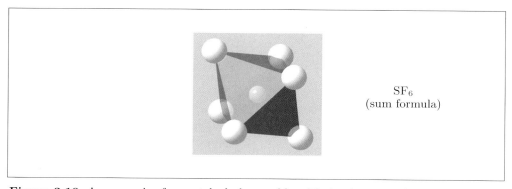

Figure 2.10. An example of an octahedral assembly with dominant covalent contributions.

The reader should know that a wavefunction ψ reflects the existence of particles such as electrons, protons, neutrons, and nuclei. In particular, if no particle is there, no wavefunction is there. In particular, the absolute amount $|\psi|^2$ of a wavefunction ψ supplies us with the probability density to measure the result "particles". The remarks that follow may supply the reader with an intuitive access to this important feature of a wavefunction. (i) Following Einstein's ideas, $-\triangle\Phi + \cdots \propto \rho c^2$ defines the interrelation between the mass energy ρc^2 and the mass potential Φ. Following Schrödinger's ideas, $-\triangle\psi + \cdots \propto E\psi$ defines the interrelation between the particle energy E and the particle function ψ. (ii) In the first case, the r.h.s. does not depend on Φ so that field properties and solid properties are separable. In the second case, the r.h.s. does depend on ψ so that field properties and solid properties are not separable. So to speak, the r.h.s. generates the principal difference. (iii) In both cases, the elongations of the field function indicate the presence of a mass, implying an across-the-board logic.

The reader should know that the wavefunction ψ can be obtained as solution of the *time-independent Schrödinger equation* and accordingly of reformulations therefrom and generalizations therefrom, while transitions of the wavefunction ψ can be calculated by applying the *time-dependent Schrödinger equation* and accordingly reformulations therefrom and generalizations therefrom. In order to gain a first insight into such types of partial differential equations, let us here have a brief look at Box 2.2, which collects some basic specifications of the time-independent Schrödinger equation. (i) In order to understand the meaning of this partial differential equation, dear reader, please think about the threads that follow. For example, for the sake of simplicity, considering the time-independent Schrödinger equation (2.13) and considering the wavefunctions $\psi(\boldsymbol{x}) \propto \exp(-\mathrm{i}\boldsymbol{k}\boldsymbol{x})$ implying eigenvalues $E = \boldsymbol{p}^2/2m_0$ with $\boldsymbol{p} = \hbar\boldsymbol{k}$, we realize that the product term $E\psi(\boldsymbol{x})$, combining energies E and wavefunctions $\psi(\boldsymbol{x})$, reflects the entity of microscopic systems, namely that wave and particle properties occur coexistently and in an inseparable manner. In particular, we realize that the product term $E\psi(\boldsymbol{x})$ concatenates a wave-type expression $\exp(-\mathrm{i}\boldsymbol{k}\boldsymbol{x})$ and a particle-type expression $\boldsymbol{p}^2/2m_0$, and we realize that this concatenation is unravelable since the momentum vector $\boldsymbol{p}$ is not specified as a particle-type quantity $m_0\boldsymbol{v}$, but as a wave-type quantity $\hbar\boldsymbol{k}$, so to speak, exemplarily telling us that microscopic systems are *wave–particle systems*. In particular, we realize that the sum of all terms defines an energy balance which consists of both wave contributions and particle contributions, so to speak, exemplarily telling us that such *eigenvalue equations* are *wave–particle energy balance equations*. (ii) In order to understand the buildup of this partial differential equation, dear reader, please think about the threads that follow. Firstly, we realize that each energy eigenvalue E includes a kinetic energy contribution T evoked by kinetic energy operators $\hat{T}_i = -\hbar^2 \triangle_i / 2m_i$, where $\hbar$ is Planck's constant anf m_i is the mass of an electron e_i^- or a nucleus n_i^+, and includes a potential energy contribution V evoked by potential energy operators $\hat{V}_i = V_i$, where ϵ_0 is the dielectric constant and e is the elementary charge. Secondly, we realize that the number of kinetic energy operators and potential energy operators depends on the number of particles. Thirdly, we realize that additional approximations such as the Born–Oppenheimer approximation, which separates the electrons from the nuclei, are readily introduced in order to reduce the operating expense.

Box 2.2 (Important formulae: on the time-independent Schrödinger equation).

Time-independent Schrödinger equation, free particle without spin:

$$-\tfrac{1}{2}\triangle\psi(\boldsymbol{x}) = \tfrac{m_0}{\hbar^2} E\psi(\boldsymbol{x}) \ , \ \ \triangle = \tfrac{\partial^2}{\partial x^2} + \tfrac{\partial^2}{\partial y^2} + \tfrac{\partial^2}{\partial z^2} \ . \tag{2.13}$$

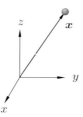

Time-independent Schrödinger equation, hydrogen atom without spin:

$$\left[-\tfrac{1}{2}\triangle + \tfrac{m_e}{\hbar^2} V(r)\right]\psi(\boldsymbol{x}) = \tfrac{m_e}{\hbar^2} E\psi(\boldsymbol{x}) \ , \ \ \triangle = \tfrac{\partial^2}{\partial x^2} + \tfrac{\partial^2}{\partial y^2} + \tfrac{\partial^2}{\partial z^2} \ ,$$
$$V(r) = -\tfrac{1}{4\pi\epsilon_0}\tfrac{e^2}{r} \ , \ r = \sqrt{x^2+y^2+z^2} \ . \tag{2.14}$$

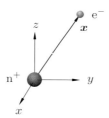

Time-independent Schrödinger equation, hydrogen ion molecule without spin:

$$\left[-\tfrac{1}{2}\triangle_1 - \tfrac{1}{2}\tfrac{m_e}{m_n}\triangle_a - \tfrac{1}{2}\tfrac{m_e}{m_n}\triangle_b + \tfrac{m_e}{\hbar^2}\left(V_{a1}(r_{a1}) + V_{b1}(r_{b1}) + V_{ab}(r_{ab})\right)\right]\psi(\boldsymbol{x}_a, \boldsymbol{x}_b, \boldsymbol{x}_1) =$$
$$= \tfrac{m_e}{\hbar^2} E\psi(\boldsymbol{x}_a, \boldsymbol{x}_b, \boldsymbol{x}_1) \ ,$$
$$\triangle_1 = \tfrac{\partial^2}{\partial x_1^2} + \tfrac{\partial^2}{\partial y_1^2} + \tfrac{\partial^2}{\partial z_1^2} \ , \ \triangle_a = \tfrac{\partial^2}{\partial x_a^2} + \tfrac{\partial^2}{\partial y_a^2} + \tfrac{\partial^2}{\partial z_a^2} \ , \ \triangle_b = \tfrac{\partial^2}{\partial x_b^2} + \tfrac{\partial^2}{\partial y_b^2} + \tfrac{\partial^2}{\partial z_b^2} \ ,$$
$$V_{a1}(r_{a1}) = -\tfrac{1}{4\pi\epsilon_0}\tfrac{e^2}{r_{a1}} \ , \ V_{b1}(r_{b1}) = -\tfrac{1}{4\pi\epsilon_0}\tfrac{e^2}{r_{b1}} \ , \ V_{ab}(r_{ab}) = \tfrac{1}{4\pi\epsilon_0}\tfrac{e^2}{r_{ab}} \ . \tag{2.15}$$

Continuation of Box.

Time-independent Schrödinger equation, hydrogen molecule without spin:

$$\hat{H}_{H_2}\psi(\boldsymbol{x}_a, \boldsymbol{x}_b, \boldsymbol{x}_1, \boldsymbol{x}_2) = E\psi(\boldsymbol{x}_a, \boldsymbol{x}_b, \boldsymbol{x}_1, \boldsymbol{x}_2),$$

$$\hat{H}_{H_2} = \hat{T}_{H_2} + \hat{V}_{H_2},$$

$$\hat{T}_{H_2} = \hat{T}_e + \hat{T}_n,$$

$$\hat{T}_e = -\frac{\hbar^2}{2m_e}\triangle_1 - \frac{\hbar^2}{2m_e}\triangle_2, \quad \hat{T}_n = -\frac{\hbar^2}{2m_n}\triangle_a - \frac{\hbar^2}{2m_n}\triangle_b,$$

$$\hat{V}_{H_2} = V_{a1}(r_{a1}) + V_{b1}(r_{b1}) + V_{a2}(r_{a2}) + V_{b2}(r_{b2}) + V_{ab}(r_{ab}) + V_{12}(r_{12}),$$

$$\triangle_{1/2} = \frac{\partial^2}{\partial x_{1/2}^2} + \frac{\partial^2}{\partial y_{1/2}^2} + \frac{\partial^2}{\partial z_{1/2}^2},$$ (2.16)

$$\triangle_a = \frac{\partial^2}{\partial x_a^2} + \frac{\partial^2}{\partial y_a^2} + \frac{\partial^2}{\partial z_a^2}, \quad \triangle_b = \frac{\partial^2}{\partial x_b^2} + \frac{\partial^2}{\partial y_b^2} + \frac{\partial^2}{\partial z_b^2},$$

$$V_{a1/2}(r_{a1/2}) = -\frac{1}{4\pi\epsilon_0}\frac{e^2}{r_{a1/2}}, \quad V_{b1/2}(r_{b1/2}) = -\frac{1}{4\pi\epsilon_0}\frac{e^2}{r_{b1/2}},$$

$$V_{ab}(r_{ab}) = \frac{1}{4\pi\epsilon_0}\frac{e^2}{r_{ab}}, \quad V_{12}(r_{12}) = \frac{1}{4\pi\epsilon_0}\frac{e^2}{r_{12}}.$$

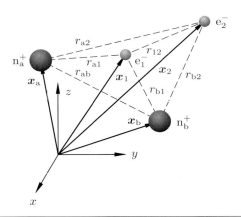

> Continuation of Box.

Time-independent Schrödinger equation, hydrogen molecule with spin:

$$\hat{H}_{\mathrm{H}_2}\psi(\boldsymbol{x}_\mathrm{a},\boldsymbol{x}_\mathrm{b},\boldsymbol{x}_1,\boldsymbol{x}_2,\sigma_\mathrm{a},\sigma_\mathrm{b},\sigma_1,\sigma_2) = E\psi(\boldsymbol{x}_\mathrm{a},\boldsymbol{x}_\mathrm{b},\boldsymbol{x}_1,\boldsymbol{x}_2,\sigma_\mathrm{a},\sigma_\mathrm{b},\sigma_1,\sigma_2),$$

$$\hat{H}_{\mathrm{H}_2} = \hat{T}_{\mathrm{H}_2} + \hat{V}_{\mathrm{H}_2} + \hat{T}'_{\mathrm{H}_2},\ \hat{T}_{\mathrm{H}_2} = \hat{T}_\mathrm{e} + \hat{T}_\mathrm{n},$$

$$\hat{T}_\mathrm{e} = -\frac{\hbar^2}{2m_\mathrm{e}}\triangle_1 - \frac{\hbar^2}{2m_\mathrm{e}}\triangle_2,\ \hat{T}_\mathrm{n} = -\frac{\hbar^2}{2m_\mathrm{n}}\triangle_\mathrm{a} - \frac{\hbar^2}{2m_\mathrm{n}}\triangle_\mathrm{b},$$

$$\hat{V}_{\mathrm{H}_2} = V_{\mathrm{a}1}(r_{\mathrm{a}1}) + V_{\mathrm{b}1}(r_{\mathrm{b}1}) + V_{\mathrm{a}2}(r_{\mathrm{a}2}) + V_{\mathrm{b}2}(r_{\mathrm{b}2}) + V_{\mathrm{ab}}(r_{\mathrm{ab}}) + V_{12}(r_{12}),$$

$$V_{\mathrm{a}1/2}(r_{\mathrm{a}1/2}) = -\frac{1}{4\pi\epsilon_0}\frac{e^2}{r_{\mathrm{a}1/2}},\ V_{\mathrm{b}1/2}(r_{\mathrm{b}1/2}) = -\frac{1}{4\pi\epsilon_0}\frac{e^2}{r_{\mathrm{b}1/2}},$$

$$V_{\mathrm{ab}}(r_{\mathrm{ab}}) = \frac{1}{4\pi\epsilon_0}\frac{e^2}{r_{\mathrm{ab}}},\ V_{12}(r_{12}) = \frac{1}{4\pi\epsilon_0}\frac{e^2}{r_{12}},$$

$$\triangle_{1/2} = \frac{\partial^2}{\partial x_{1/2}^2}+\frac{\partial^2}{\partial y_{1/2}^2}+\frac{\partial^2}{\partial z_{1/2}^2},\ \triangle_\mathrm{a} = \frac{\partial^2}{\partial x_\mathrm{a}^2}+\frac{\partial^2}{\partial y_\mathrm{a}^2}+\frac{\partial^2}{\partial z_\mathrm{a}^2},\ \triangle_\mathrm{b} = \frac{\partial^2}{\partial x_\mathrm{b}^2}+\frac{\partial^2}{\partial y_\mathrm{b}^2}+\frac{\partial^2}{\partial z_\mathrm{b}^2}.$$

(2.17)

Applying the self-consistent network of model conceptions which is introduced in the framework of more far-reaching ideas in Chapter 5 already in advance at this point, we here quote the following advanced spin model
(the advanced kinetic operators $\hat{T}'$ are kinetic spin energy operators
that are suggested by the self-consistent network of model conceptions):

$$\hat{T}'_{\mathrm{H}_2} = \hat{T}'_\mathrm{e} + \hat{T}'_\mathrm{n},\quad \begin{aligned}\hat{T}'_\mathrm{e} &= -\frac{\hbar^2}{2m_\mathrm{e}}\left(\boldsymbol{\theta}^{\ni}_{1,\mathrm{spin}}\nabla_1\right)\nabla_1 - \frac{\hbar^2}{2m_\mathrm{e}}\left(\boldsymbol{\theta}^{\ni}_{2,\mathrm{spin}}\nabla_2\right)\nabla_2,\\ \hat{T}'_\mathrm{n} &= -\frac{\hbar^2}{2m_\mathrm{n}}\left(\boldsymbol{\theta}^{\ni}_{\mathrm{a},\mathrm{spin}}\nabla_\mathrm{a}\right)\nabla_\mathrm{a} - \frac{\hbar^2}{2m_\mathrm{n}}\left(\boldsymbol{\theta}^{\ni}_{\mathrm{b},\mathrm{spin}}\nabla_\mathrm{b}\right)\nabla_\mathrm{b}.\end{aligned}$$

(2.18)

The reader may already here compare with Box 2.3, which compares the operators $\hat{T}$ and $\hat{T}'$ with the classical energies (2.32) and (2.33). We already here realize that the replacement $p_i \to -\mathrm{i}\hbar\partial/\partial x^i$ recasts (2.32) into $\hat{T}$. Therefore, is it a miracle that we expect that the same replacement $p_i \to -\mathrm{i}\hbar\partial/\partial x^i$ recasts the rewritten formulation (2.36) of (2.33) into $\hat{T}'$?

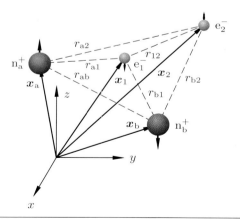

> **Continuation of Box.**

Born–Oppenheimer separation:

$$\psi(\boldsymbol{x}_a, \boldsymbol{x}_b, \boldsymbol{x}_1, \boldsymbol{x}_2, \sigma_a, \sigma_b, \sigma_1, \sigma_2) = \psi_n(\boldsymbol{x}_a, \boldsymbol{x}_b, \sigma_a, \sigma_b)\psi_e(\boldsymbol{x}_1, \boldsymbol{x}_2, \sigma_1, \sigma_2; \boldsymbol{x}_a, \boldsymbol{x}_b)$$

$$\Downarrow$$

$$\frac{\psi_e \hat{T}_n \psi_n}{\psi_e \psi_n} + \underbrace{\frac{\psi_n \hat{T}_n \psi_e}{\psi_n \psi_e}}_{\approx 0} + \frac{\psi_e \hat{T}'_n \psi_n}{\psi_e \psi_n} + \underbrace{\frac{\psi_n \hat{T}'_n \psi_e}{\psi_n \psi_e}}_{\approx 0} + \underbrace{\frac{\hat{T}_e \psi_e}{\psi_e} + \frac{\hat{T}'_e \psi_e}{\psi_e} + \hat{V}_{H_2}}_{E_e(\boldsymbol{x}_a, \boldsymbol{x}_b)} = E \quad (2.19)$$

$$\Downarrow$$

$$\hat{T}_e \psi_e + \hat{T}'_e \psi_e + \hat{V}_{H_2}\psi_e = E_e(\boldsymbol{x}_a, \boldsymbol{x}_b)\psi_e \;,\; \hat{T}_n \psi_n + \hat{T}'_n \psi_n + E_e(\boldsymbol{x}_a, \boldsymbol{x}_b)\psi_n = E\psi_n \;,$$

$$E_e(\boldsymbol{x}_a, \boldsymbol{x}_b) = E_e(|\boldsymbol{x}_b - \boldsymbol{x}_a|) = E_e(r_{ab}) = E_e(r_r) \;.$$

Barycentric coordinates
(nuclear spin properties here are neglected):

$$\boldsymbol{x}_a, \boldsymbol{x}_b \to \boldsymbol{x}_{cg}, \boldsymbol{x}_r \;,$$

$$\boldsymbol{x}_{cg} = \frac{m_a \boldsymbol{x}_a + m_b \boldsymbol{x}_b}{m_a + m_b} \stackrel{m_a = m_b = m_n}{=} \frac{\boldsymbol{x}_a + \boldsymbol{x}_b}{2} \;,\quad \boldsymbol{x}_r = \boldsymbol{x}_b - \boldsymbol{x}_a \;, \quad (2.20)$$

$$\frac{1}{m_a}\triangle_a + \frac{1}{m_b}\triangle_b = \frac{1}{m_a+m_b}\triangle_{cg} + \frac{m_a+m_b}{m_a m_b}\triangle_r \stackrel{m_a=m_b=m_n}{=} \frac{1}{2m_n}\triangle_{cg} + \frac{2}{m_n}\triangle_r \;,$$

$$\hat{T}_n(\boldsymbol{x}_a, \boldsymbol{x}_b)\psi_n(\boldsymbol{x}_a, \boldsymbol{x}_b) + E_e(\boldsymbol{x}_a, \boldsymbol{x}_b)\psi_n(\boldsymbol{x}_a, \boldsymbol{x}_b) = E\psi_n(\boldsymbol{x}_a, \boldsymbol{x}_b)$$
$$\Downarrow \qquad (2.21)$$
$$\hat{T}_n(\boldsymbol{x}_{cg}, \boldsymbol{x}_r)\psi_n(\boldsymbol{x}_{cg}, \boldsymbol{x}_r) + E_e(r_r)\psi_n(\boldsymbol{x}_{cg}, \boldsymbol{x}_r) = E\psi_n(\boldsymbol{x}_{cg}, \boldsymbol{x}_r) \;,$$

$$\hat{T}_n(\boldsymbol{x}_a, \boldsymbol{x}_b) = -\frac{\hbar^2}{2m_n}\triangle_a - \frac{\hbar^2}{2m_n}\triangle_b \;,\quad \hat{T}_n(\boldsymbol{x}_{cg}, \boldsymbol{x}_r) = -\frac{\hbar^2}{4m_n}\triangle_{cg} - \frac{\hbar^2}{m_n}\triangle_r \;.$$

Chain rule
(transformation of $\triangle_a$ and $\triangle_b$):

$$\frac{\partial}{\partial x_a}F(\boldsymbol{x}_{cg}, \boldsymbol{x}_r) = \left(\frac{\partial x_{cg}}{\partial x_a}\frac{\partial}{\partial x_{cg}} + \frac{\partial x_r}{\partial x_a}\frac{\partial}{\partial x_r}\right)F(\boldsymbol{x}_{cg}, \boldsymbol{x}_r) \;,$$
$$(2.22)$$
$$\frac{\partial^2}{\partial x_a^2}F(\boldsymbol{x}_{cg}, \boldsymbol{x}_r) = \left(\frac{\partial x_{cg}}{\partial x_a}\frac{\partial}{\partial x_{cg}} + \frac{\partial x_r}{\partial x_a}\frac{\partial}{\partial x_r}\right)^2 F(\boldsymbol{x}_{cg}, \boldsymbol{x}_r) \; \text{etc.}$$

Separation
(motion of the center of gravity, relative motion of the nuclei):

$$\psi_n(\boldsymbol{x}_a, \boldsymbol{x}_b) = \psi_n(\boldsymbol{x}_{cg}, \boldsymbol{x}_r) = \varphi(\boldsymbol{x}_{cg})\varphi(\boldsymbol{x}_r) \;,\; E = E_{cg} + E_r \;,$$
$$(2.23)$$
$$-\frac{\hbar^2}{4m_n}\triangle_{cg}\varphi(\boldsymbol{x}_{cg}) = E_{cg}\varphi(\boldsymbol{x}_{cg}) \;,\; -\frac{\hbar^2}{m_n}\triangle_r\varphi(\boldsymbol{x}_r) + E_e(r_r)\varphi(\boldsymbol{x}_r) = E_r\varphi(\boldsymbol{x}_r) \;.$$

Box 2.3 (Important formulae: on the advanced spin model).

$\hat{T}$, eigenvalue equation, eigenvalues:

$$\hat{T} = -\frac{\hbar^2}{2m_0}\left(\nabla\right)\nabla = -\frac{\hbar^2}{2m_0}\triangle \,, \qquad (2.24)$$

$$-\frac{1}{2}\triangle\psi = -\frac{1}{2}\left(\sum_{i=1}^{3}\frac{\partial^2}{\partial x^{i2}}\right)\psi = \chi_L\psi \,,$$

$$\chi_L = \frac{m_0}{\hbar^2}E_L \qquad (2.25)$$

(compare with (2.13)),

$$\psi \propto \exp\left[\pm i(\boldsymbol{kx} + \varphi)\right] \,, \qquad (2.26)$$

$$\chi_L = \frac{1}{2}\boldsymbol{k}^2 = \frac{1}{2}\sum_{i=1}^{3}k_i^2 = \frac{1}{2}\left(k_1^2 + k_2^2 + k_3^2\right) \qquad (2.27)$$

(spherical wave vector ($\boldsymbol{k}$) geometry).

$\hat{T}'$, eigenvalue equation, eigenvalues
(we here focus on $\boldsymbol{\theta}^{\ni} \neq \boldsymbol{\theta}^{\ni}(\boldsymbol{x})$):

$$\hat{T}' = -\frac{\hbar^2}{2m_0}\left(\boldsymbol{\theta}^{\ni}\nabla\right)\nabla = -\frac{\hbar^2}{2m_0}\triangle' \,, \qquad (2.28)$$

$$-\frac{1}{2}\triangle'\psi = -\frac{1}{2}\left(\sum_{i,j=1}^{3}\gamma^{ij}\frac{\partial}{\partial x^i}\frac{\partial}{\partial x^j}\right)\psi = \chi_S\psi \,,$$

$$\chi_S = \frac{m_0}{\hbar^2}E_S \,, \qquad (2.29)$$

$$\psi \propto \exp\left[\pm i(\boldsymbol{kx} + \varphi)\right] \,, \qquad (2.30)$$

$$\chi_S = \frac{1}{2}\boldsymbol{k}\boldsymbol{\theta}^{\ni}\boldsymbol{k} = \frac{1}{2}\sum_{i,j=1}^{3}\gamma^{ij}k_ik_j =$$
$$= \frac{1}{2}\left(\gamma^{11}k_1^2 + \gamma^{22}k_2^2 + \gamma^{33}k_3^2 + 2\gamma^{12}k_1k_2 + 2\gamma^{13}k_1k_3 + 2\gamma^{23}k_2k_3\right) \qquad (2.31)$$

(ellipsoidal wave vector ($\boldsymbol{k}$) geometry).

> **Continuation of Box.**
>
> Kinetic energy of classical mechanics:
>
> $$\frac{\boldsymbol{p}^2}{2m_0} = \frac{(m_0\boldsymbol{v})^2}{2m_0} = \frac{1}{2m_0}\left(p_1^2 + p_2^2 + p_3^2\right) \qquad (2.32)$$
>
> (spherical momentum vector ($\boldsymbol{p}$) geometry).
>
> Kinetic spin energy (Poinsot ellipsoid) of classical spin mechanics:
>
> $$\frac{1}{2}\boldsymbol{\omega}\boldsymbol{L} = \frac{1}{2}\boldsymbol{\omega}\boldsymbol{\theta}\boldsymbol{\omega} =$$
>
> $$= \frac{1}{2}\left(\theta^{11}\omega_1^2 + \theta^{22}\omega_2^2 + \theta^{33}\omega_3^2 + 2\theta^{12}\omega_1\omega_2 + 2\theta^{13}\omega_1\omega_3 + 2\theta^{23}\omega_2\omega_3\right) \qquad (2.33)$$
>
> (ellipsoidal frequency vector ($\boldsymbol{\omega}$) geometry).
>
> With the mass m_0 and the radius R of the solid body, the tensor of inertia $\boldsymbol{\theta}$ can be rewritten as
>
> $$\boldsymbol{\theta} = \boldsymbol{\theta}' m_0 R^2 \,. \qquad (2.34)$$
>
> Applying (2.34), we obtain
>
> $$\frac{1}{2}\boldsymbol{\omega}\boldsymbol{L} = \frac{1}{2}\boldsymbol{\omega}\boldsymbol{\theta}\boldsymbol{\omega} =$$
>
> $$= \frac{m_0 R\boldsymbol{\omega}\,\boldsymbol{\theta}'\,m_0 R\boldsymbol{\omega}}{2m_0} = \frac{\boldsymbol{p}\,\boldsymbol{\theta}'\,\boldsymbol{p}}{2m_0}\,, \qquad (2.35)$$
>
> $$\boldsymbol{p} = m_0 R\boldsymbol{\omega}\,,$$
>
> and thus
>
> $$\frac{1}{2}\boldsymbol{\omega}\boldsymbol{L} = \frac{1}{2}\boldsymbol{\omega}\boldsymbol{\theta}\boldsymbol{\omega} =$$
>
> $$= \frac{1}{2m_0}\left(\theta^{11'}p_1^2 + \theta^{22'}p_2^2 + \theta^{33'}p_3^2 + 2\,\theta^{12'}p_1p_2 + 2\,\theta^{13'}p_1p_3 + 2\,\theta^{23'}p_2p_3\right) \qquad (2.36)$$
>
> (ellipsoidal momentum vector ($\boldsymbol{p}$) geometry).
>
> **Notes:**
> We note that the operator $\hat{T}'$ which is given by (2.28) includes much more than kinetic spin energies. As it is discussed in [63], it includes a whole class of inner energies of particles.

Applying such time-independent Schrödinger equations, reformulations thereof, or generalizations thereof, or in most practical cases rather departing from approximations thereof, numerical calculations can be accomplished, supplying us with an impression of wavefunctions ψ and probability densities $|\psi|^2$, directly giving us an insight into bond orientations, bond symmetries, and bond strengths. Going beyond it, by applying time-dependent extensions of such time-independent Schrödinger equations such as the time-dependent Schrödinger equation or the linear and nonlinear time-dependent extensions that are introduced in the framework of more far-reaching ideas in Chapter 5, numerical calculations can be accomplished, supplying us with an impression of the time-dependent changes of wavefunctions ψ and probability densities $|\psi|^2$, finally giving us an insight into time-dependent changes of bond orientations, bond symmetries, and bond strengths, offering us basic ways of analyzing time-dependent processes such as the crack evolution in materials and the phase transitions of materials.

Applying the time-independent Schrödinger equation (2.14), the energy eigenvalues E_i and the wavefunctions ψ_i covering the states of the electron of the hydrogen atom, and if e^2 is formally replaced by ne^2, where n is the nuclear charge number, also covering the states of the electron of hydrogen-like atoms, are immediately derived. Based upon the wavefunctions ψ_i, depending on the quantum numbers n, l, and m usually termed s orbitals, p orbitals, d orbitals and so forth, hybrid orbitals allowing us to model the connections of atoms are immediately constructed. We should know that these are only simple approximations, namely these contain cardinal properties of actually existing molecular orbitals, allowing us to use these as simple models of actually existing molecular orbitals. We should also know that these are conveniently named sp ($\longrightarrow$ linear bond orientation), sp^2 ($\longrightarrow$ planar–trigonal bond orientation), sp^3 ($\longrightarrow$ tetrahedral bond orientation), sp^{3}d^2 ($\longrightarrow$ octahedral bond orientation) and so forth, formally reflecting that superpositions of s orbitals, p orbitals, d orbitals and so forth are at the botttom of these hybrid orbitals.

Dealing with ceramic materials, in particular, the sp^3 orbitals should be known. As it is sketched in Figure 2.17, such sp^3 orbitals are obtained as linear combinations of the s orbital and the p orbitals. As it is sketched in Figure 2.17, focussing on the electron configurations of carbon (C) and silicon (Si), the construction principle of such sp^3 orbitals is the following. Do consider the ground state. Raise one of the two s electrons of the completely populated s state into the non-populated p state. Construct linear combinations of the s orbital and the p orbitals to obtain the four sp^3 orbitals. As it is sketched in Figure 2.17, this evokes hybrid orbitals containing cardinal properties of actuallay existing molecular orbitals forming tetrahedral bond structures, and we arrive at tetrahedron that is shown at the bottom of Figure 2.17 if we bring out the tetrahedral bond directions. Maybe we should make a mental note that we certainly construct superpositions based upon the same principal quantum number n defining the period in the periodic table of the elements, but in special cases superpositions based upon different principal quantum numbers n might be the better choice. In all cases, however, the wavefunctions that are illustrated by Figures 2.11–2.16 supply us with the mathematical basis.

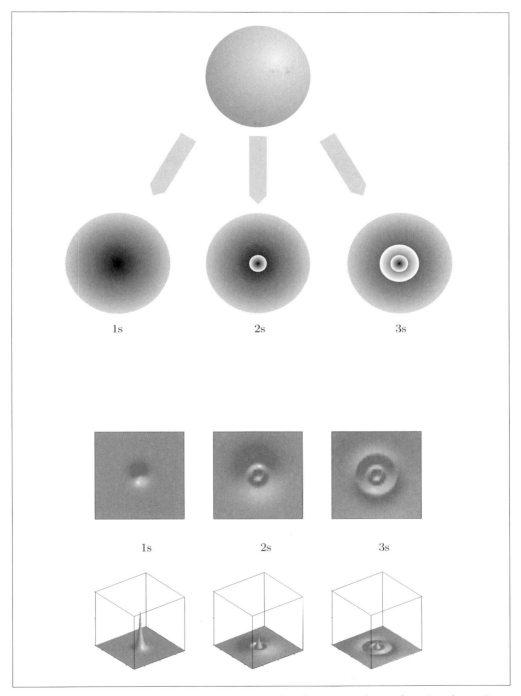

Figure 2.11. The 1s, 2s, and 3s orbitals. Probability densities and wavefunction elongations.

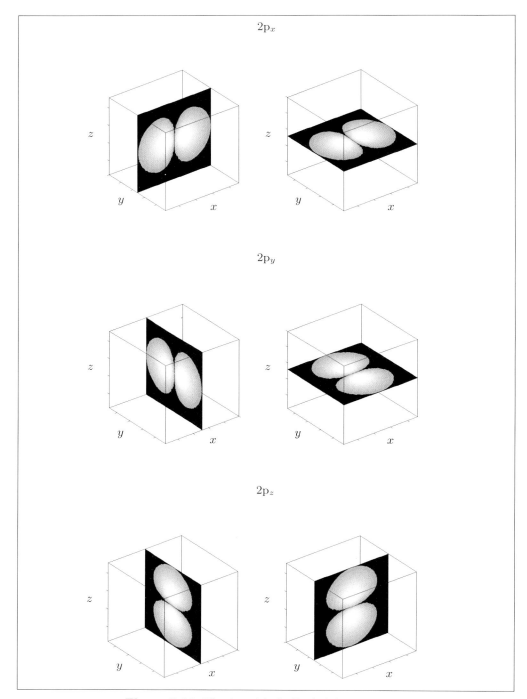

Figure 2.12. The 2p orbitals. Probability densities.

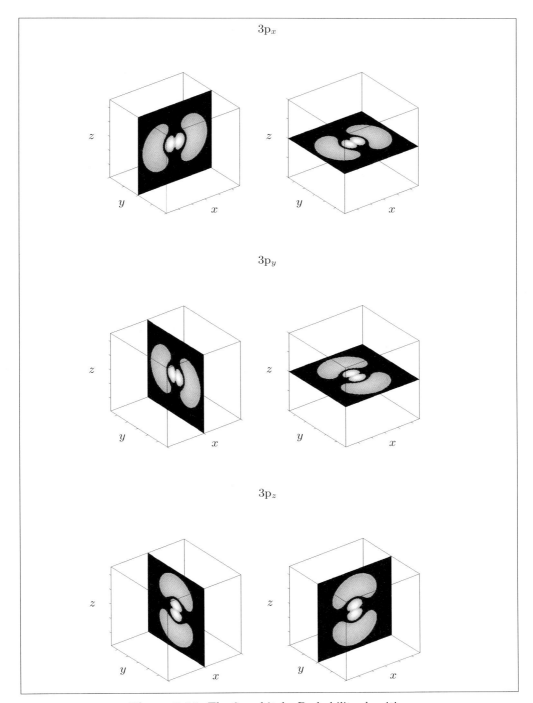

Figure 2.13. The 3p orbitals. Probability densities.

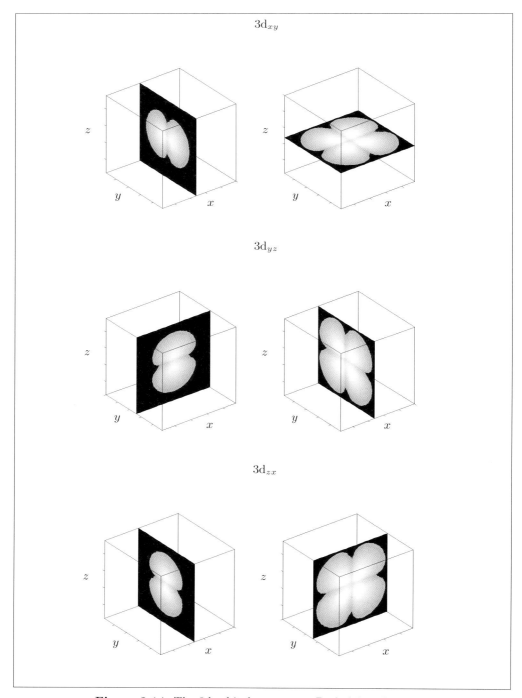

Figure 2.14. The 3d orbitals, part one. Probability densities.

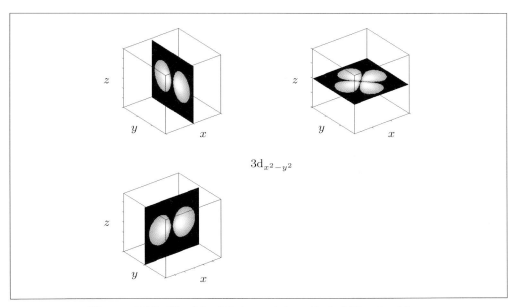

Figure 2.15. The 3d orbitals, part two. Probability densities.

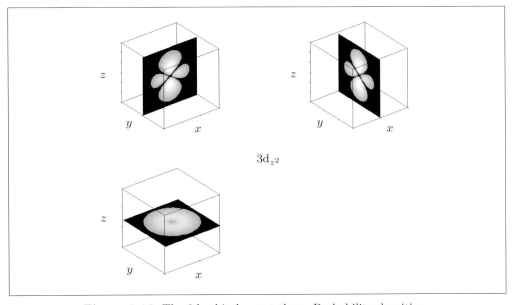

Figure 2.16. The 3d orbitals, part three. Probability densities.

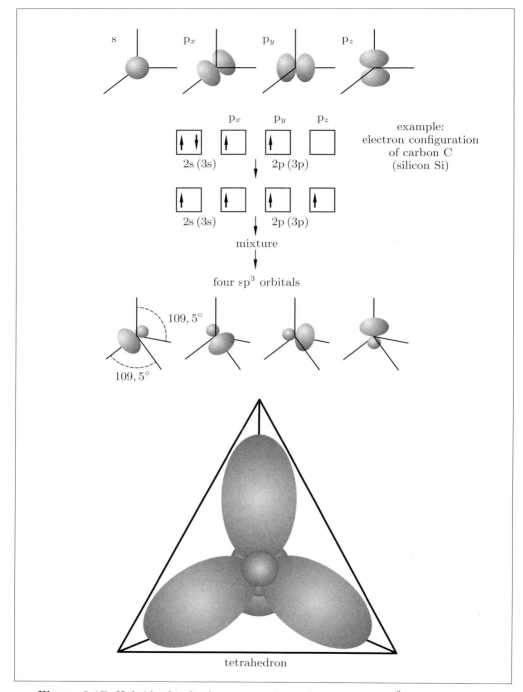

Figure 2.17. Hybrid orbitals. As an example, we here consider sp^3 hybrid orbitals.

As a complementary "black box" for students, engineers, and material scientists, on the one hand, we quote that the chemical elements of the main group IV of the periodic table of the elements, for example, C and Si, can develop four electron pairs in form of single, double, or triple bonds, in this way, establishing connections with neighbors, approaching the energetically advantageous, stable noble gas configuration, and on the other hand, we quote that for solids formed of such chemical elements, completely consistent with the notion of electron pairs mutually repelling each other, tetrahedral structures, for example, four CC or four SiO single bonds tetrahedrally arranged, are to be expected. Indeed, C-, Si-, and Ge-based ceramics exhibit such tetrahedral structures.

2.1.3 Van der Waals Bond

In good approximation, the van der Waals interaction between two atoms/molecules is described by the Lennard–Jones potential (2.37), in particular, by its specification (2.38). A modification of the Lennard–Jones potential (2.37), we compare with (2.39), makes use of an exponential function. The attractive terms, the $1/r^6$ terms, summarize the diverse contributions to the van der Waals interaction, i.e. the Keesom interaction between a dipole and another dipole, the Debye interaction between a dipole and a polarizable atom/molecule, and the van der Waals interaction in the narrower sense, namely the London dispersion interaction, the origin of which are temporarily occurring atomic/molecular dipoles inducing other atomic/molecular dipoles. Firstly, the reader should take notice that the van der Waals bond evoked by the van der Waals interaction in comparison to the ionic bond and the covalent bond is rather weak, i.e. we observe bond energies E_b of about $-0.5\ldots-5\,\mathrm{kJ/mol}$. Secondly, the reader should take notice that the van der Waals radius σ is used to define what we want to understand by the radius of atoms/molecules. Thirdly, the reader should take notice that c_1 and c_2 contain substance-specific constants, for example, the ionization energy of atoms/molecules, and that γ and ρ follow from elastic constants.

Since atoms/molecules with relatively big surface and with electrons located far away from the nucleus/nuclei are easier to polarize than atoms/molecules with relatively small surface and without electrons located far away from the nucleus/nuclei, the van der Waals interaction increases with increasing atomic/molecular mass. Since the van der Waals interation is based upon dipoles able to adjust on top of each other, whenever atoms/molecules can move freely, dense packings of atoms/molecules emerge. Firstly, the reader should take notice that the van der Waals force is responsible for the boiling point and the solubility of liquids of nonpolar compounds. Secondly, the reader should take notice that the van der Waals force is responsible for the formation and the stability of solid phases of noble gases. Thirdly, the reader should take notice that the van der Waals force is responsible for cohesion effects occurring between layers of equal solid materials or different solid materials. The basal layers of graphite, which are "glued together" by this force, define a first example, we compare with Figure 2.39. The powder suspensions, the powder particles of which are eventually "glued together" by this force, define a second example, we compare with Figure 4.14.

> **Box 2.4** (Important formulae: Lennard–Jones potential).
>
> Lennard–Jones potential $V_{\mathrm{LJ}}(n,6)$:
>
> $$V_{\mathrm{LJ}}(n,6) = \frac{c_1}{r^n} - \frac{c_2}{r^6} \;. \tag{2.37}$$
>
> Lennard–Jones potential $V_{\mathrm{LJ}}(12,6)$:
>
> $$\begin{aligned} V_{\mathrm{LJ}}(12,6) &= |E_{\mathrm{b}}| \left[\left(\frac{r_0}{r}\right)^{12} - 2\left(\frac{r_0}{r}\right)^6 \right] \\ &= 4|E_{\mathrm{b}}| \left[\left(\frac{\sigma}{r}\right)^{12} - \left(\frac{\sigma}{r}\right)^6 \right] \;. \end{aligned} \tag{2.38}$$
>
> Lennard–Jones potential $V_{\mathrm{LJ}}(\exp,6)$:
>
> $$V_{\mathrm{LJ}}(\exp,6) = \gamma \left[\exp(-r/\rho) - \left(\frac{\rho}{r}\right)^6 \right] \;. \tag{2.39}$$

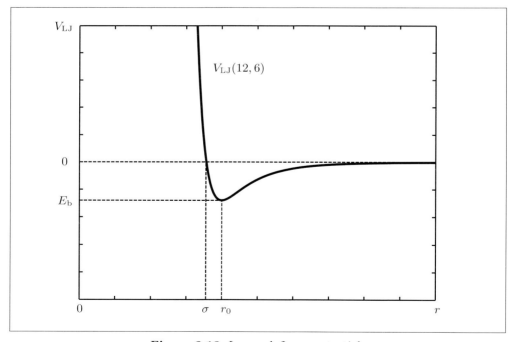

Figure 2.18. Lennard–Jones potential.

2.1.4 Metallic Bond

Depending on further constraints such as pressure, with respect to electric conductivity, solid materials can be *conductors, semi-conductors*, or *insulators*. Here only considering metallic conductors, i. e. here not considering ionic conductors or superconductors, we state that electrons populating quasi-continuous energy domains (*"energy bands"*) are the reason for it, we compare with Figures 2.19–2.21. (i) In particular, insulators show the following band structure. A series of energy bands is completely populated. Between the highest populated energy band, i. e. the *valence band*, and the first unpopulated energy band, i. e. the *conduction band*, a relatively large band gap exists so that neither thermal nor electromagnetic or sound excitations could make the band gap conquerable without destroying the material and no electric conductivity is observable since electric conductivity needs partially populated energy bands. (ii) In the case of semi-conductors, a relatively small band gap between the valence band and the conduction band exists. In this case, applied excitation fields, for example, let us think of electromagnetic fields or sound fields, are able to raise electrons from the valence band into the conduction band, evoking electric conductivity. In this case, further constraints such as pressure are able to move the conduction band upon the valence band, evoking electric conductivity, or at least are able to minimize the band gap, under suitable thermal conditions, evoking electric conductivity. (iii) In the case of conductors, a partially populated highest band, a fully populated highest band and a strongly or slightly overlapping unpopulated band, or a fully populated highest band and an unpopulated band that are separated by a minimal or vanishing band gap, which under practical thermal conditions can be conquered, exists. All cases show electric conductivity. (iv) In this energy scheme, dopants are represented as energy states located below the conduction band edge or above the valence band edge. In the first case, these act as donors, i. e. electrons can be released into the conduction band, and in the second case, these act as acceptors, i. e. electrons can be taken from the valence band, leaving holes in the valence band, in both cases, eventually setting up a conduction mechanism.

The reader should know that the analysis of the band structure is used to subdivide metals into *semi-metals* and *metals in the closer sense*. As shown in Figure 2.19, if there is a partially populated highest band, or if there is a fully populated highest band and a strongly overlapping unpopulated band, we then call the material *metallic*, but if there is a fully populated highest band and a slightly overlapping unpopulated band, evoking electric conductivity, or if there is a fully populated highest band and an unpopulated band separated by a minimal or vanishing band gap, under practical thermal conditions, evoking electric conductivity. we then call the material *semi-metallic*.

The reader should know that typical ceramic materials applied as insulators are Al_2O_3 and MgO, and that typical ceramic materials applied as semi-conductors are AlN, GaAs, and SiC. Mind you, we should appreciate that AlN, GaAs, and SiC only under normal conditions are semi-conductors. Applying relatively high pressures and/or relatively high temperatures, AlN, GaAs, and SiC become conductors. We here note in passing that we also know a lot of organic semi-conductors such as phthalocyanines and perylene-tetracarboxylic-dianhydride (PTCDA). We here also note in passing that the metalloids (B, Si, As, Te, Ge, Sb) are semi-conductors or semi-metals.

50   2. Structures

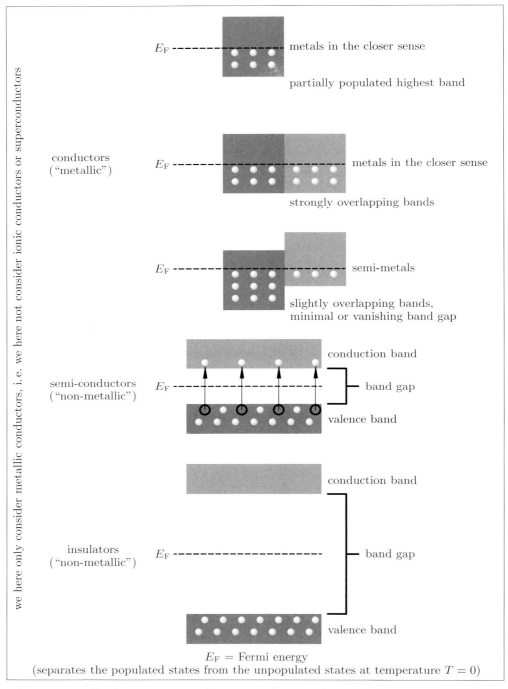

Figure 2.19. Conductors, semi-conductors, insulators. Valence band and conduction band.

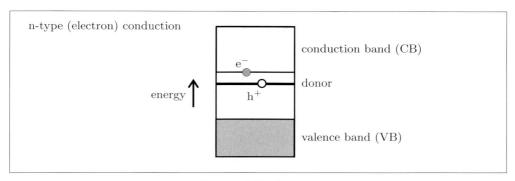

Figure 2.20. Donors.

For the reader it might be interesting to know that there are direct semi-conductors and indirect semi-conductors. For example, silicon (Si) is an indirect semi-conductor, namely the maximum of the valence band and the minimum of the conduction band are not located one upon the other. For example, gallium arsenide (GaAs) is a direct semi-conductor, namely the maximum of the valence band and the minimum of the conduction band are located one upon the other. It might be helpful to throw a glance at Figure 2.22, which supplies the reader with a graphic notion of the evolution of band structures. For example, bringing two Li atoms together, a new electronic state evolves. On the one hand, the inner electrons form two energy levels each populated with two electrons distinguished by antiparallel oriented spins. On the other hand, the $2s^1$ valence electrons form a bonding MO which without excitations is populated with two electrons distinguished by antiparallel oriented spins and an antibonding MO which without excitations is not populated by electrons. Bringing n Li atoms together, band structures evolve, in particular, the $2s^1$ valence electrons form the valence band, which without excitations is half populated and half unpopulated. It is remarkable that one mole (i.e. N_A) atoms, each with X atomic orbitals in the valence shell, evokes X moles of atomic orbitals, leading to X moles of molecular orbitals.

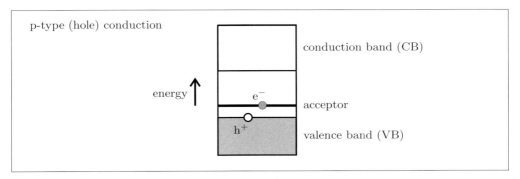

Figure 2.21. Acceptors.

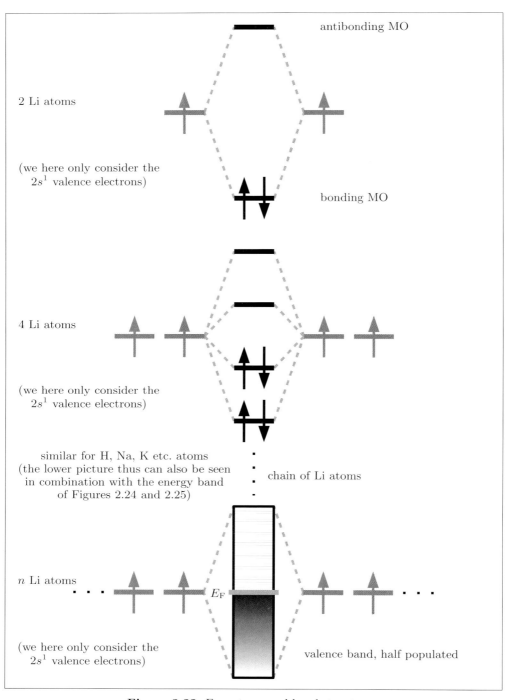

Figure 2.22. Free atoms and band structures.

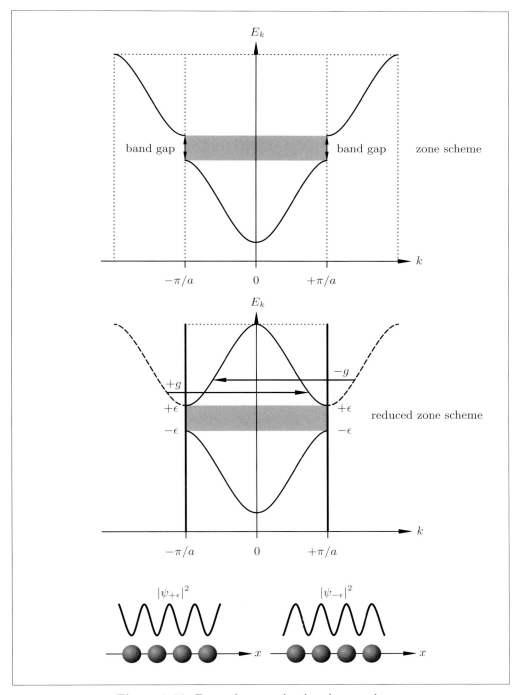

Figure 2.23. Zone scheme and reduced zone scheme.

For the reader it might be interesting to know that Schrödinger's theory supplies us with a direct access to band structures. As it is summarized in Box 2.5, in combination with the Bloch theorem, the time-independent Schrödinger equation (2.40) defines the energy eigenvalues $E = E(\boldsymbol{k})$ of an individual electron in a periodic potential, where $\boldsymbol{k}$ represents nothing else but the wave vectors of the wave functions $\psi(\boldsymbol{x}) = \psi_{\boldsymbol{k}}(\boldsymbol{x})$ of the individual electron in the periodic potential. As it is indicated in Figure 2.23 (top) a zone scheme with a band gap separating different energy bands comes into being, which is readily restructured as it is indicated in Figure 2.23 (middle), in this way, obtaining a reduced zone scheme of the type usually used for the visualization of band structures. We note that the flattening of the first energy band at the band gap reflects a situation that is described by the *Bragg condition*, which states that for certain specific sets $\{\lambda, \theta\}$ of wave lengths λ and angles θ between reflecting planes ("Bragg planes") and wave directions, depending on the given lattice parameters, waves are reflected. We point out that the minimum of the first energy band reflects the particle properties and the wave properties of a nearly free electron which moves in a cubic crystal with lattice spacing a and, on the one hand, can show the angles $\theta = \pm 90°$ between the reflecting planes and the wave directions and, on the other hand, can show the momenta $p \propto k$, in particular, the property $E_k \propto k^2$. We point out that approaching the domain of the Bragg condition, however, the "electron waves" are reflected, as it is outlined in Figure 2.23 (bottom), leading to two different "electron waves", relating to two different energies making up the band gap. As it is outlined in Box 2.5, the Bragg condition in this case reduces to $k = \pm n\pi/a$, for the first energy band indeed requiring the corner points $-\pi/a$ and $+\pi/a$ that apply in Figure 2.23, marking off the so-called first *Brillouin zone* of the cubic crystal, the primitive Wigner–Seitz cell of the *reciprocal lattice* of the cubic cystal. As it is outlined in Box 2.5, it can be shown that $E(\boldsymbol{k}) = E(\boldsymbol{k} + \boldsymbol{g})$ and $\psi_{\boldsymbol{k}}(\boldsymbol{x}) = \psi_{\boldsymbol{k}+\boldsymbol{g}}(\boldsymbol{x})$, where $\boldsymbol{g}$ is nothing else but the lattice spacing of the reciprocal lattice, and this justifies the transition from the zone scheme to the reduced zone scheme via the application of the lattice spacing g of the reciprocal lattice of the cubic crystal that is carried out in Figure 2.23.

For the reader's delight, in Figure 2.24, we outline the link between crystal structure, band structure, Bloch functions, and density of states (DOS), and in Figure 2.25, we additionally outline the link between crystal structure, band structure, Bloch functions, and crystal orbital overlap population (COOP), for the sake of simplicity, in both cases considering the simple example of a H chain. On the one hand, let us here remark that the crystal orbital (CO) theory, generalizing the molecular orbital (MO) theory, suggests different types of superpositions of H wavefunctions for different states of the energy band. We firstly note that a strong curvature of the energy band implies a high density of states dm/dE. We secondly note that the energy band is half populated as it is indicated by the Fermi energy E_F. On the other hand, let us here remark that negative values of the crystal orbital overlap population characterize nothing else but antibonding shares, while positive values of the crystal orbital overlap population characterize nothing else but bonding shares. Therefore, the sign that is obtained by integration up to the Fermi energy E_F defines a measure that is characteristic for the bond order of the crystal structure.

Talking about the Bragg condition, the following more far-reaching annotations should be made. Aiming at analyzing the structure of crystals, the X-ray diffraction, the electron scattering, and the neutron scattering come into play. We annotate that the Bragg diffraction that is illustrated in Box 2.5 explains the diffraction patterns which are produced by these important scattering techniques. So to speak, in the directions of constructive interference where the Bragg condition is valid, intensity maxima of the scattered beams occur, and in the directions of destructive interference where the Bragg condition is not valid, intensity minima of the scattered beams occur, together producing the diffraction patterns. We additionally annotate that the position of the maxima ("spots", "reflections") provides us with an insight into the lattice structure. In particular, utilizing (2.42), we gain the access to the lattice parameters. We additionally annotate that a three-dimensional picture of the electron density within a crystal is obtained if the crystal is placed on a goniometer. In particular, gradually rotating the crystal collecting the diffraction patterns for each orientation of the crystal, after applying the method of Fourier transformation, we gain the access to the distribution of the electrons, the positions of the nuclei, and the orientations of the bonds.

Talking about the Bragg condition, the following more far-reaching annotations should also be made. We annotate that scattering techniques like the X-ray diffraction are applied to monocrystals, polycrystals, and powders. We additionally annotate that such scattering techniques applied to monocrystals lead to diffraction spots, whereas such scattering techniques applied to polycrystals/powders lead to diffraction rings, reflecting the randomness of the orientations of the Bragg planes of a crystalline unit in a polycrystal/powder. We additionally annotate that such scattering techniques in combination with temperature and pressure control allow the tracking of changes of the lattice parameters caused by changes of temperature and pressure, supplying us with an insight into expansion tensors and bulk moduli. Last but not least, isn't it an interesting aspect of scattering phenomena that phase transitions become manifest in changes of the diffraction patterns, for example, Laue spots vanish and Laue spots emerge, lines split and lines coalesce, or amorphous diffraction patterns come into being depending on the nature of the diverse crystalline phases?

Talking with physicists, we learn that the entirety of freely moving electrons, called *electron gas* or *Fermi gas*, interacts with the underground of metal ions, finally evoking the metallic bond. On the one hand, we here point out that these electrons reside within partially populated energy bands. On the other hand, we here point out that these electrons are responsible for the electric conductivity and the thermal conductivity of metallic objects as well as for more far-reaching macroscopic characteristics observable for metallic objects, for example, for the ductility of metallic objects. We additionally take notice that not only ceramic substances such as SiC show metallic traits, even organic substances such as polymers show metallic traits provided these are suitably manufactured. For example, delocalized electrons evoking metallic traits are created by the integration of double bonds.

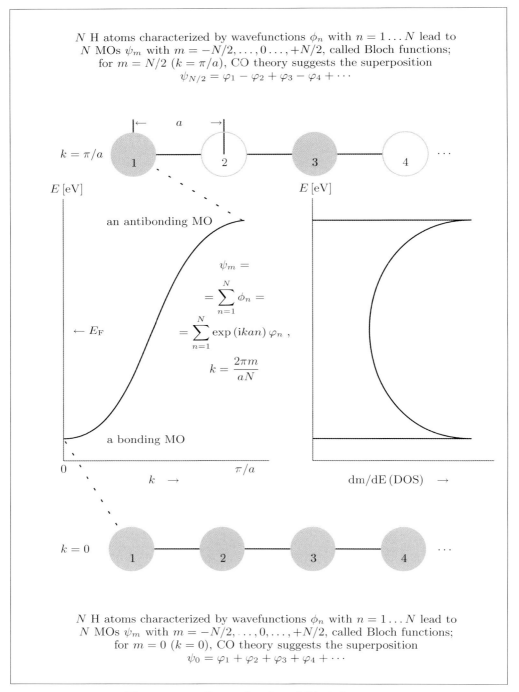

Figure 2.24. Energy bands and Bloch functions.

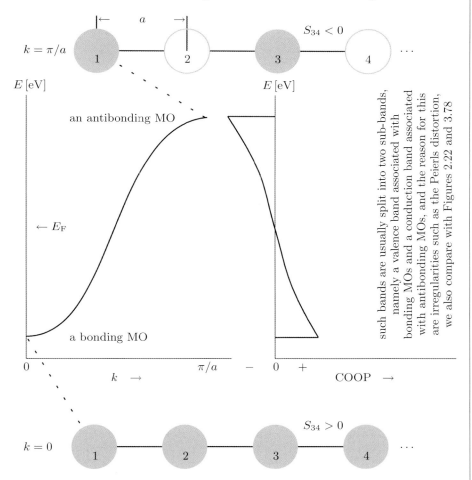

Figure 2.25. Energy bands and overlap population.

Box 2.5 (Important formulae: periodic potentials, Bloch theorem, Bragg condition).

Time-independent Schrödinger equation for periodic potentials
(one particle; lattice constants a_1, a_2, a_3):

$$\hat{H}\psi(\boldsymbol{x}) = E\psi(\boldsymbol{x}) \ , \quad \hat{H} = -\frac{\hbar^2}{2m_0}\triangle + V(\boldsymbol{x}) \ , \quad V(\boldsymbol{x}) = V(\boldsymbol{x}+\boldsymbol{x}_n) \ ,$$
$$\boldsymbol{x}_n = n_1\boldsymbol{a}_1 + n_2\boldsymbol{a}_2 + n_3\boldsymbol{a}_3 \ . \tag{2.40}$$

Bloch theorem:

$$\psi_{\boldsymbol{k}}(\boldsymbol{x}) = u_{\boldsymbol{k}}(\boldsymbol{x})\exp\left[\mathrm{i}\boldsymbol{k}\boldsymbol{x}\right] \ , \quad u_{\boldsymbol{k}}(\boldsymbol{x}) = u_{\boldsymbol{k}}(\boldsymbol{x}+\boldsymbol{x}_n)$$
$$(\psi(\boldsymbol{x}) = \psi_{\boldsymbol{k}}(\boldsymbol{x}) \ , \quad \psi_{\boldsymbol{k}}(\boldsymbol{x}) = \psi_{\boldsymbol{k}+\boldsymbol{g}}(\boldsymbol{x}) \ , \quad E = E(\boldsymbol{k}) \ , \quad E(\boldsymbol{k}) = E(\boldsymbol{k}+\boldsymbol{g})) \ . \tag{2.41}$$

Bragg condition:

$$2d\sin\theta = n\lambda \ . \tag{2.42}$$

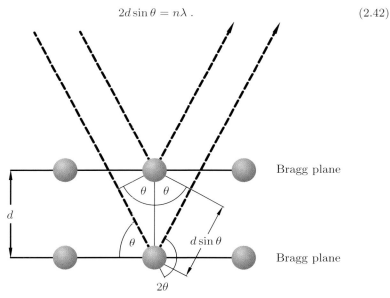

Interplanar distance d for a cubic crystal with lattice spacing a
(h, k, and l are the Miller indices; $n = 1, 2, 3, \ldots$):

$$d = \frac{a}{\sqrt{h^2+k^2+l^2}} \ ,$$
$$2d\sin\theta = n\lambda \rightarrow \frac{\lambda}{2a} = \frac{\sin\theta}{\sqrt{h^2+k^2+l^2}} \quad \text{or} \quad k(\lambda) = \frac{\sqrt{h^2+k^2+l^2}}{\sin\theta}\pi/a \ . \tag{2.43}$$

Allowing $\theta = \pm 0 \ldots \pm 180°$, in this way, imprinting $\pm$ orientations:

$$k(\lambda) = \pm n\pi/a \quad \text{for} \quad \theta = \pm 90° \ . \tag{2.44}$$

2.2 Crystal Structures

We all know it, our lifes show elements of regularity and irregularity, order and chaos. Moreover, we all know it, mental dynamics just like social dynamics, organic dynamics just like inorganic dynamics, animated compositions just like inanimate compositions, gases just like liquids, all shows regularity and irregularity, order and chaos. Naturally, we all know it, this applies for atomic and molecular assemblies as well. For instance, SiO_2 shows unsymmetric, amorphous patterns, we compare with Figures 2.26 and 2.28, and symmetric, crystalline patterns, we compare with Figure 2.27, but all are based upon regular [SiO_4] tetrahedrons.

We are now ready to study crystalline patterns by considering crystal structures of ceramic substances. Before we go into details, however, we point out the criteria that govern the formation of crystal structures. (i) *Valency* and *stoichiometry*. Closely following dictionaries, we define the term "valency" as the number of hydrogen atoms that an atom or a chemical group can combine or substitute in forming compounds, and we define the term "stoichiometry" as the relation between the quantities of substances that take part in a reaction or form a compound. The formation of crystal structures only is possible in accordance with these chemical properties. (ii) *Bond type*. Closely following dictionaries, we define "bond type" as the type of force of attraction that holds atoms or ions together in a molecule or a crystal and we note that bonds are usually created by transfering or sharing one or more electrons in form of single, double, or triple bonds, we also compare with Section 2.1. The formation of crystal structures depends on the bond type, in particular, if only (directional) covalent contributions are present, the coordination number 4 associated with tetrahedral geometry is observed, but if also (non-directional) ionic contributions are present, other coordination numbers associated with other coordination geometries are observed, we also compare with Section 2.1. (iii) *Radius ratio*. Closely following dictionaries, we define "radius ratio" as the relative magnitude of two radii expressed as a quotient, we also compare with Section 2.1. The formation of crystal structures consisting of more than one species can be understood as the development of a superior (space filling) lattice consisting of a relatively big kind of particles (which are usually provided by anions) concatenated with inferior (hole filling) lattices each consisting of a relatively small kind of particles (which are usually provided by cations).

We again take notice that structures of complex ionic crystals obey *Pauling's rules* [42]. (1) Around each cation, a coordination polyhedron of anions is formed. The sum of radii of ions determines the ion distance, i.e. the distance between cations and anions. The radius ratio determines the coordination number. (2) The charge of an anion is approximately or exactly equal to the negative sum of the electrostatic bond strengths of the cations that surround the anion. (3) The stability of an ionic structure decreases if edges and particularly faces are shared by two anion polyhedra. For cations with high charge and low coordination number, this effect is particularly large. (4) In a crystal with different kinds of cations, those of high valency and small coordination number tend not to share polyhedrons with one another. (5) The number of different kinds of constituents in a crystal tends to be small.

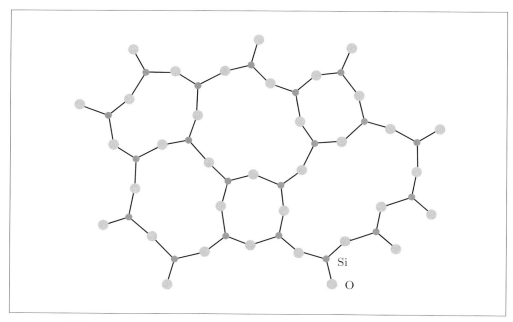

Figure 2.26. Amorphous SiO$_2$ (quartz glass), planar representation.

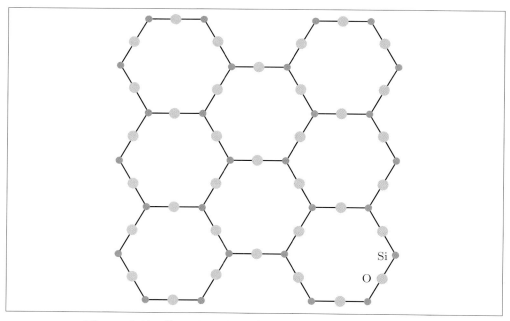

Figure 2.27. Crystalline SiO$_2$ (quartz), planar representation.

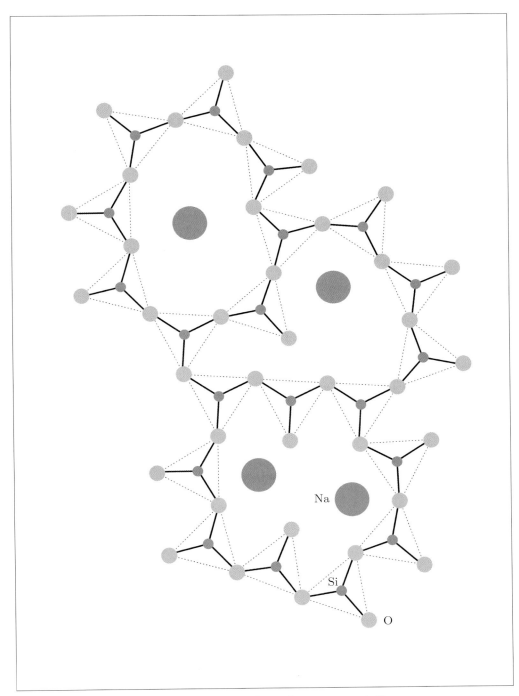

Figure 2.28. Silicate glass, planar representation.

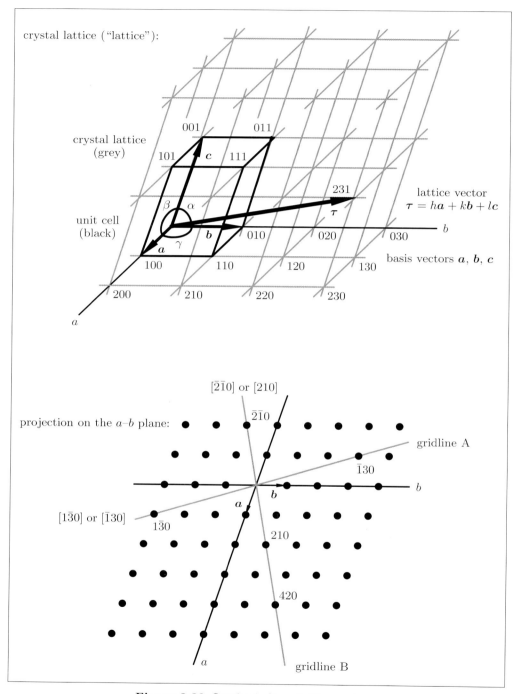

Figure 2.29. Lattice indices, lattice vectors.

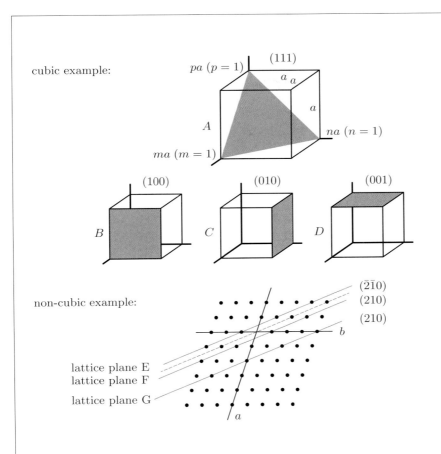

Figure 2.30. Miller indices, lattice planes.

2.2.1 Lattice Types

Cooling down liquefied ceramic material, letting stream solutions of ceramic material over appropriate layers, or letting condense vapors of ceramic material on appropriate substrates, crystal structures emerge. (i) We here note that crystallization processes are exergonic (exothermal) processes i.e. energy is released. (ii) We here also note that a crystalline solid can be monocrystalline or polycrystalline. For example, falling below the melting point, at the beginning of the crystallization, seed crystals which eventually establish centers for the evolution of crystallites evolve. If the crystallization starts at one location, a *monocrystal* evolves. If the crystallization starts at more than one location, a *polycrystal* evolves. In most cases, a crystalline solid is a polycrystal composed of many crystallites that are separated by grain boundaries. (iii) We here also note that crystal structures depend on further constraints such as temperature and pressure. If the constraints are altered, *phase transformations* may take place, evoking different crystalline modifications. (iv) It is a known fact that crystalline states are characterized by a periodic formation of their basis elements (ions, atoms, molecules) in the three-dimensional space. It is also a known fact that the crystalline structures that come into being are described by space lattices, the lattice points of which are placeholders for individual particles or particle formations. Following the explanations of the preceding sections, we understand that the principal influence parameters for the evolution of crystalline states are the ratios of the particle radii, the numbers of bonds, the directions of bonds, and the pursuit of relatively low internal energy.

2.2.1.1 Lattice Indices, Lattice Vectors, Lattice Planes

As it is shown in Figure 2.29, for each *lattice* a *unit cell* that is characterized by the *lattice constants* a, b, c, α, β, and γ, i.e. the basis vectors and the angles between them, can be defined from which the lattice can be reconstructed via translational operations. We note that such a unit cell can be a primitive (the smallest possible) unit cell or not, however, in any case such a unit cell should reflect the complete symmetry of the crystalline structure. As it is shown in Figure 2.29, each lattice point is conventionally described by integers h, k, and l, and negative values are conventionally outlined by an additional bar. We note that the *lattice vector* $\boldsymbol{\tau}$ then is readily expressed by the basis vectors $\boldsymbol{a}$, $\boldsymbol{b}$, and $\boldsymbol{c}$ as $\boldsymbol{\tau} = h\boldsymbol{a} + k\boldsymbol{b} + l\boldsymbol{c}$, and a *gridline* then is readily expressed as a value triple $[hkl]$. As it is shown in Figure 2.30, each *lattice plane* is conventionally described by integers h, k, and l, too, and negative values are conventionally outlined by an additional bar, too. However, in the first case the h, k, and l are *lattice indices* that are counted starting from a given origin, but in the second case the h, k, and l are *Miller indices* that come into being as follows. In a first step, we have to determine the intersection points of a lattice plane with the a, b, and c axes, evoking the values m, n, and p. In a second step, we construct the reciprocal form $1/m$, $1/n$, and $1/p$ and multiply with the lowest common multiple f of the denominator. The gridlines, i.e. the lattice directions, are indicated by brackets $[hkl]$, while the lattice planes are indicated by parenthesis (hkl), and for the description of hexagonal and rhombohedral systems also the four Bravais–Miller indices h, k, i, and l are used.

2.2.1.2 Crystal Systems, Bravais Lattices, Space Groups

Leafing through the textbooks of crystallography, in particular, we learn that there are seven *crystal systems* – the cubic, the rhombohedral, the tetragonal, the hexagonal, the orthorhombic, the monoclinic, and the triclinic system – which can be characterized by 14 unit cells, termed *Bravais lattices*, collected in Figures 2.31 and 2.32. Leafing through the textbooks of group theory, in particular, we learn that there are 32 point groups and 230 space groups characterizing the symmetry of the seven crystal systems.

The reader should realize that the Bravais lattices shown in Figures 2.31 and 2.32 vary in the length of the unit cell axes and in the angles of the unit cell axes, arranged back-to-back, finally forming translational lattices. The reader should also realize that in all cases where more than one type of individual atoms or atom groups establish a translational lattice, nested sub-lattices are applied to characterize the arrangement of the individual atoms or atom groups. As a general rule, the first sub-lattice is a close-packed (more or less dense-packed) lattice composed of relatively bigger anions such as O^{2-}, which essentially determine the substantial filling of the lattice, and the second sub-lattice is composed of relatively smaller cations placed at the spacings between the anions. If the lattice points are only placed on the cell corners, one speaks of "primitive centering" (often indicated by the symbol P), if one additional lattice point is placed in the center of the cell, one speaks of "body-centering" (often indicated by the symbol I), if one additional lattice point is placed in the center of each of the faces of the cell, one speaks of "face-centering" (often indicated by the symbol F), and if one additional lattice point is placed in the center of one of the faces of the cell, one speaks of "single-face-centering" (depending on the face orientation, often indicated by the symbols A, B, and C). We realize that with the hexagonal system two ways of description of the unit cell are common. On the one hand, we consider the complete hexagonal element. On the other hand, we consider one-third of it. We compare with Figure 2.32, where these circumstances are illustrated.

The reader should realize that the *cubic close-packed (ccp) lattice*, we compare with Figure 2.33, and the *hexagonal close-packed (hcp) lattice*, we compare with Figure 2.34, are of particular interest. Looking through the cubic close-packed (ccp) lattice along those diagonal direction where a 3–6–3 polyhedron instead of a 4–4–4 polyhedron is visible, a comparison of both lattice types is directly possible, i.e. we then realize that the principal difference is that a cubic close-packed (ccp) lattice is characterized by an A–B–C layering, while a hexagonal close-packed (hcp) lattice is characterized by an A–B–A layering, where A, B, and C indicate layers of close-packed atom planes that are placed differently, we compare with Figures 2.35 and 2.36. In both lattices, there are tetragonal gaps, namely gaps in which the center is surrounded by four atoms, and octahedral gaps, namely gaps in which the center is surrounded by six atoms. There are eight tetrahedral gaps and four octahedral gaps within the unit cell of a cubic close-packed lattice, and ten tetrahedral gaps and six octahedral gaps within the unit cell of a hexagonal close-packed lattice, we compare with Figures 2.37 and 2.38. Naturally, consistent with the stoichiometry of the compounds, more or less of these gaps are populated.

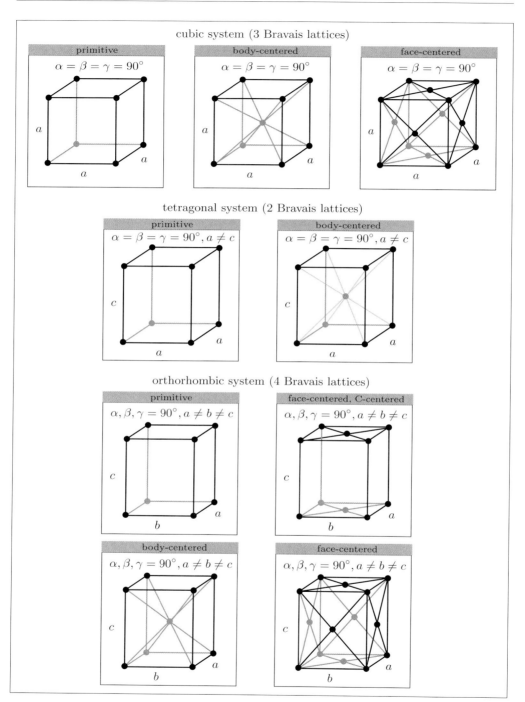

Figure 2.31. Bravais lattices, part one.

hexagonal system (1 Bravais lattice)

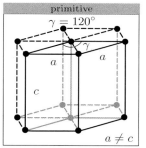

annotation:
in accordance to
other publications,
in special cases,
we also want to consider
an individual block,
e. g. the drawn through block,
as the hexagonal unit cell

rhombohedral system (1 Bravais lattice)

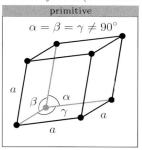

annotation:
a rhombohedral object
is a special trigonal object;
this special trigonal object
and the special trigonal object
that is defined by a block
of a hexagonal object
define the trigonal system

monoclinic system (2 Bravais lattices)

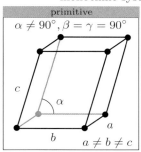
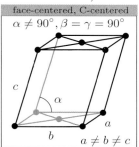

triclinic system (1 Bravais lattice)

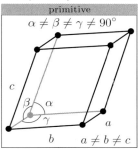

Figure 2.32. Bravais lattices, part two.

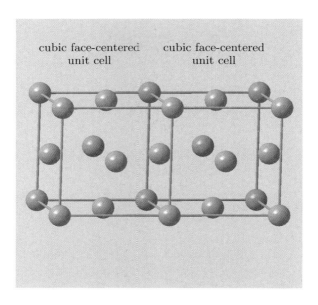

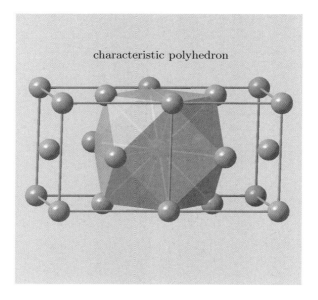

Figure 2.33. Cubic close-packed (ccp) lattice. Unit cell and characteristic polyhedron.

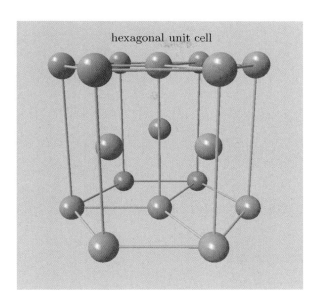

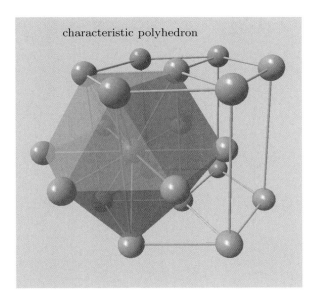

Figure 2.34. Hexagonal close-packed (hcp) lattice. Unit cell and characteristic polyhedron.

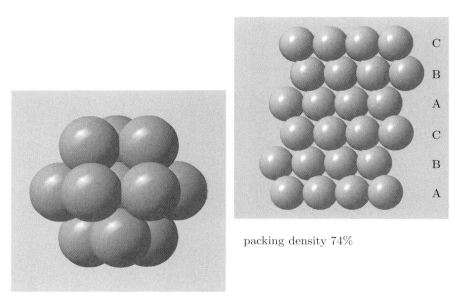

packing density 74%

note that A, B, and C here do not indicate single-phase-centering, but special position shifts of the layers

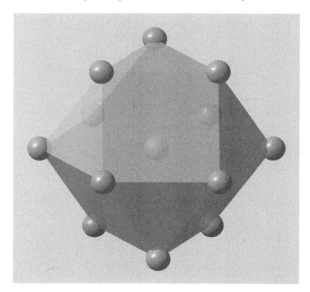

Figure 2.35. Cubic close-packed lattice. A–B–C layering.

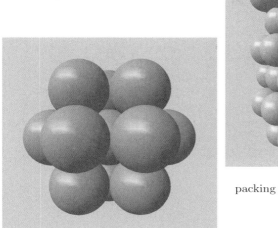
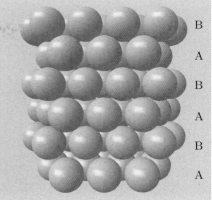

packing density 74%

note that A and B here do not indicate single-phase-centering, but special position shifts of the layers

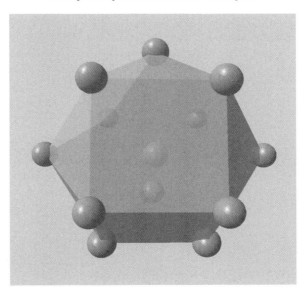

Figure 2.36. Hexagonal close-packed lattice. A–B–A layering.

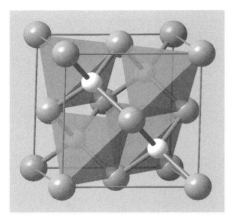

eight tetrahedral gaps here are shown,
four gaps are populated
(white spheres)
and four gaps are unpopulated
(centers of polyhedra),
all eight gaps are attributed to the unit cell

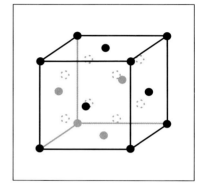

the drawn through circles outline
gaps on the cell edges,
while the dotted circles outline
gaps inside the cell

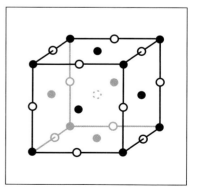

thirteen octahedral gaps here are shown,
we indicate all thirteen gaps
(centers of polyhedra),
four gaps are attributed to the unit cell

Figure 2.37. Cubic close-packed lattice. Tetrahedral gaps *versus* octahedral gaps.

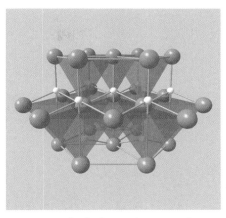

twenty tetrahedral gaps here are shown,
we show ten populated gaps
(white spheres)
and ten unpopulated gaps
(centers of polyhedra),
ten gaps are attributed to the unit cell

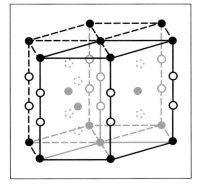

the drawn through circles outline
gaps on the cell edges,
while the dotted circles outline
gaps inside the cell

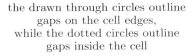

six octahedral gaps here are shown,
we indicate all six gaps
(centers of polyhedra),
all six gaps are attributed to the unit cell

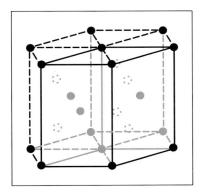

Figure 2.38. Hexagonal close-packed lattice. Tetrahedral gaps *versus* octahedral gaps.

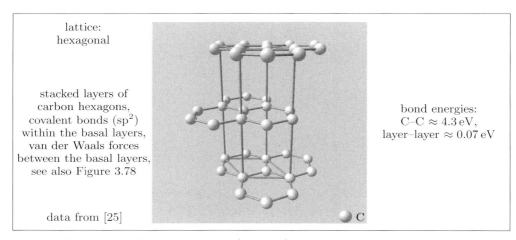

Figure 2.39. Cristalline carbon (graphite). The box indicates the unit cell.

2.2.2 Crystal Types

Leafing through the publications of advanced ceramics, in particular, we find a lot of ceramic exponents of the seven crystal systems described above. In Box 2.6, selected ceramic exponents are collected together with *structure prototypes* such as rock salt, cesium chloride, and zinc blende. We take notice that Box 2.6 provides information about the distribution of the individual constituents at the individual sub-lattices (see Figures 2.41–2.50). We also take notice that structure prototypes are compounds after which a set of compounds with the same crystal lattice is named each. We further take notice that carbon (C) materials such as graphite (see Figure 2.39), diamond (see Figure 2.40), lonsdalite, and the fullerenes are structure prototypes of the ceramic kind which consist of only one sort of constituents.

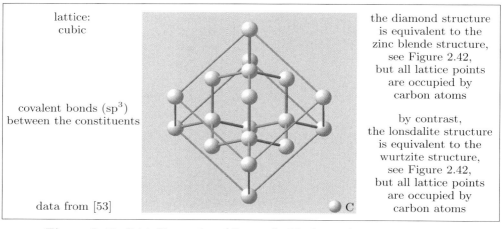

Figure 2.40. Cristalline carbon (diamond). The box indicates the unit cell.

Box 2.6 (Important data: ceramic lattices).

The compounds that follow show a more or less ionic character.
We thus distinguish between cations and anions.
(A = cations, X = anions, cn = coordination number.)

AX lattices (anions and cations have equal valency).

Rock salt (NaCl, lattice: cubic, see Figure 2.41):

anions (Cl$^-$) form a cubic close-packed lattice (cn: 6A + 12X),
cations (Na$^+$) populate all octahedral gaps of the anion lattice (cn: 6X),
radius cation/radius anion $r_c/r_a|_{stable} \geq 0.414$ (octahedron).

Some compounds with the same crystal lattice:

NaCl(0.563), MnO(0.57), FeO(0.54), CoO(0.51), NiO(0.5), MgO(0.46),
CaO(0.71), SrO(0.8), BaO(0.96), CdO(0.69).

Caesium chloride (CsCl, lattice: cubic, see Figure 2.41):

anions (Cl$^-$) form a cubic primitive lattice (cn: 8A + 6X),
cations (Cs$^+$) populate all cubic gaps of the anion lattice (cn: 8X),
radius cation/radius anion $r_c/r_a|_{stable} \geq 0.732$ (cube).

Some compounds with the same crystal lattice:

CsCl(0.95), CsBr(0.88), CsJ(0.78).

Zinc blende (α–ZnS, lattice: cubic, see Figure 2.42):

anions (S^{2-}) form a cubic close-packed lattice (cn: 4A + 12X),
cations (Z^{2+}) populate 1/2 of the tetrahedral gaps of the anion lattice (cn: 4X),
radius cation/radius anion $r_c/r_a|_{stable} \geq 0.225$ (tetrahedron).

Some compounds with the same crystal lattice:

α–ZnS(0.33), GaAs, β–SiC, β–BeO.

Wurtzite (β–ZnS, lattice: hexagonal, see Figure 2.42):

anions (S^{2-}) form a hexagonal close-packed lattice (cn: 4A + 12X),
cations (Z^{2+}) populate 1/2 of the tetrahedral gaps of the anion lattice (cn: 4X),
radius cation/radius anion $r_c/r_a|_{stable} \geq 0.225$ (tetrahedron).

Some compounds with the same crystal lattice:

β–ZnS(0.33), AlN, α–SiC, α–BeO, ZnO.

> **Continuation of Box.**
>
> The compounds that follow show a more or less ionic character.
> We thus distinguish between cations and anions.
> (A = cations, X = anions, cn = coordination number.)
>
> **AX_2 lattices (anions and cations have different valency).**
>
> *Fluorite* (CaF_2, lattice: cubic, see Figure 2.43):
> anions (F^-) form a cubic primitive lattice (cn: 4A + 6X),
> cations (Ca^{2+}) form a cubic close-packed lattice (cn: 8X + 12A),
> radius cation/radius anion $r_c/r_a|_{stable} \geq 0.732$ (cube).
>
> Some compounds with the same crystal lattice:
> CaF_2, CeF_2, ThO_2, CeO_2, UO_2, γ–ZrO_2, HfO_2, PuO_2.
>
> Note that exchanging the anions and the cations, we obtain the antifluorite strcuture.
> For the fluorite structure: $r_c > r_a$. For the antifluorite strcuture: $r_c < r_a$.
>
> *Low quartz* (α–quartz, lattice: rhombohedral, see Figure 2.44):
> radius cation/radius anion $r_c/r_a|_{stable} \geq 0.225$ (tetrahedron).
>
> *Low cristobalite* (α–cristobalite, lattice: tetragonal, see Figure 2.45):
> radius cation/radius anion $r_c/r_a|_{stable} \geq 0.225$ (tetrahedron).
>
> *Rutile* (TiO_2, lattice: tetragonal, see Figure 2.46):
> anions (O^{2-}) form a disturbed hexagonal close-packed lattice (cn: 3A),
> cations (Ti^{4+}) form a tetragonal body-centered lattice (cn: 6X),
> radius cation/radius anion $r_c/r_a|_{stable} \geq 0.414$ (octahedron).
>
> Some compounds with the same crystal lattice:
> TiO_2, GeO_2, PbO_2, SnO_2, TeO_2, MoO_2, MnO_2.
>
> **A_2X_3 lattices (anions and cations have different valency).**
>
> *Corundum* (α–Al_2O_3, lattice: hexagonal, see Figure 2.47):
> anions (O^{2-}) form a hexagonal close-packed lattice (cn: 4A + 12X),
> cations (Al^{3+}) populate 2/3 of the octahedral gaps of the anion lattice (cn: 6X).
>
> Some compounds with the same crystal lattice:
> α–Al_2O_3, α–Fe_2O_3, Cr_2O_3, α–GaO_3.

> **Continuation of Box.**
>
> The compounds that follow show a more or less ionic character.
> We thus distinguish between cations and anions.
> (A, B = cations, X = anions, cn = coordination number.)
>
> **ABX_2 lattices (more than one cation).**
>
> *Perovskite* ($CaTiO_2$, lattice: cubic, see Figure 2.48):
>
> cations A (Ti^{4+}) form a cubic primitive lattice (cn: 6X),
> cations B (Ca^{2+}) form a cubic primitive lattice (cn: 12X),
> anions (O^{2-}) are arranged cubic face-centered.
>
> Some compounds with the same crystal lattice:
> $CaTiO_2$, $SrTiO_2$, $BaTiO_2$, $CaZrO_2$, $KNbO_2$, $LaAlO_2$, $YAlO_2$.
>
> **AB_2X_4 lattices (more than one cation).**
>
> *Spinel* ($MgAl_2O_4$, lattice: cubic, see Figure 2.49):
>
> all cations A (Mg^{2+}) populate tetrahedral gaps (cn: 4X),
> all cations B (Al^{3+}) populate octahedral gaps (cn: 6X).
>
> Some compounds with the same crystal lattice:
> $MgAl_2O_4$, $FeAl_2O_4$, $ZnAl_2O_4$.
>
> Inverse spinel:
>
> all cations A (Mg^{2+}) and 1/2 of the cations B (Al^{3+}) populate octahedral gaps,
> 1/2 of the cations B (Al^{3+}) populate tetrahedral gaps.
>
> Some compounds with the same crystal lattice:
> $MgTiMgO_4$, $FeMgFeO_4$, $FeTiFeO_4$, $GaMgGaO_4$, $InMgInO_4$,
> $FeFeFeO_4$, $FeNiFeO_4$, $FeCoFeO_4$.
>
> *Nickel potassium fluoride* (NiK_2F_4, lattice: tetragonal, see Figure 2.50):
> combination of perovskite structure and rock salt structure.
>
> Some compounds with the same crystal lattice:
> $La_{2-x}Sr_xCuO_4$, $YBa_2Cu_3O_{7-x}$, $Bi_2Sr_2CaCu_2O_8$, $Bi_2Sr_2Ca_2Cu_3O_{10}$.
>
> (Superconducting substances, we compare with Section 3.4.2.)

Some additional details compactly follow here. (i) Let us consider the *AX lattices*. For the reader it should be interesting to know that the covalent contributions to the bonds from cesium chloride (CsCl) via rock salt (NaCl) and wurtzite (β–ZnS) to zinc blende (α–ZnS) become much stronger. We take notice that the difference between the zinc blende structure and the diamond ("cubic diamond") structure only is that zinc blende is made up of two chemical elements (S, Zn), while diamond is made up of one chemical element (C). We take notice that the same applies for the difference between the wurtzite structure and the lonsdalite ("hexagonal diamond") structure, i.e. the difference between wurtzite and lonsdalite only is that wurtzite is made up of two chemical elements (S, Zn), while lonsdalite is made up of one chemical element (C). At this point, we also note in passing that a lot of compounds can crystallize into the zinc blende lattice or/and the wurtzite lattice. For example, GaAs can crystallize into the zinc blende lattice, AlN can crystallize into the wurtzite lattice, and SiC is observed in zinc blende form (β–SiC) and in wurtzite form (α–SiC). (ii) Let us consider the AX_2 *lattices*. For the reader it should be interesting to know that fluorite nowadays is used to produce high performance telescopes and high performance camera lenses. Due to its high transparency for ultraviolet wavelengths of about 157 nm, it is used to produce optical elements for such ultraviolet wavelengths. Due to its low dispersion, it is used to produce optical elements with low chromatic aberration. For the reader it should be also interesting to know that rutile nowadays is used to produce pigments for paints that call for a bright white color. Since it absorbs ultraviolet light, it is used to produce sunscreens. Since the UV absorption of nano-scale particles is blue shifted in comparison to the UV absorption of the bulk material, nano-scale particles are of particular interest. (iii) Let us consider the A_2X_3 *lattices*. For the reader it should be interesting to know that the hardness of corundum (absolute hardness 400) nearly enters the territory of the hardness of diamond (absolute hardness 1500). Therefore, it is used as abrasive, for example, on sandpaper. (iv) Let us consider the ABX_2 *lattices*. For the reader it should be interesting to know that perovskite-type materials, i.e. materials showing the perovskite structure or including the perovskite structure as partial elements, show a wealth of interesting electromagnetic properties, in particular, charge ordering capabilities, spin ordering capabilities, and interplay mechanisms which concatenate electric properties and magnetic properties with structural properties and transport properties. We note that these are used as sensors, as catalyst electrodes in special types of fuel cells, as ferroelectrics, as thermoelectrics, and as superconductors. (v) Let us consider the AB_2X_4 *lattices*. For the reader it should be interesting to know that spinel-type materials, i.e. materials showing the spinel structure or including the spinel structure as partial elements, are flexible regarding the cations A and B, in particular, the cations A and B can mix, leading to different chemical compounds. We note that these are used for paints, solid-state lasers, and magnetic storage devices. (vi) In addition, for the reader it should be interesting to know that spinel-type materials can show a wealth of vacancies as integral part. We note that the deeper reason for these so-called "defect spinels" is that cations (for example, Fe^{2+} ions) can be oxidized (for example, Fe^{2+} ions can be oxidized to Fe^{3+} ions), leaving a charge imbalance, which is compensated by the release of atoms (for example, Fe atoms can be released), finally evoking the vacancies mentioned above.

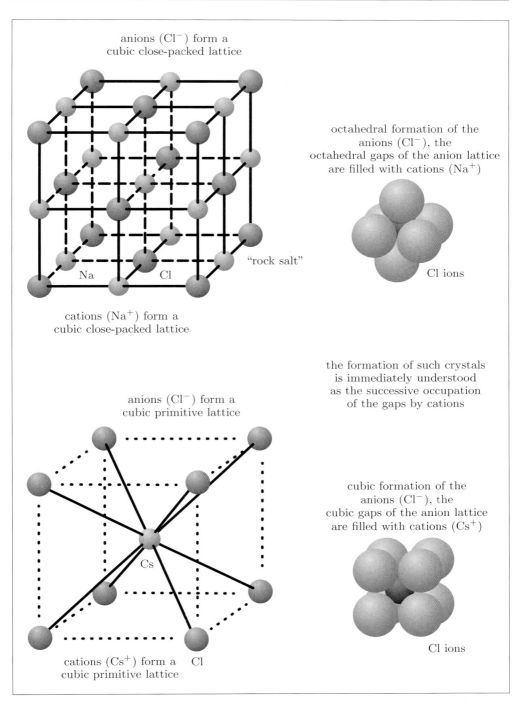

Figure 2.41. Crystalline NaCl (rock salt), crystalline CsCl (cesium chloride).

80   2. Structures

anions (S^{2-}) form a
cubic close-packed lattice

tetrahedral formation of the
anions (S^{2-}), 1/2 of the
tetrahedral gaps of the anion lattice
are filled with cations (Zn^{2+})

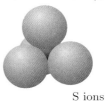

S ions

"zinc blende"   data from [68]

anions (S^{2-}) form a
hexagonal close-packed lattice

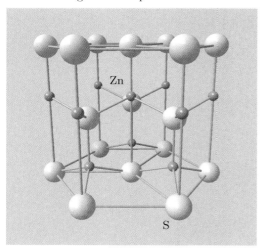

tetrahedral formation of the
anions (S^{2-}), 1/2 of the
tetrahedral gaps of the anion lattice
are filled with cations (Zn^{2+})

S ions

"wurtzite"   data from [60]

Figure 2.42. Crystalline ZnS (zinc blende, wurtzite).

cations (Ca^{2+}) form a
cubic close-packed lattice

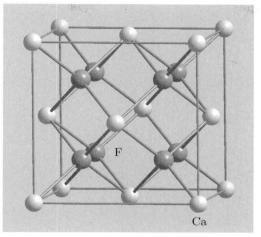

(anions (F^-) form a
cubic primitive lattice)

"fluorite", "fluorspar"

cubic formation of the
anions (F^-), the
cubic gaps of the anion lattice
are filled with cations (Ca^{2+})

F ions

the above unit cell can be decomposed
into cubic formations of the anions

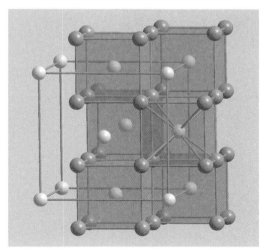

data from [5]

Figure 2.43. Crystalline CaF_2 (fluorite, fluorspar).

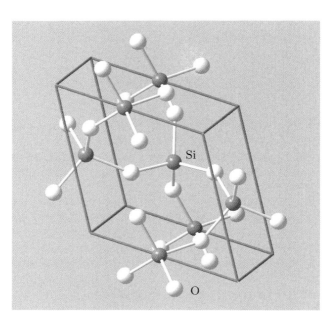

AX$_2$ lattice, rhombohedral, Si^{4+}, O^{2-}

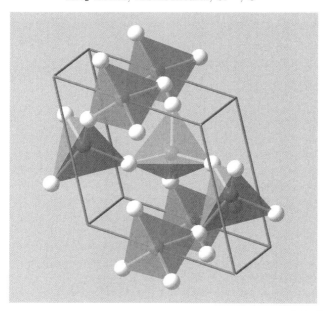

data from [69]

Figure 2.44. Crystalline SiO$_2$ (low quartz).

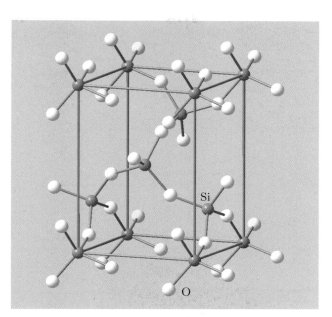

AX$_2$ lattice, tetragonal, Si^{4+}, O^{2-}

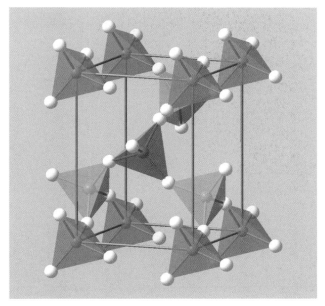

data from [45]

Figure 2.45. Crystalline SiO$_2$ (low cristobalite).

cations (Ti^{4+}) form a
tetragonal body-centered lattice

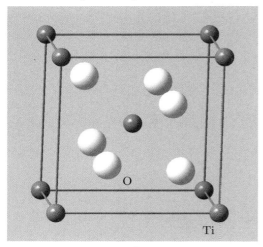

(anions (O^{2-}) form a
disturbed hexagonal close-packed lattice)

"rutile"

octahedral formation of the
anions (O^{2-}), 1/2 of the
octahedral gaps of the anion lattice
are filled with cations (Ti^{4+})

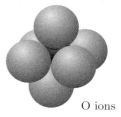

O ions

the octahedra share corners,
leading to a long-range structure

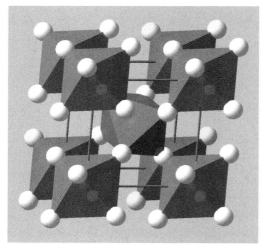

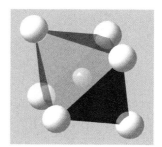

data from [21]

Figure 2.46. Crystalline TiO_2 (rutile).

anions (O^{2-}) form a
hexagonal close-packed lattice

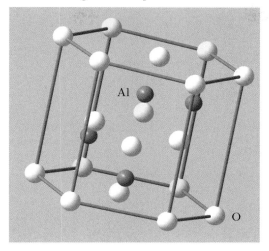

"corundum"

octahedral formation of the
anions (O^{2-}), 2/3 of the
octahedral gaps of the anion lattice
are filled with cations (Al^{3+})

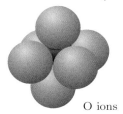

O ions

the octahedra share corners,
leading to a long-range structure

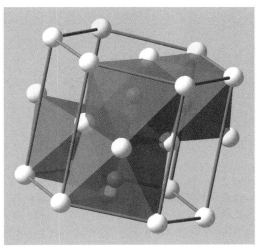

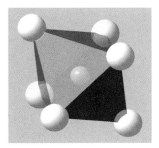

data from [16]

Figure 2.47. Crystalline Al_2O_3 (corundum).

cations A (Ti^{4+}) form a
cubic primitive lattice

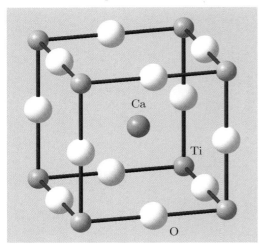

(cations B (Ca^{2+}) form a cubic primitive lattice,
anions (O^{2-}) are arranged cubic face-centered)

we here note that the
face-centered arrangement of the
anions O^{2-} becomes directly visible
by constructing a unit cell with the
cations Ca^{2+} as corners
and a cation Ti^{4+} as center

"perovskite"

octahedral formation of the
anions (O^{2-}), 1/4 of the
octahedral gaps of the anion lattice
are filled with Ti cations (Ti^{4+})

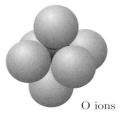

O ions

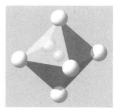

the octahedra share corners,
leading to a long-range structure

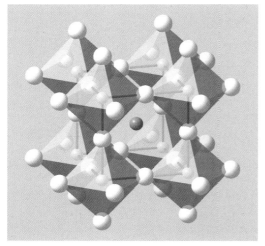

data from [46]

Figure 2.48. Crystalline $CaTiO_3$ (perovskite).

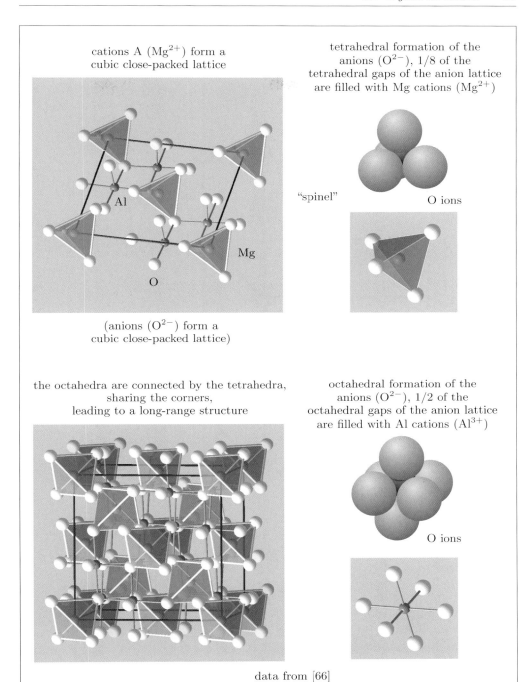

Figure 2.49. Crystalline $MgAl_2O_4$ (spinel).

88   2. Structures

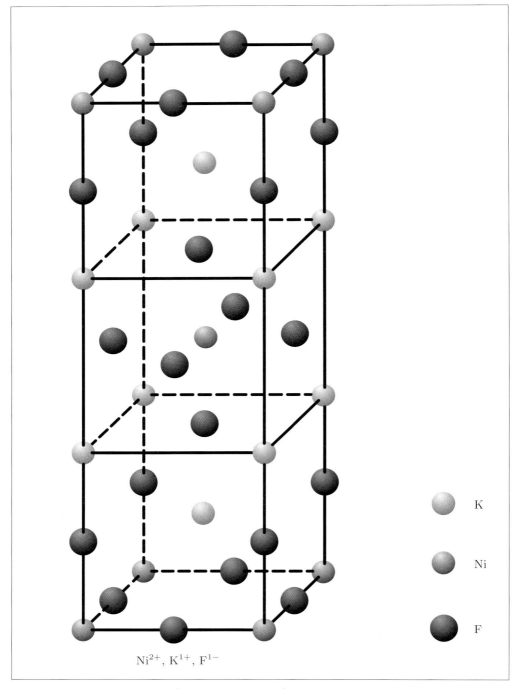

Figure 2.50. Crystalline NiK$_2$F$_4$ (nickel potassium fluoride).

Due to their importance and their rich variety of forms of appearance, we consider *silicates* as an own class of compounds. We point out that silicates can be defined as compounds that contain silicon atoms surrounded by oxygen to form anions such as $[SiO_4]^{4-}$. (i) On the one hand, one then can conveniently classify the structural patterns by means of the most important chemical assemblies, i.e. one distinguishes between arrangements based upon lone tetrahedrons $[SiO_4]^{4-}$ ("nesosilicates"), based upon double tetrahedrons $[Si_2O_7]^{6-}$, $[Si_2O_6]^{4-}$, $[Si_2O_5]^{2-}$ ("sorosilicates"), based upon rings $[Si_nO_{3n}]^{2n-} = [SiO_3^{2-}]_n$ ("cyclosilicates"), based upon single chains or double chains $[Si_nO_{3n}]^{2n-} = [SiO_3^{2-}]_n$ or $[Si_{4n}O_{11n}]^{6n-} = [Si_4O_{11}^{6-}]_n$ ("inosilicates"), based upon sheets $[Si_{2n}O_{5n}]^{2n-} = [Si_2O_5^{2-}]_n$ ("phyllosilicates"), and based upon frameworks $[Al_nSi_mO_{2(n+m)}]^{n-}$ ("tectosilicates"). (ii) On the other hand, one then can conveniently classify the structural patterns by means of numbers. (1) The coordination number cn. (2) The number of linkedness L. $L = 0$: isolated polyhedra. $L = 1$: polyhedra linked via corner points. $L = 2$: polyhedra linked via edges. $L = 3$: polyhedra linked via faces. (3) The number of connectedness Q. Q^0: isolated polyhedron. Q^1: polyhedron linked to one other polyhedron. Q^2: polyhedron linked to two other polyhedra. Q^3: polyhedron linked to three other polyhedra. Q^4: polyhedron linked to four other polyhedra. (4) We may also encounter the number of branchedness B, which characterizes the ramification of chains, rings, sheets etc., the number of multiplicity M, which counts the chains, rings, sheets etc. that build up a structural pattern, and the number of periodicity P, which counts the chains, rings, sheets etc. after which a theme recurs. (iii) We note that further schemes for the classification of silicates are found in literature. For instance, extended schemes for the classification of silicates include structural units such as $[SiF_6]^{2-}$ so that further numbers, for example, the number of different anions, are needed.

In the above sequence of arrangements, the silicon-to-oxygen ratio decreases from $1:4$ in the case of the $[SiO_4]^{4-}$-type compounds (with four charges per unit) to $1:2$ in the case of the SiO_2-type compound *cristobalite* (with no charges per unit). We note that these can be considered as end-members of arrangements of the type shown in Figures 2.51–2.54. We also note that the SiO_2-type compound *cristobalite* is illustrated in Figure 2.45. Departing from arrangements of the type shown in Figures 2.51–2.54, silicious networks such as those shown in Figures 2.55–2.64, are easily constructed. On the one hand, let us here point out that silicious networks can be considered as assemblies of polyhedra sharing corners, edges, and planes. On the other hand, let us here point out that phyllosilicates can show a single-layer structure, but also a two-layer structure (kaolinite etc.), a three-layer structure (montmorillonite etc.), and even a four-layer structure (donbassite etc.). For the convenience of the reader, in Figure 2.61, we illustrate the crystal structure of talc, which is a three-layer silicate where two tetrahedral sheets formed of SiO_4 tetrahedrons occur combined with one octahedral layer formed of $Mg(OH)_6$ octahedrons. We here especially mention that this sandwich structure is responsible for the water storage capacity of phyllosilicates as well as for the plasticity of phyllosilicates. We here additionally mention that the water storage capacity of phyllosilicates can lie above 700% and that the plasticity of phyllosilicates can be directly led back to the capability of phyllosilicates to shear along the layers, and this is the basis for forming clay.

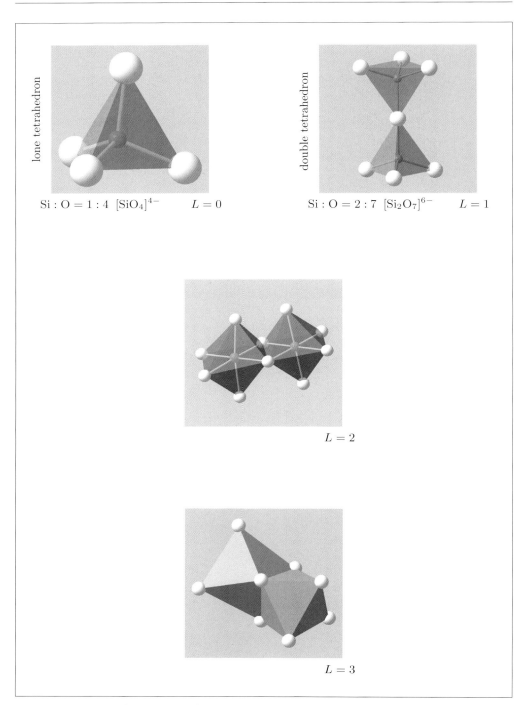

Figure 2.51. On the Bragg–Liebau symbols, part one.

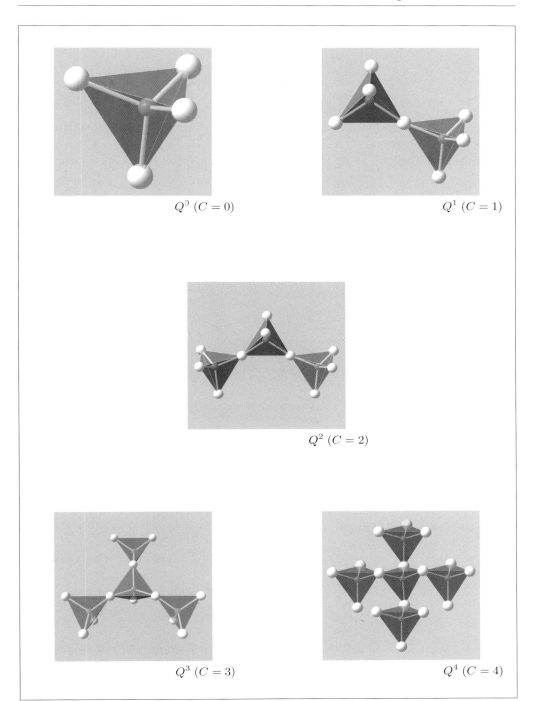

Figure 2.52. On the Bragg–Liebau symbols, part two

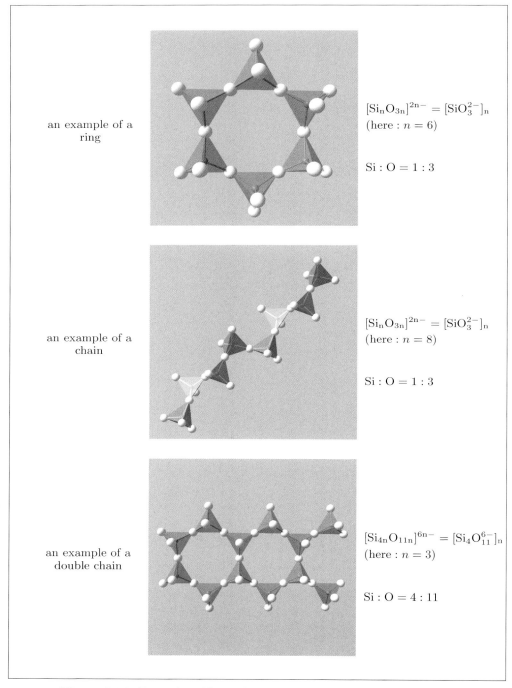

Figure 2.53. Examples of basis elements of siliceous networks, part one.

an example of a sheet

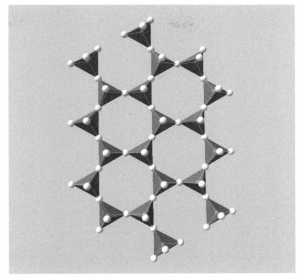

$[Si_{2n}O_{5n}]^{2n-}$
$=$
$[Si_2O_5^{2-}]_n$
(here : $n = 10$)

an example of a framework

$[Al_nSi_mO_{2(n+m)}]^{n-}$
(here : $[AlSi_2O_6^-]_6$)

Figure 2.54. Examples of basis elements of siliceous networks, part two.

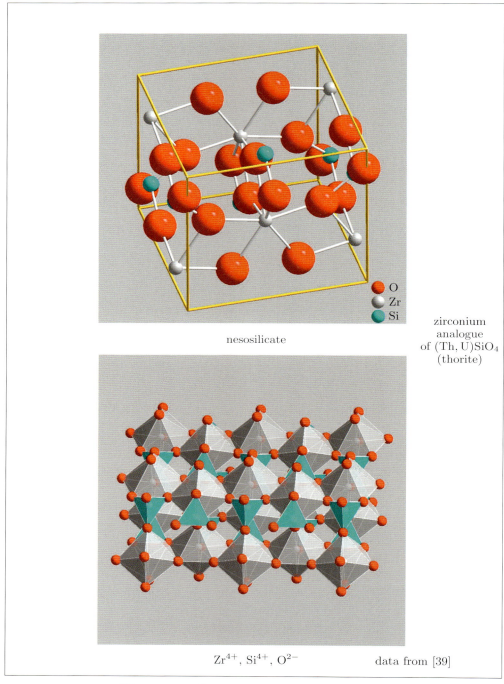

Figure 2.55. Zirconium silicate ZrSiO$_4$ (zircon).

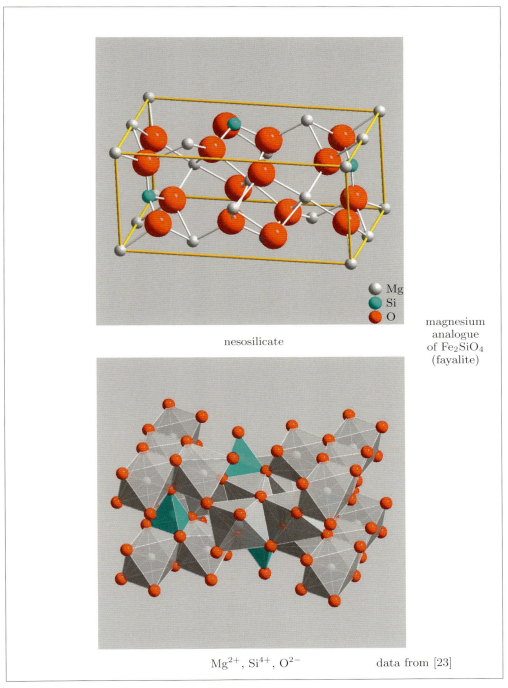

Figure 2.56. Magnesium silicate Mg_2SiO_4 (forsterite).

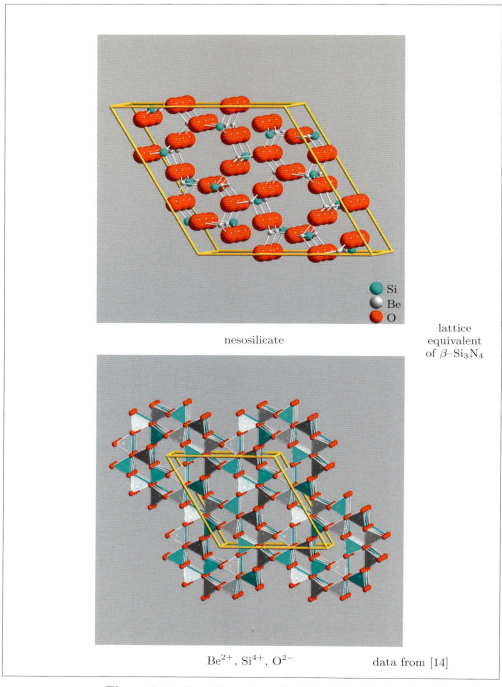

Figure 2.57. Beryllium silicate Be$_2$SiO$_4$ (phenakite).

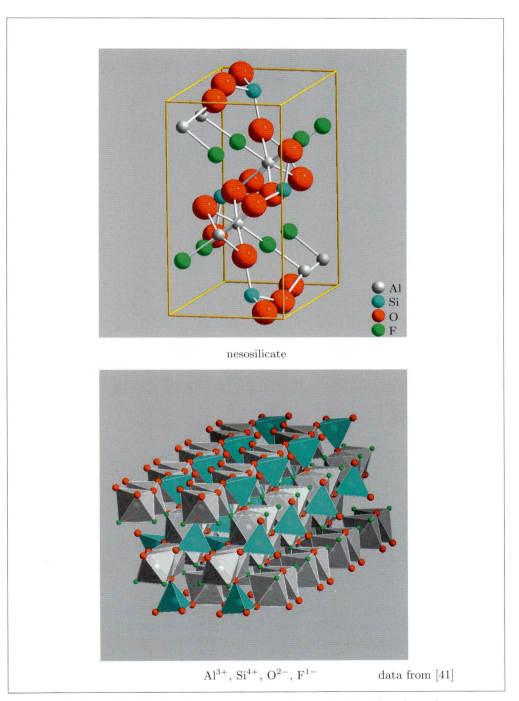

Figure 2.58. Dialuminum silicate with fluoride $Al_2(SiO_4)F_2$ (topaz).

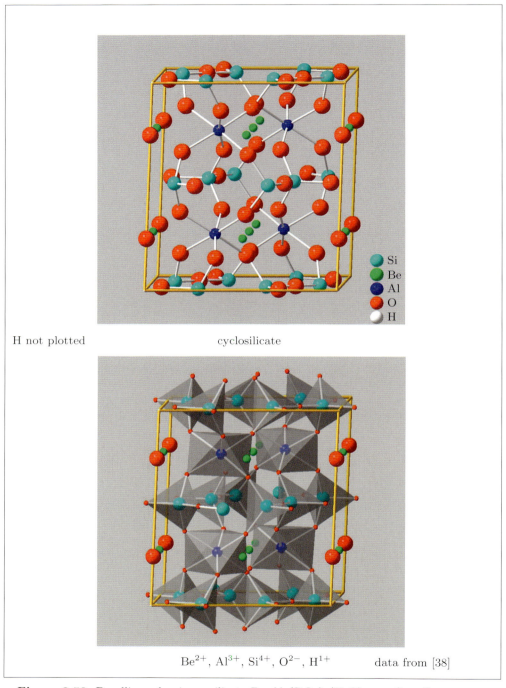

Figure 2.59. Beryllium aluminum silicate $Be_3Al_2(SiO_3)_6(H_2O)_{0.0991}$ (beryl), part one.

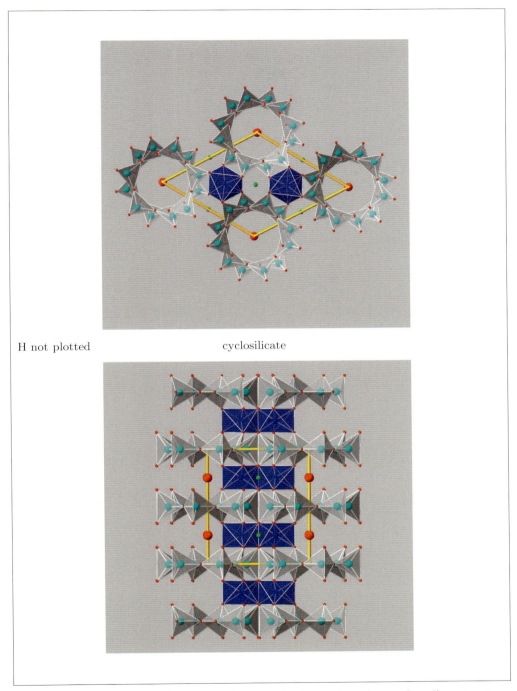

Figure 2.60. Beryllium aluminum silicate $Be_3Al_2(SiO_3)_6(H_2O)_{0.0991}$ (beryl), part two.

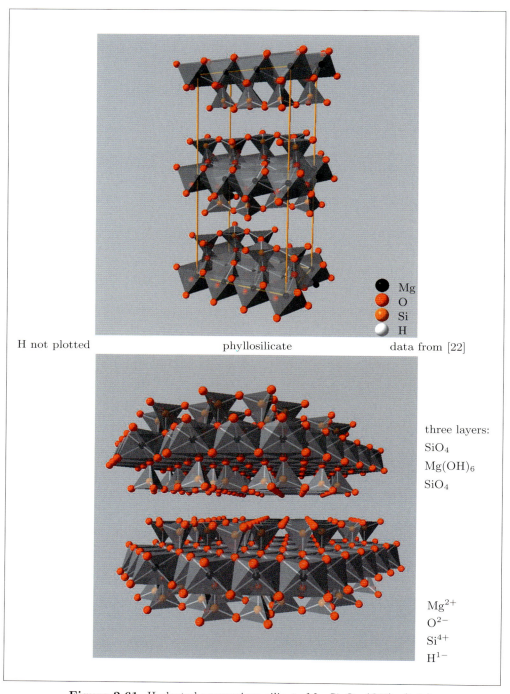

Figure 2.61. Hydrated magnesium silicate $Mg_3Si_4O_{10}(OH)_2$ (talc).

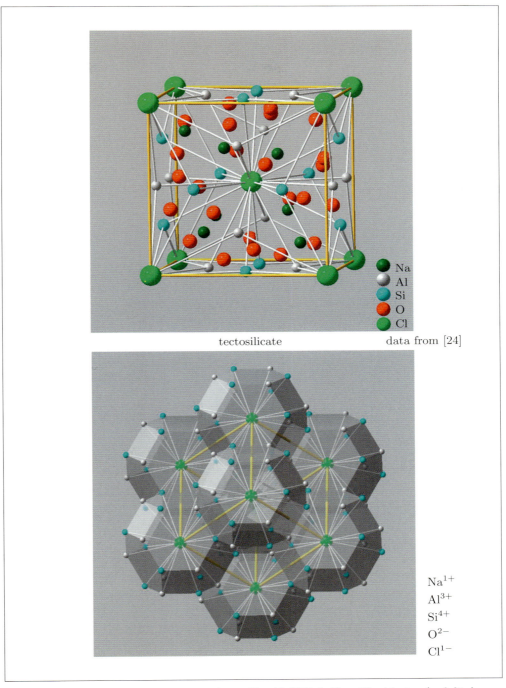

Figure 2.62. Sodium aluminum silicate $Na_8Al_8(SiO_4)_6Cl_2$ with chlorine (sodalite).

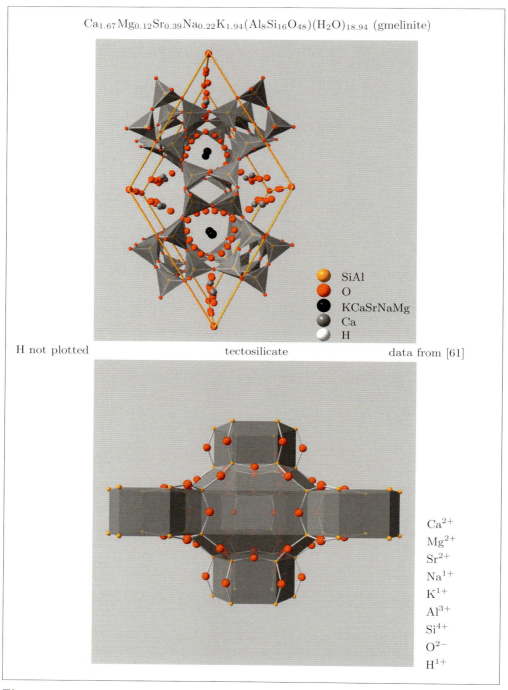

Figure 2.63. Hydrated calcium magnesium strontium sodium potassium aluminum silicate.

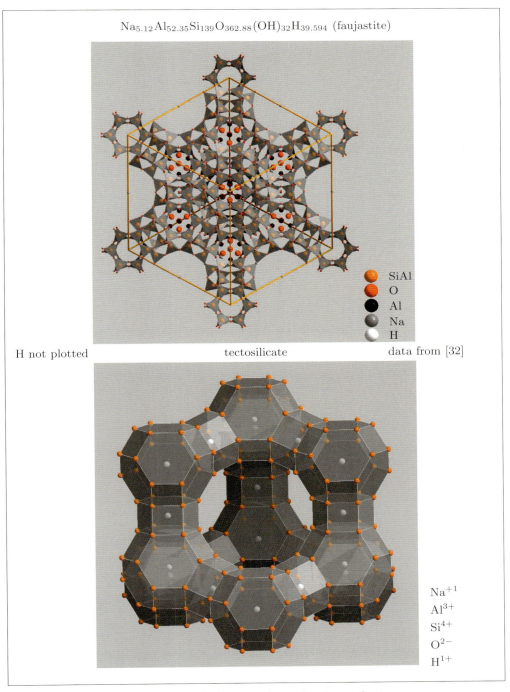

Figure 2.64. Hydrogen sodium aluminum silicate.

The term *isomorphic* means nothing else but "having the same form". Therefore, the term *isomorphism* catches the one-to-one correspondence between different systems. Consequently, if we speak of isomorphic compounds, we mean that the crystal lattices of the compounds have the same form. Examples are provided by Box 2.6. Following Box 2.6, for example, we realize that β–SiC is a zinc blende isomorph and that α-SiC is a wurtzite isomorph.

In crystallography and mineralogy, extending the term *allomorphism* which catches the property of certain chemical elements of crystallizing in two or more different forms, the property of certain compounds of crystallizing in two or more different forms is called *polymorphism*. Therefore, if we speak of a *polymorphic* compound, we mean that the crystal lattice of the compound can show different modifications. As a special example, let us consider the ZrO_2 polymorphs shown in Box 2.7. As it is outlined in Box 2.7, cooling down the liquefied ZrO_2 material, a cubic polycrystalline phase evolves after falling below the melting point, followed by a tetragonal polycrystalline phase and a monoclinic polycrystalline phase after falling below further critical temperatures. The cubic, tetragonal, and monoclinic ZrO_2 polymorphs that come into being including some data are collected in Box 2.7.

Talking about ZrO_2 polymorphs, the following more far-reaching annotations should be made. It is not possible to use ZrO_2 powder for the production of ceramic products because the drastic changes of the volume that occur during the phase transition from the tetragonal to the monoclinic modification would cause extensive cracking in the ceramic products during the cooling down period that follows after the sintering period. In order to avoid this problem, one adds dopants that stabilize the cubic or tetragonal polymorph of the ceramic material that comes into being, in particular, the dopants CaO, MgO, and Y_2O_3, but also the dopants CeO_2, ScO_3, and YbO_3. The following mechanism is responsible for this effect. The transformation temperature is shifted to lower temperatures. In the case of the tetragonal-to-monoclinic phase transition, the transformation temperature is lowered to temperatures where the phase transition is kinetically hindered. In the case of the cubic-to-tetragonal phase transition, provided the concentration of the dopants is high enough, the cubic phase even can be stabilized to room temperature. Firstly, we note that *FSZ* (fully stabilized zirconia) points at polycrystalline ZrO_2 fully stabilized by the integration of oxides of these types into the microstructure, whereas *PSZ* (partly stabilized zirconia) points at polycrystalline ZrO_2 partly stabilized by the integration of oxides of these types into the microstructure. Secondly, we note that *TZP* (tetragonal zirconia polycrystal), a slightly stabilized form of zirconia, in particular Y-TZP, for technological purposes is most important since microcracks that occur in such ceramic materials call for phase transitions, converting metastable tetragonal areas in stable monoclinic areas, leading to the sealing of the microcracks due to the increase of the volume of these areas ("self-healing effect"). This toughening mechanism is also used to enforce ceramic materials such as alumina or mullite by embedding small particles of tetragonal zirconia. We also compare with the microstructure images presented in Figures 2.83 and 2.84.

Box 2.7 (Important data: ZrO_2 polymorphs).

ZrO_2 polymorphs:

liquefied material $\overset{2680°C}{\Longrightarrow}$ c–ZrO_2 $\overset{2370°C}{\Longrightarrow}$ t–ZrO_2 $\overset{1170°C}{\Longrightarrow}$ m–ZrO_2,

liquefied material $\overset{2680°C}{\Longrightarrow}$ cubic $\overset{2370°C}{\Longrightarrow}$ tetragonal $\overset{1170°C}{\Longrightarrow}$ monoclinic

(melting point: 2680°C).

Densities:

liquefied material $\overset{2680°C}{\Longrightarrow} \approx 6.27\,\mathrm{g/cm}^3$ $\overset{2370°C}{\Longrightarrow} \approx 6.1\,\mathrm{g/cm}^3$ $\overset{1170°C}{\Longrightarrow} \approx 5.6\,\mathrm{g/cm}^3$

(molar mass: 123.22 g/mol).

We also compare with Figure 2.65.

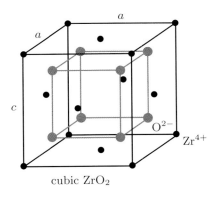

cubic ZrO_2

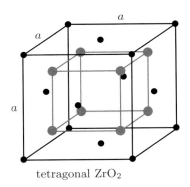

tetragonal ZrO_2

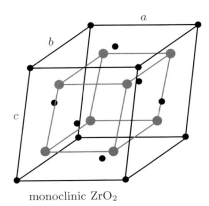

monoclinic ZrO_2

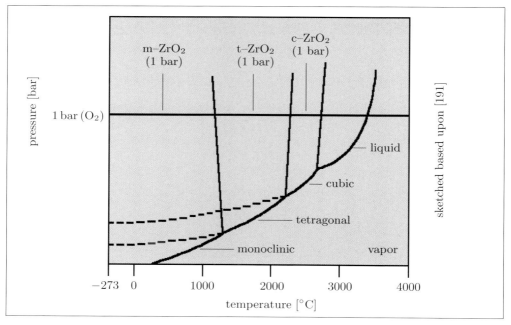

Figure 2.65. Transformations of ZrO$_2$ polymorphs.

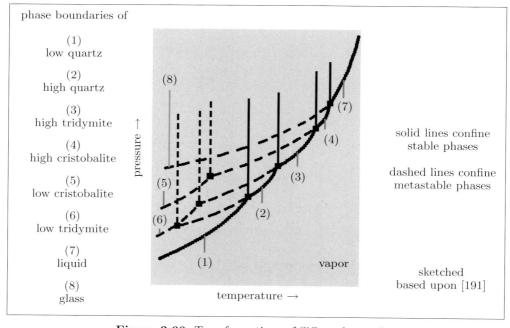

Figure 2.66. Transformations of SiO$_2$ polymorphs.

Box 2.8 (Important data: SiO₂ phases).

SiO$_2$ melts at 1710° C.

Transformation temperatures of SiO$_2$ polymorphs:

```
                    867° C                    1470° C
high quartz (2)  ◄──────►  high tridymite (3)  ◄──────►  high cristobalite (4)
       ▲                          ▲                              ▲
                                160° C
                                  ▼
     573° C              intermediate tridymite              200–270° C
                                  ▲
                                105° C
       ▼                          ▼                              ▼
 low quartz (1)             low tridymite                  low cristobalite
```

Dependence of SiO$_2$ polymorphs on temperature and pressure
(only stable phases are shown, we compare with Figure 2.66):

[Phase diagram: pressure (kbar) vs temperature (°C), showing regions for stishovite, coesite, liquid, and phases (1), (2), (3), (4). Note: "vapor data were not incorporated". sketched based upon [191]]

> **Box 2.9** (Important data: phase transitions initiated by changes of temperature).
>
> **Polymorphisms.**
>
> Silicon carbide (SiC):
> $$\beta\text{–SiC (cubic)} \overset{2100°C}{\Longrightarrow} \alpha\text{–SiC (hexagonal)}.$$
>
> Silicon nitride (Si$_3$N$_4$):
> $$\alpha\text{–Si}_3\text{N}_4\text{ (hexagonal)} \overset{1300°C}{\Longrightarrow} \beta\text{–Si}_3\text{N}_4\text{ (hexagonal)}.$$
>
> Barium titanium oxide (BaTiO$_3$):
> $$\text{m–BaTiO}_3 \overset{-100°C}{\Longrightarrow} \text{o–BaTiO}_3 \overset{0°C}{\Longrightarrow} \text{t–BaTiO}_3 \overset{100°C}{\Longrightarrow} \text{c–BaTiO}_3.$$

It is a well-known fact that phase transitions connecting polymorphs, on the one hand, can be initiated by the change of temperature, and on the other hand, can be initiated by the change of pressure. Box 2.7 and Box 2.9 show examples of the first case, while Box 2.10 shows examples of the second case. It is a well-known fact that phase transitions connecting polymorphs, on the one hand, can be reversible, and on the other hand, can be irreversible. We note that the phase transition of SiC shown in Box 2.9 is an example of the first case, while the phase transition of Si$_3$N$_4$ shown in Box 2.9 is an example of the second case. We also note that the reason for such irreversibilities varies. For instance, in many cases, the reason is that foreign ions have stabilized the original modification. For instance, in many cases, the reason is that the original modification has represented a metastable state. The p–T diagrams shown in Figures 2.65 and 2.66, together with Box 2.8, extensively illustrate the dependence on temperature and pressure in the context of ZrO$_2$ polymorphs and SiO$_2$ polymorphs.

> **Box 2.10** (Important data: phase transitions initiated by changes of pressure).
>
> **Allomorphisms.**
>
> Carbon (C):
> $$\text{graphite (hexagonal)} \overset{50\text{ kbar }(1500°C)}{\Longrightarrow} \text{diamond (cubic)}.$$
>
> **Polymorphisms.**
>
> Boron nitride (BN):
> $$\text{h–BN} \overset{60\text{ kbar }(1200°C)}{\Longrightarrow} \text{c–BN}.$$

It is a well-known fact that the condensation from the vapor phase ("vapor") as well as the solidification from the melt ("liquid"), on the one hand, can lead to a stable modification (within a certain temperature range), and on the other hand, for various reasons, can lead to a metastable modification. It is also a well-known fact that the transition from a metastable modification into a stable modification is called *monotropic transformation*. The p–T diagram which is presented in Box 2.8 exhibits stable modifications of SiO_2, while the p–T diagram which is presented in Figure 2.66 exhibits stable modifications as well as metastable modifications of SiO_2. Observing the p–T diagram which is presented in Figure 2.66, we firstly conceive that the lowering of the temperature of a melt can lead to solidification into a metastable glass phase ("silicate glass") and we secondly conceive that a further lowering of the temperature can lead to metastable low tridymite or stable low quartz. Observing the p–T diagram which is presented in Figure 2.66, we also conceive that impurities applied to glasses (for example, NaCl applied to glasses with the sweat of hands), can evoke the fracture of glasses according to the following mechanism. The typical operating temperature of technical glasses is about 1500°C. At this operating temperature, impurities at the technical glasses diffuse into the technical glasses. Finally cooling down the impurified technical glasses, starting from a certain temperature, initial nuclei ("seed crystals") develop which destabilize the technical glasses, after application of stress, leading to fracture. We here firstly note that this is a problem for laboratories as well as for the processing industry. We here secondly note that this is the reason why one likes holding technical glasses at a certain temperature.

Polytypism is a special type of polymorphism to be explained as follows. Very often, crystals can be regarded as built up by stacking layers, stacking blocks, or stacking rods each with exactly the same structure or at least with nearly the same structure. If there are modifications that differ in the stacking sequence, block sequence, or rod sequence, one speaks of *polytypism*. As a special example, let us consider the SiC polytypes shown in Figure 2.67. (i) The additional letter C and H, respectively, in 3C SiC and 2H SiC, respectively, indicates the stacking geometry, namely the letter C ("cubic") is used to indicate a cubic A–B–C stacking sequence tracing the cubic zinc blende structure and the letter H ("hexagonal") is used to indicate a hexagonal A–B stacking sequence tracing the hexagonal wurtzite structure. The number 3 and 2, respectively, indicates the respective stacking layer number within a stacking period. Of course, in other cases, analogous "letter + number" combinations are used. (ii) The 2H SiC polytype and the 3C SiC polytype define the simplest stacking sequences of the SiC structure family, in the sections before, termed β-SiC (zinc blende isomorph) or α-SiC (wurtzite isomorph), respectively. More than 220 SiC polytypes are known today. The 4680R SiC polytype even shows a stacking periodicity of about 100 μm, a length which is visible to the naked eye. (iii) The lattice spacings "layer–to–layer" in the examples quoted here as well as in all other cases vary from polytype to polytype and therefore are indicatives for the stacking periods. The physical properties in this example as well as in all other cases vary from polytype to polytype and this supplies us with a starting point for the fine tuning of physical properties of ceramic materials.

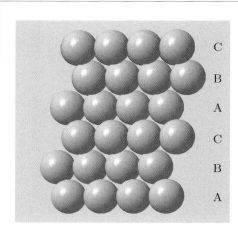
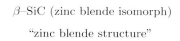
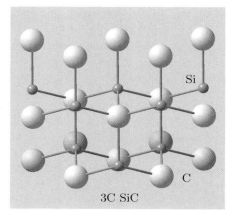
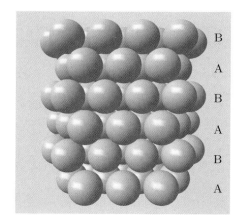
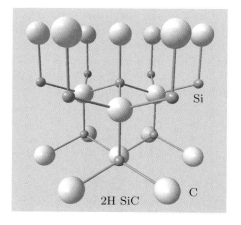

Figure 2.67. SiC polytypes.

2.2.3 Defect Types

Irregularities in crystal structures may evoke charateristics that are most welcome and may evoke charateristics that are not welcome. On all accounts, irregularities decisively dictate the characteristics of crystal structures. On all accounts, one distinguishes between *point defects*, *line defects*, *planar defects*, and *volume defects* ("bulk defects"). (i) The class of *point defects* covers all point-like defects in a crystal structure. Typically, missing particles, misplaced particles, and foreign particles define representatives of this class of defects. *Schottky defects* and *Frenkel defects* define further examples of this class of defects. (ii) The class of *line defects* covers all line-like defects in a crystal structure. Typically, *edge dislocations* and *screw dislocations* define representatives of this class of defects. Mixed types of both of them define further examples of this class of defects. (iii) The class of *planar defects* covers all area-related defects in a crystal structure. Typically, *grain boundaries* evoked by "jumps" of the crystallographic direction and *stacking faults* evoked by "jumps" of the stacking sequence are representatives of this class of defects. Twin boundaries that come into being if two crystals of the same nature finally grow together are further examples of this class of defects. (iv) The class of *volume defects* ("bulk defects") covers voids and inclusions.

The first two classes are shown in Figures 2.68 and 2.69. (i) We here take notice that Schottky defects are combinations of cation vacancies and anion vacancies congruent to the valency of the compounds of an ionic crystal. The anions and cations taken out usually are displaced to dislocations, grain boundaries, and material surfaces. We here take notice that Frenkel defects are combinations of an anion (a cation) vacancy and an anion (a cation) each. The anions (cations) taken out usually populate interstitials. Going beyond it, it should be clear that anions usually are too big for interstices and thus cause too high energy shifts at interstices so that interstitial anions are rare. (ii) We here take notice that inserted half-planes of particles lead to the edge dislocations, whereas shifted half-planes of particles forming a spiral lead to the screw dislocations. It is an interesting fact that such topological defects show characteristics known from "wave–particle systems". In particular, such dislocations can move about without losing their identity, leaving behind an undistorted crystal. In particular, two dislocations of opposite orientation may annihilate each other within the crystal. Consequently, one also uses the notation *solitons* for such dislocations. We note that such dislocations are characterized by two geometrical terms, the dislocation line and the Burgers vector b, which describes the magnitude and the direction of the distortion. Furthermore, we note that in an edge dislocation, the Burgers vector is perpendicular to the direction of the dislocation line, while in a screw dislocation, the Burgers vector is parallel to the direction of the dislocation line. Moreover, we note that in compound crystals (ionic crystals or covalent crystals) the formation of dislocations can be quite complex. For example, for an edge dislocation to form in a NaCl crystal, in order to maintain charge neutrality, to half-plains have to be inserted, and this leads to a Burgers vector that is in general somewhat twice in size. For example, for covalent crystals showing rigid tetrahedral bonds, the formation of dislocations is associated with rather high energy, and this makes them extremly resistant to shear. It holds $U \propto \mu b^2$ and $\sigma \propto \mu b$ with $U =$ internal energy, $\sigma =$ shear stress, and $\mu = G =$ shear modulus.

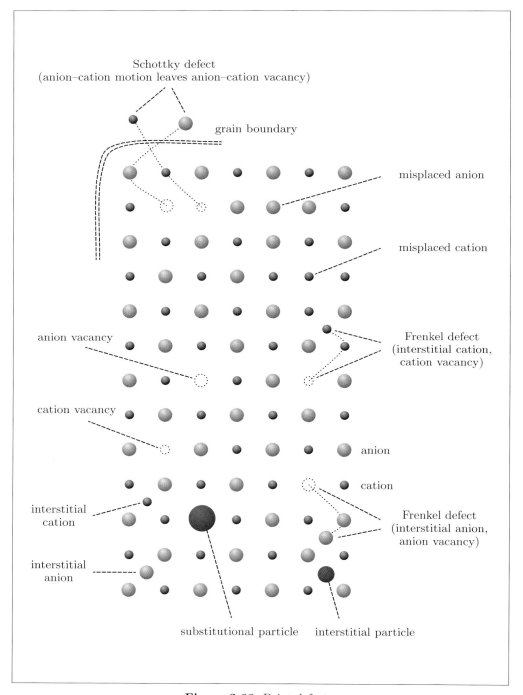

Figure 2.68. Point defects.

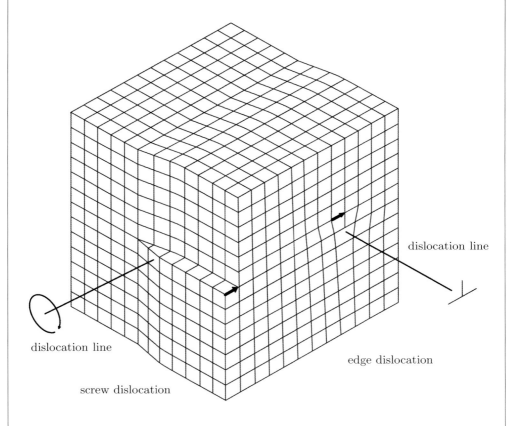

Figure 2.69. Line defects.

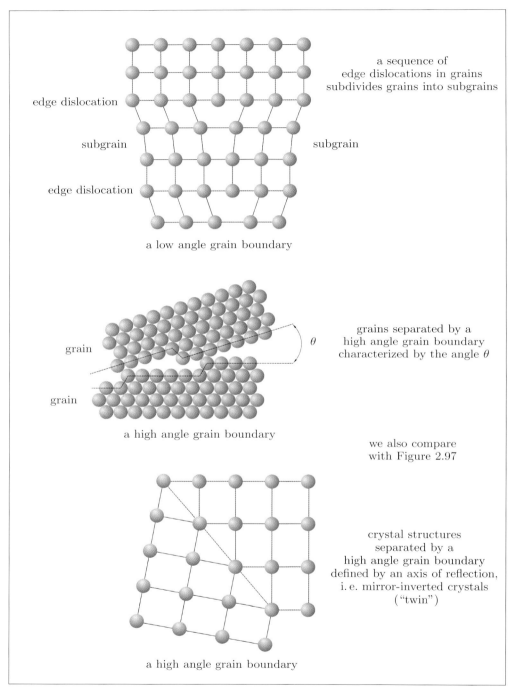

Figure 2.70. Planar defects.

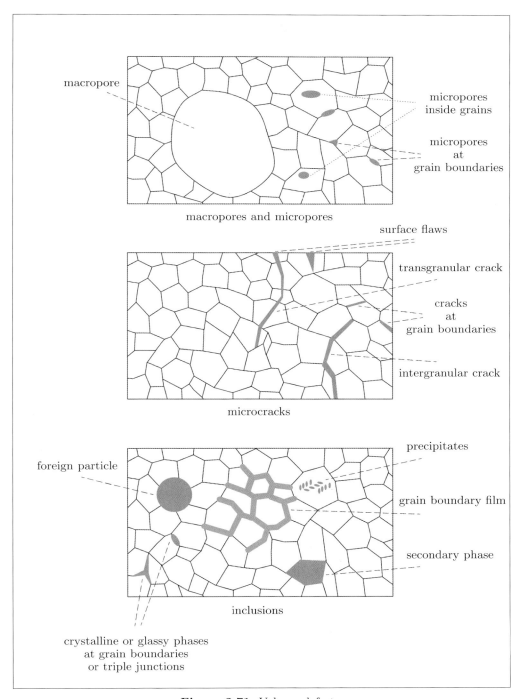

Figure 2.71. Volume defects.

The second two classes are shown in Figures 2.70 and 2.71. (i) We remark that one distinguishes between low angle grain boundaries and high angle grain boundaries depending on the angle that characterizes the different crystalline orientations around the grain boundary. For example, low angle grain boundaries are caused by a sequence of edge dislocations in grains, leading to orientation differences of only few degrees, finally subdividing grains into subgrains. For example, high angle grain boundaries separate grain from grain, showing relatively bigger orientation differences. (ii) We remark that one distinguishes between macropores and micropores subject to the size range of the grain sizes. Macropores are typically much larger than grains. Micropores can be located at grain boundaries or inside the grains, we compare with Figure 2.71, top. We also remark that one distinguishes between macrocracks and microcracks. Microcracks can be surface flaws, cracks in the crystal, and cracks at grain boundaries, we compare with Figure 2.71, middle. Beyond that, volume defects can be grain boundary films, crystalline phases or glassy phases located at grain boundaries or triple junctions, secondary phases, precipitates, and foreign particles, all originating from impurities or added substances, we compare with Figure 2.71, bottom. Beyond that, we point at Figures 2.72 and 2.73, supplying us with some complementary images.

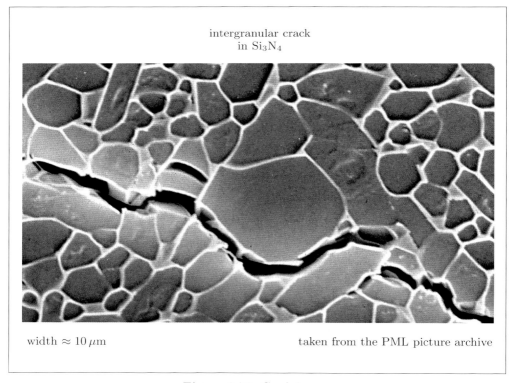

Figure 2.72. Crack image.

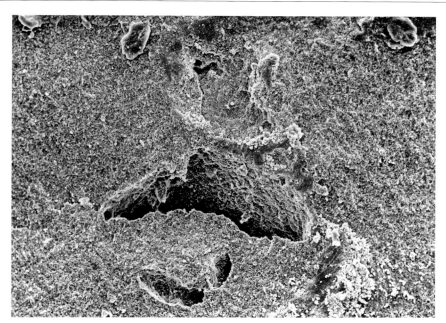
taken from the PML picture archive

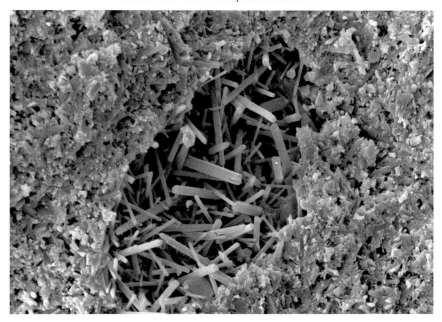
whiskers inside a pore, pore length $\approx 40\,\mu$m

Figure 2.73. Pore images.

The reader should know that one distinguishes between *intrinsic point defects* and *extrinsic point defects*. We speak of *intrinsic defects* if the defects already can be found in the pure crystal in thermodynamic equilibrium, but we speak of *extrinsic defects* if the defects are caused by the presence of second elements. For example, Schottky defects and Frenkel defects are classified as intrinsic defects, but substitutional particles that occur after a doping procedure are classified as extrinsic defects. The reader should also know that intrinsic point defects are subdivided into *stoichiometric defects* and *non-stoichiometric defects*. We speak of *stoichiometric defects* if the ratio of cations to anions does not change, but we speak of *non-stoichiometric defects* if cations (anions) are finally added or cations (anions) are finally removed. For example, Schottky defects and Frenkel defects are classified as stoichiometric defects, but interstitial particles are classified as non-stoichiometric defects. We compare with Box 2.11, which collects important defect reactions, and with Figure 2.74 and Figure 2.75, which illustrate these important defect reactions. It should be clear that such defect reactions are governed by the law of conservation of mass, i.e. $\sum m_{initial} = \sum m_{final}$, as well as by the law of electroneutrality, i.e. $\sum q^+ = \sum q^-$. It should also be clear that one speaks of a *substitutionally mixed crystal* if at least some particles of at least one sub-lattice filled with a specific type of particles are replaced by particles of another type, and that one speaks of an *interstitially mixed crystal* if at least some lattice gaps of one type of lattice gaps are filled by particles of another type.

In Box 2.11 and Figures 2.74 and 2.75, we apply the *Kröger–Vink notation*, which describes defect reactions as explained in Box 2.11. Following Box 2.11 and Figure 2.74, we especially realize that the emission of atomic oxygen $\frac{1}{2}O_2(g)$ out of an ionic crystal containing ionic oxygen O^{2-} as regular contribution, in a first step, is described as the emergence of a vacancy with effective zero charge at the oxygen position ($\to V_O^x$), and in a second step, is described as the release of two electrons into the conduction band where each electron shows a single effective negative charge ($\to 2e'$), thus generating a vacancy with double effective positive charge ($\to V_O^{\bullet\bullet}$). Following Box 2.11 as well as Figure 2.74, we especially realize that the absorption of atomic oxygen $\frac{1}{2}O_2(g)$ into an ionic crystal containing ionic oxygen O^{2-} as regular contribution, in a first step, is described as the emergence of interstitial oxygen with effective zero charge ($\to O_i^x$), and in a second step, is described as the absorption of two electrons from the conduction band where each electron shows a single effective negative charge, therefore leaving two holes in the conduction band where each hole shows a single effective positive charge ($\to 2h^\bullet$), therefore generating interstitial oxygen with double effective negative charge ($\to O_i''$). Following Box 2.11 as well as Figure 2.75, on the one hand, we additionally realize that particles of the surrounding of a crystal can be absorbed, leading to a surplus of cations with effective positive charge populating interstitials, and on the other hand, we additionally realize that particles of a crystal can be released, leading to a deficiency of cations, i.e. to vacancies with effective negative charge. It is a remarkable fact that the latter mechanism leads to the alteration of the valency of cations, for instance, releasing electrons into the conduction band enforcing the refilling of the holes that are generated during the release of Fe(g) particles, Fe^{2+} cations are finally transformed (oxidized) in Fe^{3+} cations.

> **Box 2.11 (Important formulae: defect reactions).**

Kröger–Vink notation:

E_s^c stands for an atom (Si, O etc.), a vacancy V, an electron e, or a hole h, where s indicates the site that is occupied by E (i indicates a lattice interstice), and where c stands for the relative electric charge
(a single • indicates a single positive effective charge, a single / indicates a single negative effective charge, and x indicates an effective zero charge).

Defect reactions (stoichiometric imperfections).

Schottky defects (MgO crystal):

$$Mg_{Mg}^x + O_O^x \rightarrow V_{Mg}'' + V_O^{\bullet\bullet} + Mg_{surface} + O_{surface} \; . \tag{2.45}$$

Frenkel defects (AgCl crystal):

$$Ag_{Ag}^x \rightarrow V_{Ag}' + Ag_i^{\bullet} \; . \tag{2.46}$$

Anti-structure defects (SiC crystal):

$$Si_{Si} + C_C \rightarrow Si_C + C_{Si} \; . \tag{2.47}$$

Defect reactions (non-stoichiometric imperfections).

Oxygen emission:

$$O_O^x \rightarrow \tfrac{1}{2}O_2(g) + V_O^{\bullet\bullet} + 2e' \; . \tag{2.48}$$

Oxygen absorption:

$$\tfrac{1}{2}O_2(g) \rightarrow O_i'' + 2h^{\bullet} \; . \tag{2.49}$$

Surplus of cations (deficiency of anions, ZnO crystal):

$$Zn(g) \rightarrow Zn_i^{\bullet} + e' \Rightarrow Zn_{1,00033}O \; . \tag{2.50}$$

Deficiency of cations (surplus of anions, FeO crystal):

$$3Fe_{Fe}^x \rightarrow 2Fe_{Fe}^{\bullet} + V_{Fe}'' + Fe(g) \Rightarrow Fe_{0,9}O \; . \tag{2.51}$$

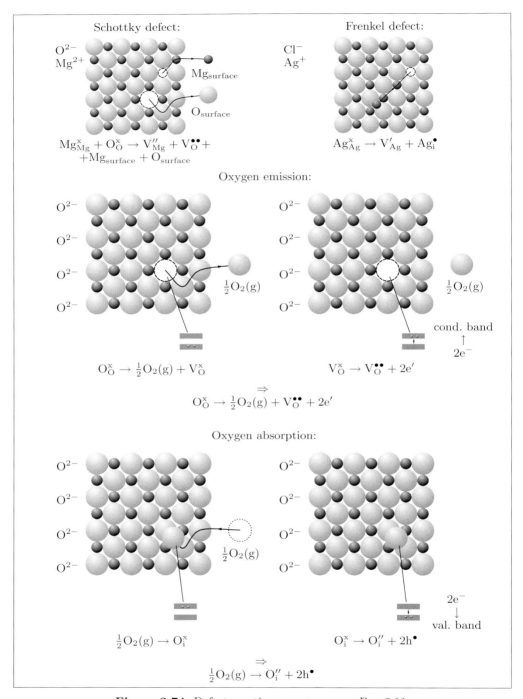

Figure 2.74. Defect reactions, part one: see Box 2.11.

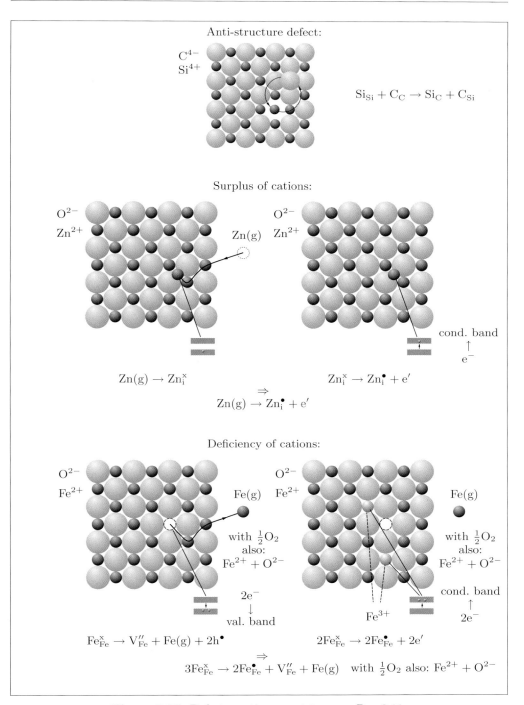

Figure 2.75. Defect reactions, part two: see Box 2.11.

The reader should realize that the result of defect reactions, for example, could be crystal structures of the type shown in Figure 2.76. (i) For example, crystal structures showing structural vacancies, on the one hand, leading to stoichiometric imperfections expressed by decimal numbers, and on the other hand, leading to the alteration of the valency of cations, could come into being. The $Fe_{0,9}O$ structure, which is characterized by some Fe aroms being oxidized into Fe^{3+} cations creating oxygen vacancies to save neutrality and resulting in the decimal number 0,9, is an example. (ii) For example, as a consequence of doping procedures, substitutionally mixed crystals could come into being. The Y-doped ZrO_2 structure is an example for a substitutionally mixed crystal where structural oxygen vacancies compensate the charge mismatch caused by the dopant, and this is manifested by an increase if the ionic conductivity due to the increase of the mobility of oxygen via these vacancies. This type of compensation is also called *anion vacancy model*. (iii) For example, as a consequence of doping procedures, interstitially mixed crystals could come into being. The Mg-doped Al_2O_3 structure is an example for an interstitially mixed crystal where interstitial cations compensate the charge mismatch caused by the dopant. This type of compensation is also called *cation interstitial model*. (iv) An anion interstitial model and a cation vacancy model exist, too. However, an anion interstitial model is rather exceptional since anions usually are rather large and then hardly fit into tetrahedral, octahedral, or cubic lattice gaps. The substitution of Zr^{4+} in Y_2O_3 is a rare case where the additional charge of Zr^{4+} is compsensated by oxygen interstitials. (v) Which defect model applies is checked by density measurements, we compare with Figure 2.77. (vi) Which "solid solution" comes into being is governed by basic rules, we compare with Box 2.12.

2.2.3.1 Thermodynamics: A Toolkit of Thermodynamic Formulae

In order to prepare the reader to what follows, let us here additionally present a toolkit of thermodynamic formulae in Boxes 2.13 and 2.14, together with Examples 2.1 and 2.2, also supplying us with first simple applications. Applying thermodynamic formulae, we should appreciate the issues that follow. (i) With respect to certain variables allowing to describe proceeding evolution processes, the *entropy S* in thermodynamic equilibrium shows a maximum, but the *Gibbs free energy G* in thermodynamic equilibrium shows a minimum. (ii) If $\Delta G < 0$ for a chemical reaction, the reaction proceeds from the reactant (educt) side to the product side. If $\Delta G > 0$ for a chemical reaction, the reaction proceeds from the product side to the reactant (educt) side. (iii) ΔG is fixed by an interplay of *reaction enthalpy* ΔH and *reaction entropy* ΔS, in the case of the formation of compounds usually termed *formation enthalpy* and *formation entropy*. A chemical reaction can be exothermic ($\Delta H < 0$), namely energy is released, or a chemical reaction can be endothermic ($\Delta H > 0$), namely energy is needed. On the one hand, if the product of temperature T and entropy gain ΔS is relatively small, an endothermic situation ($\Delta H > 0$) implies an endergonic situation ($\Delta G > 0$), i.e. a reaction does not start. On the other hand, if the product of temperature T and entropy gain ΔS is relatively big, an endothermic situation ($\Delta H > 0$) implies an exergonic situation ($\Delta G < 0$), i.e. a reaction does start. (iv) We here assume that p and T are constants. If V and T are constants, A replaces G and U replaces H.

> **Box 2.12** (Important data: solid solutions).

Types of solid solutions:

cation solutions or anion solutions,
interstitial solid solutions (interstitially mixed crystal) or
substitutional solid solutions (substitutionally mixed crystal).

Rules for wide/total solubility.

Parameters:

ion size, valency, isomorphy, chemical affinity.

Rules:

wide/total solubility requires a difference of radii $\leq 15\%$ ("criterion of ion size");
total solubility requires equal valency ("criterion of valency");
total solubility requires equal crystal structure ("criterion of isomorphy");
high chemical affinity implies high chemical reactivity,
leading to new phases ("criterion of chemical affinity").

(If the valency is not equal, additional alterations come into being.
In addition to the above examples, we here quote the SiAlON (Si_3N_4–Al_2O_3) example.
For each Al atom that populates an Si site, an O atom is embedded in an N site.)

Examples for wide/total solubility.

Wide solubility:

ZrO_2–Y_2O_3, ZrO_2–CaO, Si_3N_4–Al_2O_3.

Total solubility:

MgO–NiO, MgO–FeO, Al_2O_3–Cr_2O_3, Gd_2O_3–Y_2O_3.

Examples for formation energies.

ZnO ($O_O^x \to O_i'' + V_O^{\bullet\bullet}$): $2.5\,\text{eV}$.

MgO (Null $\to V_{Mg}'' + V_O^{\bullet\bullet}$): $7.7\,\text{eV}$. Al_2O_3 (Null $\to 2V_{Al}''' + 3V_O^{\bullet\bullet}$): $\approx 25\,\text{eV}$.

(These examples show that usually relatively high formation energies
are required for the formation of defects.)

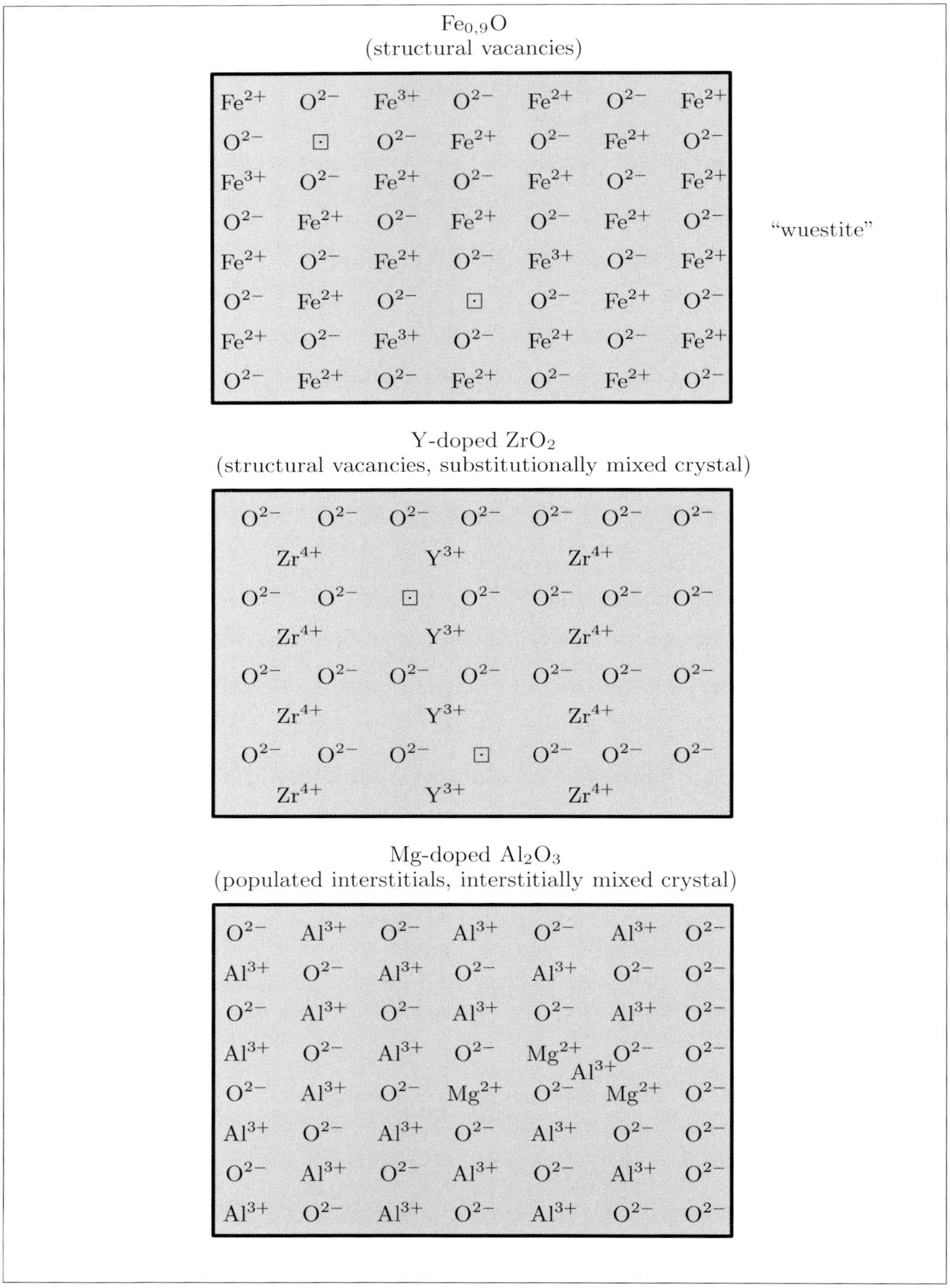

$Fe_{0.9}O$
(structural vacancies)

Fe^{2+}	O^{2-}	Fe^{3+}	O^{2-}	Fe^{2+}	O^{2-}	Fe^{2+}
O^{2-}	□	O^{2-}	Fe^{2+}	O^{2-}	Fe^{2+}	O^{2-}
Fe^{3+}	O^{2-}	Fe^{2+}	O^{2-}	Fe^{2+}	O^{2-}	Fe^{2+}
O^{2-}	Fe^{2+}	O^{2-}	Fe^{2+}	O^{2-}	Fe^{2+}	O^{2-}
Fe^{2+}	O^{2-}	Fe^{2+}	O^{2-}	Fe^{3+}	O^{2-}	Fe^{2+}
O^{2-}	Fe^{2+}	O^{2-}	□	O^{2-}	Fe^{2+}	O^{2-}
Fe^{2+}	O^{2-}	Fe^{3+}	O^{2-}	Fe^{2+}	O^{2-}	Fe^{2+}
O^{2-}	Fe^{2+}	O^{2-}	Fe^{2+}	O^{2-}	Fe^{2+}	O^{2-}

"wuestite"

Y-doped ZrO_2
(structural vacancies, substitutionally mixed crystal)

O^{2-}	O^{2-}	O^{2-}	O^{2-}	O^{2-}	O^{2-}	O^{2-}
	Zr^{4+}		Y^{3+}		Zr^{4+}	
O^{2-}	O^{2-}	□	O^{2-}	O^{2-}	O^{2-}	O^{2-}
	Zr^{4+}		Y^{3+}		Zr^{4+}	
O^{2-}	O^{2-}	O^{2-}	O^{2-}	O^{2-}	O^{2-}	O^{2-}
	Zr^{4+}		Y^{3+}		Zr^{4+}	
O^{2-}	O^{2-}	O^{2-}	□	O^{2-}	O^{2-}	O^{2-}
	Zr^{4+}		Y^{3+}		Zr^{4+}	

Mg-doped Al_2O_3
(populated interstitials, interstitially mixed crystal)

O^{2-}	Al^{3+}	O^{2-}	Al^{3+}	O^{2-}	Al^{3+}	O^{2-}
Al^{3+}	O^{2-}	Al^{3+}	O^{2-}	Al^{3+}	O^{2-}	O^{2-}
O^{2-}	Al^{3+}	O^{2-}	Al^{3+}	O^{2-}	Al^{3+}	O^{2-}
Al^{3+}	O^{2-}	Al^{3+}	O^{2-}	Mg^{2+} Al^{3+}	O^{2-}	O^{2-}
O^{2-}	Al^{3+}	O^{2-}	Mg^{2+}	O^{2-}	Mg^{2+}	O^{2-}
Al^{3+}	O^{2-}	Al^{3+}	O^{2-}	Al^{3+}	O^{2-}	O^{2-}
O^{2-}	Al^{3+}	O^{2-}	Al^{3+}	O^{2-}	Al^{3+}	O^{2-}
Al^{3+}	O^{2-}	Al^{3+}	O^{2-}	Al^{3+}	O^{2-}	O^{2-}

Figure 2.76. Examples for crystal structures with defects.

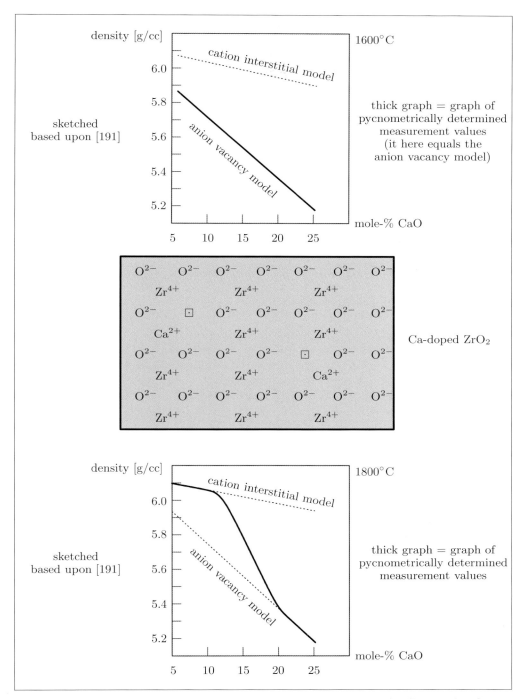

Figure 2.77. Cation interstitial model and anion vacancy model for Ca-doped ZrO_2.

Regarding Box 2.13, which shows a collection of thermodynamic relations, the remarks that follow should be made. (i) If energy/matter/particle exchange with the surrounding takes place, the thermodynamic system is called an *open system*, but if this is not the case, the thermodynamic system is called a *closed system*. If a process in a total thermodynamic system can be reversed without leaving modifications of any kind in the total thermodynamic system, the process is called *reversible*, but if this is not the case, the process is called *irreversible*. (ii) If the pressure p is a constant of evolution during a process, the *heat capacity* C_p, which is a measure for the heat energy dQ that is required to increase the temperature T of a substance about dT provided the pressure p is a constant of evolution, is needed to compute the alteration of enthalpy H and entropy S with the temperature T. (2.72), and (2.73) define "Ansätze" that are frequently applied for the heat capacity C_p, and (2.69) and (2.70), in close combination with Examples 2.1 and 2.2, illustrate the principle of computation. (iii) Based upon the concentrations of the reactants (educts) A and B and the products C and D of a chemical reaction at temperature T, the alteration of the Gibbs free energy G can be computed according to (2.71). Based upon (2.71), in close combination with (2.57) specified by (2.69), (2.70), (2.72), and (2.73), the alteration of the Gibbs free energy G at another temperature T also can be computed following the principle that is launched by (2.78) and (2.83) of Examples 2.1 and 2.2.

Dealing with ceramic substances, the pressure p usually is a constant of evolution, and G usually is the thermodynamic potential adapted best, for the case of a reaction proceeding from the reactant (educt) side to the product side displaying $\Delta G < 0$, and for the case of a reaction proceeding from the product side to the reactant (educt) side displaying $\Delta G > 0$. If the temperature T is a constant, $\Delta G = \Delta H - T\Delta S$, but if the temperature T is a variable, $\Delta G = \Delta H - T\Delta S - S\Delta T$, which follows from (2.57). (i) From a physical point of view, ΔH is the energy that is used in bond breaking actions minus the energy that is released in bond making actions. (ii) From a physical point of view, the heat capacity C_p reflects the vibrational behavior of bonds. However, not all types of bonds are equivalent carriers of the vibrational behavior and thus not all types of bonds equivalently contribute to the heat capacity C_p. For example, keeping binary compounds together, the metal–O bonds in good approximation determine the vibrational behavior in the high-temperature limit, and if the number of metal–O bonds during a chemical reaction in total does not change, the heat capacity C_p in total in good approximation is the same before and after a chemical reaction. (iii) Speaking about the physical background of the thermodynamic formulae presented below, we here should point out that (2.76) is a differential specification of ΔG for the case of a sintering process. We note that the reduction of the volume of the ceramic substance during the sintering process evokes a negative term $p\Delta V$, and that the aggregation of powder particles, on the one hand, leads to the reduction of the total surface of the ceramic substance thus evoking a negative term $\gamma_s \Delta A_s$, and on the other hand, leads to the evolution of a grain boundary surface within the ceramic substance thus evoking a positive term $\gamma_b \Delta A_b$, together generating a negative contribution so that in total $\Delta G < 0$. We also note that the differential specification of ΔG that is quoted here covers each point of the sintering process.

Regarding Box 2.14, which shows a collection of statistical thermodynamic relations, the remarks that follow should be made. (i) If we speak of a "microcanonical ensemble", we mean that the constraints that restrict the particles do not allow the exchange of energy/matter/particles with the environment, i.e. the total energy is kept constant, the total energy does not fluctuate. If we speak of a "canonical ensemble", we mean that the constraints that restrict the particles do allow the exchange of heat energy with the environment at given temperature, volume, and particle numbers. If we speak of a "grand canonical ensemble", we mean that the constraints that restrict the particles do allow the exchange of heat energy and the exchange of particles with the environment at given temperature, volume, and chemical potentials. (ii) It is noteworthy that (2.86) is determined by the number Ω_m of micro-states that realize the same macro-state. Specifying Ω_m as the number of particle distributions in energy space that realize the same total energy, (2.86) reduces to the configurational entropy of the total energy, namely the well-known statistical representation of the thermal entropy, namely the well-known Boltzmann entropy. Specifying Ω_m as the number of particle distributions in crystalline structure space that realize the same total crystalline structure, (2.86) reduces to the configurational entropy of the total crystalline structure. (iii) It is also noteworthy that in the case of "canonical ensembles", thermodynamic quantities such as the entropy and the internal energy can be expressed by the partition function Z_c, the normalization function of the Boltzmann distribution $\rho = \exp\left(-\beta E_k\right)/Z_c$, and in the case of "grand canonical ensembles", thermodynamic quantities such as the entropy and the internal energy can be expressed by the partition function Z_{gc}, the normalization function of $\rho = \exp\left[-\beta\left(-\sum_l \mu_l N_{kl} + E_k\right)\right]/Z_{gc}$. (iv) It is also noteworthy that the total entropy of a ceramic material, on the one hand, consists of the thermal entropy, reflecting the entirety of population possibilities of particle motion states, and on the other hand, consists of the configurational entropies of crystalline structures, reflecting the entirety of population possibilities of particle lattice positions, and is completed by second-level modes such as electronic modes, for instance, reflecting the entirety of population possibilities of states of electrons within partially populated energy bands, and magnetic modes, for instance, reflecting the entirety of population possibilities of states of magnetic moments of electrons within partially populated energy bands. (v) The alteration of the configurational entropy of a crystalline structure in the framework of a thermodynamic process and the difference of the configurational entropies of a crystalline structure 1 and a crystalline structure 2, for example, is calculated as follows. Let M sites in a crystal that can be populated by N_1 ions/atoms/molecules and N_1' gaps with $M = N_1 + N_1'$ be given. Let us compare this scenario (a) with a scenario where M sites are populated by N_2 ions/atoms/molecules and N_2' gaps with $M = N_2 + N_2'$ and (b) with a scenario where M' sites are populated by N_1 ions/atoms/molecules and N_1'' gaps with $M' = N_1 + N_1''$. Using these data in order to determine Ω_1 and Ω_2 of (2.93), according to (2.94) in the case (a) and according to (2.95) in the case (b), the alteration (a) of the configurational entropy of the crystalline structure in the framework of a thermodynamic process and the difference (b) of the configurational entropies of the crystalline structure 1 and the crystalline structure 2 are readily calculated. (vi) In those cases where the gaps can be led back to different types of ions/atoms/molecules, according to (2.96), Ω products replace Ω terms [50].

> **Box 2.13 (Important formulae: a first glance at thermodynamic relations).**
>
> d = not path-dependent infinitesimal changes, δ = path-dependent infinitesimal changes, U = internal energy, H = enthalpy, G = free enthalpy, A = free energy, Ω = grand potential, S = entropy, Q = heat, W = work, m = mass of the material, p = pressure, V = volume, T = absolute temperature, μ = chemical potential of chemical species, N = particle number of chemical species, σ_{ij} = stress tensor, ϵ_{ij} = strain tensor, C = heat capacity, c = specific heat capacity, Q_A = concentration of substance A, Q_r = reaction quotient, R = molar gas constant, γ_s = specific surface energy, γ_b = specific grain boundary surface energy, ν = number of moles, $dA_{s,b}$ = surface reductions.
>
> First law of thermodynamics:
>
> $$dU = \delta Q - \delta W + \sum_i \mu_i dN_i + \cdots \qquad (2.52)$$
>
> (δW = work that is evoked by the system, reducing the internal energy of the system).
>
> Second law of thermodynamics:
>
> $$dS \geq \frac{\delta Q}{T} \qquad (2.53)$$
>
> (for reversible processes "=" and for irreversible processes ">").
>
> Third law of thermodynamics:
>
> $$\lim_{T \to 0} S = \text{constant} \quad \text{(closed systems)}. \qquad (2.54)$$
>
> Thermodynamic potentials:
>
> $$U = Q - W + \sum_i \mu_i N_i + \cdots \qquad (2.55)$$
> (internal energy),
>
> $$H = U + pV \qquad (2.56)$$
> (enthalpy),
>
> $$G = H - TS = U + pV - TS \qquad (2.57)$$
> (Gibbs free energy, also free enthalpy),
>
> $$A = U - TS \qquad (2.58)$$
> (Helmholtz free energy, also free energy),
>
> $$\Omega = U - TS - \sum_i \mu_i N_i \qquad (2.59)$$
> (Landau potential, also grand potential).

Continuation of Box.

Mechanical work, compressible substance:
$$\delta W = dW_{\text{comp}} = p dV . \tag{2.60}$$

Mechanical work, elastic substance, stress and strain gradients:
$$\delta W = dW_{\text{elast}} = \sum_{ij} V \sigma_{ij} d\epsilon_{ij} = \sum_{ijkl} V C_{ijkl} \epsilon_{kl} d\epsilon_{ij} . \tag{2.61}$$

Fundamental relations
$(\delta Q \leq T dS , \quad \delta W = dW_{\text{mech}} = p dV)$:

$$\begin{aligned}
dU &\leq +T dS - p dV + \sum_i \mu_i dN_i + \cdots , \\
dH &\leq +T dS + V dp + \sum_i \mu_i dN_i + \cdots , \\
dG &\leq -S dT + V dp + \sum_i \mu_i dN_i + \cdots , \\
dA &\leq -S dT - p dV + \sum_i \mu_i dN_i + \cdots , \\
d\Omega &\leq -S dT - p dV + \cdots .
\end{aligned} \tag{2.62}$$

Note that the linkage of these thermodynamic relations is managed by Legendre transformations.

Additional relations
(the lower indices p, V, T, S, and N_i, indicate constant quantities p, V, T, S, and N_i):

$$\begin{aligned}
-p &= \left(\frac{\partial U}{\partial V}\right)_{S,\{N_i\}} = \left(\frac{\partial A}{\partial V}\right)_{T,\{N_i\}} , \\
+V &= \left(\frac{\partial H}{\partial p}\right)_{S,\{N_i\}} = \left(\frac{\partial G}{\partial p}\right)_{T,\{N_i\}} , \quad +T = \left(\frac{\partial U}{\partial S}\right)_{V,\{N_i\}} = \left(\frac{\partial H}{\partial S}\right)_{p,\{N_i\}} , \\
-S &= \left(\frac{\partial G}{\partial T}\right)_{p,\{N_i\}} = \left(\frac{\partial A}{\partial T}\right)_{V,\{N_i\}}
\end{aligned} \tag{2.63}$$

(state relations) ,

$$\begin{aligned}
\left(\frac{\partial T}{\partial V}\right)_{S,\{N_i\}} &= -\left(\frac{\partial p}{\partial S}\right)_{V,\{N_i\}} , \quad \left(\frac{\partial T}{\partial p}\right)_{S,\{N_i\}} = +\left(\frac{\partial V}{\partial S}\right)_{p,\{N_i\}} , \\
\left(\frac{\partial T}{\partial V}\right)_{p,\{N_i\}} &= -\left(\frac{\partial p}{\partial S}\right)_{T,\{N_i\}} , \quad \left(\frac{\partial T}{\partial p}\right)_{V,\{N_i\}} = +\left(\frac{\partial V}{\partial S}\right)_{T,\{N_i\}}
\end{aligned} \tag{2.64}$$

(Maxwell relations) .

Note that these relations do not include changes of the particle numbers N_i. Observing changes of the particle numbers N_i, complementary relations are needed.

Continuation of Box.

Chemical potential:

$$\mu_i = \left(\frac{\partial U(V, S, N_i)}{\partial N_i}\right)_{V,S,\{N_{j\neq i}\}} = \left(\frac{\partial H(p, S, N_i)}{\partial N_i}\right)_{p,S,\{N_{j\neq i}\}} \qquad (2.65)$$
$$= \left(\frac{\partial G(p, T, N_i)}{\partial N_i}\right)_{p,T,\{N_{j\neq i}\}} = \left(\frac{\partial A(V, T, N_i)}{\partial N_i}\right)_{V,T,\{N_{j\neq i}\}}.$$

Heat capacity:

$$C_{\mathrm{V}} = \left(\frac{\delta Q}{\mathrm{d}T}\right)_V = T\left(\frac{\partial S}{\partial T}\right)_V = \left(\frac{\partial U}{\partial T}\right)_V ,$$
$$C_{\mathrm{p}} = \left(\frac{\delta Q}{\mathrm{d}T}\right)_p = T\left(\frac{\partial S}{\partial T}\right)_p = \left(\frac{\partial H}{\partial T}\right)_p ,$$
(2.66)

$$c_{\mathrm{V}} = \left(\frac{\partial C}{\partial m}\right)_V , \quad c_{\mathrm{p}} = \left(\frac{\partial C}{\partial m}\right)_p \quad \text{(specific heat capacity)} ,$$
$$C_{\mathrm{V,m}} = \frac{1}{\nu}\left(\frac{\delta Q}{\mathrm{d}T}\right)_V , \quad C_{\mathrm{p,m}} = \frac{1}{\nu}\left(\frac{\delta Q}{\mathrm{d}T}\right)_p \quad \text{(molar heat capacity)} .$$
(2.67)

For liquids and solids, the difference between $C_{\mathrm{V}}, c_{\mathrm{V}}$ and $C_{\mathrm{p}}, c_{\mathrm{p}}$ often is neglectable and we then may apply the following relations:

$$C = \frac{\delta Q}{\mathrm{d}T} , \quad c = c_m = \frac{C}{m} = \frac{C}{\rho V} , \quad \rho = m/V . \qquad (2.68)$$

Heat capacity and entropy
($\mathrm{d}S = \delta Q/T$, $\delta Q = C_{\mathrm{p}}\mathrm{d}T$, see also Example 2.2):

$$\Delta S = \int_{T_1}^{T_2} \mathrm{d}S = \int_{T_1}^{T_2} \frac{1}{T}\delta Q = \int_{T_1}^{T_2} \frac{C_{\mathrm{p}}}{T}\mathrm{d}T , \quad S_{T_2} = S_{T_1} + \int_{T_1}^{T_2} \frac{C_{\mathrm{p}}}{T}\mathrm{d}T . \qquad (2.69)$$

Heat capacity and enthalpy
($\mathrm{d}p = 0$, $\mathrm{d}N_i = 0$, $\mathrm{d}H = T\mathrm{d}S$, $\mathrm{d}S = \delta Q/T$, $\delta Q = C_{\mathrm{p}}\mathrm{d}T$, see also Examples 2.1 and 2.2):

$$\Delta H = \int_{T_1}^{T_2} \mathrm{d}H = \int_{T_1}^{T_2} T\mathrm{d}S = \int_{T_1}^{T_2} \delta Q = \int_{T_1}^{T_2} C_{\mathrm{p}}\mathrm{d}T ,$$
$$H_{T_2} = H_{T_1} + \int_{T_1}^{T_2} C_{\mathrm{p}}\mathrm{d}T .$$
(2.70)

Continuation of Box.

Resulting Gibbs free energy ΔG of a chemical reaction at temperature T showing reactants (educts) A and B and products C and D:

$$\begin{aligned}\Delta G &= G_C + G_D - G_A - G_B = \\ &= G_{0,C} + RT\ln Q_C(T) + G_{0,D} + RT\ln Q_D(T) - \\ &\quad - G_{0,A} - RT\ln Q_A(T) - G_{0,B} - RT\ln Q_B(T) = \\ &= \Delta G_0 + RT\ln Q_r(T) , \end{aligned} \quad (2.71)$$

$$\Delta G_0 = G_{0,C} + G_{0,D} - G_{0,A} - G_{0,B}, \quad Q_r(T) = \frac{Q_C Q_D}{Q_A Q_B}.$$

Kubaschewski "Ansatz" for the heat capacity C_p:

$$C_p = a + b_1 T + b_2 T^2 + b_{-1} T^{-1} . \quad (2.72)$$

Extended Kubaschewski "Ansatz" for the heat capacity C_p:

$$\begin{aligned} C_p &= a + b_1 T + b_2 T^2 + \cdots + b_{-1} T^{-1} + b_{-2} T^{-2} + \cdots \\ &\cdots + c_{1/2} T^{1/2} + c_{1/3} T^{1/3} + \cdots + c_{-1/2} T^{-1/2} + c_{-1/3} T^{-1/3} + \cdots \\ &\cdots + d_{2/3} T^{2/3} + \cdots + d_{-2/3} T^{-2/3} + \cdots \\ &\qquad \cdots . \end{aligned} \quad (2.73)$$

Internal energy balance equation for open systems ($dU_{\text{in}}/dU_{\text{out}}$ = internal energy entering/leaving the system, δW = work done by the system):

$$\frac{dU}{dt} = \frac{\delta Q}{dt} + \frac{dU_{\text{in}}}{dt} - \frac{dU_{\text{out}}}{dt} - \frac{\delta W}{dt} . \quad (2.74)$$

Entropy balance equation for open systems ($\dot{m}_j$ = change of mass associated with species j, S_j = entropy per mass unit, $\dot{S}_{\text{internal}}$ internal production of entropy):

$$\frac{dS}{dt} = \frac{\dot{Q}}{T} + \sum_j \dot{m}_j S_j + \dot{S}_{\text{internal}} . \quad (2.75)$$

Reduction of free enthalpy (due to surface reductions dA_s (total surface) and dA_b (grain boundary surface) and due to volume reduction dV):

$$dG = \gamma_s dA_s + \gamma_b dA_b + p dV \text{ ("sintering equation")}. \quad (2.76)$$

2. Structures

Example 2.1 (An additional tutorial example: reaction enthalpy).

Calculate the reaction enthalpy (heat of reaction) ΔH_T of the reaction FeO(s)+H$_2$(g) at 500°C with the help of (2.70). For the formation enthalpy (heat of formation) of FeO(s) and H$_2$O(g), respectively, use the value -265.7 kJ/mol and -241.8 kJ/mol, respectively, at 25°C. For the heat capacities $C_{p,m}$ of the products and educts in the relevant range of temperature use the "Ansatz"

$$\begin{aligned}
\text{FeO(s)}: & \quad C_{p,m} = 51.81 + 06.78 \cdot 10^{-3} T - 1.59 \cdot 10^5 T^{-2} \; [\text{J/mol K}], \\
\text{Fe(s)}: & \quad C_{p,m} = 17.49 + 24.78 \cdot 10^{-3} T \qquad\qquad\qquad\quad [\text{J/mol K}], \\
\text{H}_2\text{(g)}: & \quad C_{p,m} = 27.29 + 03.26 \cdot 10^{-3} T - 0.50 \cdot 10^5 T^{-2} \; [\text{J/mol K}], \\
\text{H}_2\text{O(g)}: & \quad C_{p,m} = 30.01 + 10.71 \cdot 10^{-3} T - 0.33 \cdot 10^5 T^{-2} \; [\text{J/mol K}].
\end{aligned} \qquad (2.77)$$

Solution.

Setup of the reaction scheme:

$$\begin{array}{ccccc}
\text{FeO} + \text{H}_2 & \xrightarrow{500°\text{C}, \Delta H} & \text{Fe} + \text{H}_2\text{O} \\
\uparrow \quad \uparrow & & \uparrow \quad \uparrow \\
\text{FeO} + \text{H}_2 & \xrightarrow{25°\text{C}, \Delta H_{298}} & \text{Fe} + \text{H}_2\text{O}
\end{array} \qquad (2.78)$$

Application of (2.70) (we resort to the rule "products minus educts"):

$$\begin{aligned}
\Delta H = & \\
= \Delta H_{298} + \int_{298\,\text{K}}^{773\,\text{K}} & \left[C_{p,m}(\text{Fe}) + C_{p,m}(\text{H}_2\text{O}) - C_{p,m}(\text{FeO}) - C_{p,m}(\text{H}_2) \right] dT,
\end{aligned} \qquad (2.79)$$

$$\Delta H_{298} = -241.8 \,\text{kJ/mol} + 265.7 \,\text{kJ/mol} = 23.9 \,\text{kJ/mol}, \qquad (2.80)$$

$$\begin{aligned}
\Delta H = & \\
= 23.9 \,\text{kJ/mol} + & \left[17.49T + \tfrac{24.78}{1000}\tfrac{T^2}{2} \right]_{298\,\text{K}}^{773\,\text{K}} + \\
+ & \left[30.01T + \tfrac{10.71}{1000}\tfrac{T^2}{2} + 0.33 \cdot 10^5 \tfrac{1}{T} \right]_{298\,\text{K}}^{773\,\text{K}} - \\
- & \left[51.81T + \tfrac{6.78}{1000}\tfrac{T^2}{2} + 1.59 \cdot 10^5 \tfrac{1}{T} + 27.29T + \tfrac{3.26}{1000}\tfrac{T^2}{2} + 0.50 \cdot 10^5 \tfrac{1}{T} \right]_{298\,\text{K}}^{773\,\text{K}} = \\
= 23.9 \,\text{kJ/mol} + & (14\,610.9 + 16\,910.9 - 26\,006.5 - 13\,688.9) \,\text{J/mol} = \\
= 23.9 \,\text{kJ/mol} - & 8.2 \,\text{kJ/mol} = 15.7 \,\text{kJ/mol}.
\end{aligned} \qquad (2.81)$$

Example 2.2 (An additional tutorial example: formation enthalpy/entropy).
Calculate the change of the formation enthalpy/entropy $\Delta H_T/\Delta S_T$ for Si_3N_4 from 25°C to 427°C with the help of (2.70)/(2.69). The heat capacities $C_{p,m}$ in the range of temperature 298–900 K shall be given by

$$C_{p,m} = a + bT \begin{cases} Si: & a = 25.61\,\text{J/mol K},\ b = 0\,\text{J/mol K}^2, \\ N_2: & a = 27.87\,\text{J/mol K},\ b = 4.27\cdot 10^{-3}\,\text{J/mol K}^2, \\ Si_3N_4: & a = 70.42\,\text{J/mol K},\ b = 98.75\cdot 10^{-3}\,\text{J/mol K}^2. \end{cases} \qquad (2.82)$$

Solution.

Setup of the reaction scheme:

$$\begin{array}{ccc} 3Si + 2N_2 & \xrightarrow{427°C, \Delta H, \Delta S} & Si_3N_4 \\ \uparrow \qquad \uparrow & & \uparrow \\ 3Si + 2N_2 & \xrightarrow{25°C, \Delta H_{298}, \Delta S_{298}} & Si_3N_4 \end{array} \qquad (2.83)$$

Application of (2.70) (we resort to the rule "products minus educts"):

$$\Delta H = (\Delta H_{298} := 0+) \int_{298\,\text{K}}^{700\,\text{K}} \left[C_{p,m}(Si_3N_4) - 2C_{p,m}(N_2) - 3C_{p,m}(Si) \right] dT =$$

$$= \int_{298\,\text{K}}^{700\,\text{K}} \left[-62.15 + 90.21\cdot 10^{-3}T \right] dT = \qquad (2.84)$$

$$= \left[-62.15\,T + 90.21\cdot 10^{-3}\frac{T^2}{2} \right]_{298\,\text{K}}^{700\,\text{K}} = -6888.35\,\text{J/mol}\,.$$

Application of (2.69) (we resort to the rule "products minus educts"):

$$\Delta S = (\Delta S_{298} := 0+) \int_{298\,\text{K}}^{700\,\text{K}} \frac{1}{T}\left[C_{p,m}(Si_3N_4) - 2C_{p,m}(N_2) - 3C_{p,m}(Si) \right] dT =$$

$$= \int_{298\,\text{K}}^{700\,\text{K}} \frac{1}{T}\left[-62.15 + 90.21\cdot 10^{-3}T \right] dT = \qquad (2.85)$$

$$= \left[-62.15\ln T + 90.21\cdot 10^{-3}T \right]_{298\,\text{K}}^{700\,\text{K}} = -16.81\,\text{J/mol K}\,.$$

Box 2.14 (Important formulae: a first glance at statistical thermodynamic relations).

Ω_m = number of micro-states that realize the same macro-state,
Z = partition function, E_k = total energy of particles in micro-state k,
N_{kl} = number of particle species l in micro-state k, k_B = Boltzmann constant.

(In special cases, the sums must be replaced by integrals.
In case of degenerations, degeneration factors must be inserted.)

Entropy of a closed system
(entropy of a "microcanonical ensemble"):

$$S = k_\mathrm{B} \ln \Omega_\mathrm{m} . \tag{2.86}$$

Entropy of an open system, particle numbers preserved
(entropy of a "canonical ensemble"):

$$S = k_\mathrm{B}(\ln Z_\mathrm{c} + \beta U) , \quad Z_\mathrm{c} = \sum_k \exp(-\beta E_k) , \quad \beta = \frac{1}{k_\mathrm{B} T} . \tag{2.87}$$

Entropy of an open system, particle numbers not preserved
(entropy of a "grand canonical ensemble"):

$$S = k_\mathrm{B} \left(\ln Z_\mathrm{gc} + \beta U - \beta \sum_l \mu_l N_l \right) ,$$

$$Z_\mathrm{gc} = \sum_k \exp \left[\beta \left(\sum_l \mu_l N_{kl} - E_k \right) \right] , \quad \beta = \frac{1}{k_\mathrm{B} T} . \tag{2.88}$$

Internal energy, canonical ensemble:

$$U = -\left(\frac{\partial \ln Z_\mathrm{c}}{\partial \beta} \right)_{V, \{N_i\}} . \tag{2.89}$$

Internal energy, grand canonical ensemble:

$$U = -\left(\frac{\partial \ln Z_\mathrm{gc}}{\partial \beta} \right)_{\{\mu_i\}} + \sum_i \frac{\mu_i}{\beta} \left(\frac{\partial \ln Z_\mathrm{gc}}{\partial \mu_i} \right)_\beta . \tag{2.90}$$

Pressure, canonical ensemble:

$$p = \frac{1}{\beta} \left(\frac{\partial \ln Z_\mathrm{c}}{\partial V} \right)_{T, \{N_i\}} . \tag{2.91}$$

Pressure, grand canonical ensemble:

$$p = \frac{1}{\beta} \left(\frac{\partial \ln Z_\mathrm{gc}}{\partial V} \right)_{\{\mu_i\}} - \sum_i \frac{\mu_i}{\beta} \left(\frac{\partial}{\partial V} \frac{\partial \ln Z_\mathrm{gc}}{\partial \mu_i} \right)_\beta . \tag{2.92}$$

Continuation of Box.

Alteration of the total entropy:

$$\Delta S_{\text{tot}} = \Delta S_{\text{thermal}} + \Delta S_{\text{configurational}} + \Delta S_{\text{further modes}},$$

$$\Delta S_{\text{thermal}} = \frac{\delta Q}{T}, \qquad (2.93)$$

$$\Delta S_{\text{configurational}} = k_B (\ln \Omega_1 - \ln \Omega_2) = k_B \left(\ln \frac{\Omega_1}{\Omega_2} \right).$$

(S_{thermal} is the entropy introduced in Box 2.13; S_{thermal} can also be modeled by (2.86); S_{thermal} can only be modeled by (2.86).)

Alteration of the configurational entropy:

(a) population of crystal sites, equal number M of crystal sites
$(M = N_1 + N_1' = N_2 + N_2' = \text{constant})$

$$\Rightarrow$$

$$\Omega_1 := \frac{M!}{N_1! N_1'!} = \frac{(N_1 + N_1')!}{N_1! N_1'!},$$

$$\Omega_2 := \frac{M!}{N_2! N_2'!} = \frac{(N_2 + N_2')!}{N_2! N_2'!}$$

Stirling's formula for large x ($\ln x! = x \ln x - x$)
$$\Rightarrow$$
$$\qquad (2.94)$$

$$\Delta S_{\text{configurational}} =$$

$$= -k_B \left(N_1 \ln \frac{N_1}{N_1 + N_1'} + N_1' \ln \frac{N_1'}{N_1' + N_1} - N_2 \ln \frac{N_2}{N_2 + N_2'} - N_2' \ln \frac{N_2'}{N_2' + N_2} \right);$$

(b) population of crystal sites, equal number N_1 of ions/atoms/molecules
$(N_1 = \text{constant})$

$$\Rightarrow$$

$$\Omega_1 := \frac{M!}{N_1! N_1'!} = \frac{(N_1 + N_1')!}{N_1! N_1'!},$$

$$\Omega_2 := \frac{M'!}{N_1! N_1''!} = \frac{(N_1 + N_1'')!}{N_1! N_1''!}$$

Stirling's formula for large x ($\ln x! = x \ln x - x$)
$$\Rightarrow$$
$$\qquad (2.95)$$

$$\Delta S_{\text{configurational}} =$$

$$= -k_B \left(N_1 \ln \frac{N_1}{N_1 + N_1'} + N_1' \ln \frac{N_1'}{N_1' + N_1} - N_1 \ln \frac{N_1}{N_1 + N_1''} - N_1'' \ln \frac{N_1''}{N_1' + N_1} \right);$$

Continuation of Box.

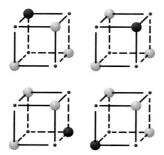

example for Ω_1

$M_{1,A} = M_{1,X} = 4,$
$M_{1,A} + M_{1,X} = 8$

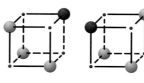

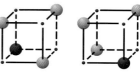

$N_{1,A} = 3,\ N'_{1,A} = 1,$
$\Omega_{1,A} = 4$

$\Omega_1 = \Omega_{1,A}\Omega_{1,X} = 16$

$N_{1,X} = 3,\ N'_{1,X} = 1,$
$\Omega_{1,X} = 4$

(c) AX crystal, equal number $M = M_{1,A} = M_{1,X} = M_{2,A} = M_{2,X}$ of crystal sites (populated sites $N_{i,A} = N_{i,X}$, defects $N'_{i,A} = N'_{i,X}$, $i = 1, 2$)

$$\Rightarrow$$

$$\Omega_i = \Omega_{i,A}\Omega_{i,X} =$$

$$= \frac{M!}{N_{i,A}!N'_{i,A}!}\frac{M!}{N_{i,X}!N'_{i,X}!} = \frac{(N_{i,A} + N'_{i,A})!}{N_{i,A}!N'_{i,A}!}\frac{(N_{i,X} + N'_{i,X})!}{N_{i,X}!N'_{i,X}!}. \qquad (2.96)$$

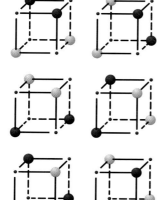

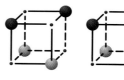

example for Ω_2

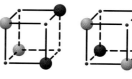

$M_{1,A} = M_{1,X} = 4,$
$M_{1,A} + M_{1,X} = 8$

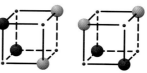

$N_{1,A} = 2,\ N'_{1,A} = 2,$
$\Omega_{2,A} = 6$

$\Omega_2 = \Omega_{2,A}\Omega_{2,X} = 36$

$N_{1,X} = 2,\ N'_{1,X} = 2,$
$\Omega_{2,X} = 6$

Box 2.15 (Important formulae: a second glance at statistical thermodynamic relations).

K = constant to be chosen suitably, k_B = Boltzmann constant,
p_i = probability of a system state i, where i covers all possible system states,
$\hat{\rho}$ = quantum-mechanical density matrix,
M = total number of atoms/molecules,
N_j = number of atoms/molecules within an element j ($j = 1 \ldots s$) of the phase space,
Ω = "thermodynamic probability" =
= number of micro-states that realize the same macro-state.

Information entropy:

$$I = -K \sum_i p(x_i) \log_b p(x_i) \tag{2.97}$$

(in special cases, the sum must be replaced by an integral)

$\Downarrow$

Gibbs entropy
("Γ space entropy", i.e. particle systems are considered):

$$S = -k_B \sum_i p_i \ln p_i \tag{2.98}$$

(in special cases, the sum must be replaced by an integral)

$\Downarrow$

Boltzmann entropy
("μ space entropy", i.e. individual particles are considered):

$$S = k_B \ln \Omega, \quad \Omega = \frac{M!}{N_1! N_2! \ldots N_s!}, \tag{2.99}$$

$$S \stackrel{\text{Stirling's formula}}{=} k_B \left(M \ln M - \sum_{j=1}^{s} N_j \ln N_j \right) = -k_B \sum_{j=1}^{s} N_j \ln \frac{N_j}{M} \tag{2.100}$$

$\Uparrow$

Gibbs entropy
("Γ space entropy", i.e. particle systems are considered):

$$S = -k_B \sum_i p_i \ln p_i \tag{2.101}$$

(in special cases, the sum must be replaced by an integral)

$\Uparrow$

Von Neumann entropy:

$$S = -k_B \operatorname{tr}(\hat{\rho} \ln \hat{\rho}) \tag{2.102}$$

On the one hand, we here note that the total entropy is a measure for the notion *information content of a thermodynamic system*. On the other hand, we here note that the increase or decrease of the notion *information content of a thermodynamic system* usually means the increase or decrease of the disorder of a thermodynamic system.

It might be interesting for the readers to know that it was C. Shannon who put the notion *information* into concrete terms. It might be interesting for the readers to know, too, that it was C. Shannon who established a mathematical expression specifying the notion *information* [49]. Let us here quote that this mathematical expression is given by (2.97), where K is a constant to be chosen suitably, and where the $p(x_i)$ record the probabilities for the occurrence of special values of the random variables x_i. Let us here quote that this mathematical expression can be understood as follows. Suppose that 10 bits (10 0/1 digits within a 0/1 chain) are transmitted via a transmission channel. If the receiver knows the 0/1 chain already before the transmission has taken place, 0 bits of information are transmitted. If the receiver knows the 0/1 chain only once the transmission has taken place, 10 bits of information are transmitted, provided the 0 outcome and the 1 outcome are equally likely. Setting in (2.97) $K = 1$ bit and using in (2.97) the binary logarithm $\log_b = \log_2 = \text{lb}$, in the first case, the $p(x_i)$ are 0 or 1, with $\lim_{p \to 0} p \log_b p = 0$ and $\log_b 1 = 0$, leading to the final result "information 0 bit", and in the second case, the $p(x_i)$ are $1/2$, with $\text{lb}\, 1/2 = -1$, leading to the final result "information 10 bit", supplying us with a mathematical wrapping for the notion *information*, also called *information entropy*, *Shannon entropy*, or *uncertainty*.

As it is outlined in Box 2.15, the notion *information* is carried into the domain of thermodynamics as follows. On the one hand, the Gibbs entropy (2.98) is obtained as specification of the information entropy (2.97), and the Gibbs entropy (2.98) in turn leads to the Boltzmann entropy (2.99), which directly establishes the connection to the entropy formulae of Box 2.14. On the other hand, the Gibbs entropy (2.101) is obtained as specification of the von Neumann entropy (2.102), which eventually carries the notion *information* into the domain of conventional quantum mechanics, and the Gibbs entropy (2.101) in turn leads to the Boltzmann entropy (2.99), which directly establishes the connection to the entropy formulae of Box 2.14. Since we do not require that the $p(x_i)$ of the information entropy (2.97) are restricted in any way, in particular, since we do not require that the $p(x_i)$ of the information entropy (2.97) have to be defined in the "bit and byte sense" mentioned above, such specifications are possible. Since such specifications are possible, we finally realize that the Gibbs entropy, the Boltzmann entropy, and the entropy formulae of Box 2.14 indeed are carrying the notion *information* into the domain of thermodynamics. We note in passing that the Gibbs entropy is a Γ space entropy, namely particle systems are compared, while the Boltzmann entropy is a μ space entropy, namely individual particles are compared [50]. We note in passing that Γ space relations are the better choice if interactions that are strong between individual particles are present, while μ space relations are the better choice if interactions that are weak between individual particles are present.

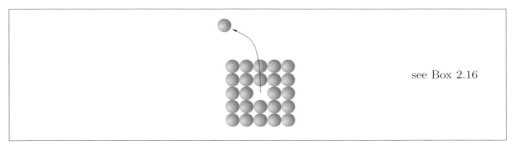

Figure 2.78. Formation of a vacancy.

2.2.3.2 Thermodynamics: Defect Models

The reader should realize that the toolkit of thermodynamic formulae introduced in the preceding section also supplies us with a mathematical access to crystal structures with defects including the defect models described above. A simple vacancy model is developed in Box 2.16. Above all, we point out that ΔG acccording to (2.112) records the change of free enthalpy due to the occurrence of vacancies. Above all, we point out that ΔG acccording to (2.113) leads to the equilibrium concentration of vacancies, i.e. $n_v/N = \exp(-\Delta g_v/k_B T)$ with $\Delta g_v = h_v - T\Delta s_{\mathrm{vib}}$, where h_v is the enthalpy needed to create a vacancy, and where Δs_{vib} is the change of thermal entropy evoked by the occurrence of a vacancy provided we here only do consider vibrational motions of identical quantum-mechanical particles defined by vibrational frequencies ω and ω'. Comparing with Examples 2.3 and 2.4, we realize that this simple vacancy model is immediately adapted to similar situations such as the generation of vacancy pairs of Schottky defects or anion/cation Frenkel pairs of Frenkel defects. We realize that the concentrations $n_{v,\mathrm{anion}}/N_{\mathrm{anion}}$ (anion vacancies) and $n_{v,\mathrm{cation}}/N_{\mathrm{cation}}$ (cation vacancies) are governed by the change of free enthalpy $\Delta g_{\mathrm{vacancy\ pair}}$ which is associated with the occurrence of a vacancy pair (in the case of Schottky defects) or are governed by the changes of free enthalpy $\Delta g_{\mathrm{anion\ Frenkel\ pair}}$ and $\Delta g_{\mathrm{cation\ Frenkel\ pair}}$ which are associated with the occurrence of an anion Frenkel pair and a cation Frenkel pair (in the case of Frenkel defects), where $n_{v,\mathrm{anion}}$ and $n_{v,\mathrm{cation}}$ are the numbers of anion vacancies and cation vacancies, respectively, and N_{anion} and N_{cation} are the numbers of anion sites and cation sites, respectively.

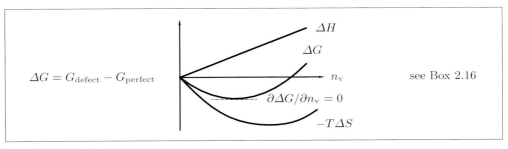

Figure 2.79. Progress of the free enthalpy due to the formation of vacancies.

Box 2.16 (Important formulae: a simple vacancy model).

G = free enthalpy, H = enthalpy, S = entropy,
k_B = Boltzmann constant, $\hbar$ = Planck's constant; ω = vibrational frequency,
n_v = number of vacancies, χ = weight factor,
h_v = enthalpy needed to create a vacancy,
Δs_vib = change of thermal entropy evoked by the occurrence of a vacancy,
N = number of particles, M = number of crystal sites.

$$G = H - TS = H - T\bigl[S_\mathrm{thermal} + S_\mathrm{configurational} + S_\mathrm{further\ modes}\bigr]. \tag{2.103}$$

Perfect crystal ($N' = M$):

$$H = H_\mathrm{perfect}, \tag{2.104}$$

$$S_\mathrm{thermal} = N' k_\mathrm{B} \left(\ln \frac{k_\mathrm{B} T}{\hbar \omega} + 1 \right) \tag{2.105}$$

(since we here do only consider vibrational motions of
N' identical quantum-mechanical particles),

$$S_\mathrm{configurational} = k_\mathrm{B} \ln \Omega = 0$$

(since there is only one possibility to distribute
N' identical quantum-mechanical particles on $N' = M$ crystal sites
$\Rightarrow \Omega = 1 \Rightarrow \ln \Omega = 0$), \hfill (2.106)

$S_\mathrm{further\ modes} = 0$ (since we here do not consider further modes). \hfill (2.107)

Defect crystal ($N + n_\mathrm{v} = M$):

$$H = H_\mathrm{perfect} + n_\mathrm{v} h_\mathrm{v}, \tag{2.108}$$

$$S_\mathrm{thermal} = (N - \chi n_\mathrm{v}) k_\mathrm{B} \left(\ln \frac{k_\mathrm{B} T}{\hbar \omega} + 1 \right) + \chi n_\mathrm{v} k_\mathrm{B} \left(\ln \frac{k_\mathrm{B} T}{\hbar \omega'} + 1 \right) \tag{2.109}$$

(since we here do assume that χn_v particles around n_v vacancies
show a different vibrational frequency ω'),

$$S_\mathrm{configurational} = k_\mathrm{B} \ln \Omega = -k_\mathrm{B} \left(N \ln \frac{N}{N + n_\mathrm{v}} + n_\mathrm{v} \ln \frac{n_\mathrm{v}}{N + n_\mathrm{v}} \right) \tag{2.110}$$

(since we here do follow (2.94) and (2.95)),

$S_\mathrm{further\ modes} = 0$ (since we here do not consider further modes). \hfill (2.111)

> **Continuation of Box.**
>
> Equilibrium concentration of vacancies:
>
> $$\Delta G = G_{\text{defect}} - G_{\text{perfect}} =$$
>
> $$= H_{\text{perfect}} + n_{\text{v}} h_{\text{v}} - H_{\text{perfect}} -$$
>
> $$-T(N - \chi n_{\text{v}})k_{\text{B}}\left(\ln \frac{k_{\text{B}} T}{\hbar \omega} + 1\right) - T\chi n_{\text{v}} k_{\text{B}} \left(\ln \frac{k_{\text{B}} T}{\hbar \omega'} + 1\right) +$$
>
> $$+T(N + n_{\text{v}})k_{\text{B}}\left(\ln \frac{k_{\text{B}} T}{\hbar \omega} + 1\right) + \qquad (2.112)$$
>
> $$+T k_{\text{B}} \left(N \ln \frac{N}{N + n_{\text{v}}} + n_{\text{v}} \ln \frac{n_{\text{v}}}{N + n_{\text{v}}}\right) \approx$$
>
> $$\approx n_{\text{v}} h_{\text{v}} + T k_{\text{B}} \chi n_{\text{v}} \ln \frac{\omega'}{\omega} + T k_{\text{B}} \left(N \ln \frac{N}{N + n_{\text{v}}} + n_{\text{v}} \ln \frac{n_{\text{v}}}{N + n_{\text{v}}}\right),$$
>
> $$\frac{\partial \Delta G}{\partial n_{\text{v}}} = 0$$
> (equilibrium condition)
>
> $$\Rightarrow$$
>
> $$\ln \frac{n_{\text{v}}}{N + n_{\text{v}}} \approx \ln \frac{n_{\text{v}}}{N} = -\frac{h_{\text{v}} - T \Delta s_{\text{vib}}}{k_{\text{B}} T} = -\frac{\Delta g_{\text{v}}}{k_{\text{B}} T}, \qquad (2.113)$$
>
> $$\Delta g_{\text{v}} = h_{\text{v}} - T \Delta s_{\text{vib}}, \quad \Delta s_{\text{vib}} = k_{\text{B}} \chi \ln \frac{\omega}{\omega'}$$
>
> $$\Rightarrow$$
>
> $$\frac{n_{\text{v}}}{N} = \exp\left(-\frac{\Delta g_{\text{v}}}{k_{\text{B}} T}\right),$$
>
> $$\Delta G = \Delta H - T \Delta S,$$
>
> $$\Delta H = n_{\text{v}} h_{\text{v}}, \qquad (2.114)$$
>
> $$-T \Delta S \approx T k_{\text{B}} \chi n_{\text{v}} \ln \frac{\omega'}{\omega} + T k_{\text{B}} \left(N \ln \frac{N}{N + n_{\text{v}}} + n_{\text{v}} \ln \frac{n_{\text{v}}}{N + n_{\text{v}}}\right).$$
>
> We here compare with Figures 2.78 and 2.79.

Example 2.3 (An additional tutorial example: concentration of Schottky defects).

Set up the defect reaction for a Schottky defect in an NaCl crystal. How does the proper equilibrium concentration of the vacancies look like if we adapt the vacancy model introduced Box 2.16 to this situation?

Solution.

$$\text{Na}_{\text{Na}}^{\text{x}} + \text{Cl}_{\text{Cl}}^{\text{x}} \rightarrow \text{V}_{\text{Na}}' + \text{V}_{\text{Cl}}^{\bullet} + \text{Na}_{\text{surface}}^{\text{x}} + \text{Cl}_{\text{surface}}^{\text{x}} \,, \tag{2.115}$$

$$\frac{n_{\text{v,anion}}}{N_{\text{anion}}} = \frac{n_{\text{v,cation}}}{N_{\text{cation}}} = \exp\left(-\frac{\Delta g_{\text{vacancy pair}}}{2k_\text{B}T}\right) =$$
$$= \exp\left(-\frac{h_{\text{vacancy pair}}}{2k_\text{B}T}\right) \exp\left(+\frac{\Delta s_{\text{vacancy pair}}}{2k_\text{B}}\right) \,, \tag{2.116}$$

$$\frac{n_{\text{v,anion}}}{N_{\text{anion}}} \frac{n_{\text{v,cation}}}{N_{\text{cation}}} = \exp\left(-\frac{\Delta g_{\text{vacancy pair}}}{k_\text{B}T}\right) \,. \tag{2.117}$$

Example 2.4 (An additional tutorial example: concentration of Frenkel defects).

Set up the defect reactions for Frenkel defects in an NaCl crystal. How does the proper equilibrium concentration of the vacancies look like if we adapt the vacancy model introduced Box 2.16 to this situation? How can the interstitial cations and the interstitial anions be included?

Solution.

$$\text{Na}_{\text{Na}}^{\text{x}} \rightarrow \text{V}_{\text{Na}}' + \text{Na}_{\text{i}}^{\bullet} \,, \quad \frac{n_{\text{v,cation}}}{N_{\text{cation}}} = \exp\left(-\frac{\Delta g_{\text{cation Frenkel pair}}}{2k_\text{B}T}\right) \,, \tag{2.118}$$

$$\frac{n_{\text{v,cation}} n_{\text{i,cation}}}{N_{\text{cation}} N_{\text{i}}} = \exp\left(-\frac{\Delta g_{\text{cation Frenkel pair}}}{k_\text{B}T}\right) \,; \tag{2.119}$$

$$\text{Cl}_{\text{Cl}}^{\text{x}} \rightarrow \text{V}_{\text{Cl}}^{\bullet} + \text{Cl}_{\text{i}}' \,, \quad \frac{n_{\text{v,anion}}}{N_{\text{anion}}} = \exp\left(-\frac{\Delta g_{\text{anion Frenkel pair}}}{2k_\text{B}T}\right) \tag{2.120}$$

$$\frac{n_{\text{v,anion}} n_{\text{i,anion}}}{N_{\text{anion}} N_{\text{i}}} = \exp\left(-\frac{\Delta g_{\text{anion Frenkel pair}}}{k_\text{B}T}\right) \,. \tag{2.121}$$

2.3 Microstructures

The chemical, physical, and mechanical properties of materials are no fixed quantities, but very much dependent on the *microstructure* of a material and therefore on the kind and on the number of defects of the crystals present in the material. Certainly, in special cases, individual properties of a microstructured material can be on a par with those of a monocrystal. However, in most cases, already polycrystals consisting of more or less defect-free, twisted and shifted crystals reveal quite different properties. The divergence will be even increased when the crystals of a polycrystal contain a lot of defects. In order to express the structural difference, the crystals of a polycrystal are commonly notated as *grains* and the boundaries that separate the grains as *grain boundaries*. The divergence between the properties of a microstructured material and a monocrystal will be even increased when pores or further crystalline and non-crystalline phases are present in the microstructured material. This all is especially true for ceramics. Especially, all kinds of point defects are important for the functional properties of ceramics, and features such as grain size, pores, cracks, and secondary phases are of special importance with respect to the mechanical stability of ceramics. However, in many cases, this differentiation simplifies the complex microstructure–property relation of ceramics already too much.

2.3.1 Microstructure Features

Typical features of ceramic microstructures are sketched in Figure 2.80 and illustrated by several examples shown in Figures 2.81–2.86. From Figure 2.80–2.86, we read off that different grains usually show different cell orientations, sizes, and shapes, that pores within grains and at grain boundaries can occur, that symmetric domains such as mirror-symmetrical twins or ferroelectric domains can occur, and that secondary phases due to impurities or additives can occur, which can be crystalline or non-crystalline, for example, as glass phases located in triple junctions of grains or as more or less continuous phases located at grain boundaries. From Figure 2.80–2.86, we read off that the grain shape ranges from polyeder-like to column-like and plate-like forms. Let us here note that the distribution of grain sizes can show more than one maximum. The term "duplex microstructure" circumscribes such a special case. Let us here also note that pores, on the one hand, can be channels that are connected with the atmospherical surrounding, and on the other hand, can be isolated in the ceramic microstructure, in the latter case, located either within grains or in between grains at grain boundaries or triple junctions. The terms "open pores" and "closed pores" define such special cases. Let us here also note that technical ceramics are typically fine-grained ceramics in order to achieve high mechanical performance. In general, coarse-grained ceramics and broad grain size distributions are only acceptable if functional properties are dominant and/or weak mechanical performance can be tolerated.

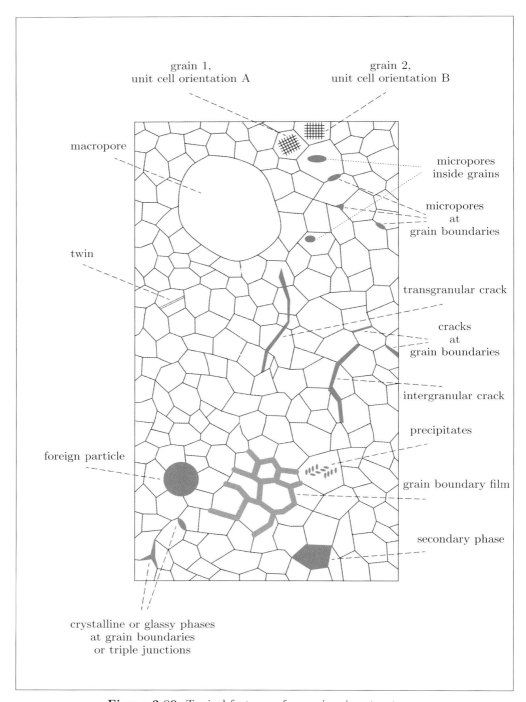

Figure 2.80. Typical features of ceramic microstructures.

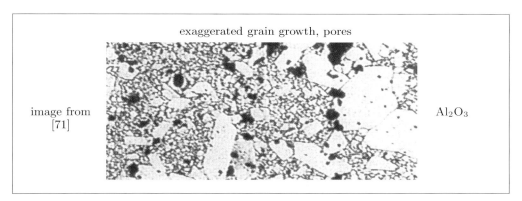

Figure 2.81. Al_2O_3 microstructure image.

The reader should particularly realize that the morphology of grain boundaries ranges from occasionally very thin (some atomic distances) grain boundary layers that consist of a glass phase that continuously runs through the ceramic material to totally isolated grain pendentives which are located at grain edges of the ceramic material, in particular, we compare with Figure 2.86 and Boxes 2.17 and 2.18. From Figure 2.86, we read off that the "wetting angle" of contact θ distinguishes the diverse forms of grain boundaries from each other. From Box 2.17, we read off that the "wetting angle" of contact θ according to (2.122) is determined by both the interfacial (surface) tension "grain-to-grain", which is here denoted as γ_{gg}, and the interfacial (surface) tension "grain-to-boundary", which is here denoted as γ_{gb}. From Box 2.17, we also read off that the interfacial (surface) tension according to (2.123) is defined as the quotient of the free enthalpy dG (which is needed to generate the surface enlargement dA) and the surface enlargement dA, provided pressure, temperature, and particle numbers are constants. Box 2.18 finally concatenates the values for the quotient γ_{gg}/γ_{gb} with the values for the "wetting angle" of contact θ.

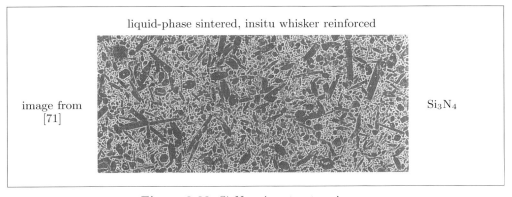

Figure 2.82. Si_3N_4 microstructure image.

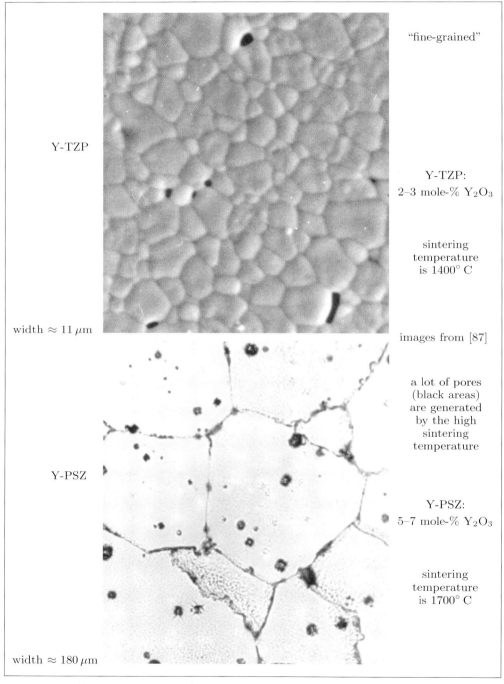

Figure 2.83. ZrO$_2$ (Y-TZP and Y-PSZ) microstructure images.

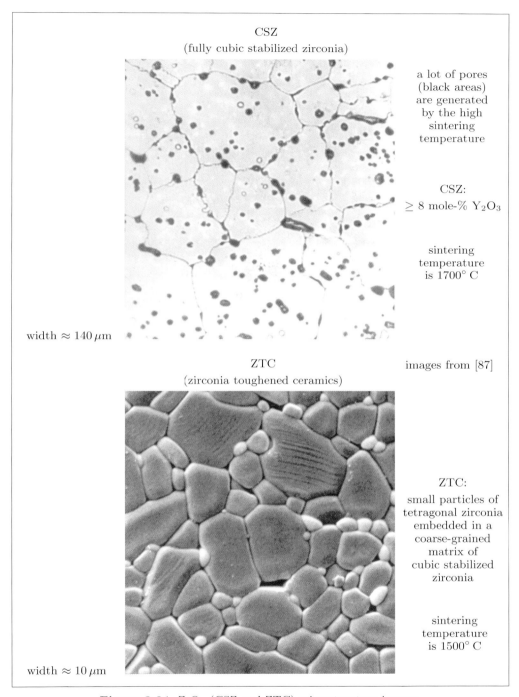

Figure 2.84. ZrO_2 (CSZ and ZTC) microstructure images.

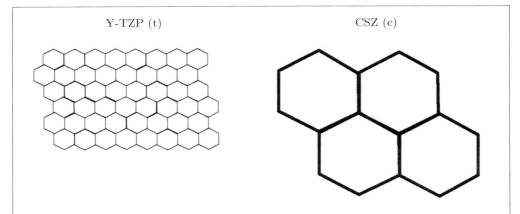

diagramed grain structures according to Figures 2.83 and 2.84

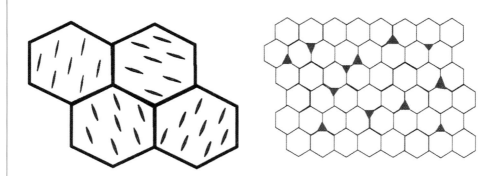

finely dispersed, lense-shaped t precipitationes within c grains

Figure 2.85. Grain structures.

2.3 Microstructures

Box 2.17 (Important formulae: grain boundary geometry and interfacial (surface) tensions).

"Wetting angle" of contact θ:

$$\cos\frac{\theta}{2} = \frac{1}{2}\frac{\gamma_{\text{gg}}}{\gamma_{\text{gb}}}. \tag{2.122}$$

Interfacial (surface) tension γ:

$$\gamma = \left(\frac{\partial G}{\partial A}\right)_{p,T,\{N_i\}}. \tag{2.123}$$

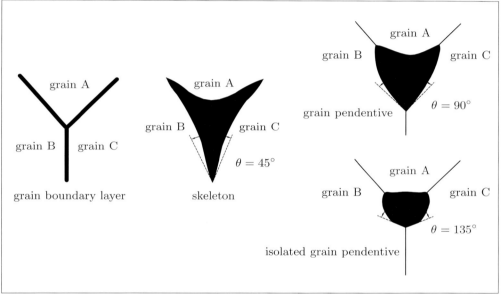

Figure 2.86. Grain boundary structures.

Box 2.18 (Important data: grain boundary geometry and interfacial (surface) tensions).

$$\begin{aligned}
\frac{\gamma_{\text{gg}}}{\gamma_{\text{gb}}} &\gg 1 & &: \quad \text{grain boundary layer}, \\
\frac{\gamma_{\text{gg}}}{\gamma_{\text{gb}}} &> \sqrt{3} & &: \quad \text{skeleton } ((2.122) \to \theta < 60°), \\
\frac{\gamma_{\text{gg}}}{\gamma_{\text{gb}}} &= 1\ldots\sqrt{3} & &: \quad \text{grain pendentive } ((2.122) \to \theta = 60\ldots 120°), \\
\frac{\gamma_{\text{gg}}}{\gamma_{\text{gb}}} &< 1 & &: \quad \text{isolated grain pendentive } ((2.122) \to \theta > 120°).
\end{aligned} \tag{2.124}$$

2.3.2 Microstructure Design

The characteristics of ceramic materials, for example, hardness, fracture toughness, thermal conductivity, electric characteristics, magnetic characteristics, and creep under compression or tension, are particularly determined by the microstructures of the ceramic materials. Therefore, designing microstructures of ceramic materials means designing ceramic materials with special characteristics, in last consequence, meeting special technical demands. However, since the microstructures of ceramic materials strongly depend on the production processes, different production processes or less controlled production processes may result in ceramic materials with quite different characteristics. Therefore, designing ceramic materials, on the one hand, particularly means modifying existing production processes, and on the other hand, particularly means getting production processes more controlled.

The following design elements are important. (i) Above all, the *grain size* is an important design element. On the one hand, small-scale grains may noticeably enhance the strength. On the other hand, functional properties strongly rely on the grain size. For example, the dielectric constant and thereby the capacitance per unit of volume increase with decreasing grain size down to grain sizes below which ferroelectric domains can hardly develop. For example, in ferrites smaller grain sizes are advantageous since the motion of domain walls then is reduced. (ii) Furthermore, the *grain shape* is an important design element. For example, microstructures containing platelet-like grains or needle-like grains show an appreciable higher fracture toughness since the toothing of grains and the deflection of cracks at grains thereby is improved. (iii) Furthermore, the *secondary phases* are an important design element. For example, glass phases occurring in liquid-phase-sintered ceramic materials, on the one hand, noticeably enhance the fracture toughness of ceramic materials, but on the other hand, noticeably reduce the strength at high termperatures. (iv) Last but not least, the *porosity* is an important design element. For example, on the one hand, small rounded pores may blunt cracks and thus may enhance fracture toughness, but on the other hand, "sharp pores" may further reduce the fracture toughness of notch-sensitive ceramics.

Wanting to design microstructures, we must not undervalue the impact of porosity. (i) Firstly, big pores (even single ones) reduce the strength of ceramics dramatically. Thus, wanting to meet reliable mechanical performance of ceramics, the densification during production is essential. Naturally, reducing the porosity in ceramics could also mean the reduction of the scattering of light in optical ceramics, for example, thereby increasing light transmittance and efficiency of sodium discharge lamps. (ii) Secondly, increased porosity can be beneficial when the pores are well-distibuted and uniform in size. On the one hand, the thermal shock resistance of ceramics can be increased in this manner, thereby providing increased thermal cycling capability, for example, of honeycomb-structured cordierite automotive catalytic converters. On the other hand, well-distributed pores introduced into ferrites act as pinning sites for domain walls, thereby preventing the motion of domain walls in ferrites. This design element is also applied in the production of aluminum titanate parts for the metallurgical processing of molten metals. (iii) Thirdly, pores may serve as reservoirs for lubricants.

2.3.2.1 Process Constraints

Wanting to design microstructures, we must evaluate the influence of process type and process parameters. (i) For example, we consider the SiC microstructures shown in Figures 2.87–2.90. On the one hand, we read off from these figures that the process type (pressureless sintering, hot-pressed SiC, liquid-phase-sintering, Si-infiltration) enables us to control the grain structure of the SiC microstructure and thereby their cardinal properties. On the other hand, we read off from these figures that the choice of the process parameters (additives parameters, sintering parameters) enables us to control the grain structure of the SiC microstructure and thereby their characteristics. (ii) For example, the application of pressure has an influence on the sintering process. In particular, the gas pressure sintering of Si_3N_4 or SiC facilitates the production of extensively defect-free and extensively homogeneous microstructures. (iii) For example, the stream velocity of a flowing solution has an influence on the deposition process of a ceramic layer which eventually evolves on a monolayer out of aqueous solutions during bio-inspired mineralization procedures. In particular, a flowing solution, in comparison to a stagnant solution, clearly increases the deposition rate.

2.3.2.2 Advanced Constraints

Wanting to design microstructures, going one step further, we must be aware of the circumstances that interior and/or exterior patterns in the end can be imprinted in the crystal structure and/or the microstructure of ceramic films (TiO_2, ZrO_2 etc.) evolving on a substrate out of aqueous solutions during bio-inspired mineralization procedures by the application of technical constraints such as rotations of the deposition tank and/or stream differentiations of the flowing solution. For example, slightly modifying the deposition tank and the feed pipe of the technical equipment that is presented in Figure 4.64 shown in Section 4.2.3 in such a way that it carries out rotations around the centrally located vertical tank axis and that stream patterns $j(x)$ of the flowing solution can be implemented, depending on the rotation frequency ω, the stream velocity v of the flowing solution, and the stream patterns $j(x)$ of the flowing solution, we expect that regular patterns and chaotic patterns can be imprinted in the crystal structure and/or the microstructure of a ceramic film (TiO_2, ZrO_2 etc.). Furthermore, we expect that the application of technical constraints such as *acoustic fields* or *electromagnetic fields* has a structuring effect on a ceramic film (TiO_2, ZrO_2 etc.), provided the wave structure, the wave lengths, and the oscillation frequencies approach the vibrational behavior of the crystalline structure, and provided these are applied during the deposition period of the ceramic film (TiO_2, ZrO_2 etc.). Therefore, we expect that we will be placed in a position to control the evolution of ceramic materials during bio-inspired mineralization procedures in finest details by utilizing this type of advanced constraints, approaching the idea of *self-evolutional ceramic nanostructures*. Beyond that, we already note here that the incorporation of synthetic polymers in bio-inspired mineralization procedures defines a sub-class of this type of advanced constraints, in last consequence, setting control parameters, evoking a variety of ceramic materials structured in finest details. We come back to this powerful technique in Section 4.2.3.

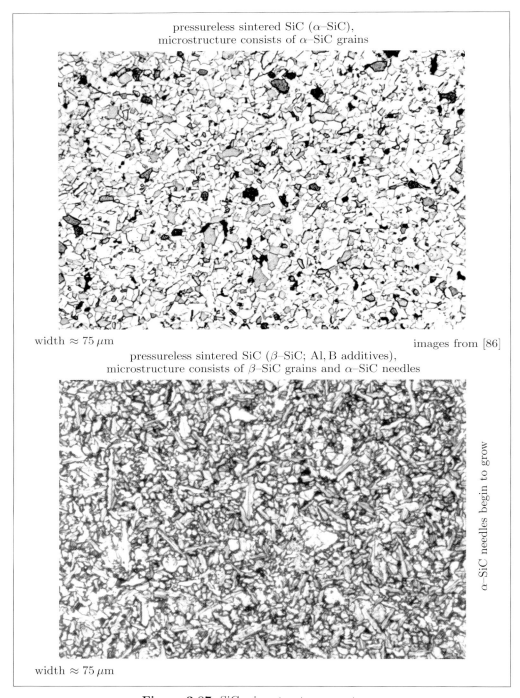

Figure 2.87. SiC microstructures, part one.

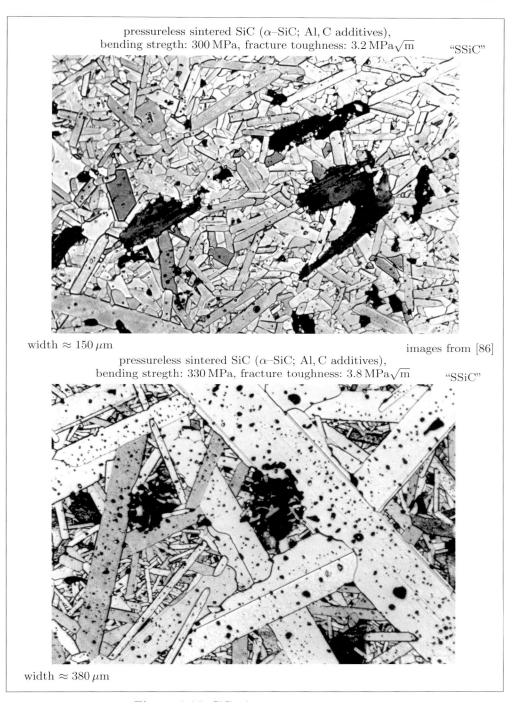

Figure 2.88. SiC microstructures, part two.

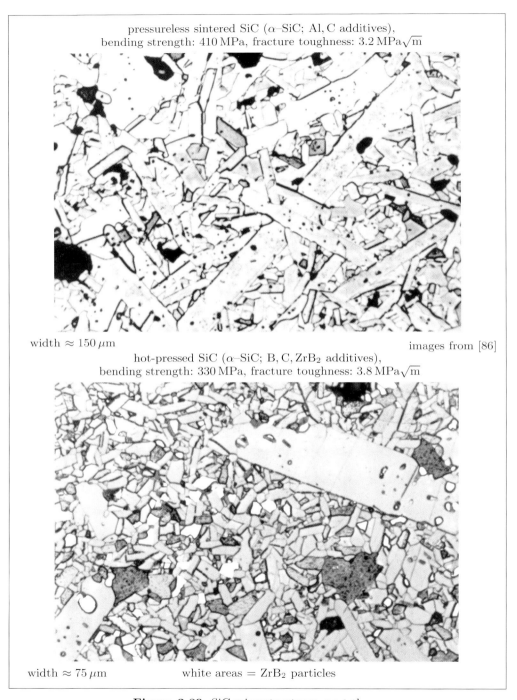

Figure 2.89. SiC microstructures, part three.

2.3 *Microstructures* 155

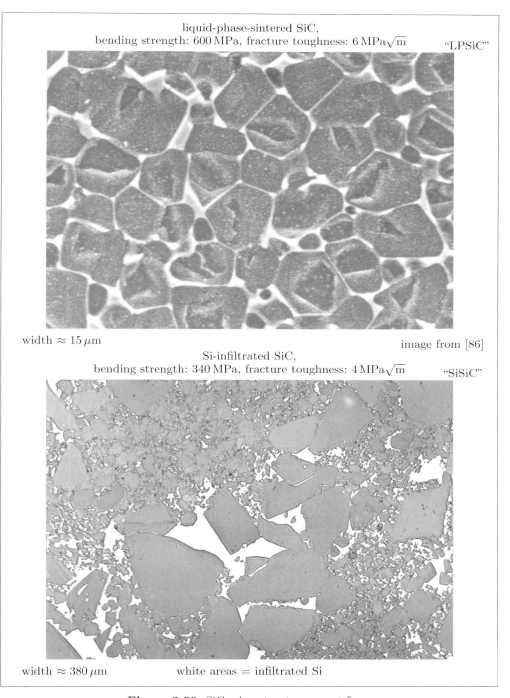

liquid-phase-sintered SiC,
bending strength: 600 MPa, fracture toughness: 6 MPa$\sqrt{\text{m}}$ "LPSiC"

width ≈ 15 μm   image from [86]

Si-infiltrated SiC,
bending strength: 340 MPa, fracture toughness: 4 MPa$\sqrt{\text{m}}$ "SiSiC"

width ≈ 380 μm   white areas = infiltrated Si

Figure 2.90. SiC microstructures, part four.

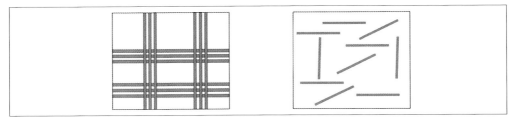

Figure 2.91. Typical structures of composite materials: fiber/whisker reinforcement.

2.3.2.3 Composite Materials

Combining ceramic and/or metallic components, organic and/or inorganic components, in last consequence, we are led to composite materials. Let us here quote the examples that follow. (i) Whisker reinforced Al_2O_3. This composite material is especially used for cutting knifes. (ii) Carbon fiber reinforced SiC. This composite material is especially used for wheel disk brakes. (iii) Nicalon® fiber reinforced Si–B–C–N ceramics and carbon fiber reinforced Si–B–C–N ceramics. An example of a duroplastic ceramics of the second type is shown in Figure 2.93. We note that such a composite material is generated via resin transfer molding (RTM). After infiltration of the starting substances into a carbon fiber mat, in a first heating step at about 200°C, a cross-linking of the starting substances is established, and in a second heating step at about 1400°C, the composite material (sometimes called "infiltrated medium") is established – a variety of chemical procedures to be explained in Section 4.2.

Combining ceramic components with metallic alloys, in last consequence, we are led to metal matrix composite microstructures. We note that such composite materials, for example, are generated via suitable preforms. Producing a preform, for example, by the method of freeze casting, after the freeze drying period, the voids are filled with the metallic alloys. We also note that such composite materials can show nearly any tribology, any tensile strength, and any creep stability. Examples where silicon (Si) and/or aluminum oxide (Al_2O_3) are/is combined with the metallic alloy $AlSi_9Cu_3$ are shown in Figure 2.94. The readers might find it very interesting that the tensile strength increases with decreasing grain size of the aluminum oxide particles, provided we fix the grain size of the silicon particles. The readers might find it also very interesting that a bigger fraction of silicon calls for an increase of the tribology, whereas a bigger fraction of aluminum oxide calls for an increase of the tensile strength.

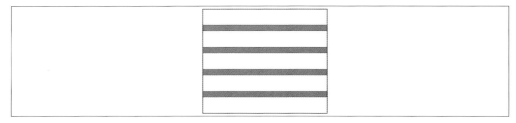

Figure 2.92. Typical structures of composite materials: sandwich structure.

2.3.3 Microstructure Visualization

Wanting to visualize microstructures, one resorts to light microscope techniques or electron microscope techniques. We here point at two electron microscope techniques relevant in this context. (i) The scanning electron microscope (SEM) technique. This special electron microscope technique is especially used for the evaluation of surfaces and fractures at surfaces. (ii) The transmission electron microscope (TEM) technique. This special electron microscope technique especially supplies us with a detailed insight into the "atomic layout" of thin layers. We compare with Figures 2.95–2.98, which supply us with TEM variants and complementary HRTEM images.

Wanting to visualize microstructures, the sample needs an appropriate preparation. The reader should conceive that this includes mechanical preparation methods such as polishing with diamond paste, chemical preparation methods such as etching with acids, electro-chemical preparation methods, and electro-mechanical preparation methods. In passing, we want to record this information. (i) The samples that are shown in Figure 2.100 were processed by mechanical grinding (waxed SiC paper) and additionally polished with diamond paste (through 7, 3, and 1 μm) and electro-mechanical means (820 ml H_2O, 50 g $Na_2S_2O_3 \cdot 5H_2O$, 70 g $K_3(Fe(CN)_6)$, 60 g NH_4Cl, 1.5 A/cm^2, 3 min). (ii) The samples that are shown in Figure 2.102 were processed by mechanical grinding (1000 grit SiC paper), electrolytic polishing (100 ml CH_3OH, 10 ml H_2SO_4, 12°C, 25 V), and etching (100 ml H_2O, 20 g $K_2S_2O_5$, 15–45 min). (iii) The SiC structure that is shown in Figure 2.87, after a grinding and lapping period, has been made visible by fused salt with the parameters 90 g KOH, 10 g KNO_3, 480°C, 2 min. (iv) The LPSiC structure that is shown in Figure 2.90, after a grinding and lapping period, has been made visible by plasma-etching with the parameters $CF_4 + O_2$, 1 : 1, 4 min.

The reader should pick up that Figures 2.99–2.102 deal with metallic materials. However, the reader should not be surprised about the fact that we on and off consider metallic materials. In fact, the reader should awaken to the fact that many methods (preparation methods, powder technologies etc.) applicable to ceramic materials are also applicable to metallic materials. In passing, we want to record this information. (i) In Figure 2.99, we show Ag–Cu–Sn–Hg microstructures. The pictures illustrate the different evolution stages of the hardening process of Ag55Cu15Sn30+50Hg (amalgam) starting from a pre-alloyed Ag55Cu15Sn30 powder visualized at the top, left. (ii) In Figure 2.100, we show Pd–In microstructures. The Pd–In microstructure that comes into being depends on the Pd–In ratio. In Figure 2.100, top, it is Pd–33 at.-% In, and in Figure 2.100, bottom, it is Pd–37.5 at.-% In. Figure 2.100, top, corresponds to a situation where Pd_2In crystallizes, after a martensitic transformation, leading to the mosaic-like structure. Figure 2.100, bottom, corresponds to a situation where Pd_2In through slow cooling experiences a strong grain growth with the result that a subsequent martensitic transformation leads to the feather-like structure and further cooling produces a Pd_2In–PdIn eutectic in the surrounding regions. (iii) In the end, in Figures 2.101 and 2.102, we show Fe–Ni–W systems. The Fe–Ni–W microstructure that comes into being depends on the Fe–Ni–W ratio. The topological connection of Fe–Ni–W microstructure and Fe–Ni–W ratio is presented in Figure 2.101, top – a variety of phase diagrams to be explained in Section 3.1.

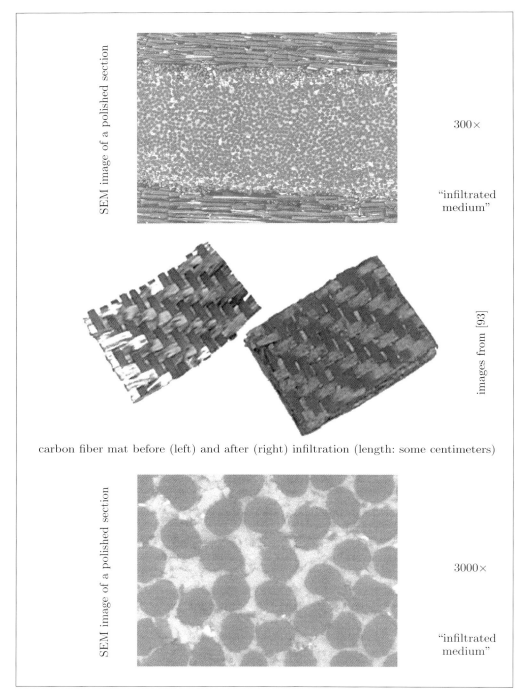

Figure 2.93. Carbon fiber reinforced Si–B–C–N ceramics.

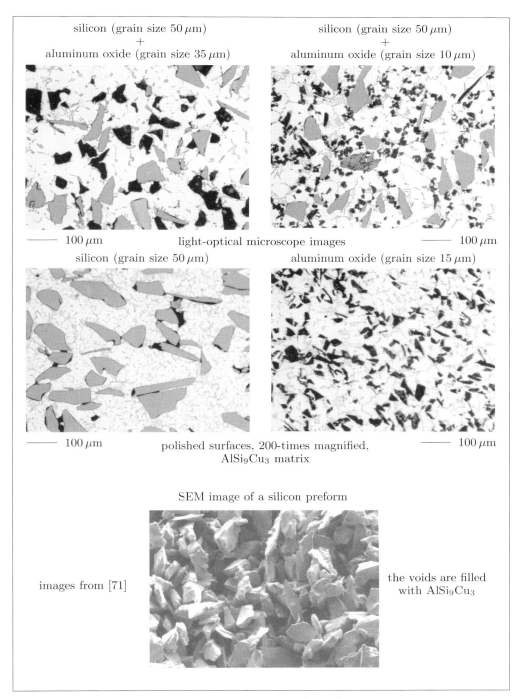

Figure 2.94. Metal matrix composite (MMC) microstructures.

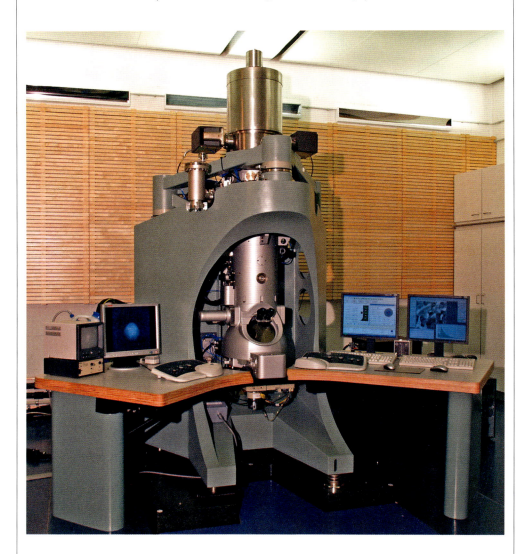

Figure 2.95. Stuttgart Center for Electron Microscopy (StEM): "SESAM".

Figure 2.96. Stuttgart Center for Electron Microscopy (StEM): "JEOL ARM 1250 KV".

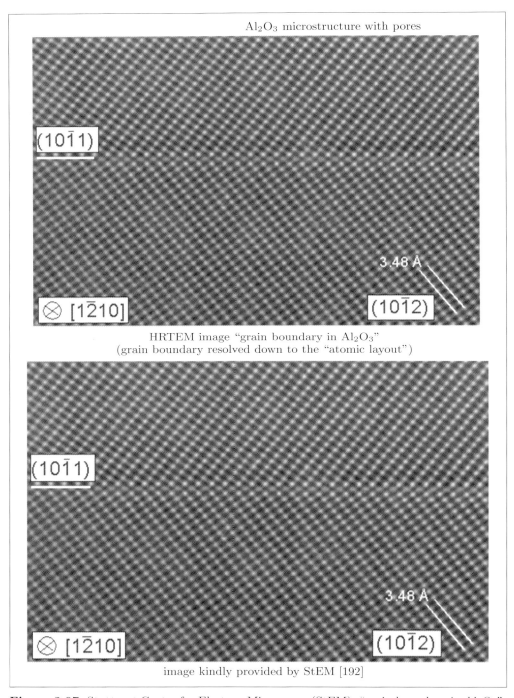

Figure 2.97. Stuttgart Center for Electron Microscopy (StEM): "grain boundary in Al_2O_3".

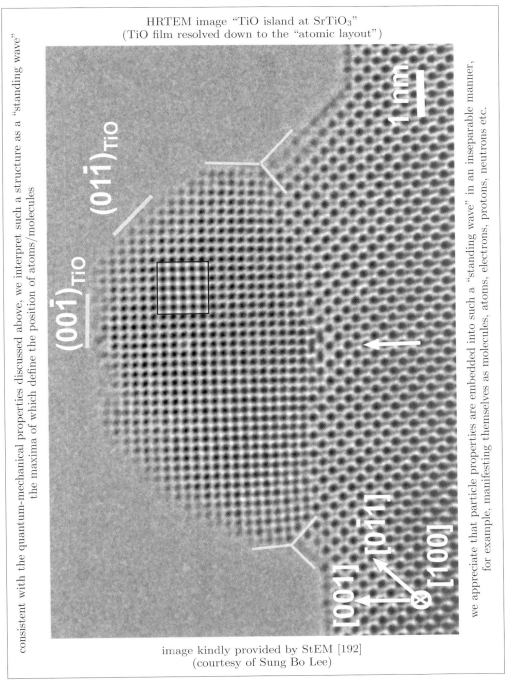

Figure 2.98. Stuttgart Center for Electron Microscopy (StEM): "TiO island at SrTiO$_3$".

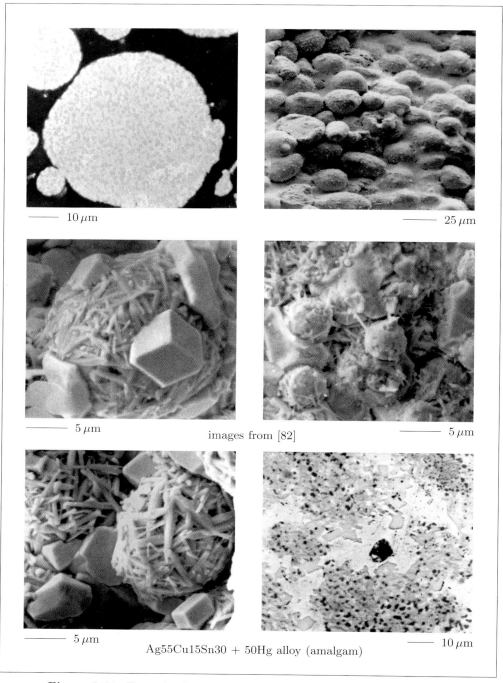

Figure 2.99. Examples for metallic microstructures: Ag–Cu–Sn–Hg alloy.

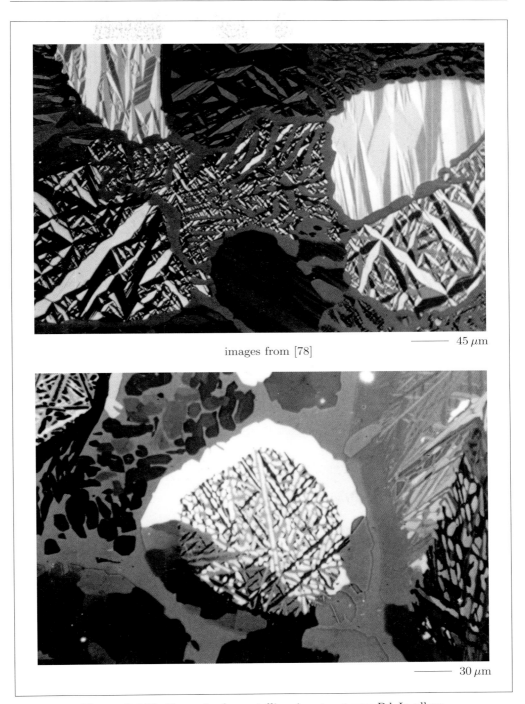

images from [78]

———— 45 μm

———— 30 μm

Figure 2.100. Examples for metallic microstructures: Pd–In alloys.

166 2. Structures

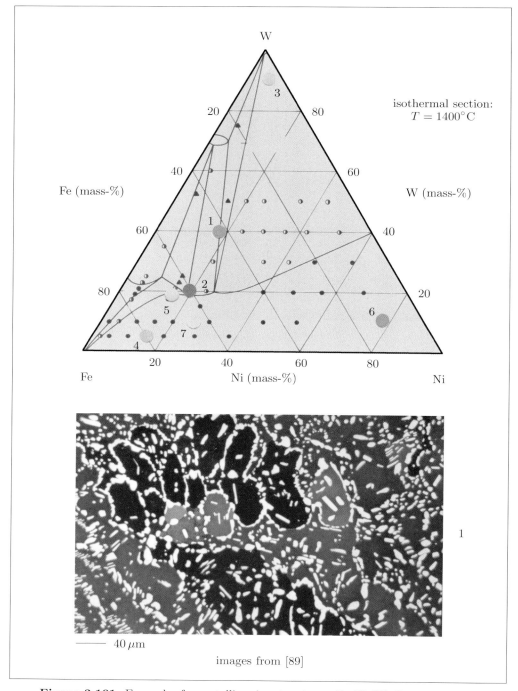

Figure 2.101. Examples for metallic microstructures: Fe–Ni–W alloys, part one.

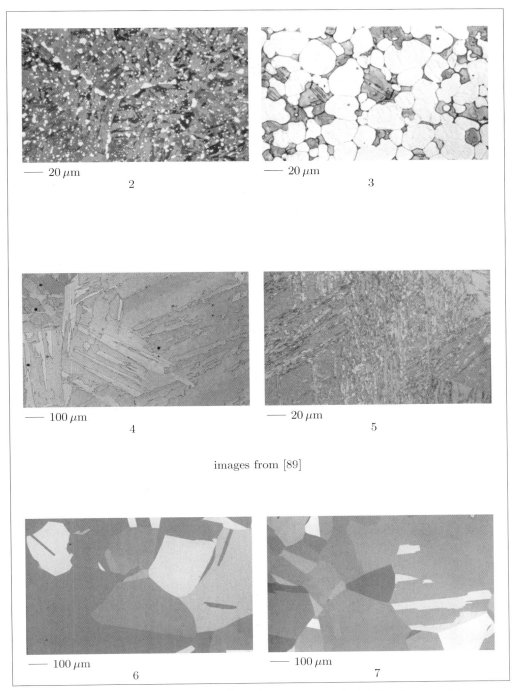

Figure 2.102. Examples for metallic microstructures: Fe–Ni–W alloys, part two.

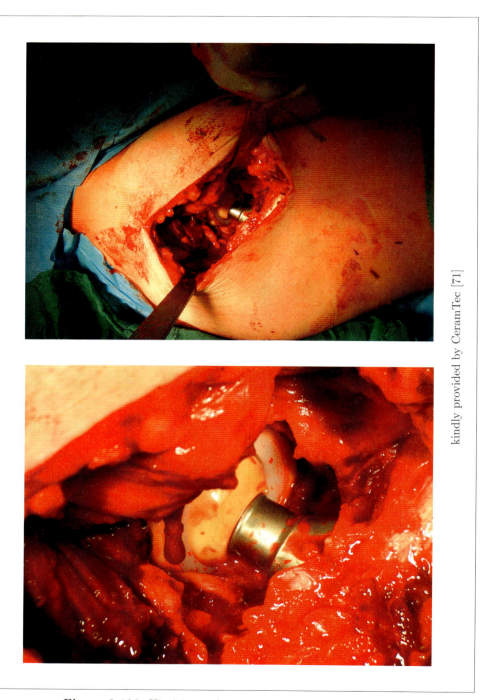

Figure 2.103. Hip joint endoprosthesis in the human body.

Let us finish this chapter with some brief remarks concerning the bio-compatibility of ceramic materials. In contrast to many metallic materials, ceramic materials normally exhibit an extremely high bio-compatibility. For example, as the clinical practice reveals again and again, hip joint endoprostheses in most cases can be implanted without harm for the human body. For example, as the clinical practice reveals over and over, dental prostheses in most cases fulfil their biting task for decades without harm for the human body. By contrast, in contact with the human body, many metallic materials may cause damage to the human body. In particular, in contact with the human body, metallic alloys such as those shown in Figures 2.100 and 2.101 may release metal ions into the human body, which are accumulated in the Central Nervous System (CNS) and other organs such as kidneys, liver, and brain eventually evoking toxic effects [10, 44, 47, 48, 55]. Nevertheless, also in the context of ceramic materials, it should be always a good choice to budget for bio-compatibility investigations, searching for possible impacts on humans, fauna, and flora!

Observing Ceramics

3. Properties

The spectrum of properties of ceramics reaches from excellent chemical properties such as corrosion resistance via excellent physical properties such as insulating properties with respect to thermal conductivity and electric conductivity or such as mechanical high temperature stability to high hardness and hight abrasion resistance. Therefore, ceramics can be used under conditions where metals or polymers rapidly break down. For example, ceramics can be smoothly used inside chemically aggressive environments, where metals would rapidly corrode. For example, ceramics can be smoothly used inside thermally aggressive high temperature environments, where polymers would rapidly disintegrate. By the way, without the great chemical resistance of laboratory glassware, laboratory research scarcely would be imaginable.

3.1 Chemical Properties

Speaking of *chemical properties*, we mean the stability, the instability, and the reactivity of materials under the influence of constraints such as pressure, temperature and/or severe atmospheres. Let us get a general idea of the main features of this topic with special respect to ceramics in the sections that follow.

3.1.1 Phase Equilibria

How does a substance behave if temperature and/or pressure are altered? For example, given the temperature and the pressure, a substance could be in a solid state, but increasing the temperature, the substance could become liquid or gaseous. How do two or more substances behave if these are alloyed depending on the temperature and the pressure? For example, given the pressure, within a given temperature/composition domain, two substances could become manifest as one new substance, but changing the composition interval, two substances also could become manifest as two new substances within one substance. In order to make such *thermochemical topologies* comprehensible, phase diagrams are extremely helpful means. These not only supply us with a wealth of theoretical information, these also supply us with a wealth of practical information, in particular, with detailed information about the parameter windows within which production processes, for example, needing a mixture of fused and solid substances, have to be established, and with detailed information about the parameter windows within which the components would be stable during service.

3.1.1.1 Phase Diagrams

The reader should know that such phase diagrams describe how ceramic compounds respond to constraints such as temperature, pressure, and composition. Since we treat p–T diagrams in Section 2.2.2, in the sections that follow, we want to concentrate on phase diagrams including compositions of ceramic compounds. The following notes should be helpful. (i) Binary compounds such as SiO_2, ZrO_2, Al_2O_3, Si_3N_4, and SiC can be considered as unary systems since in terms of stoichiometrical compounds the amounts of the two constituents of such a binary compound are fixed, i. e. the amounts of the two constituents of such a binary compound are not independent, so that the smallest number of independent constituents is reduced to one and the two constituents form an unary system – in order to express that certainly two constituents are present, however, which are not independent – called a *pseudo-unary system*. The extension of this notation is straightforward. For example, phase diagrams that are dealing with two pseudo-unary compounds are called *pseudo-binary*, and phase diagrams that are dealing with three pseudo-unary compounds are called *pseudo-ternary*. Nevertheless, among ceramists, pseudo-unary systems and pseudo-binary systems are simply denoted as "unery systems". (ii) Phase diagrams including compositions of ceramic compounds conveniently are computed via nonlinear regression algorithms enabling us to reduce measurement data to mean value curves, and this has two important consequences. On the one hand, the phase transition points to be analyzed experimentally may not coincide with the phase transition curves that are shown in such phase diagrams. On the other hand, the phase transition curves that are shown in such phase diagrams depend on the measurement data that are eventually taken as the calculative basis, which by the way explains the existence of different phase diagrams of the same type for a special ceramic compound.

The reader should further know that ceramic materials seldom are pure compounds. Either ceramic materials contain amounts of impurities or ceramic materials contain artificially added substances. The following notes should be helpful. (i) An important example for the first category is technical alumina that depending on the source of the raw material and/or the type of manufacturing process contains impurities even high enough to form secondary phases, for example, to be seen in the microstructure as grain boundary phase. Typically, these consist of silica and calcia, together with alumina, forming low melting eutectics. At the sintering temperature of alumina, such eutectics are liquid. On the one hand, this is an advantage because it enhances the densification process. On the other hand, this is an disadvantage because such materials are creep sensitive at high temperatures. Beyond it, such materials are useless, for example, for applications in optics since secondary phases scatter light thus reducing the translucence of the materials. (ii) An important example for the second category is silicon nitride where compounds such as alumina and yttria are added to form an eutectic liquid phase at the sintering temperature in order to secure that silicon nitride can be densified, which otherwise is difficult because of the mobility of silicon and nitrogen in this mainly covalently bonded nitride. The liquid phase, which is formed at the sintering temperature, during the cooling period eventually solidifies as glass phase at the grain boundaries.

We firstly consider the binary systems which are presented below. Let us here point at the following circumstances. (i) In each phase diagram, the line of existence or the field of existence of the respective compounds is plotted. In each case, the line of existence cuts the mole fraction axes at the stoichiometrically appropriate position, i.e. 2/3 for SiO_2, 2/3 for ZrO_2, 3/5 for Al_2O_3, 4/7 for Si_3N_4, and 1/2 for SiC. In the case of ZrO_2, the field of existence widens at high temperatures, i.e. the solid compound is also existent at substantially lower oxygen contents than the oxygen content of the stoichiometrical composition manifested by oxygen vacancies in the cubic polymorph of zirconia. In the other compounds shown, the deviations (if at all present) are below the detection limit even at high temperatures. (ii) We particularly realize that the selected oxide compounds do melt congruently. SiC does melt incongruently, while Si_3N_4 does not melt under normal pressure, but dissoziates into liquid silicon and nitrogen, as it is expressed by the peritectic-type three-phase reaction $3Si + 2N_2 \leftrightarrow Si_3N_4$, at some 2100 K defining the stability limit. The SiO_2 system reveals an interesting reaction, namely the reaction of liquid Si with liquid SiO_2 at some 2125 K to form gaseous SiO, i.e. SiO degasses at relatively low temperatures. If the oxygen partial pressure is low, the ceramics disaggregates in this way. (iii) We also realize that different modifications can come into being depending on temperature and mole fraction. For example, the cubic (c), tetragonal (t), and monoclinic (m) modifications in the case of ZrO_2 or the cristobalite and tridymite modifications in the case of SiO_2. The SiC system also shows such a polymorphism, namely a high temperature phase (β) that exists besides the low temperature phase (α) stable at temperatures below 2100°C. In practice, the high temperature phase may also exist at lower temperatures since the transformation from one modification into the other is kinetically hindered. (iv) We further realize that a surplus of the one or other chemical element evokes other phases. For example, Si or graphite in the case of SiC and Si or N_2 ("gas") in the case of Si_3N_4.

We secondly consider the pseudo-binary systems which are presented below. Let us here point at the following circumstances. (i) There exist pseudo-binary systems with negligibly small solubility of the components into each other, pseudo-binary systems with some solubility of the components into each other, and pseudo-binary systems with complete solubility of the components into each other. The CaO–MgO system is a system with some solubility in the one component and a negligible small solubility in the other component. The MgO–NiO system is a system with complete solid solubility. Complete solid solubility at high temperatures, a miscibility gap below a critical point, and limited solid solubility at lower temperatures is also typical for ceramic systems. Such miscibility gaps are of special importance for glasses which consist of two phases formed via such a miscibility gap. (ii) Pseudo-binary systems very often exhibit one or more intermediate phases, representing either (strongly) stoichiometrical compounds or compounds with a homogeneity range. The Y_2O_3–Al_2O_3 system with the three congruently melting phases is an example for the first case, and the MgO–Al_2O_3 system with the congruently melting spinel phase is an example for the second case. Such intermediate phases also are formed incongruently by a peritectic reaction or in the solid state by a peritectoid reaction or in a solid solution below a critical temperature, and they can become unstable upon cooling via eutectoid reactions. (iii) Naturally, such general aspects are also valid for higher component systems.

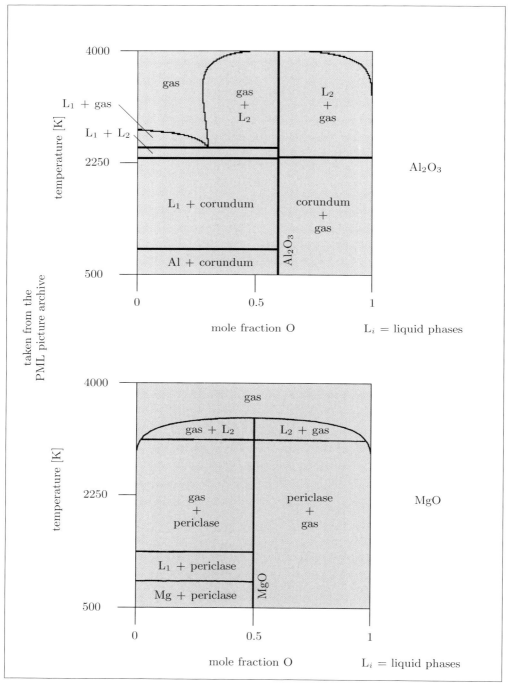

Figure 3.1. Binary systems: Al_2O_3, MgO.

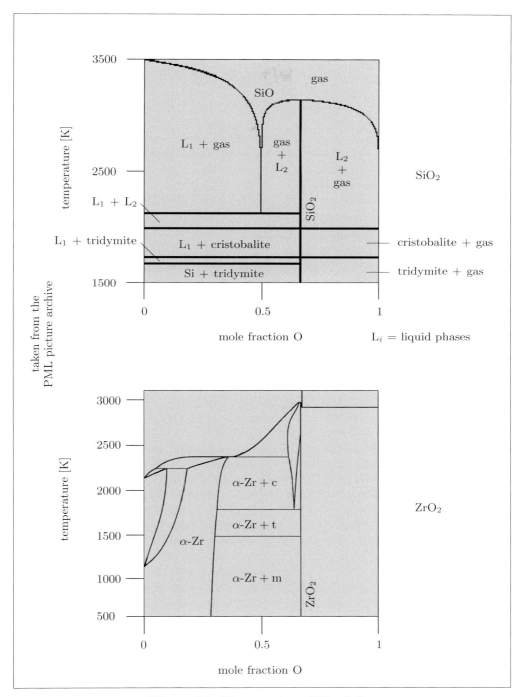

Figure 3.2. Binary systems: SiO_2, ZrO_2.

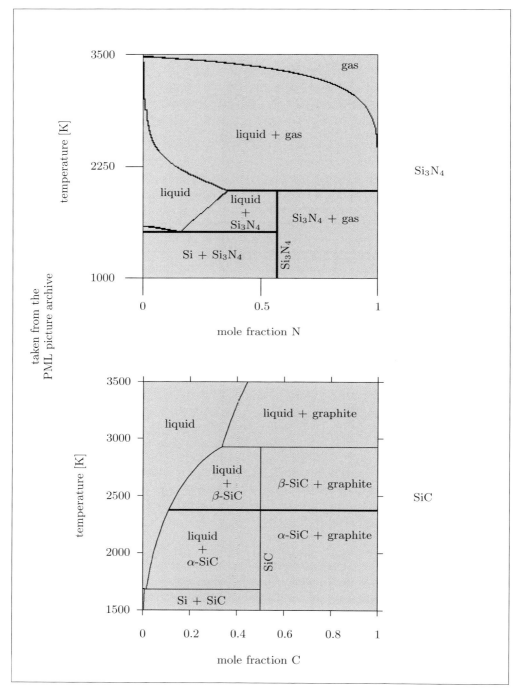

Figure 3.3. Binary systems: Si$_3$N$_4$, SiC.

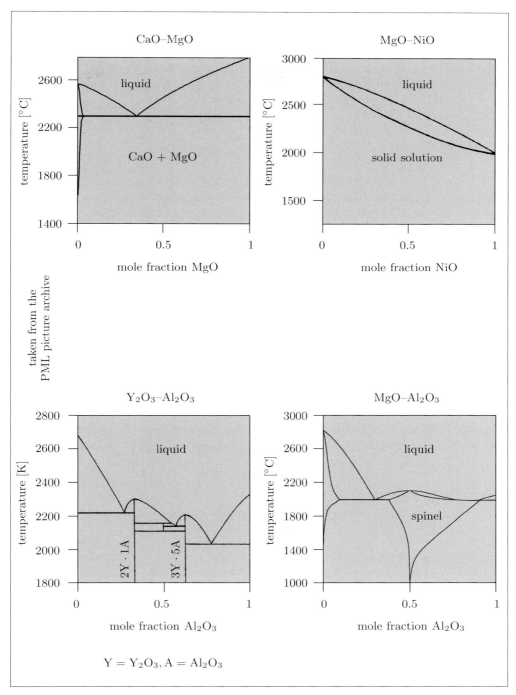

Figure 3.4. Pseudo-binary systems: CaO–MgO, MgO–NiO, Y_2O_3–Al_2O_3, MgO–Al_2O_3.

Although a pseudo-quaternary system, which comprises five different atom species, may not be adequate to be discussed in a textbook, some of its features may help to learn more about complex phase diagrams, in particular, how complex phase diagrams have to be read in order to understand the diverse states of complex materials including the dependency of complex materials on constraints. From a formal point of view, a pseudo-quaternary system is immediately decomposed into four pseudo-unary systems, six pseudo-binary systems, and four pseudo-ternary systems. The complete range of compositions is represented by a tetrahedron with the four components at the corners, the six pseudo-binary composition ranges at the edges, and the four pseudo-ternary composition ranges at the faces, while the pseudo-quaternary composition ranges are located between the four faces, i. e. within the tetrahedron.

The tetrahedron of the pseudo-quaternary system ZrO_2–Y_2O_3–Gd_2O_3–Al_2O_3 is outlined in Figure 3.5. It should be clear that the pseudo-ternary sub-systems are given by ZrO_2–Y_2O_3–Gd_2O_3 (see bottom face), Al_2O_3–Gd_2O_3–Y_2O_3 (see front face), ZrO_2–Y_2O_3–Al_2O_3 (see left hand face), and ZrO_2–Gd_2O_3–Al_2O_3 (see right hand face). It should be clear, too, that the pseudo-binary sub-systems are given by ZrO_2–Y_2O_3, ZrO_2–Gd_2O_3, Gd_2O_3–Y_2O_3, ZrO_2–Al_2O_3, Al_2O_3–Y_2O_3, and Gd_2O_3–Al_2O_3. On the one hand, this system is selected because it is a pseudo-quaternary system for which a complete set of thermodynamic data is available so that all kinds of equilibrium lines are directly computable. On the other hand, this system is selected because it reveals the stabilization of zirconia and its phase equilibria, namely the stabilization of the cubic and/or the tetragonal modification, which are of great technical importance, as already discussed in Section 2.2.2. We remark that alumina and gadolinia in combination with ZrO_2 are of great technical importance. On the one hand, alumina is added to the material since it is formed at the interface between the bond coat of zirconia during the service of zirconia-coated turbine blades at high temperatures. On the other hand, gadolinia is added to the material since it decreases the thermal conductivity thereby improving the thermal barrier performance of zirconia-coatings.

The graphs of the pseudo-binary sub-systems that are shown in Figures 3.7–3.9 are calculated lines of equilibrium ("phase boundaries"), while the clusters of symbols that are shown there indicate measurement results of various experimental groups. (i) We read off from the phase diagram of the ZrO_2–Y_2O_3 sub-system that the phase fields of tetragonal ZrO_2 and cubic ZrO_2 are stabilized to much lower temperatures. The cubic fluorite-type phase (Flu) is stable at room temperature. The tetragonal phase with high enough amount of Y_2O_3 can be cooled down to room temperature without passing into the phase field containing the monoclinic modification. (ii) Figure 2.83 shows the microstructure of the tetragonal phase stabilized with 2–3 mole-% Y_2O_3 as well as the microstructure of the cubic phase stabilized with 5–7 mole-% Y_2O_3. In the first case, the microstructure is single phase. In the second case, the microstructure is single phase only at a first glance, i. e. looking closer by TEM, the microstructure shows finely dispersed, lense-shaped precipitates of the tetragonal phase formed during heat treatment in the Tet + Flu region at about 1800 K. (iii) Figure 3.8 shows the ZrO_2–Al_2O_3 sub-system, and Figure 3.7 shows the ZrO_2–Gd_2O_3 sub-system, besides the forming of new compounds, exhibiting extensive solid solubility.

Some more details concerning the systems ZrO_2–Y_2O_3 and ZrO_2–Gd_2O_3 should be presented here. In what follows, the reader should compare with Figures 3.7 and 3.6 illustrating these systems. (i) The cubic fluorite-type phase (Flu) is stabilized over a wide range of compositions/temperatures and coexists with the tetragonal phase and the monoclinic phase on the ZrO_2-rich side. In both systems, the solubility of the dopants in the tetragonal phase does not exceed 6 mole-% and in the monoclinic phase is limited to negligibly small amounts. In both systems, one intermediate compound is confirmed. Both $Zr_3Y_4O_{12}$ and $Gd_{2x}Zr_{2x}O_{7+0.25x}$ are ordered phases, which transform to the disordered cubic solution upon heating, and the latter ordered phase is called *pyrochlor-type phase* (Pyr). The substitution of Gd^{3+} for Zr^{4+}, combined with the formation of vacancies and interstitials, eventually results in the formation of a range of non-stoichiometric compositions. (ii) An interesting observation is that there is a marked deviation ("sigmoidal (S) curve") in the lattice parameters from Vegard's slope within some 20 and 60 mole-% $GdO_{3/2}$, and this marked deviation can be led back to intermediate states residing between a random solution and a phase separation, in last consequence, evoking the formation of an order of the Flu–Pyr area, and this order is defined by a variable, temperature- and pressure-dependent long-range order parameter within a range of compositions in the system ZrO_2–Gd_2O_3.

Some more details concerning the systems Gd_2O_3–Y_2O_3 and ZrO_2–Al_2O_3 should be presented here, too. In what follows, the reader should compare with Figure 3.8 illustrating these systems. (i) In contrast to the systems ZrO_2–Y_2O_3 and ZrO_2–Gd_2O_3, which show an intermediate compound each, the system Gd_2O_3–Y_2O_3 exhibits no intermediate compound, but a complete solid solution in the bixbyite phase (C) and the hexagonal phase (Hex). (ii) Furthermore, the system ZrO_2–Al_2O_3 exhibits no intermediate compound and the solubility of the components into each other is rather limited. We note that the solubility of Al_2O_3 in cubic and teragonal zirconia is below 4–5 mole-%. We also note that the solubility of Al_2O_3 in zirconia just like the solubility of zirconia in Al_2O_3 is negligibly small.

Some more details concerning the systems Al_2O_3–Y_2O_3 and Gd_2O_3–Al_2O_3 should be presented here, too. In what follows, the reader should compare with Figure 3.9 illustrating these systems. (i) The two residual systems Al_2O_3–Y_2O_3 and Gd_2O_3–Al_2O_3 are characterized by the formation of some strongly stoichiometric compounds, namely $Y_3Al_5O_{12}$ (YAG), $YAlO_3$ (YAP), $Y_4Al_2O_9$ (YAM), $GdAlO_3$ (GAP), $Gd_4Al_2O_9$ (GAM). YAG has garnet structure. YAP and GAP have a perowskite-like structure, whereas YAM and GAM are monoclinic. The melting points of these compounds are around 2200 K. YAG, YAP, YAM, and GAP are melting congruently, whereas GAM is melting incongruently. (ii) In both of these systems, the solubility of the components into each other is negligibly small.

As a brief plug-in module, we here take notice that phase transitions that lead from one or more phases to other phases can be characterized by thermodynamic quantities such as enthalpy and entropy and their dependence on thermodynamic variables such as temperature and pressure (classification of Ehrenfest) or macroscopic variables such as crystal deformation (Landau theory of phase transitions). A structural classification (continuous, discontinuous, martensitic) and a kinetic classification (reaction velocity) are possible, too.

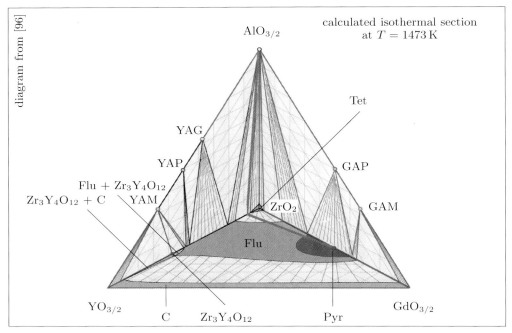

Figure 3.5. Pseudo-quaternary system: ZrO_2–Y_2O_3–Gd_2O_3–Al_2O_3.

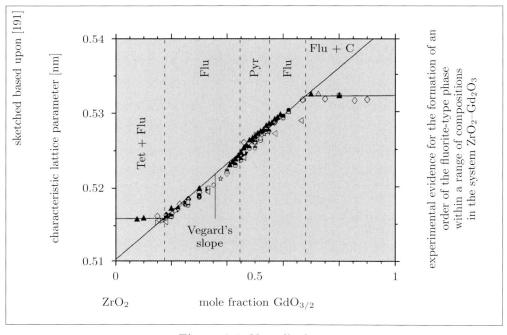

Figure 3.6. Vegard's slope.

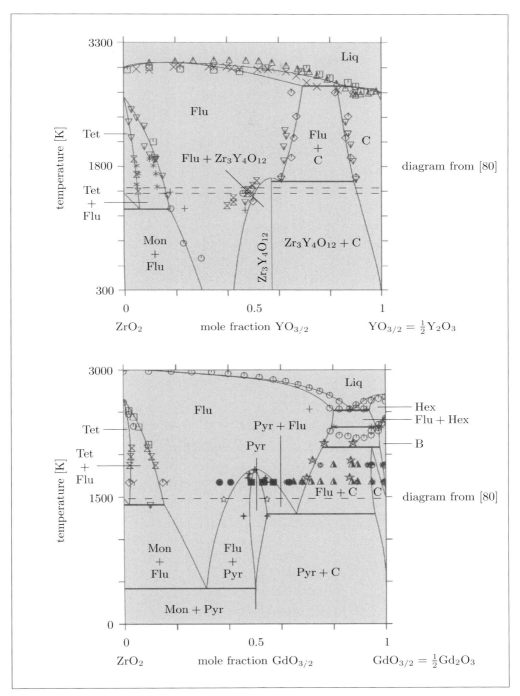

Figure 3.7. Pseudo-binary sub-systems: ZrO_2–Y_2O_3, ZrO_2–Gd_2O_3.

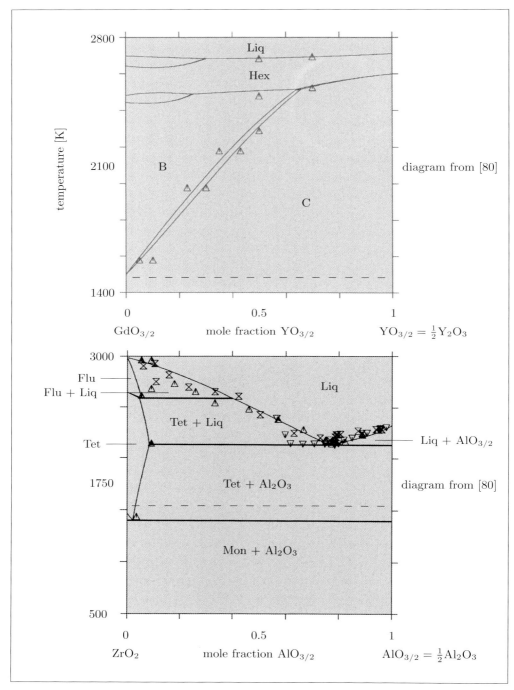

Figure 3.8. Pseudo-binary sub-systems: Gd_2O_3–Y_2O_3, ZrO_2–Al_2O_3.

3.1 *Chemical Properties* 183

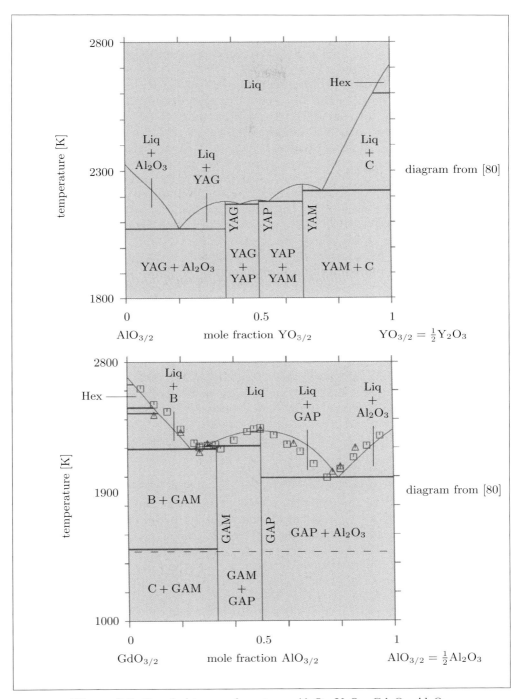

Figure 3.9. Pseudo-binary sub-systems: Al_2O_3–Y_2O_3, Gd_2O_3–Al_2O_3.

Calculations on the basis of experimental results reveal the phase boundaries of the first two pseudo-ternary systems. (i) The system ZrO_2–Y_2O_3–Gd_2O_3 reveals the numerical results on co-doping of ZrO_2, we compare with Figure 3.10, which shows an isothermal section of the system ZrO_2–Y_2O_3–Gd_2O_3 at $T = 1473\,\mathrm{K}$, and the intersections of this diagram with the respective pseudo-binary systems are indicated in Figures 3.7–3.9 as dashed lines. We point out that this isothermal section shows the wide range of compositions in which the fluorite-type phase (Flu) is stable. We also point out that this wide range of compositions in the pseudo-ternary regions is limited by the two-phase fields of phase equilibria, namely by the tetragonal phase (Tet), the bixbyite phase (C), the pyrochlor-type phase (Pyr), and the intermediate compound $Zr_3Y_4O_{12}$ (ϑ) at the ZrO_2–Y_2O_3 side. We further point out that the lines within the two-phase fields of phase equilibria are tie lines connecting the two compositions which are in equilibrium with each other. (ii) The system Al_2O_3–Gd_2O_3–Y_2O_3 reveals the alloy behavior of Al_2O_3, Gd_2O_3, and Y_2O_3, we compare with Figure 3.10, which shows an isothermal section of the system Al_2O_3–Gd_2O_3–Y_2O_3 at $T = 1473\,\mathrm{K}$, and the intersections of this diagram with the respective pseudo-binary systems are indicated in Figures 3.7–3.9 as dashed lines. In order to supply the reader with some details, we mention that the isostructural compounds YAM and GAM as well as the isostructural compounds YAP and GAP form complete solid solutions. In order to supply the reader with some details, we also mention that the compound YAG exhibits an extended range of solubility in the direction of the composition of the respective phase in the system Al_2O_3–Gd_2O_3, however, which is not stable. In order to supply the reader with some details, we further mention that the range of solubility of the compound YAG is limited by a three-phase equilibrium between the compounds YAG, GAP, and Al_2O_3. In order to supply the reader with some additional details, we finally mention that all fields of two-phase equilibria between the phases are indicated by fans of tie lines.

Calculations on the basis of experimental results also reveal the phase boundaries of the second two pseudo-ternary systems. (i) Figure 3.11 comprises the liquidus and the isothermal section of the system ZrO_2–Y_2O_3–Al_2O_3 at $T = 1523\,\mathrm{K}$. We realize that no ternary compounds or ternary solid solutions have been found in the system. We also realize that the intersections of this isothermal section with the respective pseudo-binary systems are indicated in Figures 3.7–3.9 as dashed lines, too. We further realize that this isothermal section is characterized by a couple of three-phase equilibria between the edge components and the binary phases, whereas the two-phase equilibria are again visualized by fans of tie lines when the respective phases reveal the ranges of solubility. (ii) Figure 3.12 comprises the liquidus and the isothermal sections of the system ZrO_2–Gd_2O_3–Al_2O_3 at $T = 1923\,\mathrm{K}$ and $T = 1523\,\mathrm{K}$. We realize that both isothermal sections are similar, but differ in the phase fields of GAP with the system ZrO_2–Gd_2O_3. We also realize that GAP at $T = 1923\,\mathrm{K}$ is in equilibrium with the fluorite-type phase (Flu) from 28 up to 57 mole-% Gd_2O_3. We further realize that the pyrochlor-type phase (Pyr) at $T = 1523\,\mathrm{K}$ becomes stable in the system ZrO_2–Gd_2O_3, where it strongly wedges into the two-phase field GAP + Flu by forming three new phase fields, i.e. GAP + Pyr and two three-phase fields of GAP with Pyr and two different compositions of F, a Gd-rich one and a Zr-rich one.

Regarding the latter two pseudo-ternary systems, the following additional remarks should be useful. (i) Consider the system ZrO_2–Y_2O_3–Al_2O_3. Take notice that the Y-stabilized tetragonal cubic zirconia phase and the Y-stabilized cubic zirconia phase, respectively, according to further calculations, can contain up to 4.6 and 3.5 mole-% Al_2O_3, respectively, being in equilibrium with Al_2O_3 (corundum) at 1923 K. Take also notice that the solubilities are somewhat smaller at 1523 K, but the zirconia phases are still in equilibrium with Al_2O_3. (ii) Consider the system ZrO_2–Gd_2O_3–Al_2O_3. Take notice that the solubility of Al_2O_3 in the Gd-stabilized zirconia phases is similarly small as in the Y-stabilized zirconia phases. Take also notice that all other phases reveal negligibly small solubility for the respective third component. Take further notice that a perowskite-type phase (GAP) forms if the pyrochlor-type phase (Pyr) is in contact with Al_2O_3, and this makes it impossible to use the pyrochlor-type phase as thermal barrier coating on the top of Al_2O_3 (which forms as already mentioned at the interface between the bond coat of zirconia during the service of zirconia-coated turbine blades at high temperatures) since the pyrochlor-type phase and the perowskite-type phase have different thermal expansion coefficients eventually causing cracking. (iii) Consider the system ZrO_2–Gd_2O_3–Al_2O_3. Take notice that the the liquidus surface is characterized by a wide range of compositions in which crystallization of the fluorite-type phase occurs during the cooling down of the melts. Take also notice that there are three ternary eutectic reactions denoted by E_1, E_2, and E_3. Take further notice that microstructures of solidified eutectic compositions of all three ternary eutectic reactions are presented in Figure 3.13, top, while a closer look to the microstructure evoked by E_3 is presented in Figure 3.13, bottom. The following details here should be recorded. (1) In each case, the phases are finely divided and the grain sizes are rather small. We thus expect that the casting of components of such a type of ceramics is possible. (2) A closer look to the microstructure evoked by E_3 shows that the fluorite-type phase (Flu) forms fibers embedded in a matrix of alumina.

Beyond that, it remains to be remarked that the thermodynamic description of the four pseudo-ternary systems has been combined to a thermodynamic database for the pseudo-quaternary system ZrO_2–Gd_2O_3–Y_2O_3–Al_2O_3. The database has been designed for the Thermo-Calc® software and has been used for various kinds of numerical calculations in the framework of different applications. An example is shown in Figures 3.14 and 3.15. The isoplethal sections presented in Figures 3.14 and 3.15 were calculated for industrially important compositions with a content of Y + Gd of 7.6 mole-% and 15.2 mole-% and different ratios of Gd/(Gd + Y) (0.25, 0.5, 0.75). The comparison of the calculated sections reveals that the increase of Y + Gd from 7.6 mole-% to 15.2 mole-% decreases the stability range of Flu + Tet, but increases the stability ranges of Flu + Al_2O_3 and Flu + Al_2O_3 + YAG. The comparison of the calculated sections also reveals that the increase of the ratio Y(Gd + Y) decreases the stability range of Flu + Al_2O_3, but increases that of Flu + Al_2O_3 + YAG. On the one hand, it should be recalled here that isoplethal sections do not represent tie lines between the included phases. On the other hand, it should be mentioned here that the appearance of YAG is absolutely not desirable for thermal barrier coatings because of its thermal expansion behavior.

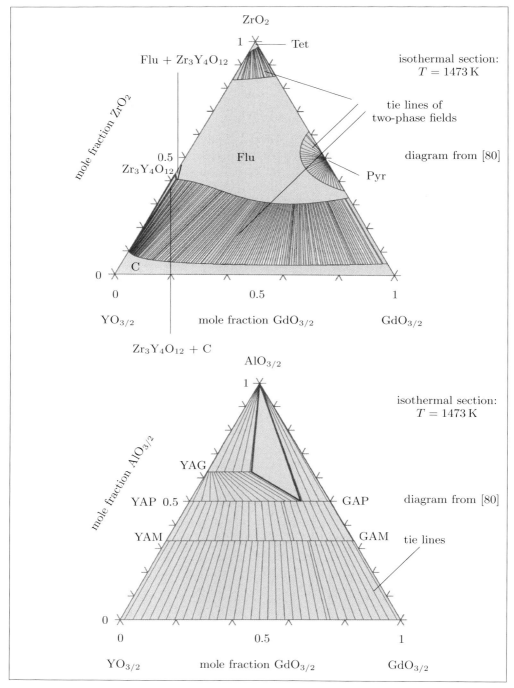

Figure 3.10. Isothermal sections: ZrO_2–Y_2O_3–Gd_2O_3, Al_2O_3–Gd_2O_3–Y_2O_3.

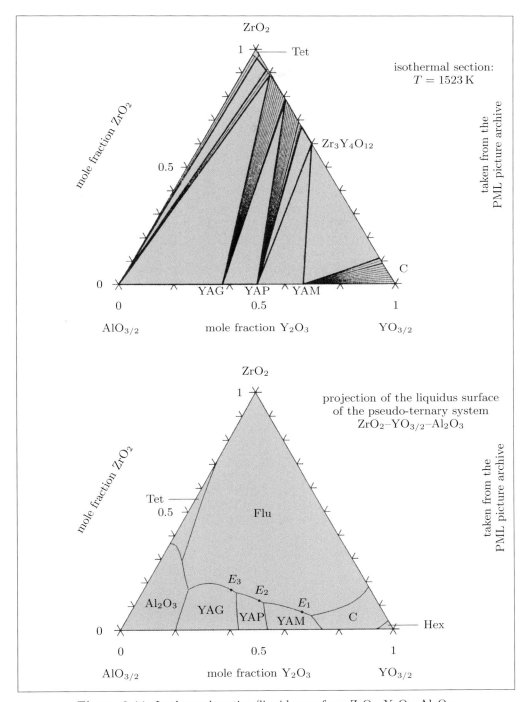

Figure 3.11. Isothermal section/liquidus surface: ZrO_2–Y_2O_3–Al_2O_3.

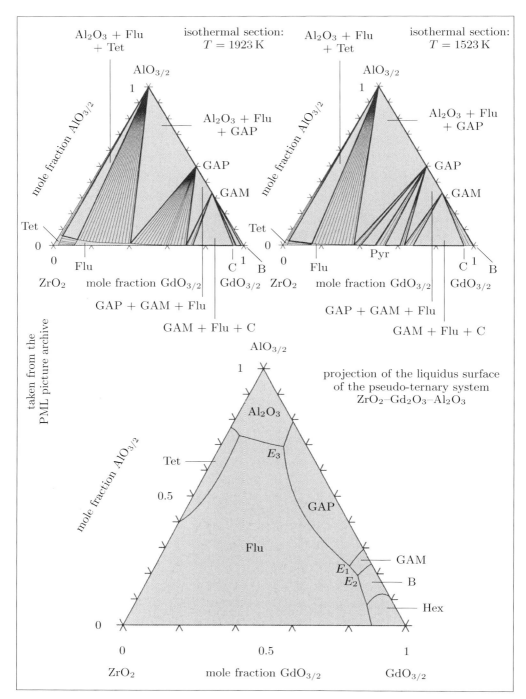

Figure 3.12. Isothermal section/liquidus surface: ZrO_2–Gd_2O_3–Al_2O_3.

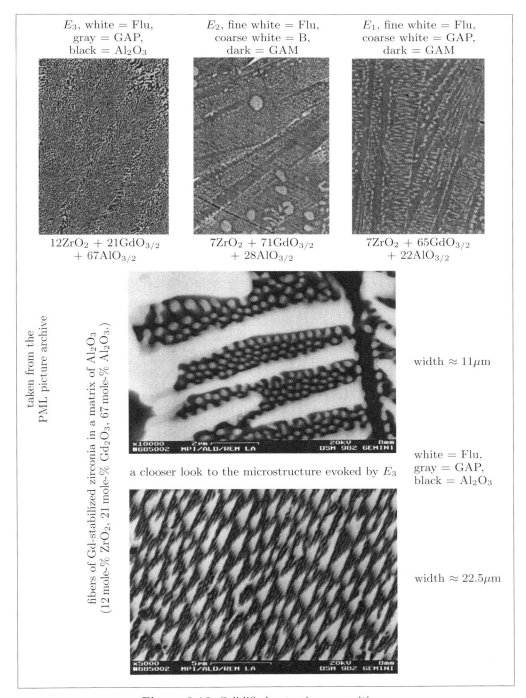

Figure 3.13. Solidified eutectic compositions.

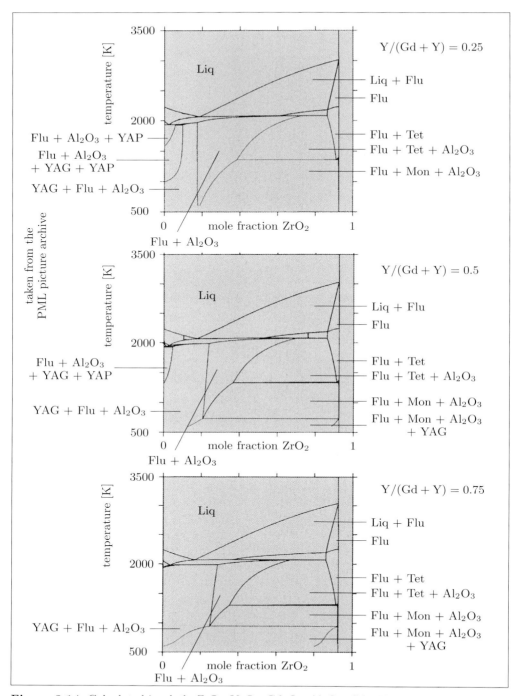

Figure 3.14. Calculated isopleth: ZrO_2–Y_2O_3–Gd_2O_3–Al_2O_3, $Gd + Y$ content of 7.6 mole-%.

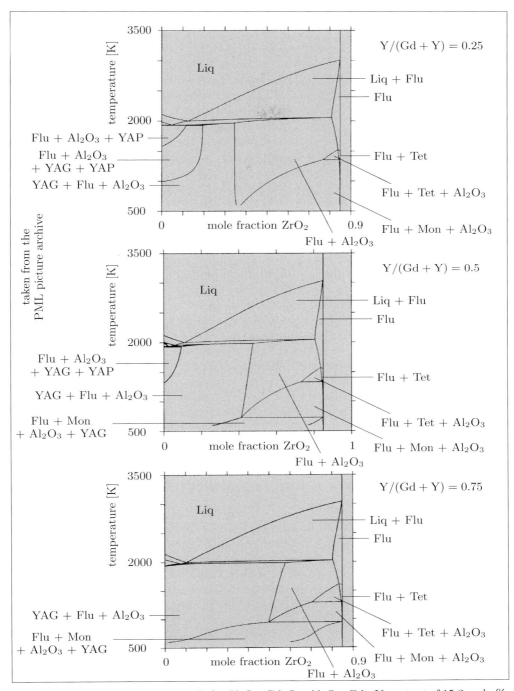

Figure 3.15. Calculated isopleth: ZrO_2–Y_2O_3–Gd_2O_3–Al_2O_3, Gd+Y content of 15.2 mole-%.

3.1.1.2 Chemical Reactions

Wanting to estimate the course direction of a chemical reaction, we are well advised to calculate the free enthalpy (Gibbs free energy) of the chemical reaction. In particular, the free enthalpy (Gibbs free energy) of a formation reaction of the binary type (3.1) is calculated according to (3.2) and (3.3). If $\Delta G_T < 0$, the reaction proceeds from the left to the right, provided no other factors (kinetic hindrances etc.) redraft the situation drastically. If $\Delta G_T > 0$, the reaction proceeds from the right to the left, provided no other factors (kinetic hindrances etc.) redraft the situation drastically. We compare with Examples 3.1 and 3.2, which both illustrate the first type of reaction, and with Example 3.3, which illustrates the first type of reaction as well as the second type of reaction. We also compare with Examples 2.1 and 2.2, which directly demonstrate how reaction/formation enthalpies/entropies and thus also free enthalpies can be calculated by comparing different temperatures and using relations of the type (3.5)–(3.7).

We here additionally note that one distinguishes between weak, medium, and strong oxides, carbides, nitrides, borides etc. depending on the ΔG_T values per O, N, C, B etc. For instance, weak oxides are characterized by $\Delta G_T > -150$ kJ/mol O. Examples are provided by CuO, HgO, and PdO. For instance, medium oxides are characterized by $\Delta G_T = -150 \ldots -350$ kJ/mol O. Examples are provided by FeO, NiO, and PbO. For instance, strong oxides are characterized by $\Delta G_T < -350$ kJ/mol O. Examples are provided by Al_2O_3, BaO, BeO, CaO, MgO, SiO_2, TiO_2, and ZrO_2. With oxides, a relatively high formation enthalpy is observable. The free formation enthalpy increases with increasing temperature, but it remains negative up to far over 3000 K. With nitrides, in comparison to oxides, the formation enthalpy is smaller. Since the increase of the free formation enthalpy with increasing temperature is also strong, such substances are unstable at higher temperatures. With carbides, in comparison to nitrides, the formation enthalpy is even smaller. Since during their decomposition no gases come into being, the changes of entropy are small, and thus the changes of free formation enthalpy are small, and therefore such substances are stable up to much higher temperatures. Regarding these circumstances, borides are comparable to carbides.

We here additionally note that chemical reactions where gases or liquids take an active part, for example, caused by particle collisions, usually are equilibrium reactions the balance point of which is more or less shifted into the right side of reaction or into the left side of reaction depending on constraints such as pressure and temperature. In Box 3.2, we show examples of equilibrium reactions/states of ceramic materials. Observing these, for example, we conclude that a BN crucible placed in a furnace with O_2 atmosphere just like a graphite heater placed in a furnace with O_2 atmosphere may change the outcome of sintering processes in this environment. Comparing these with corresponding phase diagrams, for example, comparing the equilibrium reaction $Si_3N_4 \leftrightarrow 3Si + 2N_2$ which is provided by (3.9) of Box 3.2 with the phase diagram which is presented at the top of Figure 3.3, we realize that such reactions represent points, lines, and even areas within such phase diagrams.

Chemical reactions between two solids define a very special constellation. Closely following [74], let us pose two cardinal questions. Can a reaction between two solids proceed accompanied by cooling of the system? Does it make sense to discuss a yield of a reaction between two solids of the type $m\mathrm{A(s)} + n\mathrm{B(s)} \longrightarrow \mathrm{A}_m\mathrm{B}_n\mathrm{(s)}$? (i) The answer to the first question is as follows. If only solid phases are involved, in the case of an exothermic scenario ($\Delta H < 0$), a reaction does emerge, but in the case of an endothermic scenario ($\Delta H > 0$), a reaction does not emerge. For example, the scenario (3.19) is exothermic so that a reaction does occur, but the scenario (3.20) is endothermic so that a reaction does not occur. If only solid phases are involved, in the case of an exothermic scenario as well as in the case of an endothermic scenario, entropy changes usually are small ($\Delta S \approx 0$) and thus are not considered. For example, in the case of the scenario (3.19) and in the case of the scenario (3.20) entropy changes are small and thus are not considered. By contrast, if also/only liquid phases and/or gaseous phases are involved, caused by entropy gain ($\Delta S > 0$), a reaction can evolve, depending on the temperature, also in the case of endothermic scenarios. For example, the scenario (3.21) is endothermic and is associated with entropy gain, and a reaction can evolve due to the high temperatures, and the scenario (3.22) is endothermic and is associated with entropy gain, and a reaction cannot evolve due to the low temperatures. These aspects of chemical reactions, in the general case, are formally summarized by (3.24), and in the case where the temperature does not change, are formally summarized by (3.25), telling us that the resulting Gibbs free energy of chemical reactions that lead from the reactant side (l. h. s.) to the product side (r. h. s.) has to be negative, and this is guaranteed by the teamwork of the enthalpy H, the entropy S, and the temperature T. For example, the scenario (3.21) is governed by high temperatures so that $\mathrm{d}H > 0$ and thus $\Delta H > 0$ can be compensated by $T\mathrm{d}S > 0$ involving $\Delta S > 0$, leading to $\mathrm{d}G < 0$ and thus to $\Delta G < 0$ so that a chemical reaction does emerge, but the scenario (3.22) is governed by low temperatures so that $\mathrm{d}H > 0$ and thus $\Delta H > 0$ cannot be compensated by $T\mathrm{d}S > 0$ involving $\Delta S > 0$, leading to $\mathrm{d}G > 0$ and thus to $\Delta G > 0$ so that a chemical reaction does not emerge. Having these circumstances in mind, we are able to answer the first question. Well, since entropy changes usually are small ($\Delta S \approx 0$) during a reaction between two solids, we have to conclude that changes of temperature, requiring changes of entropy, usually also do not play a crucial role ($\Delta T \approx 0$), according to (3.24) and (3.25), then requiring $\Delta G \approx \Delta H$, as pointed out by (3.26). Hence, we should not expect that a reaction between two solids can proceed accompanied by cooling of the system. Hence, this situation is different from situations where also/only liquid phases and/or gaseous phases are involved, which allow reactions where the reactants become cold upon reacting. By the way, we should appreciate that some reactions where the reactants become cold upon reacting yield instable products. For example, an instable product which is obtained by reactions where the reactants become cold upon reacting is $PbCl_4$, which is a yellow oil that tends to explode, but which can be safely prepared according to (3.23). (ii) The answer to the second question is as follows. A yield of a reaction between two solids is not reversible. Either it occurs or it does not occur, but if it occurs, then as 100%. For example, the scenario (3.19) is based upon a reaction that entirely rolls to the r. h. s., namely LiI and CsF completely react to LiF and CsI.

> **Box 3.1 (Important formulae: chemical reactions, free formation enthalpy).**
>
> M = metal, N = non-metal, X = compound,
> ΔG_T = free formation enthalpy at temperature T,
> H_T = enthalpy at temperature T, S_T = entropy at temperature T,
> C_p = heat capacity (constant pressure p).
>
> Formation reaction (binary compounds):
>
> $$m\mathrm{M} + n\mathrm{N} \rightleftharpoons x\mathrm{X}. \tag{3.1}$$
>
> Free formation enthalpy at temperature T (binary compounds):
>
> $$\Delta G_T = xG_T^\mathrm{X} - \left(mG_T^\mathrm{M} + nG_T^\mathrm{N}\right), \tag{3.2}$$
>
> $$\Delta G_T = \Delta H_T - T\Delta S_T,$$
> $$\Delta H_T = xH_T^\mathrm{X} - \left(mH_T^\mathrm{M} + nH_T^\mathrm{N}\right), \quad \Delta S_T = xS_T^\mathrm{X} - \left(mS_T^\mathrm{M} + nS_T^\mathrm{N}\right), \tag{3.3}$$
>
> $$\begin{aligned}\Delta G_T < 0 &: \text{reaction proceeds from the left to the right},\\ \Delta G_T > 0 &: \text{reaction proceeds from the right to the left}.\end{aligned} \tag{3.4}$$
>
> Temperature dependence:
>
> $$H_T = H_{T_1} + \int_{T_1}^T C_\mathrm{p}(T')\mathrm{d}T' + \cdots \tag{3.5}$$
>
> (T_1 usually is set equal to 298 K = 25°C),
>
> $$S_T = S_{T_1} + \int_{T_1}^T \frac{C_\mathrm{p}(T')}{T'}\mathrm{d}T' + \cdots \tag{3.6}$$
>
> (T_1 usually is set equal to 298 K = 25°C).
>
> Kubaschewski "Ansatz" for the heat capacity C_p:
>
> $$C_\mathrm{p} = a + b_1 T + b_2 T^2 + b_{-1}T^{-1}. \tag{3.7}$$
>
> (Generalizations are provided by (2.73).)
>
> (As always, we here resort to the rule "products minus educts".)

> **Box 3.2** (Important formulae: chemical reactions, equlibrium reactions/states).

The Si–N–O system
(for example, Si_3N_4 is oxidized in O_2 atmosphere; temperature-dependent):

$$\tfrac{1}{3}Si_3N_4 + O_2 \leftrightarrow SiO_2 + \tfrac{2}{3}N_2 \,, \quad \tfrac{1}{3}Si_3N_4 + \tfrac{1}{2}O_2 \leftrightarrow SiO + \tfrac{2}{3}N_2 \,,$$
$$SiO_2 \leftrightarrow Si + O_2 \,, \quad 2SiO \leftrightarrow SiO_2 + Si \,, \quad SiO + \tfrac{1}{2}O_2 \leftrightarrow SiO_2 \,,$$
$$Si_3N_4 + SiO_2 \leftrightarrow 2Si_2N_2O \,,$$
$$3Si + 2N_2 \leftrightarrow Si_3N_4 \,.$$
(3.8)

Remarks:

Already relatively low O_2 pressures evoke the oxydation of Si_3N_4.
The stability of Si_3N_4 is a consequence of the SiO_2 protective coating.

The Si–N system
(for example, Si_3N_4 is decomposed during vaporization; temperature-dependent):

$$Si_3N_4 \leftrightarrow 3Si + 2N_2 \quad (p_{N_2} \text{ is } 1\,\text{bar at } 2103\,\text{K}) \,. \tag{3.9}$$

The B–N–O system
(for example, BN crucible; temperature-dependent):

$$2BN + \tfrac{3}{2}O_2 \leftrightarrow B_2O_3 + N_2 \,. \tag{3.10}$$

The C–O system
(for example, graphite heater; temperature-dependent):

$$C + O_2 \leftrightarrow CO_2 \ (T > 978\,\text{K}) \,, \quad C + \tfrac{1}{2}O_2 \leftrightarrow CO \ (T > 978\,\text{K}) \,,$$
$$CO_2 + C \leftrightarrow 2CO \ (\text{Boudouard equilibrium}) \,. \tag{3.11}$$

The Si–N–Y–O system
($Si_3N_4 + SiO_2 + YN$ phase diagram):

$$Si_3N_4\text{–}Y_{10}(SiO_4)_6N_2\text{–Liq phase} \Leftrightarrow Si_3N_4, Y_{10}(SiO_4)_6N_2, \text{Liq equilibrium} \,,$$
$$Si_3N_4\text{–}Y_{10}(SiO_4)_6N_2\text{–}YSiO_2N \text{ phase} \Leftrightarrow Si_3N_4, Y_{10}(SiO_4)_6N_2, YSiO_2N \text{ equilibrium} \,,$$
$$Si_3N_4\text{–}YSiO_2N\text{–}Y_2Si_3O_3N_4 \text{ phase} \Leftrightarrow Si_3N_4, YSiO_2N, Y_2Si_3O_3N_4 \text{ equilibrium} \,.$$

($Y_{10}(SiO_4)_6N_2$ = apatite, $YSiO_2N$ = wollastonite, $Y_2Si_3O_3N_4$ = melilithe.)

Example 3.1 (An additional tutorial example: ΔG, Al_2O_3).

Calculate the free formation enthalpy at temperature $T = 298\,\text{K}$ for the reaction that leads from Al and O to Al_2O_3. For this purpose, use tabulated values.

Solution.

Setup of the reaction of formation:

$$2\text{Al} + \tfrac{3}{2}\text{O}_2 \rightarrow \text{Al}_2\text{O}_3 \,. \tag{3.12}$$

Free formation enthalpy at temperature $T = 298\,\text{K}$:

$$\Delta G_{298} =$$

$$= G_{298}^{Al_2O_3} - \left(2 G_{298}^{Al} + \tfrac{3}{2} G_{298}^{O_2}\right) = \tag{3.13}$$

$$= -1691 - (-2 \cdot 8.445 - \tfrac{3}{2} \cdot 61.15) = -1582\,\text{kJ/mol} < 0\,.$$

Example 3.2 (An additional tutorial example: ΔG, SiC).

Calculate the free formation enthalpy at temperature $T = 298\,\text{K}$ for the reaction that leads from Si and C to SiC. For this purpose, use tabulated values.

Solution.

Setup of the reaction of formation:

$$\text{Si} + \text{C} \rightarrow \text{SiC}\,. \tag{3.14}$$

Free formation enthalpy at temperature $T = 298\,\text{K}$:

$$\Delta G_{298} =$$

$$= G_{298}^{SiC} - \left(G_{298}^{Si} + G_{298}^{C}\right) = \tag{3.15}$$

$$= -78.21 - (-55.613 - 1.712) = -70.89\,\text{kJ/mol} < 0\,.$$

Example 3.3 (An additional tutorial example: ΔG, Si_3N_4).

Calculate the free formation enthalpies at temperatures $T = 298\,\text{K}$ and $T = 2200\,\text{K}$ for the reaction that leads from Si and N to Si_3N_4 by using tabulated values.

Solution.

Setup of the reaction of formation:

$$3Si + 2N_2 \rightleftharpoons Si_3N_4 \,. \tag{3.16}$$

Free formation enthalpy at temperature $T = 298\,\text{K}$:

$$\begin{aligned}\Delta G_{298} &= \\ &= G_{298}^{Si_3N_4} - \left(3G_{298}^{Si} + 2G_{298}^{N_2}\right) = \\ &= -778.8 - (-3 \cdot 5.613 - 2 \cdot 57.12) = \\ &= -647.7\,\text{kJ/mol}\,.\end{aligned} \tag{3.17}$$

$\Delta G_{298} < 0$: reaction proceeds from the left to the right.

Free formation enthalpy at temperature $T = 2200\,\text{K}$:

$$\begin{aligned}\Delta G_{2200} &= \\ &= G_{2200}^{Si_3N_4} - \left(3G_{2200}^{Si} + 2G_{2200}^{N_2}\right) = \\ &= -1327 - (-3 \cdot 121.4 - 2 \cdot 498.1) = \\ &= 33.5\,\text{kJ/mol}\,.\end{aligned} \tag{3.18}$$

$\Delta G_{298} > 0$: reaction proceeds from the right to the left.

(As always, we here resort to the rule "products minus educts".)

Box 3.3 (Important formulae: ΔH, ΔS, ΔG).

Chemical scenarios:

$$\text{LiI(s)} + \text{CsF(s)} \longrightarrow \text{LiF(s)} + \text{CsI(s)},$$
$$\Delta H < 0, \quad \Delta G < 0, \quad (3.19)$$
$$\Delta H = -33\,\text{kcal/mol} = -138\,\text{kJ/mol},$$

$$\text{Al}_2\text{O}_3(\text{s}) + \text{Al(s)} \nrightarrow 3\text{AlO(s)},$$
$$\Delta H > 0, \quad \Delta G > 0, \quad (3.20)$$

$$\text{Al}_2\text{O}_3(\text{s}) + \text{Al(l)} \xrightarrow{\text{high temperatures}} 3\text{AlO(g)}, \quad (3.21)$$
$$\Delta H > 0, \quad \Delta S > 0, \quad \Delta G < 0,$$

$$\text{Al}_2\text{O}_3(\text{s}) + \text{Al(l)} \xrightarrow{\text{low temperatures}} 3\text{AlO(g)}, \quad (3.22)$$
$$\Delta H > 0, \quad \Delta S > 0, \quad \Delta G > 0,$$

$$(\text{NH}_4)_2\text{PbCl}_6 + 2\text{H}_2\text{SO}_4 \longrightarrow \text{PbCl}_4 + 2\text{NH}_4\text{HSO}_4 + 2\text{HCl}, \quad (3.23)$$
$$\Delta H > 0, \quad \Delta S > 0, \quad \Delta T < 0, \quad \Delta G < 0$$

(mixture gets cold).

Physical/chemical laws:

$$\Delta G < 0,$$
$$dG = dH - TdS - SdT, \quad (3.24)$$
$$\Delta G = \int dG = \Delta H - \int TdS - \int SdT$$

(chemical reactions that lead from the reactant side (l. h. s.) to the product side (r. h. s.)),

$$\Delta G < 0, \quad \Delta G = \Delta H - T\Delta S \quad (3.25)$$

(chemical reactions that lead from the reactant side (l. h. s.) to the product side (r. h. s.) in the case where the temperature does not change),

$$\Delta G < 0, \quad \Delta G \approx \Delta H \quad (3.26)$$

(chemical reactions that lead from the reactant side (l. h. s.) to the product side (r. h. s.) in the case of a reaction between two solids).

As a general rule, stable ceramic materials exhibit the highest oxidation state of the respective metalloid. Regarding the stability of ceramic materials within special corrosive environments, we take notice of the following details.

3.1.2 Oxygen Environments

In contrast to oxidic ceramic materials, which apart from a few examples such as Ce and Eu materials of functional ceramics are relatively stable at high temperatures in oxygen/air environments, non-oxidic ceramic materials such as nitrides, carbides, and borides are relatively unstable at high temperatures in oxygen/air environments. Depending on additional constraints such as temperature, oxygen partial pressure, sinter additives, and impurities, a passive (O) oxidation creating a protective coating or an active (O) oxidation not creating a protective coating or even destroying a protective coating finally leaving gaseous substances may result. In all cases where a protective coating is the result, however, the relative stability increases due to the shielding effect of the protective coating. We compare with Boxes 3.4–3.7 and Figures 3.16–3.18.

3.1.3 Technical Environments

Apart from very special examples, for instance, SiO_2 in hydrofluoric acid (HF) and BeO in hydrofluoric acid (HF) or phosphoric acid (H_3PO_4), oxidic ceramic materials are stable in (even boiling) water, acids, and bases. However, many oxidic ceramic materials and many non-oxidic ceramic materials exposed to water vapor at high temperatures or exposed to salt deposits at high temperatures are unstable. It is a fact that a lot of oxidic ceramic materials are stable regarding fused non-alkaline salts, but it is also a fact that a lot of oxidic ceramic materials are unstable regarding fused alkaline salts. It is a fact, too, that many oxidic and non-oxidic ceramic materials regarding many metal melts are stable. We compare with Box 3.8 and Figure 3.17.

3.1.4 Biological Environments

Going beyond these basic types of corrosion, let us here annotate that microorganisms such as bacteria evoke further types of corrosion. However, these types of corrosion rather affect metallic materials than ceramic materials. In passing, we take notice of this metallic scenario [11, 17]. Following reactions such as $Fe + 2H^+ \rightarrow Fe^{2+} + H_2$, in contact with an H_2O environment, objects consisting of base metals such as Fe release metal cations such as Fe^{2+} into the H_2O environment, and the remaining electrons e^- enforce transitions $2H^+ \rightarrow H_2$. In a subsequent second step, following reactions such as $4H_2 + SO_4^{2-} \rightarrow HS^- + 3H_2O + OH^-$, in the presence of dissolved sulphates and sulphate-reductive bacteria such as Desulfovibrio, a redox reaction takes place, reducing the sulphates ($SO_4^{2-} \Rightarrow OH^-$) and oxidizing the elementary hydrogen ($H_2 \Rightarrow H_2O$). In a subsequent third step, following reactions such as $Fe^{2+} + HS^- \rightarrow FeS + H^+$ and $3Fe^{2+} + 6H_2O \rightarrow 3Fe(OH)_2 + 6H^+$, iron sulphide and iron hydroxide are precipitated. We then speak of "bacterial anaerobic corrosion".

Box 3.4 (Important data: corrosion in O_2 environments, silicon nitride (Si_3N_4)).

Important oxidation reactions in an O_2 environment:

$$Si_3N_4 + 5O_2 \longrightarrow 3SiO_2 + 4NO,$$
$$Si_3N_4 + 7O_2 \longrightarrow 3SiO_2 + 4NO_2.$$

Important oxidation products in an O_2 environment:

important solid reaction products are SiO_2, SiO, Si_2N_2O;
important gaseous reaction products are N_2, NO, N_2O, NO_2, $SiN(g)$, $SiO(g)$.

Reaction product SiO_2:

relatively low temperature, relatively high oxygen partial pressure
(as protective coating, passive oxidation).

Reaction product $SiO(g)$:

relatively high temperature, relatively low oxygen partial pressure
(as gaseous reaction product, active oxidation).

Reaction product Si_2N_2O:

in the presence of sinter additives.

Additional hints:

$T < 900°C \rightarrow$ formation of solid SiO, $T > 900°C \rightarrow$ formation of solid SiO_2,
$T > 1420°C \rightarrow$ dissociation of Si_3N_4 according to $Si_3N_4 \rightarrow 3Si + 2N_2$,
$Si + SiO_2 \rightarrow 2SiO(g)$, $2SiO(g) \rightarrow Si + SiO_2$;

$T < 1065°C \rightarrow$ formation of amorphous SiO_2,
$T > 1065°C \rightarrow$ formation of crystalline SiO_2 (cristobalite),
$T > 1125°C \rightarrow$ formation of crystalline SiO_2 (tridymite),
$T > 1385°C \rightarrow$ formation of crystalline SiO_2 (quartz);

in the presence of sinter additives such as MgO, ZrO_2, Al_2O_3, and Y_2O_3,
siliceous protective coatings
such as $MgSiO_3$, Mg_2SiO_4, and $Y_2Si_2O_7$ protective coatings develop.

Box 3.5 (Important data: corrosion in O_2 environments, aluminum nitride (AlN)).

Important oxidation reactions in an O_2 environment:

$$2AlN + 2O_2 \longrightarrow Al_2O_3 + N_2O, \quad 4AlN + 5O_2 \longrightarrow 2Al_2O_3 + 4NO,$$
$$4AlN + 7O_2 \longrightarrow 2Al_2O_3 + 4NO_2.$$

Important oxidation products in an O_2 environment:

important solid reaction products are Al_2O_3;
important gaseous reaction products are N_2O, NO, NO_2.

Reaction product Al_2O_3:

always as protective coating, always passive oxidation.

Additional hints:

the starting point of the reaction $2AlN + 2O_2 \longrightarrow Al_2O_3 + N_2O$
is $\approx 900\text{--}1100°C$.

Box 3.6 (Important data: corrosion in O_2 environments, boron nitride (BN)).

Important oxidation reactions in an O_2 environment:

$$2BN + 2O_2 \longrightarrow B_2O_3 + N_2O, \quad 4BN + 5O_2 \longrightarrow 2B_2O_3 + 4NO,$$
$$4BN + 7O_2 \longrightarrow 2B_2O_3 + 4NO_2.$$

Important oxidation products in an O_2 environment:

important solid reaction products are B_2O_3;
important gaseous reaction products are N_2O, NO, NO_2.

Reaction product B_2O_3:

$T < 700°C \rightarrow$ as protective coating, passive oxidation,
$T > 700°C \rightarrow$ vaporization of the B_2O_3 layer, active oxidation.

Additional hints:

the microscopic processes strongly depend on the crystalline modification of BN.

Box 3.7 (Important data: corrosion in O_2 environments, silicon carbide (SiC)).

Important oxidation reactions in an O_2 environment:

$$SiC + 2O_2 \longrightarrow SiO_2 + CO_2,$$
$$2SiC + 3O_2 \longrightarrow 2SiO_2 + 2CO,$$
$$SiC + O_2 \longrightarrow SiO(g) + CO.$$

Important oxidation products in an O_2 environment:

important solid reaction products are SiO_2;
important gaseous reaction products are CO, CO_2, SiO(g).

Reaction product SiO_2:

relatively low temperature
$\rightarrow$
$$SiC + 2O_2 \longrightarrow SiO_2 + CO_2$$
(as protective coating, passive oxidation);

relatively high temperature
$\rightarrow$
$$2SiC + 3O_2 \longrightarrow 2SiO_2 + 2CO$$
(as protective coating, passive oxidation).

Reaction product SiO(g):

$T > 1620°C$, relatively low oxygen partial pressure $\rightarrow SiC + O_2 \longrightarrow SiO(g) + CO$
(as gaseous reaction product, active oxidation).

Additional hints:

$T < 1400°C \rightarrow$ formation of amorphous SiO_2,
$T > 1400°C \rightarrow$ formation of crystalline SiO_2 (cristobalite);

$T > 1100°C \rightarrow$ formation of crystalline SiO_2 (cristobalite)
in the presence of sinter additives.

Box 3.8 (Important data: corrosion in technical environments, Si_3N_4, SiC, SiO_2).

Water vapor corrosion
(corrosion in the presence of water vapor at low and high temperatures).

Important corrosive reactions, low temperatures:

$$SiC + 2H_2O \longrightarrow SiO_2 + CH_4$$

(corrosive reaction of SiC in water vapor, passive oxidation),

$$SiC + H_2O + H_2 \longrightarrow SiO(g) + CH_4$$

(corrosive reaction of SiC in water vapor, H_2 catalyzed, active oxidation).

Important corrosive reactions, high temperatures:

$$Si_3N_4 + 3H_2O \longrightarrow 3SiO(g) + 2N_2 + 3H_2$$

(corrosive reaction of Si_3N_4 in water vapor, active oxidation),

$$SiC + 2H_2O \longrightarrow SiO(g) + CO + 2H_2$$

(corrosive reaction of SiC in water vapor, active oxidation).

Remarks:

high H_2O partial pressure $\rightarrow$ high corrosion rate;
low (high) temperature $\rightarrow$ low (high) molecular motion $\rightarrow$ passive (active) oxidation.

Hot gas corrosion
(corrosion in the presence of salt deposits at high temperatures).

Important corrosive reactions:

$$SiO_2 + Na_2SO_4 \longrightarrow Na_2OSiO_2 + SO_2 + \tfrac{1}{2}O_2$$

(corrosive reaction with the sulphate of the alkali metal Na),

$$SiO_2 + Na_2CO_3 \longrightarrow Na_2OSiO_2 + CO + \tfrac{1}{2}O_2$$

(corrosive reaction with the carbonate of the alkali metal Na).

Remarks:

such corrosive reactions generate silicate layers with relatively low melting temperature at a silicon oxide layer;

such silicate layers facilitate the diffusion of oxidants (O_2, CO_2, SO_3) into a sample, strengthening the oxidation of the sample, leading to a healing layer.

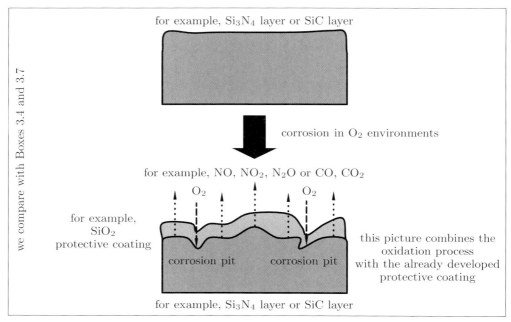

Figure 3.16. Corrosion in O_2 environments.

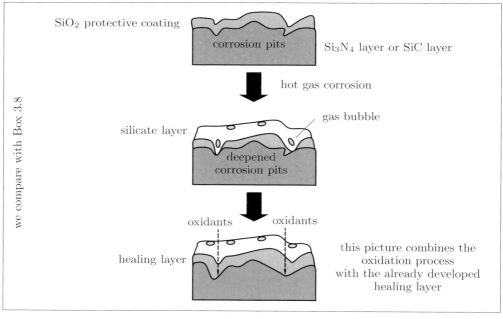

Figure 3.17. Corrosion in technical environments. Hot gas corrosion.

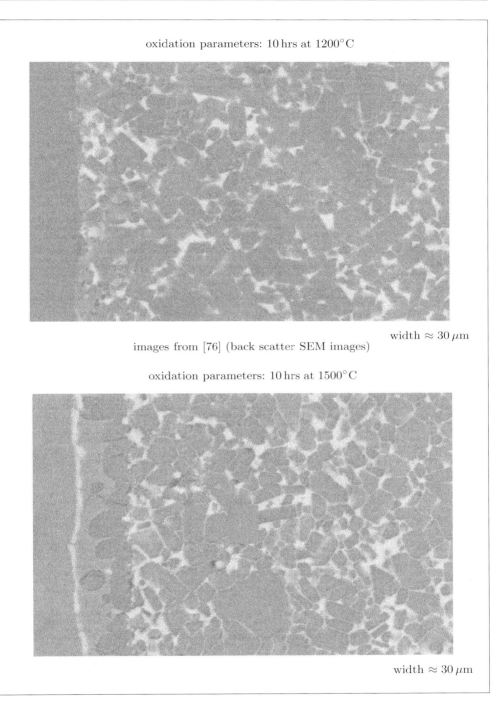

Figure 3.18. Oxidation of an SiC sample sintered with Lu_2O_3-AlN additives.

Continuation of Box.

Corrosion in environments containing fused salts.

Important corrosive reaction chains:

1.
$$2SiO_2 + Na_2SO_4 \longrightarrow Na_2Si_2O_5 + SO_2 + \tfrac{1}{2}O_2$$
(destruction of the SiO_2 protective coating),

2.
$$2Si_3N_4 + 6Na_2SO_4 + 3O_2 \longrightarrow 6Na_2SiO_3 + 4N_2 + 6SO_2$$
or
$$SiC + Na_2SO_4 + O_2 \longrightarrow Na_2SiO_3 + CO + SO_2$$
(reaction with a nitride or a carbide),

3.
diffusion of O_2 into the sample.

Corrosion in environments containing acids.

Important corrosive reactions:

$$Si_3N_4 + 16HF \longrightarrow SiF_4 + 2(NH_4)_2SiF_6$$
(at room temperature $T = 25°C$),

$$3Si_3N_4 + 16H_3PO_4 \longrightarrow 3Si_3(PO_4)_4 + 4(NH_4)_3PO_4$$
(strong reaction with H_3PO_4 for $T \geq 100°C$),

$$Si_3N_4 + 8H_2SO_4 \longrightarrow 3SiO_2 + 2(NH_4)_2SO_4 + 6SO_3$$
(strong reaction with H_2SO_4 for $T \geq 300°C$).

Remarks:

Si_3N_4 at room temperature $T = 25°C$ is resistant to hydrochloric acid (HCl), sulphuric acid (H_2SO_4), and phosphoric acid (H_3PO_4);

SiC at room temperature $T = 25°C$ is resistant to hydrochloric acid (HCl), sulphuric acid (H_2SO_4), phosphoric acid (H_3PO_4), and hydrofluoric acid (HF).

> **Continuation of Box.**
>
> Corrosion in environments containing halogens.
>
> Important corrosive reactions, reactions with chlorine:
>
> $$SiC + 2Cl_2 \longrightarrow SiCl_4(g) + C$$
> $$(T \geq 100°C),$$
>
> $$SiC + 4Cl_2 \longrightarrow SiCl_4(g) + CCl_4(g)$$
> $$(T \geq 1000°C).$$
>
> Important corrosive reactions, reactions with fluorine:
>
> $$Si_3N_4 + 6F_2 \longrightarrow 3SiF_4(g) + 2N_2$$
> (spontaneous reaction even at low temperatures),
>
> $$SiC + 4F_2 \longrightarrow SiF_4(g) + CF_4(g)$$
> (spontaneous reaction even at low temperatures).
>
> Remarks:
>
> in the presence of O_2, $SiCl_4(g)$ reacts with O_2, generating oxichlorides such as Cl_3Si–O–$SiCl_3$;
>
> oxichlorides such as Cl_3Si–O–$SiCl_3$ cannot diffuse through an SiO_2 layer, generating a passivation layer between an SiC layer and an SiO_2 layer.
>
> Corrosion in environments containing hydrogen or carbon.
>
> An important corrosive reaction, a reaction with hydrogen:
>
> $$SiO_2 + H_2 \longrightarrow SiO(g) + H_2O(g)$$
> ($T \geq 1250°C$, (O) reduction of SiO_2, (O) oxidation of H_2).
>
> An important corrosive reaction, a reaction with carbon:
>
> $$SiO_2 + 3C \longrightarrow \alpha\text{–}SiC + 2CO$$
> ($T \geq 2000°C$, the central reaction of the Acheson process).
>
> Remarks:
>
> H_2, NH_3, and CO at high temperatures act as strong reductives.

Example 3.4 (An additional tutorial example: moulded SiC padding in O_2 atmosphere).

Let a moulded SiC padding with density $\rho = 3.2\,\text{g/cm}^3$ in cylindric form ($d = 20\,\text{mm}$, $l = 60\,\text{mm}$) be given. Assume that it is oxidized by 1 litre O_2 at 900°C and 1013 mbar. Which mass change in percent is to be expected provided O_2 behaves as an ideal gas and provided the oxidation process quantitativly follows the chemical reaction $SiC + \frac{3}{2}O_2 \rightarrow SiO_2 + CO$?

Solution.

Mass of moulded SiC padding:

$$m = r^2 \pi l \rho = 60.31\,\text{g}\,. \tag{3.27}$$

Moles O_2 according to the ideal gas law:

$$n(O_2) = \frac{pV}{RT} =$$

$$= \frac{1013\,\text{mbar} \cdot 1\,\text{litre}}{8.3146\,\text{J/mol\,K} \cdot 1173.13\,\text{K}} = \frac{101300\,\text{N/m}^2 \cdot 10^{-3}\,\text{m}^3}{8.3146\,\text{J/mol\,K} \cdot 1173.13\,\text{K}} = \tag{3.28}$$

$$= 0.0104\,\text{mol}\,.$$

Moles SiC that are oxidized:

$$n(\text{SiC}) = 0.00693\,\text{mol}\ \ (2/3\ \text{of}\ 0.0104\,\text{mol}\ O_2)\,. \tag{3.29}$$

Mass SiC that is oxidized:

$$M(\text{SiC}) = 40.097\,\text{g/mol} \Rightarrow m(\text{SiC}) = M(\text{SiC})n(\text{SiC}) = 0.28\,\text{g}\,. \tag{3.30}$$

Mass SiO_2 that comes into being:

$$M(\text{SiO}_2) = 60.086\,\text{g/mol} \Rightarrow m(\text{SiO}_2) = M(\text{SiO}_2)n(\text{SiC}) = 0.42\,\text{g}\,. \tag{3.31}$$

Mass change in percent:

$$\Delta m_\%(\text{SiC}) = \frac{60.31\,\text{g} - 0.28\,\text{g} + 0.42\,\text{g}}{60.31\,\text{g}}100\% - 100\% = +0.23\%\,. \tag{3.32}$$

Example 3.5 (An additional tutorial example: precursor ceramics in N_2/O_2 atmosphere).

Let a precursor ceramics be given that is composed of Si, C, N, and B according to the gross formula $Si_{28}C_{15}N_{46}B_{10}$. Which phases are to be expected? Which reaction in an N_2 atmosphere is to be expected at 1500°C? Let us assume that 20.5 g of the precursor ceramics are completely oxidized in an O_2 atmosphere leading to the oxidation numbers +IV for Si and C and +III for N. How much oxygen is needed for this oxidation process? Which mass/mole do the oxidation products have and which mass/mole do those that are solid at room temperature have? Figures 4.44 and 4.46 might be helpful in this context!

Solution.

Phases:

$$(Si_3N_4)_9(SiC)_1(BN)_{10}C_{14} \,. \tag{3.33}$$

Reaction in N_2 atmosphere at 1500°C:

$$Si_3N_4 \rightarrow 3Si + 2N_2 \,. \tag{3.34}$$

Oxygen needed for the oxidation process:

$$28Si + 15C + 46N + 10B$$
$$\Rightarrow$$
$$28SiO_2(+IV) + 15CO_2(+IV) + 23N_2O_3(+III) + 5B_2O_3(+III) \tag{3.35}$$
(oxidation reaction);

for 1 mol of the precursor ceramics $Si_{28}C_{15}N_{46}B_{10}$ ($\rightarrow$ 1716 g/mol), 170 mol O atoms are needed;

for 20.5 g of the precursor ceramics $Si_{28}C_{15}N_{46}B_{10}$ ($\rightarrow$ 0.012 mol), 2.03 mol O atoms are needed ($\rightarrow$ 32.48 g).

Mass/mole of the oxidation products/oxidation products solid at room temperature:

$$M(Si_{28}C_{15}N_{46}B_{10}O_{170}) = 4435.9\,\text{g/mol}, \, M(Si_{28}B_{10}O_{71}) = 2028\,\text{g/mol},$$
mass loss: 45.7%.

3.2 Physical Properties

Speaking of *physical properties*, we mainly mean the thermal, electric, magnetic, and optical properties of materials. Let us get a general idea of the main features of this topic with special respect to ceramics in the sections that follow.

3.2.1 Thermal Properties

On the one hand, temperature is the expression of the existence of disordered forms of translational kinetic energy, vibrational kinetic energy, and rotational kinetic energy of particles of a plasma, a gas, a liquid, a solid, or a hybrid form such as a liquid crystal. On the other hand, temperature is the expression of the existence of thermal radiation, i.e. of the existence of infrared radiation. Focussing on the first case and focussing on solid materials, temperature is the expression of the existence of disordered forms of lattice vibrations combined with disordered forms of charge carrier motion such as free electron motion and free ion motion. It is meanwhile a well-known fact that a specific volume change comes into being if the temperature T of a solid material changes from T_1 to T_2 with $T_2 \lessgtr T_1$. It is meanwhile also a well-known fact that a specific heat flow density comes into being if the temperature T of a solid material varies from position to position, leading to a temperature gradient. In the case of thermal load, we compare with Figures 3.19–3.21, such thermal mechanisms come into play, requesting the engineer to develop materials serviceable under these conditions.

In addition to the thermodynamic relations of Box 2.13 and Box 2.14 which are presented in Section 2.2.3.1, a collection of the most important formulae which put these circumstances into concrete terms is presented in Box 3.9. Let us here particularly point at the material parameters α_L, α_A, α_V, and β. We note that these describe how solid materials react to the alteration of thermodynamic quantities such as pressure and temperature. Let us here particularly point at the Grüneisen parameter γ. We note that it describes the correlation of the volume changes δV of the solid material and the phonon frequency changes $\delta \omega$ of the phonons of the phonon branches building up the energy–momentum spectrum of the lattice vibrations of the solid material. Let us here particularly point at the thermal conductivity λ. We note that it describes the correlation of the temperature gradient $\nabla T(\boldsymbol{x})$ within the solid material and the heat flow density $\boldsymbol{j}(\boldsymbol{x})$ within the solid material. Let us here particularly point at the models λ_{ph} and λ_e for the thermal conductivity λ, specifying the two mechanisms of heat transfer in a solid material, namely a first mechanism that is based upon vibrating lattice particles exchanging phonons with other vibrating lattice particles and a second mechanism that is based upon vibrating lattice particles exchanging phonons with moving free charge carriers such as free electrons and free ions. We note that the models λ_{ph} and λ_e eventually establish the connection of the thermal conductivity with the mean phonon velocity v_{ph} and the mean charge carrier velocity v_e as well as with the mean free phonon path l_{ph} and the mean free charge carrier path l_e.

> **Box 3.9** (Important formulae: thermal properties).
>
> β = coefficient of compressibility, α = coefficients of thermal expansion,
> L_0 = original length, A_0 = original area, V_0 = original volume, ρ = mass density,
> γ = Grüneisen parameter, ω = phonon frequency,
> $\boldsymbol{j}(\boldsymbol{x})$ = heat flow density at position $\boldsymbol{x}$, $T(\boldsymbol{x})$ = temperature at position $\boldsymbol{x}$,
> λ = thermal conductivity, C = heat capacity, P = pore rate,
> $v_{\mathrm{ph(e)}}$ = mean phonon (charge carrier) velocity,
> $l_{\mathrm{ph(e)}}$ = mean free phonon (charge carrier) path.
>
> Compressibility:
>
> $$\beta = -\frac{1}{V_0}\left(\frac{\partial V}{\partial p}\right)_T \rightarrow \Delta V = -\beta V_0 \Delta p \ . \tag{3.36}$$
>
> Thermal expansion:
>
> $$\alpha_\mathrm{L} = \frac{1}{L_0}\left(\frac{\partial L}{\partial T}\right)_p \rightarrow \Delta L = \alpha_\mathrm{L} L_0 \Delta T \quad \alpha_\mathrm{A} = \frac{1}{A_0}\left(\frac{\partial A}{\partial T}\right)_p \rightarrow \Delta A = \alpha_\mathrm{A} A_0 \Delta T$$
>
> (linear thermal expansion) , (areal thermal expansion) ,
>
> $$\alpha_\mathrm{V} = \frac{1}{V_0}\left(\frac{\partial V}{\partial T}\right)_p \rightarrow \Delta V = \alpha_\mathrm{V} V_0 \Delta T \ , \quad \frac{1}{V_0}\left(\frac{\partial V}{\partial T}\right)_p = -\frac{1}{\rho}\left(\frac{\partial \rho}{\partial T}\right)_p \tag{3.37}$$
>
> (volumetric thermal expansion) .
>
> Grüneisen model:
>
> $$\alpha_\mathrm{V} = \gamma\beta\frac{C_\mathrm{V}}{V} \ , \quad \gamma = V\left(\frac{\partial p}{\partial U}\right)_V = -\frac{V}{\omega}\frac{\partial \omega}{\partial V} \ . \tag{3.38}$$
>
> Isotropic materials:
>
> $$\alpha_\mathrm{A} \approx 2\alpha_\mathrm{L} \ , \quad \alpha_\mathrm{V} \approx 3\alpha_\mathrm{L} \ . \tag{3.39}$$
>
> Heat flow density:
>
> $$\boldsymbol{j}(\boldsymbol{x}) = -\lambda \nabla T(\boldsymbol{x}) = -\lambda \operatorname{grad} T(\boldsymbol{x}) \ , \quad \lambda := \lambda_\mathrm{ph} = \frac{1}{3}C_\mathrm{V} v_\mathrm{ph} l_\mathrm{ph} \ , \quad \lambda := \lambda_\mathrm{e} = \frac{1}{3}C_\mathrm{e} v_\mathrm{e} l_\mathrm{e} \ ,$$
> $$\operatorname{Dim}[\boldsymbol{j}] = \frac{\mathrm{J}}{\mathrm{m^2\,s}} = \frac{\mathrm{W}}{\mathrm{m^2}} \ , \quad \operatorname{Dim}[\lambda] = \frac{\mathrm{J}}{\mathrm{m\,s\,K}} = \frac{\mathrm{W}}{\mathrm{m\,K}} \ . \tag{3.40}$$
>
> Thermal conductivity and pore rate:
>
> $$\lambda = \lambda_0(1-P)^b \quad (P \approx 1.5 \text{ for spherical pores}). \tag{3.41}$$

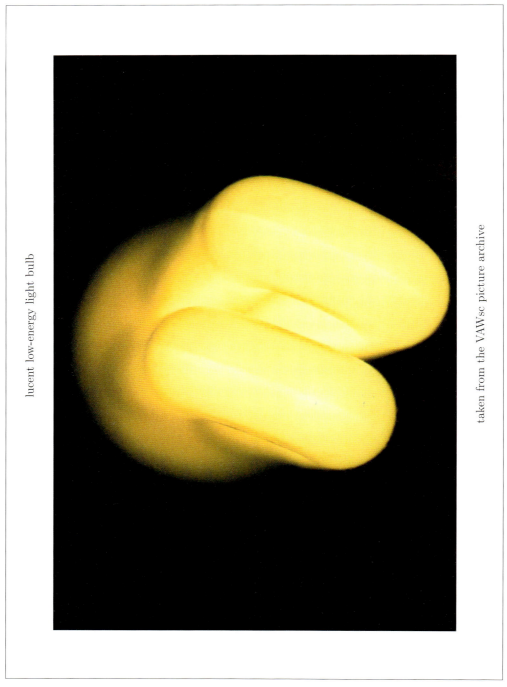

Figure 3.19. An example of thermal loads of glass, part one.

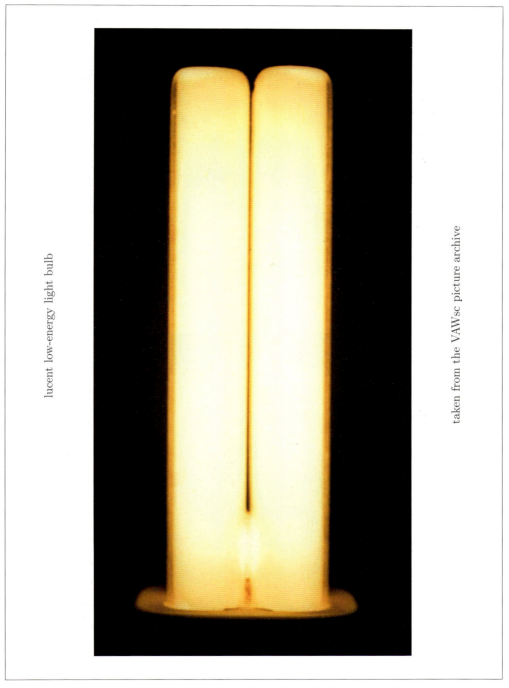

Figure 3.20. An example of thermal loads of glass, part two.

turbo charger rotor (SSiC)

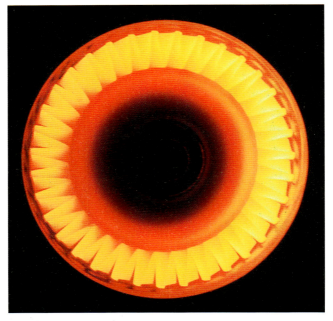

gas turbine (HIPSN)

taken from the PML picture archive

Figure 3.21. Examples for components exposed to high thermal load.

Table 3.1. Typical values for the thermal conductivity λ of ceramic materials.

ceramic material	λ [W/m K]	ceramic material	λ	ceramic material	λ
diamond	2000	Si_3N_4	20–60	TiB_2	25
BeO	300	BN	45–55	TiO_2	11.7
AlN	140–320	MgO	25–50	graphite	5
SiC	20–200	TiN	38	ZrO_2	1.5–2.5
WC	120	Al_2O_3	30	Al_2TiO_5	0.9–1.5
B_4C	30–70	TiC	30	quartz	1.3

We note that the following properties are responsible for the thermal conductivity λ of ceramic materials. (i) *Atomic mass.* Light atoms have a higher value for λ. (ii) *Anisotropy.* For anisotropic materials, λ depends on the crystal direction. For example, $\lambda \parallel c$ (SiO_2) $\approx 2\lambda \perp c$ (SiO_2). (iii) *Structure.* Materials consisting of more than two atom species have lower values for λ than materials consisting of two atom species. For example, $MgAl_2O_4$ (magnesium aluminate spinel) has a lower value for λ than Al_2O_3 (aluminum oxide). Materials showing a relatively complicated structure have lower values for λ than materials showing a relatively simple structure. For example, $Al_6Si_2O_{13}$ (mullite) has a lower value for λ than $ZnAl_2O_4$ (gahnite). Due to relatively short mean free phonon paths, all types of glassware have relatively low values for λ. Due to relatively high mean phonon velocities, all types of diamond-like materials (diamond, BeO, AlN, SiC etc.) have relatively high values for λ. (iv) *Lattice defects.* Vacancy defects, interstitial defects, and substitutional defects have an influence on λ. In particular, phonons associated with a heat flow density are scattered out of the flow direction of the heat flow density, leading to reduced values for λ. For example, O (oxygen) dissolved in AlN (aluminum nitride) leads to reduced values for λ. (v) *Pores.* Pores have an influence on λ. In particular, phonons associated with a heat flow density are stopped, deflected, and redirected by pores, leading to reduced values for λ. (vi) *Secondary phases.* Secondary phases have an influence on λ. In particular, phonons associated with a heat flow density are stopped, deflected, and redirected by grain boundaries, leading to reduced values for λ.

Table 3.2. Typical values for the coefficient of thermal expansion α_L of ceramic materials.

ceramic material	α_L [10^{-6}/K] (normal to c axis)	α_L [10^{-6}/K] (parallel to c axis)
graphite	1.0	27.0
quartz	14.0	9.0
Al_2O_3	8.3	9.0
TiO_2	6.8	8.3
Al_2TiO_5	-2.6	11.5
$CaCO_3$	-6.0	25.0

Typical values for the thermal conductivity λ of ceramic materials are shown in Table 3.1. We note that the temperature dependence of the thermal conductivity λ, apart from more far-reaching effects such as the jumps that accompany the transitions into superconducting states, typically looks like the example shown in Figure 3.22, where $T \to 0 \Rightarrow C_V \to 0 \Rightarrow \lambda = \lambda_{\text{ph}} = \frac{1}{3} C_V v_{\text{ph}} l_{\text{ph}} \to 0$ and $T \gg 0 \Rightarrow \lambda = \lambda_{\text{ph}} \sim 1/T$, exhibiting $\lambda = \lambda_e = 0$ in both limiting cases, i.e. no free charge carriers are present in the ceramic material. We also note that for other ceramic materials than diamond the values for the thermal conductivity λ are much smaller. We further note that typical values for organic materials are 0.42–0.51 (HDPE = high density polyethylene) and for metallic materials are 16-19 (steel).

Typical values for the coefficient of thermal expansion α_L of ceramic materials are shown in Table 3.2. We note that typically the coefficient of thermal expansion α_L of ceramic materials has smaller values than the coefficient of thermal expansion α_L of metallic materials. We also note that typically the coefficient of thermal expansion α_L of covalent ceramics has smaller values than the coefficient of thermal expansion α_L of ionic ceramics. In the first case, the electron gas (Fermi gas) makes the difference, i.e. the electron gas (Fermi gas) is a freely moving carrier of heat energy and thereby part of the expansion process that is induced by heat energy, in particular, then reducing the negative charge density at the positions of atoms/molecules and thereby then reducing the (negative–positive) charge compensation at the positions of atoms/molecules and thereby then enforcing a further extension of the interspace between the centers of atoms/molecules. In the second case, the packing density of atoms/molecules makes the difference, i.e. covalent ceramics are less packed ("more open") than ionic ceramics, evoking a situation where the expansion process that is induced by heat energy, on the one hand, exhibits a component breeding the expansion of ceramic materials, and on the other hand, exhibits a component not breeding the expansion of ceramic materials, but breeding the expansion of atoms/molecules into the "open space", reducing the resulting expansion of ceramic materials. We compare with Figure 3.23, which shows the interrelation "thermal expansion and anharmonic potential curve of two atoms". We read off from Figure 3.23 that the asymmetric structure ("anharmonic structure") of the potential curve of two atoms causes an increasing shift of the vibrational center away from the point of equilibrium with increasing vibrational motion that is induced by heat energy. We read off from Figure 3.23 that this effect eventually leads to an increasing mean distance between two atoms, on the macroscopic level of observation, finally manifesting itself as the expansion of ceramic materials.

We firstly acknowledge that anisotropic substances in special directions may show negative values for the coefficient of thermal expansion α_L. We secondly acknowledge that the increase in volume then is small, promising a high thermal shock resistance. We thirdly acknowledge that Al_2TiO_5, magnesium aluminosilicate (cordierite), and lithium aluminosilicate ("LAS") are examples. The reader should realize that the reason for negative thermal expansion are transverse vibrational modes, i.e. heat-induced vibrations along the diagonal of a unit cell, causing an elongation in one direction and a contraction in the perpendicular plane, rigid unit modes ("RUMs"), i.e. heat-induced vibrations in three-dimensional networks of polyhedra, causing motion of polyhedra, and phase transformations, i.e. heat-induced changes of the crystal structure.

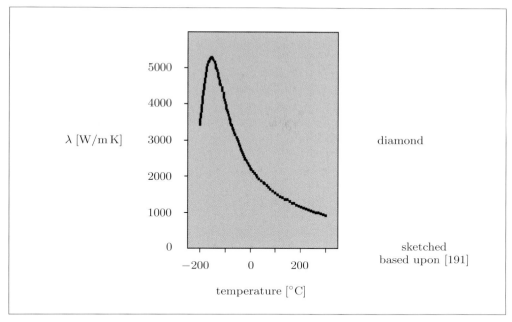

Figure 3.22. An example for the temperature dependence of thermal conductivity.

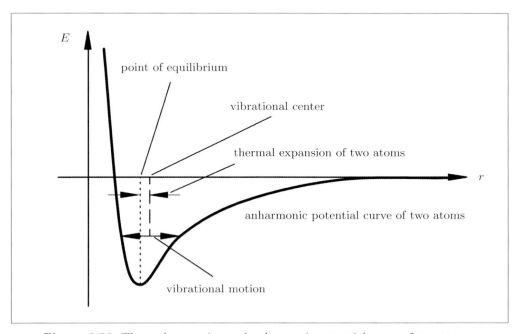

Figure 3.23. Thermal expansion and anharmonic potential curve of two atoms.

3.2.2 Electromagnetic Properties

When we touch electric devices, insulating materials protect us from electric shocks. When we switch on any devices, control any devices, and feedback control any devices, electric currents are the mediators of our wishes. Writing this book and studying this book, a wealth of electric pulses is speeding through our personal neural network, the brain, enabling us to do this. Transmitting the results of our efforts into our personal electronic network, the PC, our ideas can be easily distributed throughout the world. Observing these or similar electric scenarios, we realize that for each special scenario a material with special electric properties is the prerequisite.

We meanwhile know it, depending on constraints such as pressure and temperature, materials can be insulators, semi-conductors, or conductors. We meanwhile know it, too, depending on constraints such as pressure and temperature, conductors can be ionic conductors (with $\pm$ charges), electronic (normal) conductors (with $\pm$ charges), or superconductors. It should not be a miracle anymore that ceramics can be all of them, although ceramics mainly are considered as insulators (Al_2O_3, MgO, BeO, ZrO_2, AlN, BN, Si_3N_4, SiO_2 etc.). According to Box 3.10, briefly elucidating the electron aspect and the ion aspect, and Box 3.22, briefly elucidating the "superparticle" aspect, the electric properties of materials need the inclusion of notions like electric field intensity, electric flux density, electrical resistivity, electrical resistance, electrical conductivity, electrical conductance, electric current, electric tension (mostly termed "voltage"), permittivity, susceptibility, dielectric polarization, and electric dipole moment. From Box 3.10, we especially read off that the electrical resistivity in the case of electrons in electronic (normal) conductors, on the one hand, is modeled as electron interactions with lattice vibrations, defined as momentum exchange of electrons with phonons, and on the other hand, is modeled as electron interactions with lattice defects, defined as momentum exchange of electrons with lattice defects, eventually implying that the electrical resistivity increases with increasing T. From Box 3.10, we additionally read off that the electrical resistivity in the case of ions in ionic conductors is modeled by an exponential function depending on Q (needed to evoke ionic motion, in particular, needed to overcome the barrier of position change) and T, eventually implying that the electrical resistivity decreases with increasing T.

We meanwhile know it, insulators are called *dielectrics* if these are applied to change the electric behavior of electronic devices, for example, if these are applied to raise the capacitance of a capacitor. We meanwhile know it, too, dielectrics inside an electric field are polarized according to these mechanisms. Shift of valence electrons with respect to atomic cores, shift of ions out of their equilibrium position, orientation of already existent, permanent dipole moments, and accumulation of charges at boundary layers. We note that these mechanisms in total are captured by electric susceptibility terms and relative permittivity terms, i.e. on the one hand, by the electric susceptibility χ_e and the relative permittivity ϵ_r, and on the other hand, by their tensorial generalizations $\boldsymbol{\chi}_e$ and $\boldsymbol{\varepsilon}_r$, we compare with the relations (3.48) and (3.49). Naturally, $\boldsymbol{\chi}_e$, $\boldsymbol{\varepsilon}_r$, χ_e, and ϵ_r depend on the temperature and the frequency of the electric field intensity. Naturally, (3.48) is readily led back to (3.47), summing up the electric dipole moments $\boldsymbol{m}_{e,i}$, collecting the charge shifts in a volume V.

Box 3.10 (Important formulae: electric properties).

ρ = electrical resistivity, R = electrical resistance,
I = electric current, U = electric tension (mostly termed voltage),
L_{ec} = length of an electric conductor, A_{ec} = cross sectional area of an electric conductor,
σ = electrical conductivity, G = electrical conductance,
$\boldsymbol{E}$ = electric field intensity (electric field), $\boldsymbol{D}$ = electric flux density,
$\boldsymbol{P}$ = dielectric polarization, χ_e = electric susceptibility,
ϵ_0 = dielectric constant, ϵ_r = relative permittivity, ϵ = permittivity,
q = electric charge, $\boldsymbol{s}$ = shift vector, $\boldsymbol{m}_e$ = electric dipole moment, V = volume.

Electron aspect.

Electrical resistance R and electrical conductance G:

$$R = \rho \frac{L_{ec}}{A_{ec}} = \frac{U}{I}, \quad \text{Dim}\,[R] = \Omega, \tag{3.42}$$

$$G = 1/R. \tag{3.43}$$

Electron aspect, ion aspect.

Electrical resistivity (specific electrical resistance) ρ
and electrical conductivity (specific electrical conductance) σ:

$$\rho = \rho_{\text{phonon}}(T) + \rho_{\text{defects}}, \quad \rho_{\text{phonon}}(T) \stackrel{\Delta T = T - T_0 \to 0}{=} \rho(T_0)[1 + \alpha(T - T_0)] \tag{3.44}$$
(model for electrons in electronic (normal) conductors),

$$\rho = \rho_0 / \exp\left(-Q/RT\right) \tag{3.45}$$
(model for ions in ionic conductors, R here is the molar gas constant,
whereas Q here is the energy needed to evoke ionic motion),

$$\sigma = 1/\rho. \tag{3.46}$$

Polarization:

$$\boldsymbol{P} = \frac{1}{V} \sum_i q_i \boldsymbol{s}_i = \frac{1}{V} \sum_i \boldsymbol{m}_{e,i}. \tag{3.47}$$

Dielectric polarization:

$$\boldsymbol{P} = \begin{cases} \chi_e \epsilon_0 \boldsymbol{E} = (\epsilon_r - 1)\epsilon_0 \boldsymbol{E} & \text{(isotropic material)}, \\ \boldsymbol{\chi}_e \epsilon_0 \boldsymbol{E} = (\boldsymbol{\varepsilon}_r - 1)\epsilon_0 \boldsymbol{E} & \text{(anisotropic material)}, \end{cases} \tag{3.48}$$

$$\boldsymbol{D} = \begin{cases} \epsilon \boldsymbol{E} = \epsilon_r \epsilon_0 \boldsymbol{E} = \epsilon_0 \boldsymbol{E} + \boldsymbol{P} = \epsilon_0 \boldsymbol{E} + \chi_e \epsilon_0 \boldsymbol{E} & \text{(isotropic material)}, \\ \boldsymbol{\varepsilon} \boldsymbol{E} = \boldsymbol{\varepsilon}_r \epsilon_0 \boldsymbol{E} = \epsilon_0 \boldsymbol{E} + \boldsymbol{P} = \epsilon_0 \boldsymbol{E} + \boldsymbol{\chi}_e \epsilon_0 \boldsymbol{E} & \text{(anisotropic material)}. \end{cases} \tag{3.49}$$

We speak of "dielectric polarization" if polarization $\boldsymbol{P}$ and electric field intensity $\boldsymbol{E}$ obey the linear relation (3.48). We especially take notice that $\boldsymbol{P}$ vanishes if $\boldsymbol{E}$ vanishes. We speak of "paraelectric polarization" if polarization $\boldsymbol{P}$ and electric field intensity $\boldsymbol{E}$ are connected by a relative permittivity which depends on $\boldsymbol{E}$ in such a way that a nonlinear graph of the type shown in Figure 3.24 is generated. We especially take notice that $\boldsymbol{P}$ vanishes if $\boldsymbol{E}$ vanishes. We speak of "ferroelectric polarization" if polarization $\boldsymbol{P}$ and electric field intensity $\boldsymbol{E}$ are connected by a relative permittivity which depends on $\boldsymbol{E}$ in such a way that a nonlinear graph of the type shown in Figure 3.25 is generated. We especially take notice that a polarization remains even when the electric field is switched off, and if we apply a reverse electric field, the polarization is finally inverted, gradually running through a characteristic hysteresis curve. (i) The reader should know that a critical temperature, the *Curie temperature* T_C, is observable, which leads from ferroelectric states to paraelectric states. (ii) The reader should also know that the ferroelectric effect is evoked by well-ordered electric dipoles. (iii) The reader should also know that the electric dipoles of a sintered ceramic material form *Weiss domains*, i.e. territories delimiting sets of well-ordered electric dipoles, but the orientations of the polar axes of the Weiss domains usually are randomly distributed so that the ferroelectric effect usually is not observable. However, heating up the sintered ceramic material, then applying a static electric field to the sintered ceramic material, and then cooling down the sintered ceramic material, the Weiss domains are aligned, in last consequence, leading to a polar axis and evoking the ferroelectric effect. (iv) The reader should further know that "ferroelectricity" is named in analogy to "ferromagnetism", i.e. ferromagnetic matter states are associated with permanent magnetic moments, whereas ferroelectric matter states are associated with permanent electric moments. Straightforwardly, similar statements apply for the vocables "dielectric/diamagnetic" and the vocables "paraelectric/paramagnetic".

The reader should know that ferroelectric materials certainly show the properties pyroelectricity/piezoelectricity, but not all pyroelectric/piezoelectric materials show the property ferroelectricity. It is well-known that if the application of a mechanical force to a material evokes a current, the material is called *piezoelectric*. It is well-known, too, that if the change of the temperature of a material evokes a current, the material is called *pyroelectric*. We here should note that compressing or uncompressing a piezoelectric material, the charge centers Q^+ and Q^- of unit cells are separated, leading to electric dipoles Q^+–Q^-, eventually generating an electric potential associated with an electric tension. Contrariwise, applying an electric field instead, elongations and shortenings of the piezoelectric material are observable. We here should note that the pyroelectric effect, on the one hand, is a consequence of the lattice reorganizations due to temperature changes, and on the other hand, is a consequence of the alterations of the volume to charge ratio due to temperature changes. Contrariwise, applying an electric field instead, heatings and coolings of the pyroelectric material are observable. Important applications are to be found in the field of communications technology. Concentrating on ceramic materials, on the one hand, we quote that (piezoelectric) α-quartz plates are applied as clock generators, and on the other hand, we quote that (piezoelectric) lead zirconium titanate ("PZT materials") in tremendous amounts worldwide is used for sensors and actuators.

3.2 Physical Properties   221

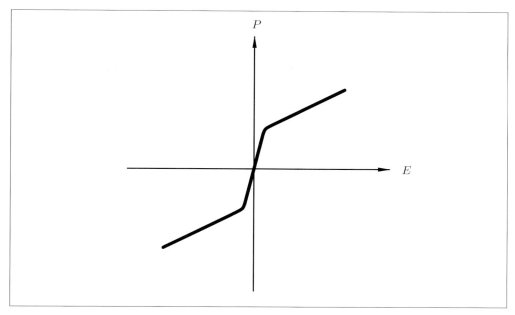

Figure 3.24. Paraelectricity (typical graph also for paramagnetism).

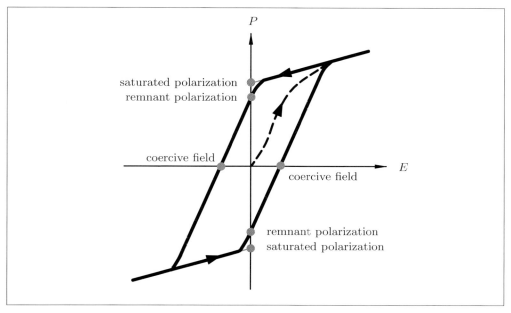

Figure 3.25. Ferroelectricity (typical graph also for ferromagnetism).

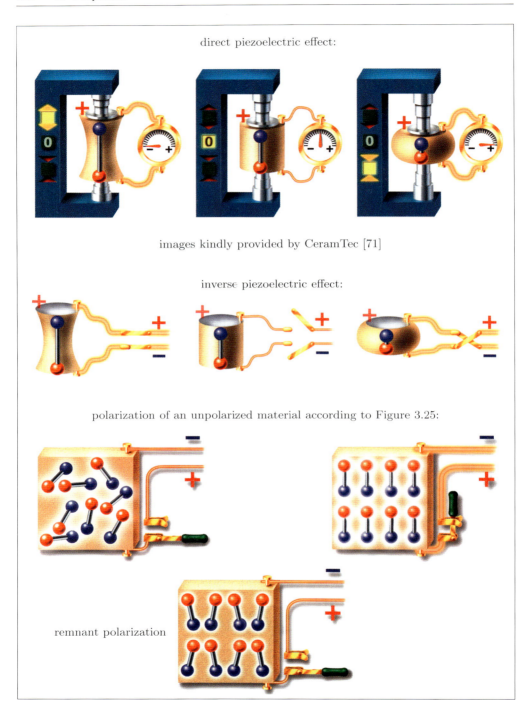

Figure 3.26. Piezoelectricity. Piezoelectric effect.

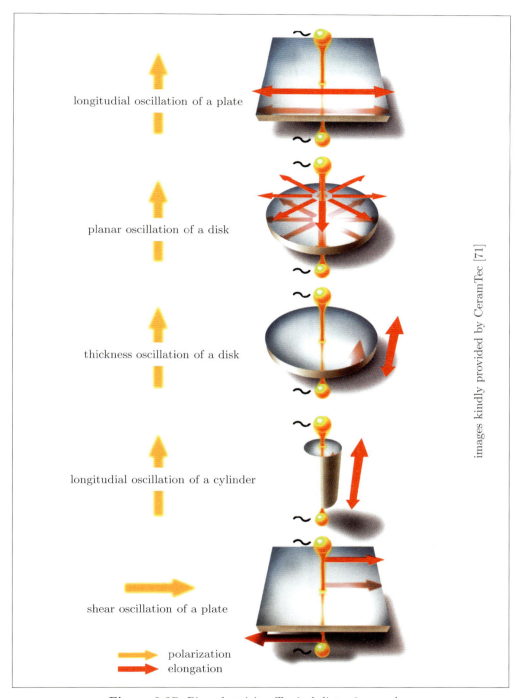

Figure 3.27. Piezoelectricity. Typical distortion modes.

Table 3.3. Relative permittivity, dielectric ceramics ($T = 298$ K, $\nu = 10^6$ Hz).

ceramic material	ϵ_r	ceramic material	ϵ_r	ceramic material	ϵ_r
TiO_2	100	porcelain	6	SiO_2	4
Al_2O_3	10	diamond	4		

In Table 3.3, the relative permittivities ϵ_r for a few selected dielectric ceramics at a special temperature and a special frequency of the electric field intensity are collected. In Table 3.4, the relative permittivities ϵ_r for a few selected ferroelectric ceramics near the Curie temperature, where the relative permittivities ϵ_r show relatively high values, are collected. In Box 3.11, areas of application of ceramic insulators, semi-conductors, and conductors are listed. Regarding Box 3.11, the following remarks should be made. (i) A thermistor is a resistor with a resistance showing an intended dependency on the temperature. Naturally, there are positive temperature coefficient (PTC) thermistors, i.e. an increasing temperature accounts for an increasing resistance. Naturally, there are negative temperature coefficient (NTC) thermistors, i.e. an increasing temperature accounts for a decreasing resistance. (ii) A varistor is a resistor with a resistance showing an abrupt reduction of its resistance above a specific threshold voltage. We annotate that varistors are abbreviated as VDR (voltage dependent resistor). We annotate, too, that metal oxide varistors are abbreviated as MOV (metal oxide varistor).

Going beyond it, for the reader it might be interesting to know that λ sensors applied in controlled catalytic converters are constructed either as resistance sensors or as Nernst sensors, in the latter case, applying ZrO_2 with Y_2O_3 as solid electrolyte, measuring the residual oxygen content of the exhaust emission with respect to an oxygen reference – a technique that is used in cars to control ignition in order to reduce gas consumption and air pollution. Going beyond it, for the reader it might be interesting to know, too, that $SrTiO_3$ formed of oxygen isotops ^{16}O (STO16) is a paraelectric substance starting from the absolute zero point 0 K, whereas $SrTiO_3$ formed of oxygen isotops ^{18}O (STO18) is a ferroelectric substance starting from the absolute zero point 0 K, but after exceeding a Curie temperature of 30 K becomes a paraelectric substance, showing us that the alteration of subtle details of ceramics can cause drastic changes of their electric properties.

Table 3.4. Relative permittivity, ferroelectric ceramics ($T \approx T_C$, T_C = Curie temperature).

ferroelectric perovskites	T_C [K]	$\epsilon_r(T \approx T_C)$	ferroelectric salts	T_C [K]	$\epsilon_r(T \approx T_C)$
$BaTiO_3$	401	14200	triglycinsulfate	323	400000
$PbTiO_3$	763	9300	K–H_2–phosphate	215	50000
$LiZrO_3$	1483	> 200	Na–K–tartrate	297	3000

Box 3.11 (Important data: electric properties).

Some areas of application of ceramic insulators:

high voltage insulators, ignition plugs, substrates for electronic circuits
(Al_2O_3, MgO, BeO, ZrO_2, AlN, BN, Si_3N_4, SiO_2 etc.),

piezoelectric elements
($BaTiO_3$, $Pb(Ti, Zr)O_3$ ("PZT materials") etc.,
we also compare with Figures 3.26, 3.27, 3.28, and 3.29).

A collection of important devices:
piezoelectric vibrators, ultrasonic transducers, surface acoustic wave devices,
piezoelectric transformers, piezoelectric actuators, capacitors,
we also compare with Figures 3.32 and 3.33.

Remarks:
steatite and quartz glass even at extremely high temperatures
are perfect insulators; this is not the case for Al_2O_3, SiO_2, BN etc.;
for example, for Al_2O_3, we measure $\rho = 10^{14}\,\Omega\,\mathrm{cm}$ at room temperature,
but $\rho = 10^5 - 10^7\,\Omega\,\mathrm{cm}$ at $1000°C$.

Some areas of application of ceramic semi-conductors:

thermistors (doped $BaTiO_3$ etc., we also compare with Figure 3.30),
varistors (doped ZnO etc., we also compare with Figure 3.31),
gas sensors (SnO_2 etc.).

Remarks:
there are type n semi-conductors and type p semi-conductors;
an example of a type n semi-conductor, i.e. electron conduction takes place,
is SiC doped by N at Si positions;
an example of a type p semi-conductor, i.e. hole conduction takes place,
is SiC doped by Al at Si positions.

Some areas of application of ceramic conductors:

heating conductors
(SiC, $MoSi_2$ etc.),

solid electrolytes for sensors
(ZrO_2 with Y_2O_3, ZrO_2 with MgO etc.),

superconducting devices
($YBa_2Cu_3O_{7-\delta}$, $Bi_2Sr_2Ca_2Cu_3O_{10}$, $HgBa_2Ca_2Cu_3O_{8+\delta}$, $Hg_{0.8}Tl_{0.2}Ba_2Ca_2Cu_3O_8$,
NbN–NbC etc.).

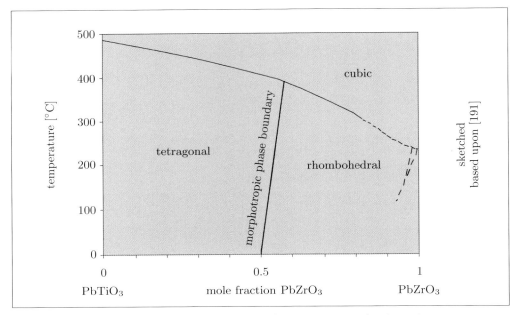

Figure 3.28. Lead zirconium titanate ("PZT materials"): phase diagram.

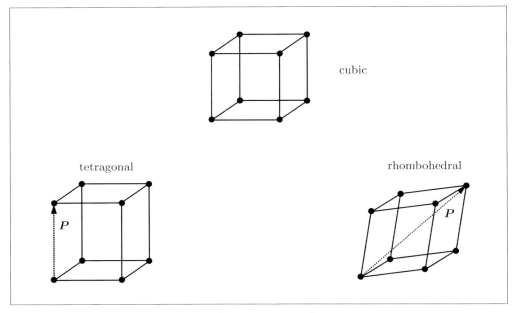

Figure 3.29. Lead zirconium titanate ("PZT materials"): vector of spontaneous polarization.

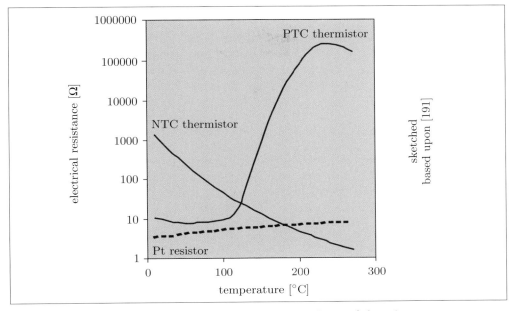

Figure 3.30. Typical temperature dependence of thermistors.

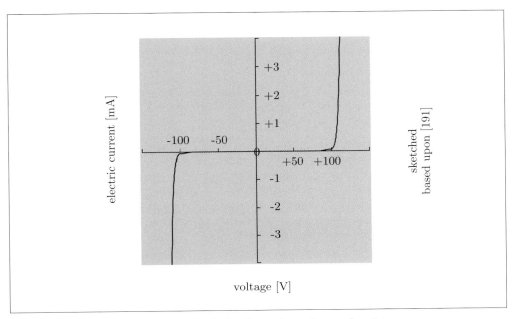

Figure 3.31. Typical voltage dependence of varistors.

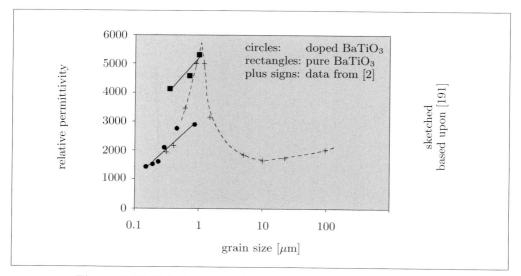

Figure 3.32. Relative permittivity: grain size dependence, BaTiO$_3$.

To the reader's delight, we also show images of ferroelectric (Weiss) domains in Figure 3.34. The following annotations should be made. The barium titanate (BaTiO$_3$) sample was created by cold isostatic pressing at 600 MPa and sintered at 1250°C in the space of 10 h in O$_2$ atmosphere. The image is a transmission optical micrograph, polarized illumination.

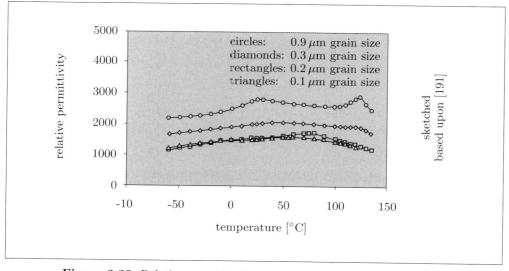

Figure 3.33. Relative permittivity: temperature dependence, BaTiO$_3$.

image of ferroelectric domains in barium titanate (BaTiO$_3$)

3000×

image taken from [90]

Figure 3.34. Ferrolectric domains, BaTiO$_3$.

The magnetic counterparts of *dielectricity*, *paraelectricity*, and *ferroelectricity* are called *diamagnetism*, *paramagnetism*, and *ferromagnetism*. We already know it, these magnetic counterparts, on the one hand, can be traced back to microscopic currents, and on the other hand, can be traced back to the magnetic dipole moments that are associated with the spins of unpaired electrons. (i) *Diamagnetism.* The reader should know that all materials show a diamagnetic polarization if an external magnetic field is applied. The reader should also know that a diamagnetic polarization is opposing to the applied external magnetic field, namely the magnetic susceptibility χ_m exhibits negative values and the relative permeability μ_r exhibits values less than one. The reader should also know that a diamagnetic material that is placed inside of a strong magnetic field is pushed out of the strong magnetic field. The reader should also know that the atomic electrons, responding to an applied external magnetic field according to Lenz's rule, which states that changes of the magnetic flux induce an electric tension generating an electric current evoking a magnetic field countervailing the changes of the magnetic flux, are responsible for the diamagnetic polarization. (ii) *Paramagnetism.* If only orbitals/bands populated with paired electrons exist, the material only shows a diamagnetic polarization. If (additionally) orbitals/bands populated with unpaired electrons exist, the material (additionally) has a paramagnetic polarization, shadowing the diamagnetic polarization. We note that the magnetic dipole moments that are associated with the spins of unpaired electrons are aligned in an external magnetic field, evoking the paramagnetic polarization and reinforcing the external magnetic field. (iii) *Ferromagnetism.* Diamagnetism and paramagnetism are magnetic properties of atoms and of material entities formed of atoms. Ferromagnetism only is a magnetic property of solids. We note that magnetic moments of unpaired electrons, coupling together and forming Weiss domains, evoke the ferromagnetism, provided the magnetic moments of statistically distributed Weiss domains are aligned via an external magnetic field, and provided the temperature does not exceed the Curie temperature T_C, which leads from the territoy of ferromagnetic states to the territory of "paramagnetic" states. (iv) The reader here should consult Boxes 3.12 and 3.22, which put some of the most important magnetic aspects of materials into concrete terms. The reader here should take notice that diamagnetic magnetization ("diamagnetic polarization") is described by the linear expressions (3.48) and (3.49). The reader here should also take notice that the typical nonlinear curve of paraelectricity which is shown in Figure 3.24 gives way to the typical nonlinear curve of paramagnetism, whereas the typical hysteresis curve of ferroelectricity which is shown in Figure 3.25 gives way to the typical hysteresis curve of ferromagnetism, in both cases replacing P (polarization) by M (magnetization) and E (electric field intensity) by H (magnetic field intensity). (v) The reader additionally should have heard that *ferrimagnetism* means that oppositely oriented ensembles of magnetic moments (for example, placed at two different sub-lattices, for instance, one consisting of Fe^{2+} ions and one consisting of Fe^{3+} ions) build up the Weiss domains and show different values and/or show different population numbers. (vi) The reader additionally should have heard that *antiferromagnetism* means that differently oriented ensembles of magnetic moments build up the Weiss domains in such a way that the ensembles of magnetic moments compensate each other. Above a critical temperature, the *Néel temperature* T_N, an antiferromagnetic material becomes paramagnetic.

> **Box 3.12** (Important formulae: magnetic properties).
>
> H = magnetic field intensity, B = magnetic flux density (magnetic field),
> M = magnetization,
> χ_m = magnetic susceptibility,
> μ_0 = magnetic constant, μ_r = relative permeability, μ = permeability,
> $\boldsymbol{m}_\mathrm{m}$ = magnetic dipole moment, V = volume,
> C = Curie constant, T_C = Curie temperature.
>
> Magnetization:
> $$M = \frac{1}{V} \sum_i \boldsymbol{m}_{\mathrm{m},i} . \tag{3.50}$$
>
> Diamagnetic magnetization ("diamagnetic polarization"):
> $$M = \begin{cases} \chi_\mathrm{m} \boldsymbol{H} = (\mu_\mathrm{r} - 1)\boldsymbol{H} & \text{(isotropic material)}, \\ \chi_\mathrm{m} \boldsymbol{H} = (\mu_\mathrm{r} - 1)\boldsymbol{H} & \text{(anisotropic material)}, \end{cases} \tag{3.51}$$
>
> $$B = \begin{cases} \mu \boldsymbol{H} = \mu_\mathrm{r}\mu_0 \boldsymbol{H} = \mu_0 \boldsymbol{H} + \mu_0 \boldsymbol{M} = \mu_0 \boldsymbol{H} + \chi_\mathrm{m}\mu_0 \boldsymbol{H} & \text{(isotropic material)}, \\ \mu \boldsymbol{H} = \mu_\mathrm{r}\mu_0 \boldsymbol{H} = \mu_0 \boldsymbol{H} + \mu_0 \boldsymbol{M} = \mu_0 \boldsymbol{H} + \chi_\mathrm{m}\mu_0 \boldsymbol{H} & \text{(anisotropic material)}. \end{cases} \tag{3.52}$$
>
> Curie's law
> (materials with a paramagnetic phase):
> $$\chi_\mathrm{m} = \frac{C}{T} . \tag{3.53}$$
>
> Curie–Weiss law
> (materials with a ferromagnetic phase and a paramagnetic phase
> separated by the Curie temperature T_C, paramagnetic phase $T > T_\mathrm{C}$):
> $$\chi_\mathrm{m} = \frac{C}{T - T_\mathrm{C}} . \tag{3.54}$$
>
> Note that critical temperatures such as the Curie temperature T_C reflect phase transitions of crystal structures, microstructures, band structures etc.

Table 3.5. Spinel ferrites. Cation distribution.

sum formula	cation structure (tetrahedral (th) site/octahedral (oh) site/O_4)
Fe_3O_4	$Fe^{3+}[Fe^{2+}Fe^{3+}]O_4$
$CoFe_2O_4$	$Fe^{3+}[Co^{2+}Fe^{3+}]O_4$
$NiFe_2O_4$	$Fe^{3+}[Ni^{2+}Fe^{3+}]O_4$
$Ni_{0.8}Zn_{0.2}Fe_2O_4$	$[Zn_{0.2}^{2+}Fe_{0.8}^{3+}][Ni_{0.8}^{2+}Fe_{1.2}^{3+}]O_4$
$Ni_{0.6}Zn_{0.4}Fe_2O_4$	$[Zn_{0.4}^{2+}Fe_{0.6}^{3+}][Ni_{0.6}^{2+}Fe_{1.4}^{3+}]O_4$
$Ni_{0.4}Zn_{0.6}Fe_2O_4$	$[Zn_{0.6}^{2+}Fe_{0.4}^{3+}][Ni_{0.4}^{2+}Fe_{1.6}^{3+}]O_4$

To the reader's convenience, we here present details relating to magnetic properties of ceramic materials. (i) Ceramic materials of technological importance which show ferrimagnetic properties are provided by ferric oxides such as magnetite (Fe_3O_4) and YIG (yttrium iron garnet). (ii) Ceramic materials of technological importance which show ferromagnetic properties are provided by spinel ferrites such as $Fe^{3+}[Fe^{2+}Fe^{3+}]O_4$, $Fe^{3+}[Co^{2+}Fe^{3+}]O_4$, and $Fe^{3+}[Ni^{2+}Fe^{3+}]O_4$. Some details relating to spinel ferrites are collected in Table 3.5 and Table 3.6. (iii) We here note that the magnetization of a material causes an alteration of the length of the material which is proportional to the magnetization of the material ("piezomagnetic effect", "magnetostriction"). The constant of proportionality has positive or negative values. For example, for Fe_3O_4, it has a positive value, but for $CoFe_2O_4$ and $NiFe_2O_4$, it has a negative value each. (iv) We here also note that one distinguishes between "hard ferrites" and "soft ferrites", indicating high ("hard") coercivity/remanence or low ("soft") coercivity/remanence, respectively. Non-permanent ferrite magnets are made of "soft ferrites". Permanent ferrite magnets are made of "hard ferrites". For example, "soft ferrites" are used in transformers. For example, "hard ferrites", also called "ceramic magnets", are used in speakers and microphones. (v) We here further note that the relative permeability just like the relative permittivity depends on the grain size, the temperature, and the driving frequency. We compare with Figures 3.35 and 3.36.

Table 3.6. Spinel ferrites. Magnetic moments.

sum formula	th site/oh site	moment (total)	moment (observed)
Fe_3O_4	$5/(4+5)$	4	4.1
$CoFe_2O_4$	$5/(3+5)$	3	3.7
$NiFe_2O_4$	$5/(2+5)$	2	2.3
$Ni_{0.8}Zn_{0.2}Fe_2O_4$	$0.8 \cdot 5/(0.8 \cdot 2 + 1.2 \cdot 5)$	3.6	3.8
$Ni_{0.6}Zn_{0.4}Fe_2O_4$	$0.6 \cdot 5/(0.6 \cdot 2 + 1.4 \cdot 5)$	5.2	5.1
$Ni_{0.4}Zn_{0.6}Fe_2O_4$	$0.4 \cdot 5/(0.4 \cdot 2 + 1.6 \cdot 5)$	6.8	5.2

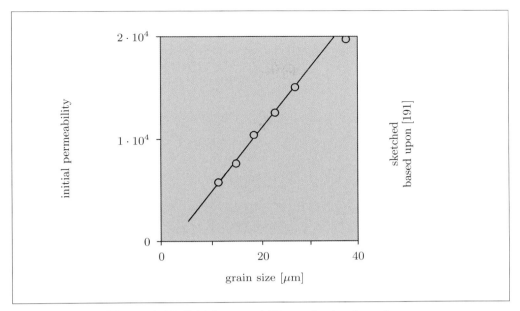

Figure 3.35. Initial permeability: grain size dependence.

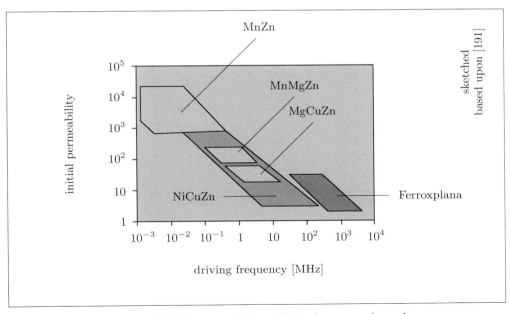

Figure 3.36. Initial permeability: driving frequency dependence.

From the basic point of view that is provided by conventional quantum mechanics, ferromagnetic and antiferromagnetic states are governed by the Hamiltonian (3.55), describing a special type of spin–spin interaction, in conventional quantum mechanics, circumscribed as spin exchange interaction. If the exchange parameters are positive, i.e. $J_{ij} > 0$, the ferromagnetic order develops. If the exchange parameters are negative, i.e. $J_{ij} < 0$, the antiferromagnetic order develops, which is immediately visualized as the superposition of antiparallely ordered, ferromagnetic sub-lattices. For example, for low Sr contents, the superconducting substance $La_{2-x}Sr_xCuO_4$ near $T=0$ exhibits antiferromagnetic order evoked by copper spins associated with unpaired electrons, we compare with Figures 5.14 and 5.15 presented in Section 5.1, and as it becomes clear from Figures 5.14 and 5.15, the antiferromagnetic order is immediately visualized as the superposition of antiparallely ordered, ferromagnetic sub-lattices. However, for a full discussion of the spin exchange interaction and the related spin interactions as applied in conventional quantum mechanics, we want to refer to [33, 36].

Applying the self-consistent network of model conceptions which is introduced in the framework of more far-reaching ideas in Chapter 5 already in advance at this point, we here quote the following advanced model for the spin exchange interaction. (i) (3.56) is a generalized energy term that covers any reference frames, while (3.57) is a generalized energy term that covers Cartesian reference frames, and both show the physical dimension of the wave vector square ($1/m^2$). (ii) (3.58) includes the term $\hbar^2/m$, in last consequence, evoked by a suitable specification of the energy matrix element of the energy momentum tensor that is at the bottom of such generalized energy terms, and thereby shows the physical dimension of the energy (J = Nm). (3.58) restricts itself to 3×3 matrices $\boldsymbol{\theta}^{\ni}$ with matrix elements γ^{ij} describing spin systems [63], which is outlined by the appendage "spin". (iii) We read off from (3.58) that a spin system implemented by $\boldsymbol{\theta}^{\ni}_{\text{spin}}$ with matrix elements $\gamma^{ij}_{\text{spin}}$ interacts with itself ($\gamma^{ij}_{\text{spin}} \leftrightarrow \gamma^{lk}_{\text{spin}}$) via a spin-specific field $\boldsymbol{\theta}^{\text{spin}}_{\ni}$ with matrix elements $\gamma^{\text{spin}}_{ij}$, which is the expression of the existence of the spin system. We also read off from (3.58) that the organization of the spin system, i.e. the "look and feel" of $\gamma^{ij}_{\text{spin}}$ and $\gamma^{\text{spin}}_{ij}$, depends on the organization of the crystal system, i.e. the "look and feel" of $\gamma_{00} = \psi$, which is the wavefunction of the electron system. We thus realize that it depends on the crystal system including constraints such as temperature and pressure whether or not long-range-type order is observable and whether this or that long-range-type order is observable.

From the advanced point of view that is provided by the self-consistent network of model conceptions which is introduced in the framework of more far-reaching ideas in Chapter 5, lower-index terms such as $\boldsymbol{\theta}^{\text{spin}}_{\ni}$ establish scalar, vectorial, "tensorial" fields the consequences of which are established by higher-index terms such as $\boldsymbol{\theta}^{\ni}_{\text{spin}}$, and the lower indices and higher indices point at their covariant origin and contravariant origin. We note that in the case of $\boldsymbol{\theta}^{\ni}_{\text{spin}}$, this is the spin property of particles, while in the case of $\boldsymbol{A}^{\ni}$, this is the precession property of spins in a magnetic field, see (5.4), and of particles in a magnetic field, see (5.3). We also note that spin exchange interactions go beyond magnetic field interactions of spins, which are too weak to be responsible for ferromagnetic and antiferromagnetic states, and advanced models for these can be easily established by departing from the generalized energy terms (5.88).

3.2 Physical Properties 235

Box 3.13 (Important formulae: spin exchange interaction).

Spin exchange interaction (conventional modeling):

$$\hat{H}_{\text{se}} = -\sum_{i,j} J_{ij} \boldsymbol{S}_i \boldsymbol{S}_j . \qquad (3.55)$$

$J_{ij} > 0$ (ferromagnetic order, "diagramed representation"):

↑ ↑ ↑ ↑
↑ ↑ ↑ ↑   "parallely ordered magnetic moments"
↑ ↑ ↑ ↑

$J_{ij} < 0$ (antiferromagnetic order, "diagramed representation"):

↑ ↓ ↑ ↓   ↑   ↑   ↓   ↓
↓ ↑ ↓ ↑ =   ↑   ↑   +   ↓   ↓   "antiparallely ordered magnetic moments"
↑ ↓ ↑ ↓   ↑   ↑   ↓   ↓

(superposition of antiparallely ordered, ferromagnetic sub-lattices).

Spin exchange interaction and much more [63], see (5.87)
(advanced modeling, any coordinates q^i, wave-vector-square representation):

$$\mathcal{S}_{00}^{(9,1)}(\boldsymbol{q}) = -\boldsymbol{X}^{(3)\,\text{T}} \hat{\boldsymbol{\mu}}_{\text{S},3} \gamma_{00} = -\frac{1}{4} \sum_{l=1}^{3} \left(\sum_{i,j,k=1}^{3} \gamma^{ij} \frac{\partial \gamma_{ij}}{\partial q^k} \gamma^{lk} \right) \frac{\partial}{\partial q^l} \gamma_{00} . \qquad (3.56)$$

Spin exchange interaction and much more [63], see (5.87)
(advanced modeling, Cartesian coordinates x^i, wave-vector-square representation):

$$\mathcal{S}_{00}^{(9,1)}(\boldsymbol{x}) = -\boldsymbol{X}^{(3)\,\text{T}} \hat{\boldsymbol{\mu}}_{\text{S},3} \gamma_{00} = -\frac{1}{4} \sum_{l=1}^{3} \left(\sum_{i,j,k=1}^{3} \gamma^{ij} \frac{\partial \gamma_{ij}}{\partial x^k} \gamma^{lk} \right) \frac{\partial}{\partial x^l} \gamma_{00} , \qquad (3.57)$$

$$\gamma_{00} = \psi(\boldsymbol{x}) .$$

Spin exchange interaction
(advanced modeling, Cartesian coordinates x^i, energy representation):

$$\hat{\mathcal{H}}_{\text{se}} \gamma_{00} = -\frac{\hbar^2}{4m} \sum_{l=1}^{3} \left(\sum_{i,j,k=1}^{3} \gamma_{\text{spin}}^{ij} \frac{\partial \gamma_{ij}^{\text{spin}}}{\partial x^k} \gamma_{\text{spin}}^{lk} \right) \frac{\partial}{\partial x^l} \gamma_{00} , \qquad (3.58)$$

$$\gamma_{00} = \psi(\boldsymbol{x}) .$$

3.2.3 Optical Properties

Meanwhile, we appreciate the relevance of laboratory glassware or fiber optic cables for the technology of nowadays. Therefore, we feel that it is necessary to supply the reader with some basics concerning optical properties of materials, in particular, of ceramic materials. The following remarks should fulfil this task. Due to interations of electromagnetic waves ("light") with matter, namely inside the object, at the surfaces of the object, and at the boundary surfaces of the object, electromagnetic waves ("light") are refracted, scattered, absorbed, and guided. (i) *Light refraction.* In Box 3.14, we present the refraction law describing how light is refracted if it crosses the boundary layer of two media. The refractive index ("refractivity"), on the one hand, increases with the number of ions per volume, and on the other hand, increases with the polarizability of the ions. (ii) *Light scattering.* Responsible for light scattering, in particular, are the roughness of the surface of the material, the grain boundaries in the material, inclusions in the material, pores in the material, and secondary phases. (iii) *Light absorption.* Responsible for light absorption, in particular, are free electrons, which interact with light of any frequency, bounded electrons, which react on visible frequencies depending on the electron states that are available, and atom groups, which react on infrared frequencies depending on the oscillation states that are available. (iv) *Light guidance.* It is no miracle anymore that light can be guided by fiber optic cables. There exist two variants, the glass variant and the plastic variant. We note that in the first case the fibers are made from fused silica, fluorozirconates, fluoroaluminates, or chalcogenides. We also note that the typical refractive index ("refractivity") for such materials is ≈ 1.5. We also note that fused silica fibers are used for long distance communication because of their ability to transmit light signals over large distances with little attenuation.

Table 3.7. Refractive index ("refractivity").

class	material	n	material	n	material	n
oxides	SiO_2	1.55	Al_2O_3	1.76	ZrO_2	2.20
	BeO	1.73	Y_2O_3	1.92	PbO	2.61
	MgO	1.74	Gd_2O_3	1.96	TiO_2	2.71
glasses	silicate glass	1.46	soda lime glass	1.51		
	Vycor glass	1.46	flint glass	1.70		
	Pyrex glass	1.47				
others	SiC	2.38				
	CdS	2.62				
	CaF_2	4.00				

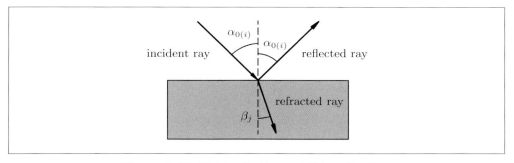

Figure 3.37. Light reflection and light refraction.

Box 3.14 (Important formulae: optical properties).

c_0 = light velocity in vacuum, λ_0 = wavelength of light in vacuum, c_j = light velocity in material j, λ_j = wavelength of light in material j, $\alpha_{0(i)}$ = angle of incidence, β_j = angle of refraction, ν = frequency.

Refractive index ("refractivity"):

$$n_j = \frac{c_0}{c_j} = \frac{\lambda_0}{\lambda_j} = \frac{\sin \alpha_0}{\sin \beta_j} = n \,. \tag{3.59}$$

Refraction law:

$$n_{ij} = \frac{n_j}{n_i} = \frac{c_i}{c_j} = \frac{\lambda_i}{\lambda_j} = \frac{\sin \alpha_i}{\sin \beta_j} \,. \tag{3.60}$$

Wavelength:

$$\lambda_j = \frac{c_j}{\nu} \,. \tag{3.61}$$

We also compare with Table 3.7 and Figures 3.37 and 3.38.

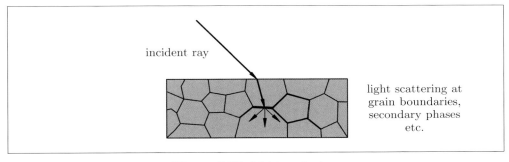

Figure 3.38. Light scattering.

3.3 Mechanical Properties

Speaking of *mechanical properties*, we mainly mean the elastic, plastic, and fracture properties of materials. Let us get a general idea of the main features of this topic with special respect to ceramics in the sections that follow.

3.3.1 Elastic Properties

As a reaction to tensile stress, compression stress, isostatic pressure, or shear stress, solid materials show reversible deformations termed *elasticity*, and these are the visible expression of the increase, the decrease, and the distortion of chemical bonds. Let us here have a brief look at Box 3.15, where a collection of important formulae making the property *elasticity* ascertainable is presented. We here especially point out that the modulus of elasticity according to (3.74) and (3.76) is a measure for the force that is needed to elongate or to compress a solid material by applying the force from one side, that the bulk modulus according to (3.80) is a measure for the force that is needed to compress a solid material by applying the force uniformly from all sides, and that the shear modulus according to (3.81) is a measure for the force that is needed to distort a solid material. We here also point out that the values for the E moduli of ceramic materials that show strong covalent bond contributions and/or strong ionic bond contributions typically lie between 100–1000 GPa, by far exceeding the typical values of materials with strong metallic bond contributions and the typical values of polymers. We here also point out that the formal structure of the E moduli of ceramic materials explicitly reflects the heterogeneous structure of ceramic materials. In particular, the E moduli of ceramic materials are functions of the pore fraction, and this is modeled by (3.83), provided the pore fraction is relatively low. In particular, the E moduli of ceramic materials can be functions of time, for example, reflecting the softening of glass phases.

Let us here point out that stresses σ_{ij} and strains ϵ_{kl} in most practical cases fulfil Hook's law (3.67). In the special case of isotropic materials, this linear physical law reduces to (3.68) provided we want to use the elastic constants λ and μ and reduces to (3.69) provided we want to use the elastic constants K and μ. In the special case of tensile stress or compression stress in one direction, the resulting lateral (transversal) contraction or expansion of the isotropic materials is described by Poisson's ratio ν, offering us the specification (3.70) of Hook's law. Let us here additionally note that stresses σ_{ij} and strains ϵ_{kl} and thus also Hook's law and its specifications in general depend on the position vector $\boldsymbol{x}$ and the time coordinate t. Let us here additionally note that this does not mean that stresses σ_{ij} and strains ϵ_{kl} and thus also Hook's law and its specifications do not include deformations on a macroscopic level of consideration. Of course, they do include deformations on a macroscopic level of consideration. Some examples of practical importance are collected in Box 3.15, namely tensile stress and compression stress applied to a cylindric rod and isostatic pressure applied to a cube as well as shear stress applied to a cube.

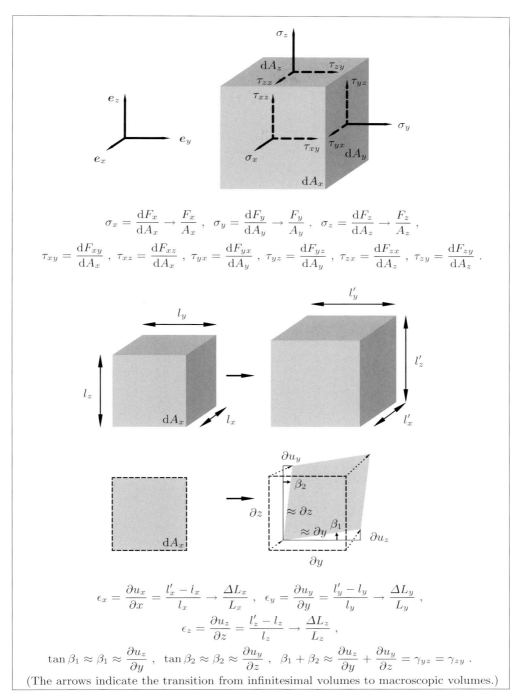

Figure 3.39. Stresses and strains.

> **Box 3.15** (Important formulae: stresses and strains).
>
> $i := 1 \equiv x$, $i := 2 \equiv y$, $i := 3 \equiv z$, $x_1 = x$, $x_2 = y$, $x_3 = z$,
> dA_i = infinitesimal cross-sectional area,
> dF_i = force normal to the infinitesimal cross-sectional area dA_i,
> dF_{ij} = force in direction j on the infinitesimal cross-sectional area dA_i,
> u_i = shift of a point under stress in direction i,
> σ_{ij} = stress tensor, ϵ_{kl} = strain tensor, C_{ijkl} = stiffness tensor,
> λ = Lamé's first parameter, $\mu = G$ = Lamé's second parameter or shear modulus,
> K = bulk modulus, E = Young's modulus or modulus of elasticity,
> ν = Poisson's ratio, δ_{ij} = Kronecker's delta.
>
> Stress tensor (compare with Figure 3.39, top):
>
> $$\{\sigma_{ij}\} = \begin{bmatrix} \sigma_{11} & \sigma_{12} & \sigma_{13} \\ \sigma_{21} & \sigma_{22} & \sigma_{23} \\ \sigma_{31} & \sigma_{32} & \sigma_{33} \end{bmatrix} = \begin{bmatrix} \sigma_{xx} & \sigma_{xy} & \sigma_{xz} \\ \sigma_{yx} & \sigma_{yy} & \sigma_{yz} \\ \sigma_{zx} & \sigma_{zy} & \sigma_{zz} \end{bmatrix} = \begin{bmatrix} \sigma_x & \tau_{xy} & \tau_{xz} \\ \tau_{yx} & \sigma_y & \tau_{yz} \\ \tau_{zx} & \tau_{zy} & \sigma_z \end{bmatrix} . \tag{3.62}$$
>
> Strain tensor (compare with Figure 3.39, bottom):
>
> $$\{\epsilon_{ij}\} = \begin{bmatrix} \epsilon_{11} & \epsilon_{12} & \epsilon_{13} \\ \epsilon_{21} & \epsilon_{22} & \epsilon_{23} \\ \epsilon_{31} & \epsilon_{32} & \epsilon_{33} \end{bmatrix} = \begin{bmatrix} \epsilon_{xx} & \epsilon_{xy} & \epsilon_{xz} \\ \epsilon_{yx} & \epsilon_{yy} & \epsilon_{yz} \\ \epsilon_{zx} & \epsilon_{zy} & \epsilon_{zz} \end{bmatrix} = \begin{bmatrix} \epsilon_x & \gamma_{xy}/2 & \gamma_{xz}/2 \\ \gamma_{yx}/2 & \epsilon_y & \gamma_{yz}/2 \\ \gamma_{zx}/2 & \gamma_{zy}/2 & \epsilon_z \end{bmatrix} \tag{3.63}$$
>
> (symmetric tensor $\rightarrow$ diagonalizable) .
>
> Relatively large infinitesimal shifts:
>
> $$\epsilon_{jk} = \frac{1}{2}\left(\frac{\partial u_j}{\partial x_k} + \frac{\partial u_k}{\partial x_j} - \sum_{i=1}^{3} \frac{\partial u_i}{\partial x_j}\frac{\partial u_i}{\partial x_k}\right) . \tag{3.64}$$
>
> Relatively small infinitesimal shifts:
>
> $$\epsilon_{jk} = \frac{1}{2}\left(\frac{\partial u_j}{\partial x_k} + \frac{\partial u_k}{\partial x_j}\right) , \tag{3.65}$$
>
> $$\begin{aligned} \epsilon_{xx} = \epsilon_x = \frac{\partial u_x}{\partial x} \, , \quad \epsilon_{yy} = \epsilon_y = \frac{\partial u_y}{\partial y} \, , \quad \epsilon_{zz} = \epsilon_z = \frac{\partial u_z}{\partial z} \, , \\ \epsilon_{xy} = \epsilon_{yx} = \frac{\gamma_{xy}}{2} = \frac{\gamma_{yx}}{2} = \frac{1}{2}\left(\frac{\partial u_x}{\partial y} + \frac{\partial u_y}{\partial x}\right) , \\ \epsilon_{yz} = \epsilon_{zy} = \frac{\gamma_{yz}}{2} = \frac{\gamma_{zy}}{2} = \frac{1}{2}\left(\frac{\partial u_y}{\partial z} + \frac{\partial u_z}{\partial y}\right) , \\ \epsilon_{xz} = \epsilon_{zx} = \frac{\gamma_{xz}}{2} = \frac{\gamma_{zx}}{2} = \frac{1}{2}\left(\frac{\partial u_z}{\partial x} + \frac{\partial u_x}{\partial z}\right) . \end{aligned} \tag{3.66}$$

Continuation of Box.

Hook's law
(the stiffness tensor C_{ijkl} summarizes 81 matrix elements,
these can be expressed by 21 independent elastic constants):

$$\sigma_{ij} = \sum_{kl=1}^{3} C_{ijkl}\epsilon_{kl} \ . \tag{3.67}$$

Hook's law for the special case of a homogeneous isotropic solid,
(only 2 elastic constants occur), for λ and μ:

$$\sigma_{ij} = \lambda \sum_{k=1}^{3} \epsilon_{kk}\delta_{ij} + 2\mu\epsilon_{ij} \ , \quad \sum_{k=1}^{3} \epsilon_{kk} = \mathrm{Tr}\,[\epsilon_{ij}] \ . \tag{3.68}$$

Hook's law for the special case of a homogeneous isotropic solid
(only 2 elastic constants occur), for K and μ:

$$\sigma_{ij} = K \sum_{k=1}^{3} \epsilon_{kk}\delta_{ij} + 2\mu \sum_{k=1}^{3} \left(\epsilon_{ij} - \frac{1}{3}\epsilon_{kk}\delta_{ij}\right) \ ,$$

$$K = \lambda + \frac{2}{3}\mu \ . \tag{3.69}$$

Hook's law for the special case of a homogeneous isotropic solid, stress in one direction
(only 2 elastic constants occur), for E and μ:

$$\epsilon_{ij} = \frac{1}{2\mu}\sigma_{ij} - \frac{\nu}{E} \sum_{k=1}^{3} \sigma_{kk}\delta_{ij} \ , \tag{3.70}$$

$$\epsilon_{11} = \frac{1}{E}\Big[\sigma_{11} - \nu\big(\sigma_{22} + \sigma_{33}\big)\Big] \ , \quad \epsilon_{22} = \frac{1}{E}\Big[\sigma_{22} - \nu\big(\sigma_{33} + \sigma_{11}\big)\Big] \ ,$$

$$\epsilon_{33} = \frac{1}{E}\Big[\sigma_{33} - \nu\big(\sigma_{11} + \sigma_{22}\big)\Big] \ ,$$

$$\epsilon_{12} = \frac{1}{2\mu}\sigma_{12} \ , \quad \epsilon_{13} = \frac{1}{2\mu}\sigma_{13} \ , \quad \epsilon_{23} = \frac{1}{2\mu}\sigma_{23} \ . \tag{3.71}$$

Relations between elastic constants and Poisson's ratio
$\left(\text{stress in } 1 \equiv x \ (2 \equiv y) \text{ direction: } \nu = -\frac{\epsilon_{22}}{\epsilon_{11}} = -\frac{\epsilon_{33}}{\epsilon_{11}} \ \left(\nu = -\frac{\epsilon_{11}}{\epsilon_{22}} = -\frac{\epsilon_{33}}{\epsilon_{22}}\right)\right)$:

$$E = 2\mu(1+\nu) = 3K(1-2\nu) = \frac{\lambda(1+\nu)(1-2\nu)}{\nu} \ , \tag{3.72}$$

$$\nu = \frac{\lambda}{2(\lambda+\mu)} \ . \tag{3.73}$$

Continuation of Box.

Example: tensile stress in y direction of a cylindric rod ($\sigma_{ij} = 0$ for $i \neq j$):

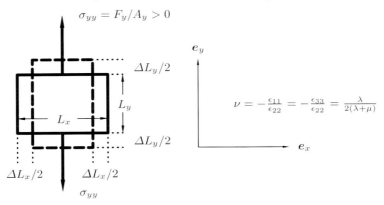

$$\sigma_{yy} = F_y/A_y > 0$$

$$\nu = -\frac{\epsilon_{11}}{\epsilon_{22}} = -\frac{\epsilon_{33}}{\epsilon_{22}} = \frac{\lambda}{2(\lambda+\mu)}$$

$$(3.70) \stackrel{\sigma_{11}=\sigma_{33}=0}{\rightarrow} \quad \epsilon_{22} = +\frac{1}{E}\sigma_{22} > 0 \,, \quad \epsilon_{11} = -\frac{\nu}{E}\sigma_{22} < 0 \,, \quad \epsilon_{33} = -\frac{\nu}{E}\sigma_{22} < 0 \,, (3.74)$$

$$\epsilon_{22} = \frac{\Delta L_y}{L_y} > 0 \,, \quad \epsilon_{11} = \frac{\Delta L_x}{L_x} < 0 \,, \quad \epsilon_{33} = \frac{\Delta L_z}{L_z} < 0 \,. \tag{3.75}$$

Example: compression stress in y direction of a cylindric rod ($\sigma_{ij} = 0$ for $i \neq j$):

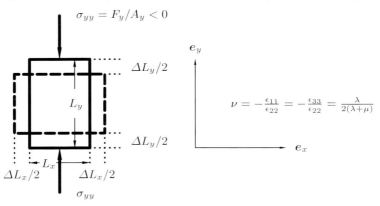

$$\sigma_{yy} = F_y/A_y < 0$$

$$\nu = -\frac{\epsilon_{11}}{\epsilon_{22}} = -\frac{\epsilon_{33}}{\epsilon_{22}} = \frac{\lambda}{2(\lambda+\mu)}$$

$$(3.70) \stackrel{\sigma_{11}=\sigma_{33}=0}{\rightarrow} \quad \epsilon_{22} = +\frac{1}{E}\sigma_{22} < 0 \,, \quad \epsilon_{11} = -\frac{\nu}{E}\sigma_{22} > 0 \,, \quad \epsilon_{33} = -\frac{\nu}{E}\sigma_{22} > 0 \,, (3.76)$$

$$\epsilon_{22} = \frac{\Delta L_y}{L_y} < 0 \,, \quad \epsilon_{11} = \frac{\Delta L_x}{L_x} > 0 \,, \quad \epsilon_{33} = \frac{\Delta L_z}{L_z} > 0 \,. \tag{3.77}$$

Continuation of Box.

Example: isostatic pressure p on a cube ($\sigma_{ij} = 0$ for $i \neq j$):

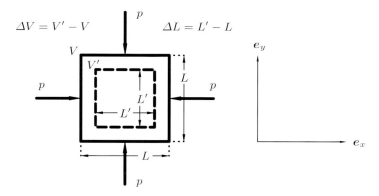

$$(3.68) \quad \overset{\epsilon_{11}=\epsilon_{22}=\epsilon_{33}=\Delta L/L}{\rightarrow} \quad \sigma_{11} = \sigma_{22} = \sigma_{33} = (2\mu + 3\lambda)\frac{\Delta L}{L} = 3K\frac{\Delta L}{L}, \quad (3.78)$$

$$\frac{\Delta V}{V} = \frac{(L+\Delta L)^3 - L^3}{L^3} = 3\frac{\Delta L}{L} + O(\Delta L), \quad (3.79)$$

$$(3.78) + (3.79) \quad \overset{|\Delta L|\ll 1, \sigma_{11}=\sigma_{22}=\sigma_{33}=-p}{\rightarrow} \quad K\frac{\Delta V}{V} = -p. \quad (3.80)$$

Example: shear stress τ_{yx} on a cube ($\sigma_{ij} = 0$ for $i = j$):

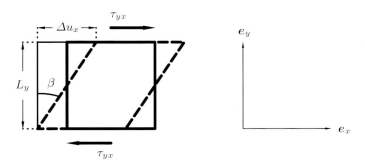

$$(3.68) \quad \overset{\sigma_{13}=\sigma_{31}=\sigma_{23}=\sigma_{32}=0}{\rightarrow} \quad \sigma_{21} = 2\mu\epsilon_{21} = 2\mu\epsilon_{12} = \mu\gamma_{yx} = \mu\gamma_{xy} = \tau_{yx}, \quad (3.81)$$

$$\gamma_{yx} = \gamma_{xy} = \frac{\Delta u_x}{\Delta y} = \frac{\Delta u_x}{L_y} = \tan\beta \approx \beta. \quad (3.82)$$

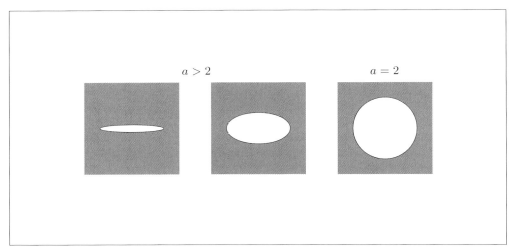

Figure 3.40. Idealized pore geometries.

Continuation of Box.

E moduli and pore fraction, low porosity, compare with Figures 3.40 and 3.41 ($E_0 = E$ modulus without pores, P = pore fraction):

$$E = E_0(1 - aP) \tag{3.83}$$

($a = 2$ for spherical pores, $a > 2$ for discoid pores).

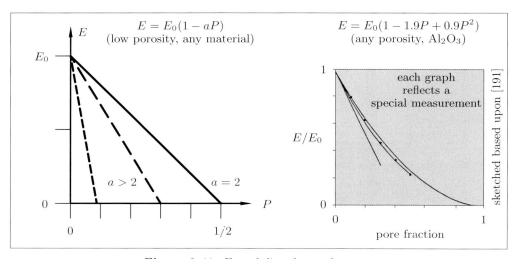

Figure 3.41. E moduli and pore fraction.

> **Mathematical post-it. Simplifications for the practitioner.**

Modulus of elasticity E as ratio of
stress σ versus strain ϵ:

$$\sigma = E\epsilon \, . \tag{3.84}$$

Metals: $E \approx \dfrac{1}{3}$. Ceramics: $E \approx \dfrac{1}{5} - \dfrac{1}{4}$.

Shear modulus $\mu = G$ as ratio of
shear stress τ versus shear strain γ:

$$\tau = \mu\gamma = G\gamma \, . \tag{3.85}$$

$\mu = G = \dfrac{E}{2(1+\nu)}$. Metals: $\mu = G \approx \dfrac{3}{8}E$. Ceramics: $\mu = G \approx \dfrac{5}{12}E$.

Bulk modulus K as ratio of
isostatic pressure p versus volume change $\Delta = \Delta V/V$:

$$p = -K\Delta = -K\dfrac{\Delta V}{V} \, . \tag{3.86}$$

$K = \dfrac{E}{3(1-2\nu)}$. Metals: $K \approx E$. Ceramics: $K \approx \dfrac{5}{9}E$.

Poisson's ratio ν as ratio of
strain ϵ_{xx} (x direction) versus strain ϵ_{yy} (y direction):

$$\nu = \dfrac{\epsilon_{xx}}{\epsilon_{yy}} \, . \tag{3.87}$$

Bond stiffness S and modulus of elasticity E
(the diverse quantities are explained via Figures 3.43 and 3.42):

$$S = \dfrac{d^2 U}{dr^2} = \dfrac{dF}{dr} \, , \quad S_0 = \left.\dfrac{d^2 U}{dr^2}\right|_{r_0} = \left.\dfrac{dF}{dr}\right|_{r_0} \, , \tag{3.88}$$

$$E \approx \dfrac{S_0}{r_0}$$

due to the approximation

$F \approx S_0(r - r_0) \Rightarrow \sigma = \dfrac{F}{A} = \dfrac{N}{A} S_0(r - r_0), \ \dfrac{N}{A} \approx \dfrac{1}{r_0^2} \Rightarrow \sigma = \dfrac{S_0}{r_0}\dfrac{r-r_0}{r_0} = E\epsilon \Rightarrow E \approx \dfrac{S_0}{r_0}$.

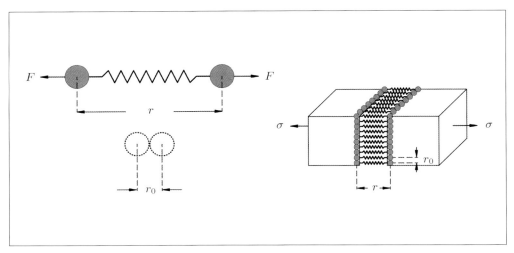

Figure 3.42. Estimation of the modulus of elasticity E: spring model.

The mathematical post-it that is attached above shows some simplifications the knowledge of which should be benefical for the practitioner. We here point at the estimation of the modulus of elasticity E according to $E \approx S_0/r_0$, which is based upon the spring model shown in Figure 3.42, generating a potential curve and a force curve of the type shown in Figure 3.43 [73]. We realize that this estimation assumes only small elongations out of the equilibrium distance r_0. We further realize that N/A together with $N/A \approx 1/r_0^2$ implements the number of bonds N per unit area A. Hence, if the potential curve/force curve or if at least r_0 and the bond stiffness S_0 at r_0 are known, an estimation of the modulus of elasticity E is possible.

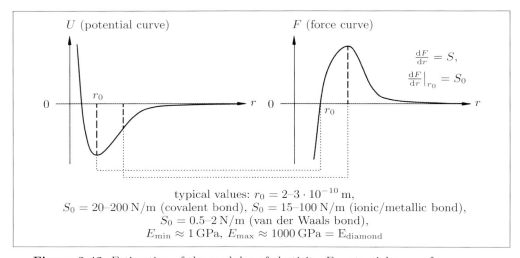

Figure 3.43. Estimation of the modulus of elasticity E: potential curve, force curve.

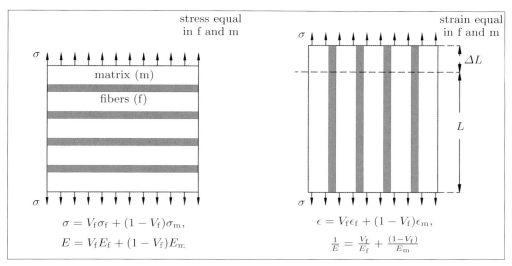

Figure 3.44. Modulus of elasticity E of composite materials: examples of rules of mixture.

The mathematical post-it that is attached above does not supply us with rules for the computation of the modulus of elasticity E of composite materials. We here thus point at Figure 3.44, which shows examples of rules of mixture, and Figure 3.45, which shows the domain of rules of mixture within which most composite materials reside [73]. We note that many other, also much more complicated rules of mixture are needed to cover composite materials.

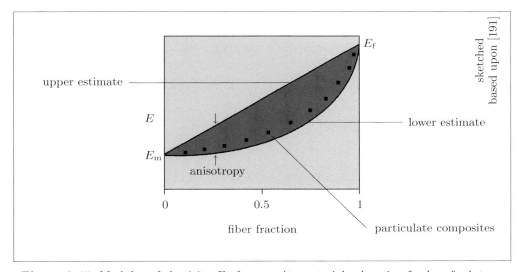

Figure 3.45. Modulus of elasticity E of composite materials: domain of rules of mixture.

3.3.2 Plastic Properties

Increasing the tensile stress, the compression stress, the isostatic pressure, or the shear stress, in each case starting from a critical point associated with a critical stress, solid materials finally leave the domain of *elasticity* and enter the domain of *plasticity*, i.e. a transition from reversible deformations to irreversible deformations, ending with fracture, takes place. This is fundamentally also true for ceramics although they are eventually mostly considered as brittle. For example, we compare with Figure 3.46, which exhibits a measurement result which illustrates the deflection of MgO under stress. In particular, we directly read off from Figure 3.46 that a region of plasticity (which is characterized by high deflection increase versus low stress increase) emerges after a region of elasticity (which is characterized by low deflection increase versus high stress increase), finally ending with fracture.

The origins of plasticity are dislocation (formation via gliding planes), twinning (formation by switching of crystal axes), phase transformation, creep, and viscous flow in the case of glasses and glass phases. (i) Observing monocrystals, strongly depending on the type of stress, in particular, this means that parts of the materials are displaced along gliding planes, that edge dislocations are switched along gliding planes, that twinnings leading to permanent changes of the crystal structure take place, or that phase transformations leading to permanent changes of the crystal structure take place. We compare with Figures 3.47–3.52, which illustrate gliding planes, gliding processes, and twinnings in a monocrystal. (ii) Observing materials showing grain structures, i.e. polycrystals, in particular, this means that parts of the materials are displaced along grain boundaries. Such grain boundary sliding processes always occur combined with grain rearrangements and grain deformations, in particular, evoked by dislocation, twinning, and phase transformation. We compare with Figure 3.63, which illustrates grain boundary sliding processes in connection with a superplastic material such as Si_3N_4 or ZrO_2 at high temperatures. (iii) We additionally note that there are usually more than one gliding (slip) system for a special material. For example, we read off from Table 3.8 that two indpendent gliding (slip) systems at low temperatures and five indpendent gliding (slip) systems at high temperatures are observed for MgO, in the latter case, representing a prerequisite for a "real" three-dimensional deformation, which is also known as the *criterion of von Mises*.

The resistance against plastic deformation is called *hardness*. The resistance against plastic deformation, for example, is measured by Brinell indentations (Brinell hardness) or Vickers indentations (Vickers hardness). For materials with high hardness such as most ceramic materials, the Vickers method is preferable because it is based upon a pyramidal indenter, allowing the imprinting of indentations in the hardest material. We compare with Figure 3.53, which illustrates the Vickers method, with Figure 3.54, which illustrates that a Vickers indentation frequently evokes crack growth around the Vickers indentation, and the table presented in Figure 3.53, collceting some examples of HV (Vickers hardness) values of ceramic materials.

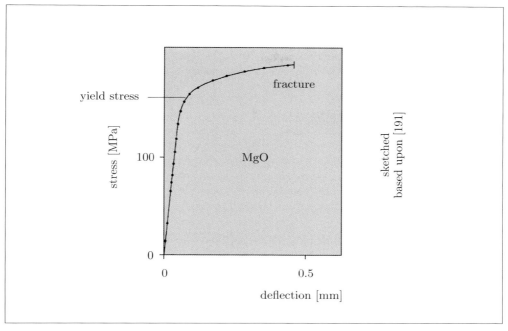

Figure 3.46. Plastic deformation: deflection.

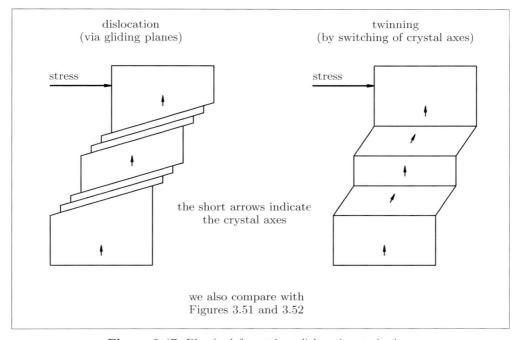

Figure 3.47. Plastic deformation: dislocation, twinning.

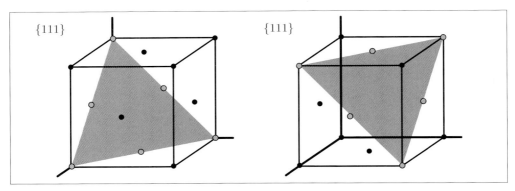

Figure 3.48. Cubic face-centered (cfc) lattices. The {111} gliding planes.

Table 3.8. Examples of gliding (slip) systems.

ceramic material	gliding system	number of independent systems	temperatures
diamond, Si, Ge	{111} $\langle 1\bar{1}0 \rangle$	5	$T > 0.5\,T_\mathrm{m}$
NaCl, LiF, MgO, NaF	{110} $\langle 1\bar{1}0 \rangle$	2	$T = $ low
NaCl, LiF, MgO, NaF	{110} $\langle 1\bar{1}0 \rangle$ {001} $\langle 1\bar{1}0 \rangle$ {111} $\langle 1\bar{1}0 \rangle$	5	$T = $ high
CaF$_2$, UO$_2$	{001} $\langle 1\bar{1}0 \rangle$	3	$T = $ low
CaF$_2$, UO$_2$	{001} $\langle 1\bar{1}0 \rangle$ {110} {111}	5	$T = $ high

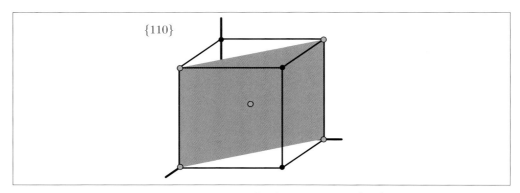

Figure 3.49. Cubic body-centered (cbc) lattices. The {110} gliding plane.

parts of the crystal are displaced along gliding planes (caused by plastic forming):

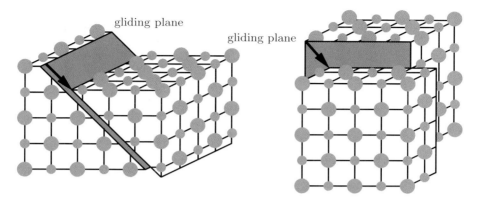

AX crystal, rock salt structure
(in the case of ionic crystals, the charge neutrality is conserved,
leading to relatively big Burgers vectors,
here pointing into the direction of the black arrows)

motion of an edge dislocation along a gliding plane (caused by plastic forming):

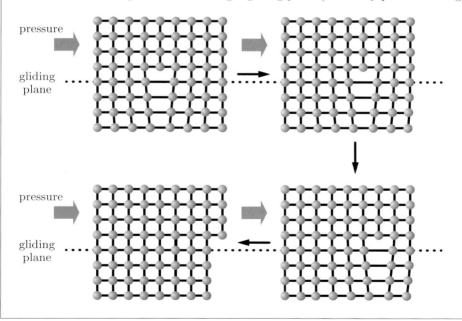

Figure 3.50. Examples of gliding processes.

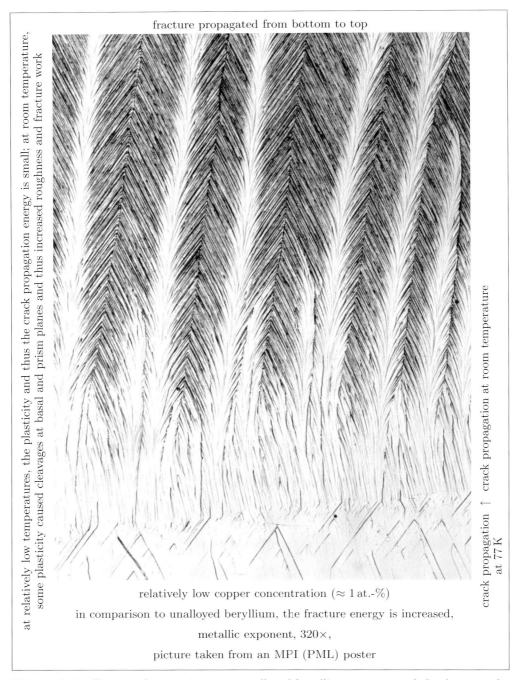

Figure 3.51. Cleavage fracture in a copper-alloyed beryllium monocrystal. In the case of a relatively low copper concentration, the fracture is controlled by gliding.

Figure 3.52. Cleavage fracture in a copper-alloyed beryllium monocrystal. In the case of a relatively high copper concentration, the fracture is controlled by twinning.

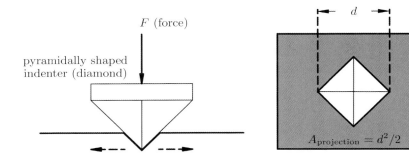

Principle of the Vickers method:

plastic material flow away from the indenter

indentation caused by the indenter

True hardness H:
$$H = \frac{F\,[\mathrm{N}]}{A_{\text{projection}}} = \frac{2\,F\,[\mathrm{N}]}{d^2\,[\mathrm{mm}^2]}\ .$$

Vickers hardness HV:
$$\mathrm{HV} = \frac{F\,[\mathrm{kg}]}{A_{\text{surface}}} = \frac{1.8544\,F\,[\mathrm{kg}]}{d^2\,[\mathrm{mm}^2]}\ ,$$
$$\mathrm{HV}\,[\mathrm{MPa}] = 9.81\,\mathrm{HV}\ .$$

According to [73].

Examples of HV (Vickers hardness) values:

ceramic material	HV value [MPa]	ceramic material	HV value [MPa]
diamond	88200	BeO	11172
B_4C	48804	MgO	6468
Al_2O_3	23226	fused SiO_2	5292
$ZrO_2 + CaO$	13818	NaCl	206

Figure 3.53. Vickers method.

3.3 *Mechanical Properties* 255

images from [91]

herringbone pattern (3), ferroelectric domains (4)

Figure 3.54. Crack growth (1) around a Vickers indentation (2). $BaTiO_3$.

The deepening remarks that follow should be useful for the reader. In crystals with strong covalent bond contributions and/or strong ionic bond contributions, due to the complicated reorganization mechanisms of such rigidly coupled material structures, gliding plane processes and edge dislocation switch processes at low temperatures only can emerge in special cases and then only occur relatively weakly developed. In crystals with strong metallic bond contributions, due to the simple reorganization mechanisms of such flexibly (electron gas) coupled material structures, gliding plane processes and edge dislocation switch processes at low temperatures can emerge in many cases and then occur relatively strongly developed. At high temperatures, crystals with strong ionic bond contributions certainly do show a considerable increase of such processes, but crystals with strong covalent bond contributions do not show a considerable increase of such processes, however, which does not mean that such processes are not possible. We again compare with Table 3.8, which reveals the clearing of gliding (slip) systems at low and high temperatures for ceramic materials with covalent and ionic character. Therefore, looking for ceramic materials with plastic properties, it is advisable to try ceramic materials with strong metallic and/or strong ionic bond contributions, but looking for ceramic materials with high-temperature stability, it is advisable to try ceramic materials with strong covalent bond contributions such as Si_3N_4 or SiC and not ceramic materials with strong ionic bond contributions such as Al_2O_3 or MgO.

The deepening remarks that follow should also be useful for the reader. Constantly applying stress to ceramic materials, one observes a plastic deformation termed *creep* dependent on stress, time, and temperature. The details that are presented here in total assume ceramic materials showing grain structures and tensile stress. (i) One separates 3 time intervals. The creep processes within the first time interval are circumscribed as primary creep, the processes of the second time interval as secondary creep, and the processes of the third time interval as tertiary creep. (ii) Phase one is predominantly characterized by grain boundary sliding processes. In the presence of a glass phase, following (3.89), a viscous flow of the grain boundary phase ("intergranular phase") is observable, whereas in the absence of a glass phase, the formation of first cavities ("pores") is observable. Since technical grades of ceramic materials typically contain a glass phase as secondary phase due to impurities of the raw materials or due to additives added to improve the densification by liquid-phase sintering, this type of viscous flow limits the mechanical stability of these materials at high temperatures. (iii) In phase two, these processes go on, namely in the presence of a glass phase, following (3.90), the viscous flow of the grain boundary phase ("intergranular phase") is observable, whereas in the absence of a glass phase, the formation of bigger cavities ("pores") is observable, and in the presence of a glass phase, first cavities ("pores") and precipitation/disolution processes develop. Furthermore, dislocation creep evoked by vacancy diffusion and climbing around obstacles comes into play. Moreover, the following diffusion mechanisms come into play: Coble creep, which means that the diffusion of atoms in a material along the grain boundaries produces a flow of material and a sliding of the grain boundaries, and Nabarro–Herring creep, which means that atoms diffuse through the grains, eventually causing elongations of the grains along the stress axis. (iv) In phase three, in particular, the growth of the cavities ("pores") leads to microcracks/macrocracks.

Box 3.16 (Important formulae: creep).

$d\epsilon/dt$ = strain rate,
σ = applied stress, σ_c = critical (threshold) stress,
η = viscosity,
d = mean diameter, D = diffusion coefficient,
A = activation energy for the creep process,
k_B = Boltzmann constant, T = absolute temperature,
t = time, t_0 = initial time.

Viscous flow, phase one:

$$\frac{d\epsilon}{dt} \propto \sigma^m \left(\frac{t_0}{t}\right)^c . \tag{3.89}$$

Viscous flow, phase two:

$$\frac{d\epsilon}{dt} \propto \frac{1}{\eta}\sigma . \tag{3.90}$$

Dislocation creep:

$$\frac{d\epsilon}{dt} \propto (\sigma - \sigma_c)^n \exp(-A_{\text{dislocation}}/k_B T) \tag{3.91}$$
$$(3 \leq n \leq 10) .$$

Coble creep:

$$\frac{d\epsilon}{dt} = \sigma \frac{1}{d^3_{\text{grain boundary}}} D_{\text{grain boundary}} \exp(-A_{\text{Coble}}/k_B T) . \tag{3.92}$$

Nabarro–Herring creep:

$$\frac{d\epsilon}{dt} = \sigma \frac{1}{d^2_{\text{grain}}} D_{\text{grain}} \exp(-A_{\text{Nabarro–Herring}}/k_B T) . \tag{3.93}$$

General creep equation:

$$\frac{d\epsilon}{dt} = \sigma^a \frac{1}{d^b} D \exp(-A/k_B T) . \tag{3.94}$$

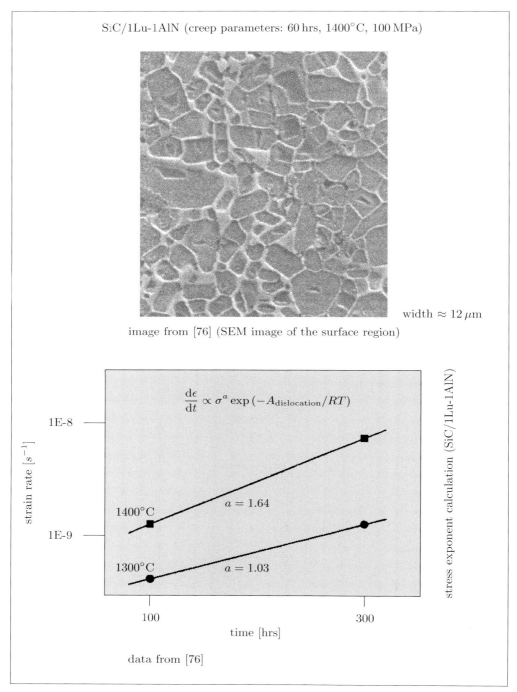

Figure 3.55. Creep under stress, liquid-phase sintered SiC samples with additives, part one.

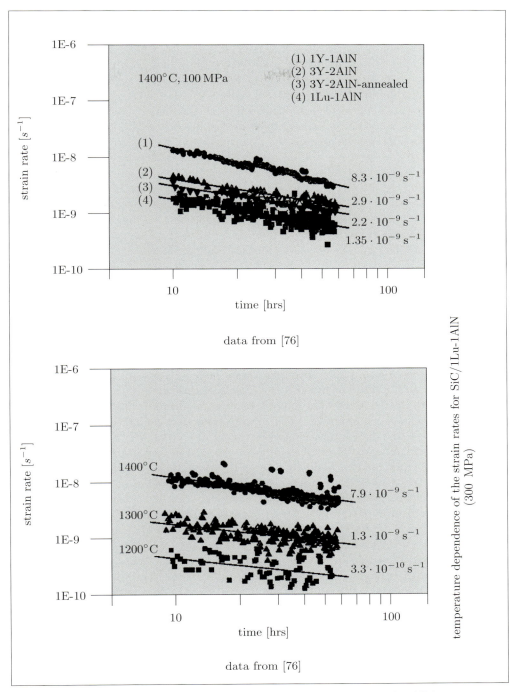

Figure 3.56. Creep under stress, liquid-phase sintered SiC samples with additives, part two.

In Figure 3.55 and Figure 3.56, a case study of liquid-phase sintered SiC with different sintering additives is shown. Figure 3.55 shows the microstructure of SiC with 10 vol-% additives consisting of 50 mol-%Lu_2O_3 and 50 mol-%AlN. The SiC grains are typically embedded in a glassy phase mainly consisting of the additives. Figure 3.55 and Figure 3.56 show diagrams making clear that strain and strain rate increase with temperature. The liquid-phase-sintered SiC samples with Lu_2O_3–AlN additives have the lowest creep rate of all tested materials, which makes this ceramic material suitable for high-temperature stability, for example, for turbine blades. In order to supply the reader with some details, we append the following remarks. (i) The powder premixes were prepared by wet attrition milling, then cold isostatically pressed at a pressure of 240 MPa, and then sintered in a gas pressure furnace in N_2 atmosphere with a first heating rate of 20 K/min to 1500°C and a second heating rate of 10 K/min to the sintering temperature. (ii) The annealing was carried out in a graphite furnace in N_2 atmosphere at ambient pressure at 1950°C with a heating/cooling rate of 10 K/min. (iii) The specification of the SiC/1Lu-1AlN powder premix was as follows: 90 vol-% SiC (α-SiC/β-SiC: 10/90 mol-%), 10 vol-% additives (50 mol-%Lu_2O_3, 50 mol-%AlN). (iv) The creep tests were performed in air on test specimens 3 mm × 4 mm × 50 mm in a testing machine (Zwick 1476, Germany) with a cross head speed of 0.002 mm/s for 60 hrs at different stresses and temperatures.

3.3.3 Strength and Fracture

Although ceramics reveal plastic behavior as we have seen above, first and foremost, they are brittle and fracture controlled by crack formation and crack growth occurs. In order to prepare the reader to what follows, we point at the following circumstances. (i) We differentiate between applied stress, i.e. the stress that is applied to ceramics, and the microscopic stress distribution caused by the applied stress. (ii) The points of origin of cracks are internal damages such as pores or external damages such as indentations. (1) *Subcritical crack growth.* Slow crack growth at $K_I \ll K_{Ic}$ under constant stress (where K_I is the stress intensity factor and K_{Ic} is the critical stress intensity factor), in particular, caused by chemical processes that reduce the formation energy of cracks, for example, evoked by H_2O molecules in the course of time diffusing into the region of the crack tip. (2) *Critical crack growth.* Fast crack growth after exceeding a critical crack length a_c, in last consequence, leading to fracture. (3) *Thermal shock.* Abrupt overload evoked by fast heating/cooling of material, in last consequence, leading to fast crack growth caused by internal tensions. (4) *Brittle fracture.* Abrupt fracture observable for materials with marginal ductility, i.e. with marginal plasticity.

Before we go into detail, for the reader's delight, in Figure 3.57, we present images of cracks, in Figure 3.58, we present images illustrating the notion *thermal shock*, and in Figure 3.59, we present an image illustrating the notion *brittle fracture*. We note that already in Figure 3.54, the crack growth around a Vickers indentation is shown. We also note that already in Figure 3.51 and Figure 3.52, fracture surfaces of special metallic crystals (copper-alloyed beryllium monocrystals) are shown.

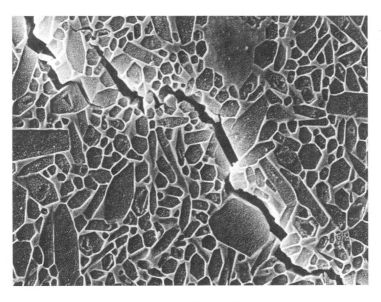

intergranular cracks — β-Si$_3$N$_4$ crystals in a glass-phase matrix

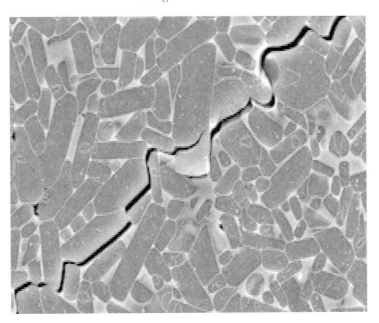

taken from the PML picture archive — β-SiC platelets in α-SiC

Figure 3.57. Cracks.

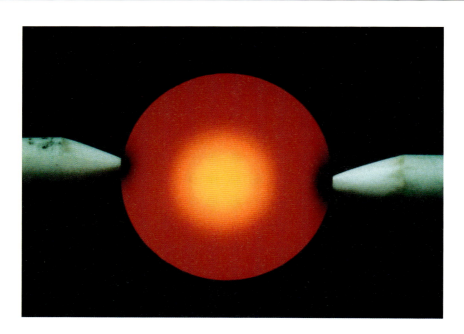

taken from the PML picture archive

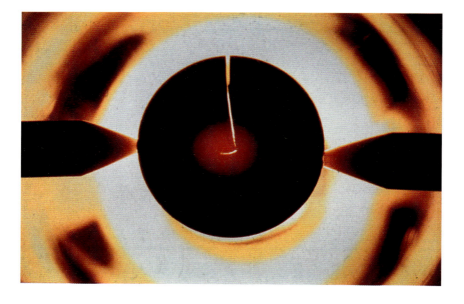

Figure 3.58. Thermal shock.

Figure 3.59. Brittle fracture.

Wanting to gain a basic access to the properties/evolution of cracks, let us resort to linear fracture mechanics supplying us with adequate means, we compare with the collection of formulae presented in Box 3.17. (i) A basic model for the fracture strength of a material is provided by (3.95). This basic model, called "theoretical strength" since it does not include perturbations, predicts high values for the fracture strength of a ceramic material. For example, for Si_3N_4, we compute $\sigma_{f,theo} \approx 55$ GPa because $E \approx 300$ GPa, $\gamma \approx 1$ J/m^2, and $d \approx 10^{-10}$ m. (ii) However, caused by perturbations such as (inhomogeneously distributed) lattice defects, (inhomogeneously distributed) grain boundaries, pores, inclusions etc. and therewith associated local stress step-up, a ceramic material usually breaks at far lower stresses depending on the type of stress. An advanced model for the fracture strength of a material that additionally includes perturbations and the type of stress is provided by (3.97). This advanced model, called "real strength" since it does include perturbations, predicts much lower values for the fracture strength of a ceramic material, indeed meeting the "notch sensitivity" of a ceramic material. For example, for Al_2O_3, we compute the values $\sigma_{f,tensile} \approx 381$ MPa and $\sigma_{f,compression} \approx 9.9$ GPa, as it is outlined in Example 3.6. (iii) The elastic stress around a crack tip is characterized by a stress step-up at the crack tip and a $\sqrt{1/r}$ decay farther away from the crack tip. The elastic stress around a crack tip is characterized by the internal loading modes $\sigma_{xx} = \sigma_x$, $\sigma_{yy} = \sigma_y$, $\sigma_{xy} = \tau_{xy}$, and $\sigma_{yz} = \tau_{yz}$. The elastic stress around a crack tip in linear fracture mechanics is modeled by (3.98), provided we consider an applied stress of type I ("loading mode I"), as it is outlined in Figure 3.60.

We note that the stress intensity factor K_I and the critical stress intensity factor K_{Ic} ("fracture toughness") characterize the material, whereas Y characterizes the type of applied stress ("loading type"). We note, too, that the fracture strength σ_f, on the one hand, depends on the type of material including the types of defects ($\rightarrow K_{Ic}$), and on the other hand, depends on the size of the defects ($\rightarrow a_c$), and the biggest values for the extent of the defects are the key values for the limitation of the breaking resistance. We note, too, that the failure conditions of Griffith ($G = G_c$) and Irwin ($K_I = K_{Ic}$), which postulate the (critical) material quantities G_c and K_{Ic}, lead from (3.96), specifying the interrelation of stress ($\rightarrow \sigma$), crack (perturbation) length ($\rightarrow a$), and material ($\rightarrow K_I$), to (3.97), specifying this interrelation at the critical point marking the failure of a specimen and telling us, on the one hand, which stress ($\rightarrow \sigma_f$) for a given material ($\rightarrow K_{Ic}$) leads to fracture if the extent of the defects of the specimen exhibits the biggest defect value a_c, and on the other hand, which biggest defect value ($\rightarrow a_c$) for a given material ($\rightarrow K_{Ic}$) leads to fracture if the stress exhibits a special value σ_f. We note, too, that K_{Ic} and a_c are statistical quantities so that it is more than wise to introduce statistical distribution functions for the description of charges of specimens. The Weibull distribution function (3.99) for the failure probability of a workpiece is applied in this statistical context, as it is outlined in Figure 3.60 and Example 3.8. The Weibull modulus m can be determined by measurements lying statistically distributed around the straight line that defines the Weibull modulus m, as it is additionally outlined in Figure 3.60.

Box 3.17 (Important formulae: linear fracture mechanics).

$\sigma_{f,theo}, \sigma_f,$ = critical crack stress (fracture strength),
σ_0 = characteristic critical crack stress (characteristic fracture strength),
K_I = stress intensity factor, K_{Ic} = critical stress intensity factor,
a = crack (perturbation) length, a_c = critical crack (perturbation) length,
Y = loading type, $f_{ij}(\theta)$ = shape factor,
G = energy release rate, G_c = critical energy release rate, γ = surface energy,
E = Young's modulus or modulus of elasticity,
F = failure probability, m = Weibull modulus,
V = volume, V_0 = normalization volume, d = atomic/molecular equilibrium distance.

Fracture strength (basic model, "theoretical strength"):

$$\sigma_{f,theo} = \sqrt{\frac{E\gamma}{d}} \, . \tag{3.95}$$

Fracture strength (advanced model, "real strength"):

$$\sigma = \frac{K_I}{Y\sqrt{\pi a}} \, , \quad G = \frac{K_I^2}{E} \, , \tag{3.96}$$

and

$G = G_c$ and thus $K_I = K_{Ic}$
(failure conditions of Griffith ($G = G_c$) and Irwin ($K_I = K_{Ic}$))
$\Downarrow$ (3.97)

$$\sigma_f = \frac{K_{Ic}}{Y\sqrt{\pi a_c}} = \frac{1}{Y}\sqrt{\frac{G_c E}{\pi a_c}} \Leftrightarrow a_c = \frac{1}{\pi Y^2}\left(\frac{K_{Ic}}{\sigma_f}\right)^2 = \frac{G_c E}{\pi Y^2 \sigma_f^2} \, .$$

(Compare with Examples 3.6 and 3.7.)

Elastic stress around a crack tip (valid for loading mode I):

$$\sigma_{ij} = \frac{K_I}{\sqrt{2\pi r}} f_{ij}(\theta) \, . \tag{3.98}$$

Failure probability
(Weibull distribution for the failure probability of a workpiece):

$$F(\sigma) = 1 - \exp\left[-\frac{V}{V_0}\left(\frac{\sigma}{\sigma_0}\right)^m\right] \, , \quad m = \frac{\ln\ln\left[(1-F)^{-V_0/V}\right]}{\ln(\sigma/\sigma_0)} \, . \tag{3.99}$$

Notes:
We note that energy release rate G means energy consumption per unit area of a newly created crack (perturbation) surface. We note that the evolution of a crack usually means the evolution of two surfaces so that $G = 2\gamma$.

Example 3.6 (An additional tutorial example: fracture strength).

Estimate the tensile fracture strength and the compression fracture strength for a charge of Al_2O_3 specimens ($K_{Ic} = 3.7\,\text{MPa}\,\text{m}^{1/2}$), which exhibits surface defects of $\approx 30\,\mu\text{m}$ depth and crack lengths of $\approx 10\,\mu\text{m}$.

Solution.

Tensile fracture strength ($1/Y \approx 1$, surface defects are decisive):

$$\sigma_{f,\text{tensile}} \approx \frac{K_{Ic}}{\sqrt{\pi a_c}} = \frac{3.7}{\sqrt{\pi\, 30 \cdot 10^{-6}}}\,\text{MPa} = 381\,\text{MPa}\,. \tag{3.100}$$

Compression fracture strength ($1/Y \approx 15$, crack lengths are decisive):

$$\sigma_{f,\text{compression}} \approx 15\frac{K_{Ic}}{\sqrt{\pi a_c}} = 15\frac{3.7}{\sqrt{\pi\, 10 \cdot 10^{-6}}}\,\text{MPa} = 9.9\,\text{GPa}\,. \tag{3.101}$$

Example 3.7 (An additional tutorial example: critical crack length).

Estimate the critcial crack length for a glass of the borosilicate type. Firstly, assume a tensile fracture strength of $\sigma_f = 102\,\text{MPa}$. Secondly, assume a surface energy of $\gamma = 1\,\text{J/m}^2$. Thirdly, assume a modulus of elaticity of $E = 70\,\text{GPa}$.

Solution.

Application of the energy criterion (3.97):

$$\sigma_f\sqrt{\pi a_c} \approx \sqrt{G_c E}\,. \tag{3.102}$$

With $G_c = 2\gamma$:

$$\sigma_f\sqrt{\pi a_c} \approx \sqrt{2\gamma E} \Rightarrow a_c \approx 2\gamma E/\pi\sigma_f^2 = 4.3\,\mu\text{m}\,. \tag{3.103}$$

3.3 Mechanical Properties

Example 3.8 (An additional tutorial example: Weibull distribution).
Calculate the maximum stress you can apply to a workpiece. For this purpose, use the Weibull distribution function (3.99). Furthermore, consider the Weibull modulus $m = 10$ and the characteristic fracture strength $\sigma_0 = 400 \text{ MPa}$. Furthermore, we require a maximum fracture rate of 1%.

Solution.

Fracture rate 1% → fracture probability $F = 0.01$.

$$F(\sigma) = 1 - \exp\left[-\frac{V}{V_0}\left(\frac{\sigma}{\sigma_0}\right)^m\right]$$
$$\Downarrow$$
$$-\ln(1-F) = \frac{V}{V_0}\left(\frac{\sigma}{\sigma_0}\right)^m$$
$$\Downarrow$$
$$\ln\left[(1-F)^{-1}\right] = \frac{V}{V_0}\left(\frac{\sigma}{\sigma_0}\right)^m \tag{3.104}$$
$$\Downarrow$$
$$\ln\left(\frac{1}{1-F}\right) = \frac{V}{V_0}\left(\frac{\sigma}{\sigma_0}\right)^m$$
$$\Downarrow$$
$$\sigma = \sqrt[m]{\ln\left(\frac{1}{1-F}\right)\frac{V_0}{V}}\sigma_0 \, ,$$

$$V = V_0$$
$$\Downarrow$$
$$\sigma = \sqrt[m]{\ln\left(\frac{1}{1-F}\right)}\sigma_0 \, . \tag{3.105}$$

With the above values:

$$\sigma = \sqrt[m]{\ln\left(\frac{1}{1-F}\right)}\sigma_0 = 400 \text{ MPa} \cdot \sqrt[10]{0.010} =$$
$$= 400 \text{ MPa} \cdot 0.631 = 252.51 \text{ MPa} \, . \tag{3.106}$$

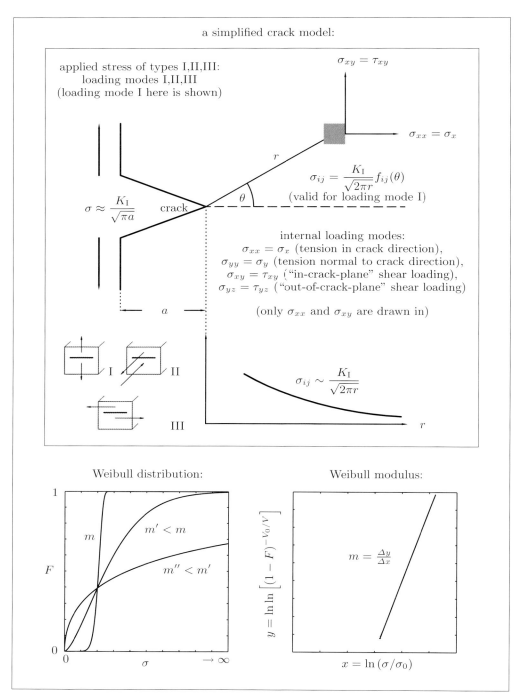

Figure 3.60. Linear fracture mechanics.

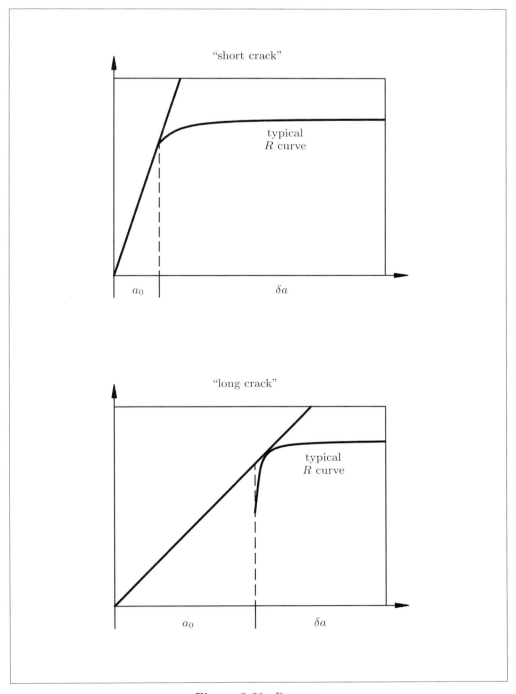

Figure 3.61. R curves.

It is remarkable that the K_{Ic} values of brittle materials (many ceramics) are much lower than the K_{Ic} values of ductile materials (many metals). It is remarkable, too, that the K_{Ic} values very often increase with increasing crack (perturbation) length, i. e. the crack resistance $R = G_c$ becomes higher. The cause for it are crack-inhibitive structures such as platelet structures or fiber structures which step by step cause direction changes of cracks, microcrack zones which step by step distribute the crack energy as well as crack flank processes such as debonding, pull-out, and bridging, i. e. energy dissipation by loss of adhesion between fibers etc. and matrix ("debonding"), energy dissipation by pulling fibers etc. out of a matrix ("pull-out"), and strengthening of crack tips by grains building bridges ("bridging"), on a macroscopic level of observation destroying the typical picture of a brittle fracture. We compare with Figure 3.61, showing typical R curves, with Figure 3.62, showing a typical fiber reinforced structure and a typical platelet reinforced structure, with Figures 2.72 and 3.57, showing direction changes of cracks evoked by debonding, pull-out, and bridging, and with Figures 3.51 and 3.52, showing two pictures of fracture surfaces of copper-alloyed beryllium monocrystals. Although beryllium is a metal, the results show that micro-plasticity at the tip of a propagating crack may hinder crack propagation by an increase of the work of fracture. We read off from Figures 3.51 and 3.52 that a drastic increase of the roughness of the fracture surfaces at room temperature against 77 K is caused by dislocation gliding and twinning depending on the copper content. Obviously, the work of fracture of beryllium is substantially increased by alloying this as such very brittle metal with copper.

Completing our consideration of "strength and fracture", we record the following items. (i) The strength of a ceramic material is limited by defects of the microstructure such as microcracks, macrocracks, micropores, macropores, surface defects, notches, and degraded grain boundaries. Naturally, such defects also can be technically utilized. For example, scoring the surface of a glass with a diamond, the glass is easily brocken along the line of scoring. (ii) The strength of a ceramic material can be increased by generating an enforced microstructure design, leading to an increased K_{Ic} value. Due to the above statements, it is clear that whiskers, fibers, and platelets integrated into a microstructure serve this purpose. Specific examples are whiskers (elongated grains) in liquid-phase sintered Si_3N_4 and SiC whiskers (elongated grains) in Al_2O_3 as well as β-SiC platelets in α-SiC. Due to the above statements, it is clear that reinforcements are evoked by the addition of substances with crystalline transformation capability. A specific example is tetragonal zirconia embedded into alumina, after the evolution of microcracks, giving way to its stable monoclinic modification, leading to the sealing of the microcracks due to the increase of the volume of its stable monoclinic modification ("self-healing effect"). We note in passing that Figure 3.57, bottom, shows the inhibition of cracks in the case of β-SiC platelets in α-SiC. We also note in passing that Box 2.7, bottom, shows the cubic, tetragonal, and monoclinic modifications of zirconia. (iii) The strength of a ceramic material can further be increased by the reduction of the defect (micropore, macropore, microcrack, macrocrack etc.) sizes. A specific example is the redensification of Si_3N_4 by hot isostatic pressing (HIP).

fiber reinforced structure (Si–B–Cl–C fibers)

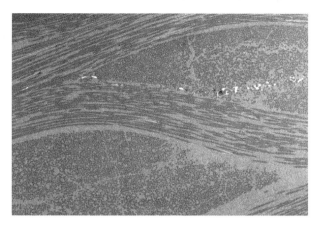

taken from the PML picture archive

platelet reinforced structure,
an example in orthopaedics: Bioloxdelta®
(SEM image of the microstructure)

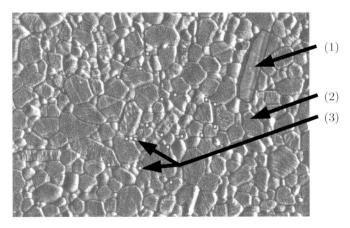

(1) oxide platelet, (2) aluminum oxide grain, (3) Y-TZP particles

image from [71]

Figure 3.62. Examples for crack-inhibitive structures.

3.4 Special Issues

3.4.1 Superplasticity

Some fine-grained materials show an unusual mechanical behavior called *superplasticity*. Superplasticity is an extreme form of plasticity. At appropriate temperatures and elongation rates, superplastic materials in a uniform manner can get extremely longer and can get extremely thinner without forming a neck or evoking internal cavities. This phenomenon, which is known with metals, was also found with ceramics. For example, ZrO_2 at the temperature $T = 1450°\,C$ and the elongation rates $1.1\ldots 5.5 \cdot 10^{-4}\,s^{-1}$ tolerates an elongation of $\approx 120\%$ without fracture. For example, Si_3N_4 at the temperature $T = 1600°\,C$ and the elongation rate $2 \cdot 10^{-5}\,s^{-1}$ tolerates an elongation of $\approx 280\%$ without fracture. We here note that the reason for the elongation and the thinning of superplastic materials of the ceramic type under applied tensile stress is the microstructure switch-over that is shown in Figure 3.63, the *Ashby–Verall mechanism*. We here also note that, caused by *grain boundary sliding*, grains are deformed without changing the volume.

3.4.2 Superconductivity

Many more or less complicated metallic materials and many more or less complicated ceramic materials show an unusual electromagnetic behavior called *superconductivity*. Superconductivity is an extreme form of conductivity. In particular, reaching a critical temperature T_c, the electrical resistance jumps to zero, accompanied by the suppression of interior magnetic fields. In particular, reaching a critical temperature T_c, the heat capacity shows a discontinuous jump. The readers should know that one distinguishes between superconductors of type I and superconductors of type II, i. e. one distinguishes between superconductors where the lines of magnetic flux apart from a thin layer at the surface of the superconductor are completely suppressed if the exterior magnetic field and the interior current density fall below critical values (type I) and superconductors where the lines of magnetic flux form so-called *vortices* if the exterior magnetic field falls below a first critical value and where the lines of magnetic flux apart from a thin layer at the surface of the superconductor are completely suppressed if the exterior magnetic field falls below a second critical value (type II). The readers should know that *Meißner–Ochsenfeld effect* means that the lines of magnetic flux apart from a thin layer at the surface of the superconductor are completely suppressed and should know that the corresponding phase and the corresponding penetration depth are called *Meißner phase* and *London penetration depth*. Special examples of ceramic materials are the perovskite-type cuprates $YBa_2Cu_3O_{7-\delta}$, $Bi_2Sr_2Ca_2Cu_3O_{10}$, $HgBa_2Ca_2Cu_3O_{8+\delta}$, and $Hg_{0.8}Tl_{0.2}Ba_2Ca_2Cu_3O_8$ with the critical temperatures $T_c \approx 93\,K$, $T_c \approx 110\,K$, $T_c \approx 133\,K$, and $T_c \approx 138\,K$. On the one hand, they are superconductors of type II, and on the other hand, they are high-temperature superconductors since their critical temperatures are above $23\,K$, by the way, which was at the time of the discovery of superconductivity with ceramics the highest critical temperature with metals.

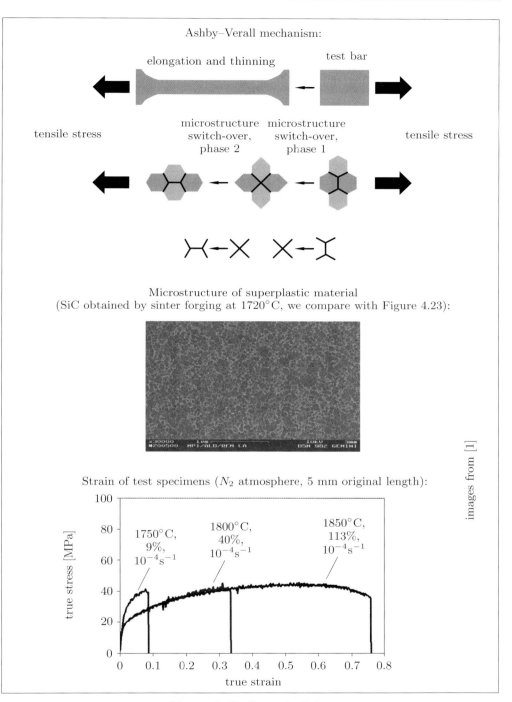

Figure 3.63. Superplasticity.

Figure 3.64 shows the crystal structure of yttrium barium cuprate $YBa_2Cu_3O_{7-\delta}$ ("YBCO", "123") in the orthorhombic (superconducting) phase. Figures 3.65 and 3.66 show microstructures of yttrium barium cuprate $YBa_2Cu_3O_{7-\delta}$. The remarks that follow should be useful. (i) YBCO has a perovskite-like, oxygen-deficient structure. YBCO undergoes a tetragonal/orthorhombic phase transformation when it is cooled down in oxygen-containing atmospheres. This phase transformation is accompanied by ordering of oxygen atoms along the crystallographic b axis. This phase transformation is accompanied by the development of stresses, which are finally relieved by twinning. This twinned structure is characteristic for the orthorhombic (superconducting) phase of YBCO. (ii) Microstructure 1 corresponds to the tetragonal (normal) phase of YBCO, microstructure 5 corresponds to the orthorhombic (superconducting) phase of YBCO, and microstructures 2–4 correspond to transition states of YBCO. (iii) Let us secure some details concerning the manufacturing process: specimens sintered at 950°C for 5 h, heating rates were 10 K/min, cooling rates were 1 K/min for 1,2, and 3, but 10 K/min for 4. (iv) Let us also secure some details concerning the metallographic preparation: specimens grinded with SiC paper, subsequently polished using (3 μm) diamond paste, and subsequently vibration-polished with colloidal silica. The microstructures were observed under polarized light. In order to enhance the contrast, the specimens were reactively sputtered using an Fe cathode in O_2 atmosphere.

Figure 3.64 particularly shows that copper–oxygen CuO_2-planes separated by an yttrium Y-spacer alternate with copper–oxygen CuO-planes. The remarks that follow should be useful. (i) The copper–oxygen CuO_2-planes are responsible for the superconductivity, namely the copper–oxygen CuO-planes attract electrons of the copper–oxygen CuO_2-planes, generating holes in the copper–oxygen CuO_2-planes, in last consequence, allowing electrons fed by power supply units to move along the copper–oxygen CuO_2-layers. (ii) Consistently, the copper–oxygen CuO_2-layers define the main direction of the current. Consistently, we always observe the anisotropy "current parallel to CuO_2-layers $\gg$ current normal to CuO_2-layers". (iii) We here yet point out that CuO_2-planes just like CuO_2-plane-CuO_2-plane spacers (Y, Ca etc.) are also the decisive crystalline structure units of other high-temperature superconductors such as Tl-1223 ($TlBa_2Ca_2Cu_3O_9$) or Tl-2223 ($Tl_2Ba_2Ca_2Cu_3O_{10}$).

Figure 3.67 shows examples for jump temperature diagrams. The remarks that follow should be useful. (i) These examples show that the critical (jump) temperature T_c that paves the way from a normal state to a superconducting state is a function of the fractions of the individual chemcial elements. (ii) In Figure 3.67, top, we show that an increasing contribution of Ca leads to a decreasing critical (jump) temperature T_c. (iii) In Figure 3.67, bottom, we show that an increasing contribution of Bi leads to a decreasing critical (jump) temperature T_c. (iv) In Figure 3.68, we additionally show that the increase of Ca (y) in the first case as well as the increase of Bi (x) in the second case involves the increase of the oxygen fraction (d). (v) Compounds of the type $Bi_{2+x}(Sr, Ca)_3Cu_2O_{8+d}$ define one important example for this effect. Compounds of the type $La_{2-x}Sr_xCuO_4$ define another important example for this effect.

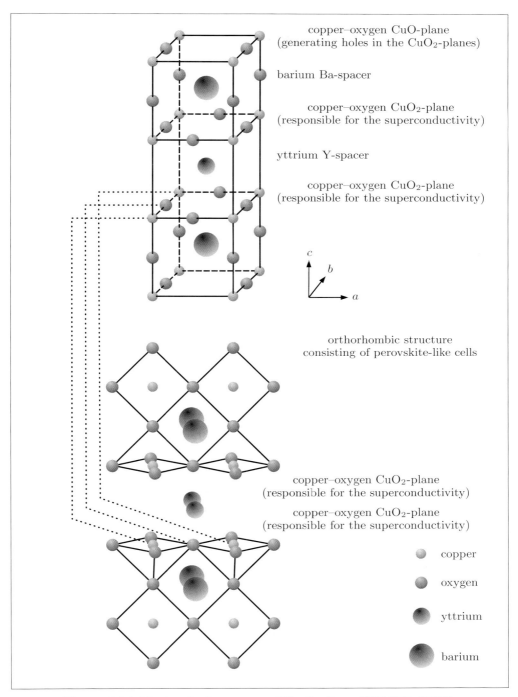

Figure 3.64. Yttrium barium cuprate $YBa_2Cu_3O_{7-\delta}$ ("YBCO", "123").

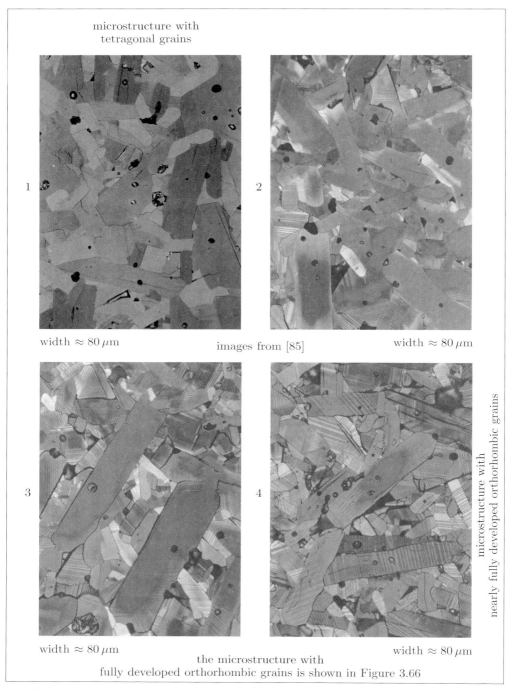

Figure 3.65. YBCO microstructure: tetragonal/orthorhombic phase transformation.

width ≈ 65 μm

the crystalline structure is shown in Figure 3.64

Figure 3.66. YBCO microstructure: orthorhombic (superconducting) phase.

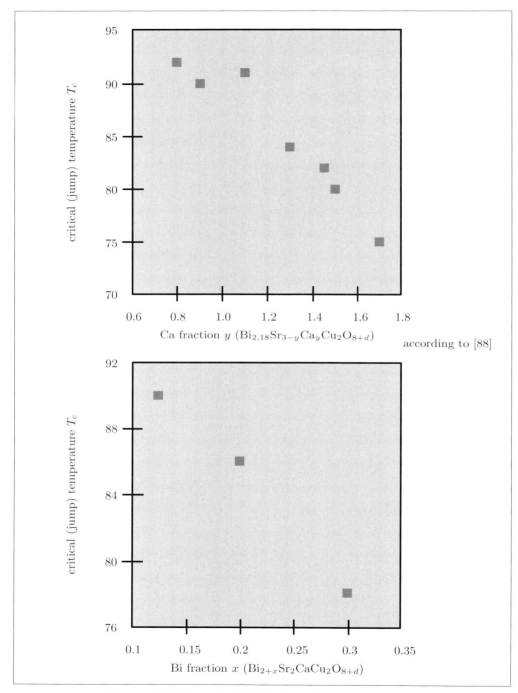

Figure 3.67. $Bi_{2+x}(Sr, Ca)_3Cu_2O_{8+d}$: jump temperatures.

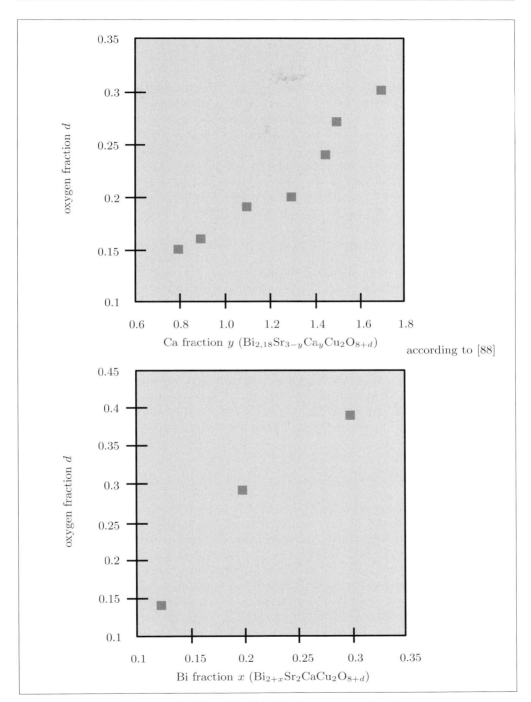

Figure 3.68. $Bi_{2+x}(Sr,Ca)_3Cu_2O_{8+d}$: oxygen fractions.

3.4.2.1 Conventional Modeling

The readers surely know that the effect that lines of magnetic flux apart from a thin layer at the surface of the superconductor are completely suppressed is conventionally called *Meißner–Ochsenfeld effect*. The readers surely also know that the corresponding phase and the corresponding penetration depth are conventionally called *Meißner phase* and *London penetration depth*, in the latter case, honoring F. and H. London who in 1935 proposed differential equations modeling the Meißner–Ochsenfeld effect, namely the *London equations* [30], evoking the *London theory of superconductivity*. An early phenomenological theory of superconductivity was proposed by Ginzburg and Landau in 1950 [20], nowadays known as the *Ginzburg–Landau theory of superconductivity*. An early microscopic theory of superconductivity was proposed by Bardeen, Cooper, and Schrieffer in 1957 [3], nowadays known as the *BCS theory of superconductivity*.

We note that the BCS theory resorts to the notion of two electrons that are bounded via lattice vibrations. So to speak, a first electron excites lattice vibrations, namely a phonon is generated, and a second electron is excited by the lattice vibrations, namely a phonon is destroyed. So to speak, an interaction between the two electrons comes into being, where the phonon acts as "exchange particle". In the BCS theory, following Pauli's principle, the two electrons of such a *Cooper pair* couple with antiparallely oriented spins, forming a boson. In the BCS theory, following Coulomb's law, namely minimizing the Coulomb repulsion, the two electrons of such a Cooper pair couple with some distance, defining a *coherence length*. In the BCS theory, the Cooper pairs condensate into a ground state leaving a band gap in the range of about one meV, explaining the critical temperatures and the critical magnetic fields. In the BCS theory, during the transition into the ground state, the properties of the Cooper pairs change, explaining the jumps of the electrical resistance and the heat capacity.

3.4.2.2 Advanced Modeling

The readers surely know that the Ginzburg–Landau theory of superconductivity [20] as well as the BCS theory of superconductivity [3] and modifications thereof [37] with considerable success are applied in many research fields to model the behavior of superconducting substances. However, we also know that the top-down approach of Ginzburg/Landau as well as the bottom-up approach of Bardeen/Cooper/Schrieffer and modifications thereof reveal limitations. For example, the interrelation between the energy alterations that occur during transitions into superconducting states, the Fermi surfaces, and the gaps in the Fermi surfaces only hardly can be explained by applying these established theories. A top-down approach to a substance-comprehensive description of superconductivity which on the one hand overcomes such limitations and on the other hand contains the Ginzburg–Landau theory of superconductivity and the BCS theory of superconductivity and modifications thereof as special limiting cases is presented in the sections that follow. We already here want to refer the reader to Figures 3.69–3.75, accompanying the next considerations.

Our starting point is given by a collection of energy terms, combined in form of an energy balance equation supplied by "generalized Einstein field theory (GEFT)", which is a "grand unified field theory" recently evoked in [63], in the limiting case of non-relativistic space time areas ("non-relativistic laboratory frames"), successively allowing the deduction of a self-consistent network of model conceptions, which some of our colleagues may consider as an extension of the Ginzburg–Landau theory of superconductivity [20] or as a semi-classical approach to quantum systems comparable to the Wunner–Main approach to quantum systems [34, 35, 64], however, which could turn out to be the true nonlinear extension of conventional quantum mechanics [63], on all accounts, which works without any known restrictions, which establishes the collection of energy terms to be applied here, and which is introduced in the framework of more far-reaching ideas in Chapter 5. In their most general formulation, these energy terms are defined by fundamental tensors, namely by the scalars $\psi = \gamma_{00}$ and $\bar{\psi} = \gamma^{00}$, the vectors $\boldsymbol{A}_\ni$ and $\boldsymbol{A}^\ni$ collecting the scalar functions $\gamma_{i0} = \gamma_{0i}$ and $\gamma^{i0} = \gamma^{0i}$, and the "tensors" $\boldsymbol{\theta}_\ni$ and $\boldsymbol{\theta}^\ni$ collecting the scalar functions $\gamma_{ij} = \gamma_{ji}$ and $\gamma^{ij} = \gamma^{ji}$, allowing us to model the scalar, the vectorial, and the "tensorial" energy aspects associated with quantum systems, including energy aspects associated with masses, charges, and more far-reaching quantum aspects such as photons, phonons, and spin [63]. In their most general formulation, these energy terms belong to the energy matrix element $\mathcal{KT}_{00}^*$ of an energy momentum tensor, after a suitable specification, launching quantum systems. For the sake of simplicity, we focus on energy relations. For the sake of simplicity, we focus on non-relativistic relations.

Applying the self-consistent network of model conceptions which is introduced in the framework of more far-reaching ideas in Chapter 5 already in advance at this point, a superconductivity model with the following properties is readily launched [75]. (i) The superconductivity model to be launched is a nonlinear model that is directly attached to linear quantum mechanics and linear quantum field theory, in special limiting cases, directly reproducing issues that are well-known from linear quantum mechanics and linear quantum field theory. (ii) The superconductivity model to be launched resorts to the notion of a crystal, on the one hand, consisting of atomic cores (in each case, summarizing a nucleus and core electrons that surround the nucleus), and on the other hand, consisting of valence electrons responsible for electric conductivity, defined by a superordinate differential equation that establishes crystal states and that works in a substance-comprehensive mode. (iii) The superconductivity model to be launched resorts to a separation of the Born–Oppenheimer type, in the case of superconductivity, generating a model where vectorial potentials $\boldsymbol{P}_s$ and $\bar{\boldsymbol{P}}_s$, implementing the influence of the crystal underground ("crystal lattice") which surrounds the valence electrons responsible for electric conductivity, via a subordinate differential equation establish superconducting-electrons–crystal-lattice states to be characterized as new entities with an eigenfield, but not as electrons that interact with each other via an interaction field attached to the crystal lattice, going beyond BCS theory and modifications thereof, explaining the jumps of the electrical resistance (eigenfield, no interaction field!) and the heat capacity (new entities!).

282 3. Properties

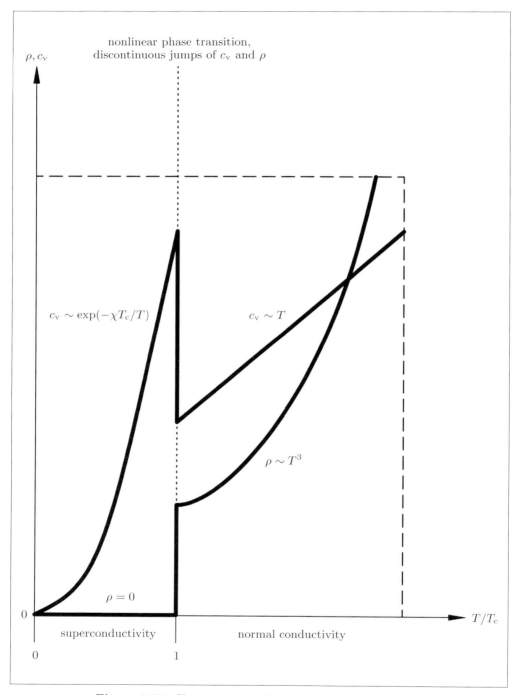

Figure 3.69. Characteristics of the superconductivity state.

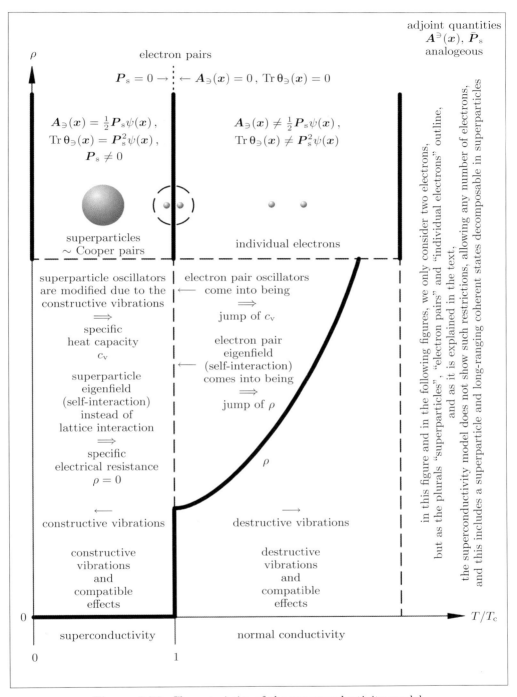

Figure 3.70. Characteristics of the superconductivity model.

We consider Figures 3.69 and 3.70. (i) The superconductivity model to be launched assumes two structural domains, a domain which we want to call *individual electrons* and a domain which we want to call *superparticles*, separated by an intermediate state which we want to call *electron pairs*, and requiring fixed compounds, the first domain covers states with normally conducting electrons, the second domain covers states with superconducting electrons, and the intermediate state connects these states, and going beyond BCS theory and modifications thereof, only the domain *individual electrons* deals with electrons that interact with each other, but the domain *superparticles*, going beyond *Cooper pairs*, deals with superconducting-electrons–crystal-lattice states to be defined as units per se showing extended eigenfields, but not as electrons that interact with each other, and the intermediate state *electron pairs* captures the transition from individual electrons to superparticles (i. e. units per se showing extended eigenfields), and allowing any number of electrons, this includes a superparticle and long-ranging coherent states decomposable in superparticles. (ii) The superconductivity model to be launched assumes that destructive vibrations and compatible effects dominate the domain *individual electrons*, whereas constructive vibrations and compatible effects dominate the domain *superparticles*, and the intermediate state *electron pairs* neither is characterized by resulting destructive influences nor is characterized by resulting constructive influences, but it is characterized by remaining influences, in particular, by electron–electron interactions defined by $\bar{\psi}(\boldsymbol{x})$ reflecting the electronic basis. (iii) The superconductivity model to be launched models the domain *individual electrons* by specifying $\boldsymbol{A}_{\ni}(\boldsymbol{x}) \neq \frac{1}{2}\boldsymbol{P}_\mathrm{s}\psi(\boldsymbol{x}) \neq 0$, $\boldsymbol{A}^{\ni}(\boldsymbol{x}) \neq \frac{1}{2}\bar{\boldsymbol{P}}_\mathrm{s}\bar{\psi}(\boldsymbol{x}) \neq 0$, $\mathrm{Tr}\,\boldsymbol{\theta}_{\ni}(\boldsymbol{x}) \neq \boldsymbol{P}_\mathrm{s}^2\psi(\boldsymbol{x}) \neq 0$ and the domain *superparticles* by the implementation of the "order parameter relations" $\boldsymbol{A}_{\ni}(\boldsymbol{x}) = \frac{1}{2}\boldsymbol{P}_\mathrm{s}\psi(\boldsymbol{x}) \neq 0$, $\boldsymbol{A}^{\ni}(\boldsymbol{x}) = \frac{1}{2}\bar{\boldsymbol{P}}_\mathrm{s}\bar{\psi}(\boldsymbol{x}) \neq 0$, $\mathrm{Tr}\,\boldsymbol{\theta}_{\ni}(\boldsymbol{x}) = \boldsymbol{P}_\mathrm{s}^2\psi(\boldsymbol{x}) \neq 0$, and we reach the intermediate state *electron pairs* from the "individual electrons side" by requiring $\boldsymbol{A}_{\ni}(\boldsymbol{x}) = 0$, $\boldsymbol{A}^{\ni}(\boldsymbol{x}) = 0$, $\mathrm{Tr}\,\boldsymbol{\theta}_{\ni}(\boldsymbol{x}) = 0$. and from the "superparticles side" by requiring $\boldsymbol{P}_\mathrm{s} = 0$, $\bar{\boldsymbol{P}}_\mathrm{s} = 0$, where the scalars $\psi(\boldsymbol{x})$ and $\bar{\psi}(\boldsymbol{x})$ are treated as wavefunction and adjoint wavefunction, respectively, and where the vectorial potentials $\boldsymbol{P}_\mathrm{s}$ and $\bar{\boldsymbol{P}}_\mathrm{s}$ shall implement constructive vibrations and compatible effects and thus shall be called *vibrational potential* and *adjoint vibrational potential*, respectively, consequently setting $\boldsymbol{P}_\mathrm{s} = 0$, $\bar{\boldsymbol{P}}_\mathrm{s} = 0$ for the intermediate state *electron pairs*. (iv) We additionally point at Figures 3.71 and 3.72, elucidating structural properties of the three areas, and we additionally point at Figures 3.73, 3.74, and 3.75, on the one hand, giving us an idea of "constructive vibrations and destructive vibrations", and on the other hand, in a decisive first step, transforming the notion of electrons that interact with each other via an interaction field attached to the crystal lattice into the notion of electron pairs (units per se showing basic eigenfields implemented by $\bar{\psi}(\boldsymbol{x})$) as well as superparticles (units per se showing extended eigenfields implemented by $\boldsymbol{P}_\mathrm{s}$, $\bar{\boldsymbol{P}}_\mathrm{s}$, and $\bar{\psi}(\boldsymbol{x})$), and in a decisive second step, finally transforming the notion of eigenfields into the notion of self-interaction fields. We additionally point out that the eigenfields are immanently bred by $\nabla\psi\nabla\psi$ terms and similar terms, enforcing the notion of self-interaction fields. (1) Self-consistently to kinetic energies established by kinetic energy terms, such terms establish the eigenfields of units per se such as electron pairs and superparticles. (2) Due to the formal structure "$\nabla\psi$ acts on $\nabla\psi$" of such terms, the eigenfields become manifest as self-interaction fields.

We here note that in the intermediate state "electron pairs", the self-interaction is of the basic type $(1-\bar{\psi})\nabla\psi\nabla\psi$ ("particle ψ with itself driven by $(1-\bar{\psi})$"), in the sum decomposition $\psi = \psi_1 + \psi_2 + \cdots$, evoking self-interactions of the type $\nabla\psi_i\nabla\psi_i$ ("particle ghost ψ_i with itself", where "particle ghost" outlines that the ψ_i at best can be interpreted as "hints of particles") as well as cross-interactions of the type $\nabla\psi_i\nabla\psi_j$ ("particle ghost ψ_i with particle ghost ψ_j", where "particle ghost" again outlines that the ψ_i at best can be interpreted as "hints of particles"), we compare with Figure 3.75. We here also note that in the domain "superparticles", the self-interaction is of the extended type $(1-\bar{\psi}_s)\nabla\psi\nabla\psi + \cdots$ with $\bar{\psi}_s = \bar{\psi}\,1 - \frac{1}{2}\bar{P}_s \cdot \bar{P}_s\,\bar{\psi}\bar{\psi}$ additionally making constructive vibrations and compatible constructive effects to inseparable parts of the self-interaction, we again compare with Figure 3.75.

It is important to know that the wavefunction $\psi(\boldsymbol{x})$ comprehensively incorporates individual electrons, electron pairs, or superparticles. It is also important to know that sum decompositions and product decompositions are applied to gain the access to sub-particle structures or sub-particle ghost structures. For example, modeling the domain "individual electrons", we apply the sum decomposition $\psi = \psi_{e_1^-} + \psi_{e_2^-}$ with $\psi_{e_i^-}$ representing electron wavefunctions, in last consequence, putting the notion of individual particles into practice, we compare with Figure 3.71. For example, modeling the domain "electron pairs", we apply $\psi = \psi_{e_1^-} + \psi_{e_2^-} + \psi_3 + \cdots$ with $\psi_{e_i^-}$ representing electron-like wavefunctions and ψ_3 and $\cdots$ representing a suitable system of functions, in last consequence, putting the notion of units per se that in some approximation however can be decomposed into individual particles into practice, we compare with Figures 3.72 and 3.75. For example, modeling the domain "superparticles", we apply $\psi = \psi_1 + \psi_2 + \psi_3 + \cdots$ with ψ_i and $\cdots$ representing a suitable system of functions, in last consequence, putting the notion of units per se that in rough approximation however can be decomposed into individual particles into practice, we again compare with Figures 3.72 and 3.75, and we refer to the next sections.

It is important to know that the mechanical approach to lattice vibrations, as shown in Figures 3.73 and 3.74, leads constructive vibrations back to mean compression stress and destructive vibrations back to mean tensile stress, involving that the transition from constructive vibrations to destructive vibrations in the mean is characterized by the absence of compression stress and the absence of tensile stress. We consider an "electron pair sphere"/a "superparticle sphere" formed of two "electron spheres", interacting with each other via one spring and connected to the "lattice" formed of two "lattice spheres" via two springs. In the case of relatively high "lattice energies" ("lattice temperatures"), incoherent oscillations such as chaotic oscillations lead to the destruction of an "electron pair sphere"/a "superparticle sphere". In the case of of relatively low "lattice energies" ("lattice temperatures"), coherent oscillations such as inversely phased oscillations or in-phase oscillations lead to the generation of an "electron pair sphere"/a "superparticle sphere". Modes that involve the absence of resulting compression stress and resulting tensile stress enter the "GEFT model" via $\boldsymbol{P}_s = 0$ and $\bar{\boldsymbol{P}}_s = 0$ (wavefunctions are not changed by lattice vibrations). Modes that involve the presence of resulting compression stress enter the "GEFT model" via $\boldsymbol{P}_s \neq 0$ and $\bar{\boldsymbol{P}}_s \neq 0$ (wavefunctions are changed by lattice vibrations).

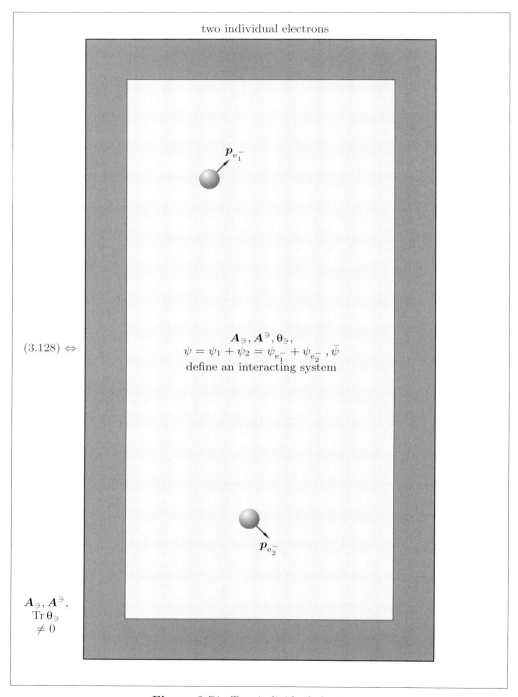

Figure 3.71. Two individual electrons.

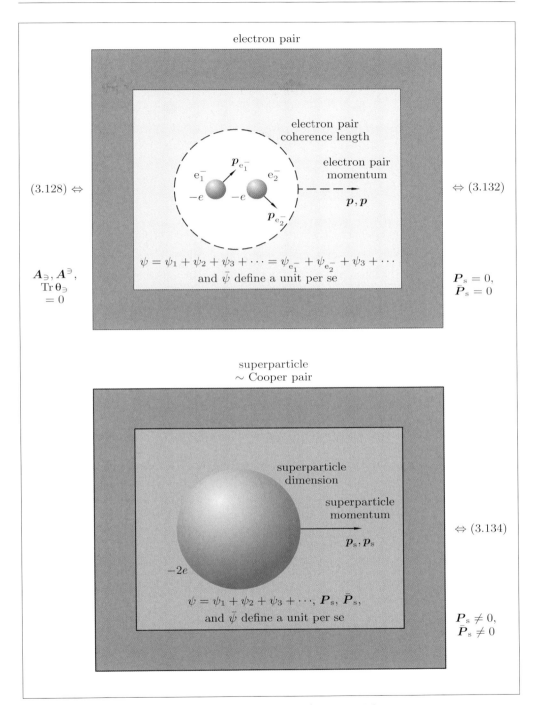

Figure 3.72. Electron pair/superparticle.

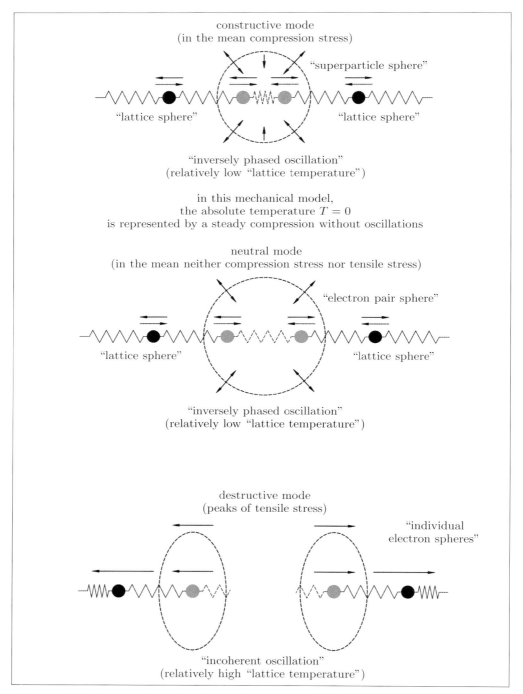

Figure 3.73. Constructive vibrations and destructive vibrations. A mechanical approach.

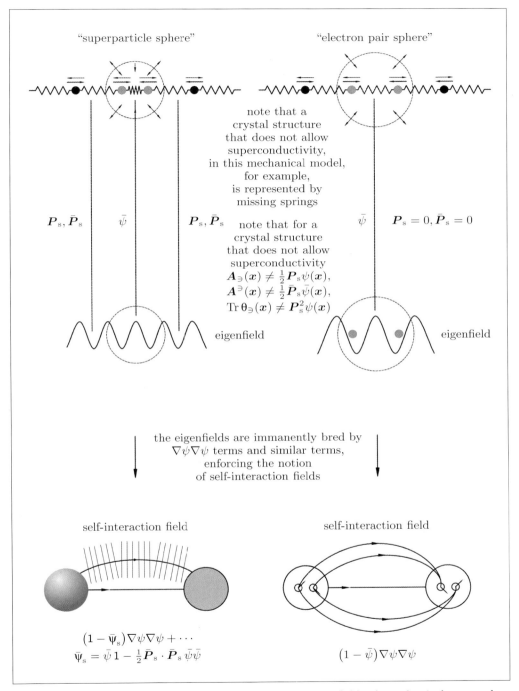

Figure 3.74. Constructive vibrations and self-interaction fields. A mechanical approach.

290   3. Properties

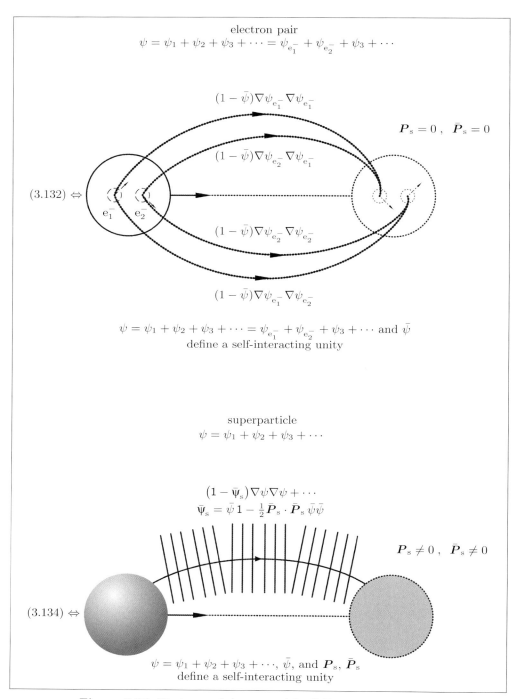

Figure 3.75. Electron pair/superparticle. Self-interaction aspects.

The vectorial potential $\boldsymbol{P}_\mathrm{s}$ is a superordinate (conjoint) potential, on the one hand, entering the differential equation that establishes electron pairs/superparticles, and on the other hand, entering the differential equation that establishes crystal states, in the first case, defined by a nonlinearly extended Hamiltonian $\hat{\mathcal{H}}_\mathrm{s}$, and in the second case, defined by a nonlinearly extended Hamiltonian $\hat{\mathcal{H}}_\mathrm{c}$. In the sections that follow, such nonlinearly extended Hamiltonians are simply called "Hamiltonians". In the sections that follow, after considering GEFT essentials, we refer to the differential equation that establishes electron pairs/superparticles and not to the differential equation that establishes crystal states or the separation of the Born–Oppenheimer type. In the boxes that follow, consistent with the above figures, but regardless of the actual situation, we refer to an electron pair/a superparticle, but not to long-ranging coherent states decomposable in electron pairs/superparticles.

GEFT Essentials

We consider the GEFT "Ansatz" (3.107). How can we interpret the three terms of the r. h. s.? Following [63], the three terms of the r. h. s. allow us to model the kinetic energy of combined particle–field systems including the "kinetic energy" of more far-reaching quantum aspects such as photon fields or phonon fields, namely in the first instance in wave-vector-square form, and specifying $\mathcal{KT}_{00}^*$ suitably, in energy form. For example, preparing ψ, $\boldsymbol{A}_\ni$, and $\boldsymbol{\theta}_\ni$ as complex exponential space time functions, these three terms in the first instance lead to a superordinate wave vector $\boldsymbol{k}_\mathrm{s}$ within a superordinate energy eigenvalue in "wave-vector-square form" $\boldsymbol{k}_\mathrm{s}^2/2$, and provided we specify $\mathcal{KT}_{00}^*$ as $m/\hbar^2 E\psi(\boldsymbol{x},t)$, where $\hbar$ is Planck's quantum of action ("Planck's constant") and m is the mass of the composite system, evoking a superordinate momentum vector $\boldsymbol{p}_\mathrm{s} = \hbar\boldsymbol{k}_\mathrm{s}$ within a superordinate energy eigenvalue $E = \boldsymbol{p}_\mathrm{s}^2/2m$, relating to a composite system that does not show interactions ("free particle–field system"), i. e. consisting of particles characterized by $\psi(\boldsymbol{x},t)$, of vectorial field contributions $\boldsymbol{A}_\ni$ concatenated with the particles, in general, enclosing transversally oscillating photon fields or phonon fields, and of "tensorial" field contributions $\boldsymbol{\theta}_\ni$ concatenated with the particles, in general, enclosing centrally oscillating photon fields or phonon fields.

We consider the GEFT "Ansätze" (3.117) and (3.122). How can we interpret the diverse terms? Following [63], the terms (7) and (9) implement self-interactions of the system, and the terms (10) and (12) implement gradiental/rotational interactions of the system, namely in the first case, due to the formal structure "$\nabla\psi$ acts on $\nabla\psi$", a microscopic unit described by ψ creates a feedback loop onto itself, namely in the second case, due to the formal structure $\boldsymbol{A}^\ni\nabla\psi/\boldsymbol{A}^\ni\times\nabla\psi$, a microscopic unit described by ψ is influenced by additional interaction fields manifesting themselves via $\boldsymbol{A}^\ni$ in a gradiental concatenation ($\boldsymbol{A}^\ni\nabla$) or in a rotational concatenation ($\boldsymbol{A}^\ni\times\nabla$). Following [63], ψ includes particles as well as "hints of particles" ("particle ghosts"), in the first case, evoking the notion of sub-particles ("sub-particle ghosts") that create a feedback loop onto themselves as well as the notion of sub-particles ("sub-particle ghosts") that interact with each other, and this is evoked by systems of functions ψ_i.

Electron Pairs/Superparticles, Kinetic Energy

The idea is to model a superparticle and long-ranging coherent states decomposable in superparticles (the extension of a Cooper pair and long-ranging coherent states decomposable in Cooper pairs) as states of a solid where electrons and phonon fields or compatible fields form a unity that apart from interaction energies (going beyond the above example) can be described by superordinate kinetic energies $E = \boldsymbol{p}_s^2/2m$, which are defined by superordinate momentum vectors $\boldsymbol{p}_s = \hbar \boldsymbol{k}_s$, which on their part are defined by superordinate wave vectors $\boldsymbol{k}_s$, and m is the mass of the composite system. For this purpose, let us here apply the GEFT "Ansatz" (3.108). In the spirit of GEFT, F_{00} covers rest-mass-less energy values, χ_{00} covers rest-mass-containing energy values, and χ_{00} is determined by (not necessarily space-independent and/or time-independent) energy values E. In the spirit of GEFT, the order parameter relations (3.115) manage the transition to the superordinate kinetic energy operator $\hat{T} = \hat{\boldsymbol{p}}_s^2/2m$, which finally evokes the desired superordinate kinetic energies $E = \boldsymbol{p}_s^2/2m$. In the spirit of GEFT, the new entities that are defined in this way explain the jump of the heat capacity, namely "indiviual electron oscillators" vanish and "superparticle oscillators" emerge, in last consequence, changing the capability of the solid to store heat energy.

So to speak, the order parameter relations (3.115) create an interrelation between the crystal-lattice-specific terms $\boldsymbol{A}_\ni$ and $\boldsymbol{\theta}_\ni$, specifying the oscillatory capabilities or compatible capabilities of the crystal underground ("crystal lattice") which surrounds the electrons causing electric conductivity, and the electron-specific term ψ, specifying the motion capabilities of the electrons causing electric conductivity, in this manner, defining states of matter where the separation of crystal lattice and electrons causing electric conductivity is resolved into new entities that apart from interaction energies are characterized by superordinate wave vectors, superordinate momentum vectors, and superordinate kinetic energies, not reflecting the notion of electrons that interact with each other via an interaction field attached to the crystal lattice, but reflecting the notion of "superordinate oscillatory states of matter", in this manner, explaining the jump of the electrical resistance, namely "interacting individual electrons" vanish and "superordinate oscillatory states of matter" emerge, in last consequence, changing the capability of the solid to breed electrical resistance.

So to speak, the order parameter relations (3.115) instal a vibrational potential $\boldsymbol{P}_s$ which describes the contribution of constructive vibrations and compatible effects to superconducting states of solid matter. In the spirit of GEFT, $\boldsymbol{P}_s \neq 0$ defines units where wavefunctions do "feel" constructive vibrations and compatible effects, in the framework of a differential treatment, leading to the alteration of wavefunctions, we compare with Figures 3.73 and 3.74. In the spirit of GEFT, $\boldsymbol{P}_s = 0$ defines units where wavefunctions do not "feel" constructive vibrations and compatible effects, in the framework of a differential treatment, not leading to the alteration of wavefunctions, we again compare with Figures 3.73 and 3.74. In the above terminology, $\boldsymbol{P}_s \neq 0$ covers "a superparticle and long-ranging coherent states decomposable in superparticles". In the above terminology, $\boldsymbol{P}_s = 0$ covers "an electron pair and long-ranging coherent states decomposable in electron pairs".

Electron Pairs/Superparticles, Interactions

The interaction energies in the case $\boldsymbol{P}_s \neq 0$ as well as in the case $\boldsymbol{P}_s = 0$, consistent with the above statements, are self-interaction energies relating to self-interaction fields defining eigenfields. Following Box 3.19, this "self-interacting structure" is generated by the "order parameter relations" (3.129), the application of which eventually leaves operative rules of the types "$\nabla \psi$ acts on $\nabla \psi$" and "ψ acts on $\nabla \psi$". Following Box 3.19, we directly arrive at the "wave–particle energy balance equation" (3.130), modeling superconducting states as "dynamic wave–particle objects" that are characterized by superordinate kinetic energies and superordinate self-interactions the carriers of which are self-interaction fields defining eigenfields, and one eigenfield contribution is defined by $\boldsymbol{P}_s$, where the adjoint $\bar{\boldsymbol{P}}_s$ defines the action of the eigenfield contribution, and one eigenfield contribution is defined by ψ itself, where the adjoint $\bar{\psi}$ defines the action of the eigenfield contribution. According to (3.132), an electron pair (just like states that are decomposable into electron pairs) does not "feel" the crystal surrounding ($\boldsymbol{P}_s, \bar{\boldsymbol{P}}_s = 0$!), thus representing a situation without resistance. According to (3.134), a superparticle (just like states that are decomposable into superparticles) certainly does "feel" the crystal surrounding ($\boldsymbol{P}_s, \bar{\boldsymbol{P}}_s \neq 0$!), but as part of itself, thus also representing a situation without resistance. On the one hand, we already here note that the differential formulation that is defined by (3.130) eventually paves the way to the Ginzburg–Landau theory of superconductivity. On the other hand, we already here note that the operative formulation that is additionally defined by (3.140) eventually paves the way to the BCS theory of superconductivity. As it is already outlined in Figure 3.71, the "wave–particle energy balance equation" (3.128) completes the considerations, i.e. it describes the situation above the phase transition point separating superconductivity from normal conductivity, i.e. it describes the situation where individual electrons and not electron pairs/superparticles are responsible for electric conductivity.

Regarding the GEFT equations below, the remarks that follow should be helpful. (i) We first remark that the notion of a forming superordinate unity involves the notion of a degenerating energy function F_{00}, i.e. the energy is shifted from the energy function F_{00} to the energy values E so that the energy function F_{00} is not needed anymore if we consider an electron pair/a superparticle. (ii) We second remark that the space derivatives in the starting equations lead to wave vector contributions $\boldsymbol{k}$, while the time derivatives in the starting equations lead to frequency contributions ω, in a direct manner, allowing the inclusion of vibrational modes needing the inclusion of wave vector contributions $\boldsymbol{k}$ and frequency contributions ω. For the sake of simplicity, neglecting higher-order terms that are given by deviations from $\exp(i\omega t)$ terms, we are led to a mathematical scheme on all levels of consideration certainly evoking time-independent measurement quantities, but nevertheless incorporating frequency contributions ω, in a direct manner, allowing the inclusion of vibrational modes. Certainly, we here could design a model on the basis of branches of discrete frequency values ω_i or continuous frequency values $\omega(i)$. However, for the sake of simplicity, we here focus on only one single frequency value ω. (iii) We leave it to the reader to philosophize about the meaning of $\exp(i\omega t)$ terms. Does this point at an inherent time lapse? Does this point at an unobservable background dynamics? And so fourth.

294   3. Properties

> **Box 3.18** (Important formulae: electron pair/superparticle, kinetic energy).
>
> ψ = wavefunction,
> $\boldsymbol{A}_{\ni}$ = general vectorial field function with components γ_{i0} ($i = 1, 2, 3$),
> $\boldsymbol{\theta}_{\ni}$ = general tensorial field function with components γ_{ij} ($i, j = 1, 2, 3$),
> x^i = Cartesian space coordinates ($x^i = x, y, z$),
> x^0 = Cartesian time coordinate ($x^0 = ct$),
> χ_{00} = rest-mass-containing energy values in wave-vector square form,
> F_{00} = rest-mass-less energy values in wave-vector square form,
> m = electron pair/superparticle rest mass, ω = circular frequency,
> c = light velocity, $\hbar$ = Planck's constant,
> $\hat{\boldsymbol{p}}_{\mathrm{s}}$ = electron pair/superparticle momentum operator,
> $\boldsymbol{P}_{\mathrm{s}}$ = vibrational potential.
>
> Implementation of
> kinetic energy terms according to (1), (3), and (5)
> of the wave-related *and* particle-related energy balance equation (5.42):
>
> $$\mathcal{KT}_{00}^{*} := F_{00}(\boldsymbol{x}, t) + \chi_{00}(\boldsymbol{x}, t)\psi(\boldsymbol{x}, t) =$$
>
> $$= -\frac{1}{2}(\nabla)^{\mathrm{T}}\nabla \psi(\boldsymbol{x},t) + \frac{1}{c}\frac{\partial}{\partial t}(\nabla)^{\mathrm{T}}\boldsymbol{A}_{\ni} - \frac{1}{2}\frac{1}{c^2}\left(1\frac{\partial^2}{\partial t^2}\right) \bullet (\boldsymbol{\theta}_{\ni}) =$$
>
> $$= -\frac{1}{2}\sum_{i=1}^{3}\frac{\partial^2 \psi(\boldsymbol{x},t)}{\partial x^i \partial x^i} + \qquad (3.107)$$
>
> $$+ \sum_{i=1}^{3}\frac{\partial^2 \gamma_{i0}}{\partial x^0 \partial x^i} - \frac{1}{2}\sum_{i=1}^{3}\frac{\partial^2 \gamma_{ii}}{\partial x^0 \partial x^0}.$$
>
> Implementation of
> (not necessarily space-independent and/or time-independent) energy values E:
>
> $$\mathcal{KT}_{00}^{*} := F_{00}(\boldsymbol{x}, t) + \chi_{00}(\boldsymbol{x},t)\psi(\boldsymbol{x},t), \quad \chi_{00}(\boldsymbol{x},t) := E\frac{m}{\hbar^2} \qquad (3.108)$$
>
> $$\Longrightarrow$$
>
> $$F_{00}(\boldsymbol{x},t) + E\frac{m}{\hbar^2}\psi(\boldsymbol{x},t) =$$
>
> $$= -\frac{1}{2}(\nabla)^{\mathrm{T}}\nabla \psi(\boldsymbol{x},t) + \frac{1}{c}\frac{\partial}{\partial t}(\nabla)^{\mathrm{T}}\boldsymbol{A}_{\ni} - \frac{1}{2}\frac{1}{c^2}\left(1\frac{\partial^2}{\partial t^2}\right) \bullet (\boldsymbol{\theta}_{\ni}) =$$
>
> $$= -\frac{1}{2}\sum_{i=1}^{3}\frac{\partial^2 \psi(\boldsymbol{x},t)}{\partial x^i \partial x^i} + \qquad (3.109)$$
>
> $$+ \sum_{i=1}^{3}\frac{\partial^2 \gamma_{i0}}{\partial x^0 \partial x^i} - \frac{1}{2}\sum_{i=1}^{3}\frac{\partial^2 \gamma_{ii}}{\partial x^0 \partial x^0}.$$

> **Continuation of Box.**

Implementation of
time-dependent complex functions, certainly evoking time-independent
measurement quantities, but nevertheless incorporating circular frequencies ω:

$$\boldsymbol{A}_{\ni}(\boldsymbol{x}, t) := \exp(\mathrm{i}\omega t)\, \boldsymbol{A}_{\ni}(\boldsymbol{x})$$
$$\Longrightarrow$$
$$+\frac{1}{c}\frac{\partial}{\partial t}(\nabla)^{\mathrm{T}} \boldsymbol{A}_{\ni} := \mathrm{i}\frac{\omega}{c} \exp(\mathrm{i}\omega t)\, \nabla \boldsymbol{A}_{\ni}(\boldsymbol{x})\;, \tag{3.110}$$

$$\mathbf{1} \bullet \boldsymbol{\theta}_{\ni}(\boldsymbol{x}, t) := \exp(\mathrm{i}\omega t)\, \mathbf{1} \bullet \boldsymbol{\theta}_{\ni}(\boldsymbol{x}) = \exp(\mathrm{i}\omega t)\, \mathrm{Tr}\,\boldsymbol{\theta}_{\ni}(\boldsymbol{x})$$
$$\Longrightarrow$$
$$-\frac{1}{c^2}\left(\mathbf{1}\frac{\partial^2}{\partial t^2}\right)\bullet(\boldsymbol{\theta}_{\ni}) := -\mathrm{i}^2\frac{\omega^2}{c^2}\exp(\mathrm{i}\omega t)\,\mathbf{1}\bullet\boldsymbol{\theta}_{\ni}(\boldsymbol{x}) = -\mathrm{i}^2\frac{\omega^2}{c^2}\exp(\mathrm{i}\omega t)\,\mathrm{Tr}\,\boldsymbol{\theta}_{\ni}(\boldsymbol{x})$$
$$\Longrightarrow$$

$$F_{00}(\boldsymbol{x},t) + E\frac{m}{\hbar^2}\psi(\boldsymbol{x},t) =$$
$$= \frac{1}{2}\left[-\nabla^2\psi(\boldsymbol{x},t) + 2\mathrm{i}\frac{\omega}{c}\exp(\mathrm{i}\omega t)\,\nabla\boldsymbol{A}_{\ni}(\boldsymbol{x}) - \mathrm{i}^2\frac{\omega^2}{c^2}\exp(\mathrm{i}\omega t)\,\mathrm{Tr}\,\boldsymbol{\theta}_{\ni}(\boldsymbol{x})\right] \tag{3.111}$$
$$\Longrightarrow$$

$$\psi(\boldsymbol{x},t) = \psi(\boldsymbol{x})\exp(\mathrm{i}\omega t)\;,\quad F_{00}(\boldsymbol{x},t) = F_{00}(\boldsymbol{x})\exp(\mathrm{i}\omega t) \tag{3.112}$$
$$\Longrightarrow$$

$$F_{00}(\boldsymbol{x}) + E\frac{m}{\hbar^2}\psi(\boldsymbol{x}) = \frac{1}{2}\left[-\nabla^2\psi(\boldsymbol{x}) + 2\mathrm{i}\frac{\omega}{c}\nabla\boldsymbol{A}_{\ni}(\boldsymbol{x}) - \mathrm{i}^2\frac{\omega^2}{c^2}\mathrm{Tr}\,\boldsymbol{\theta}_{\ni}(\boldsymbol{x})\right] \tag{3.113}$$
$$\Longrightarrow$$

$$\frac{\hbar^2}{m}F_{00}(\boldsymbol{x}) + E\psi(\boldsymbol{x}) = \frac{\hat{\boldsymbol{p}}^2}{2m}\psi(\boldsymbol{x}) - \hbar\omega\frac{\hat{\boldsymbol{p}}}{mc}\boldsymbol{A}_{\ni}(\boldsymbol{x}) + \hbar\omega\frac{\hbar\omega}{2mc^2}\mathrm{Tr}\,\boldsymbol{\theta}_{\ni}(\boldsymbol{x})\;, \tag{3.114}$$
$$\hat{\boldsymbol{p}} = -\mathrm{i}\hbar\nabla\;.$$

Implementation of
a superparticle:

$$\boldsymbol{A}_{\ni}(\boldsymbol{x}) := \frac{1}{2}\boldsymbol{P}_{\mathrm{s}}\psi(\boldsymbol{x})\;,\quad \mathbf{1}\bullet\boldsymbol{\theta}_{\ni}(\boldsymbol{x}) = \mathrm{Tr}\,\boldsymbol{\theta}_{\ni}(\boldsymbol{x}) := \boldsymbol{P}_{\mathrm{s}}^2\psi(\boldsymbol{x}) \tag{3.115}$$
$$\Longrightarrow$$

$$E\psi(\boldsymbol{x}) = \frac{\hat{\boldsymbol{p}}_{\mathrm{s}}^2}{2m}\psi(\boldsymbol{x})\;,\quad \hat{\boldsymbol{p}}_{\mathrm{s}} = -\mathrm{i}\hbar\nabla - \hbar\frac{\omega}{c}\boldsymbol{P}_{\mathrm{s}}\;, \tag{3.116}$$

$$\boldsymbol{P}_{\mathrm{s}} = 0 \Longrightarrow \text{``electron pair''}\;,\quad \boldsymbol{P}_{\mathrm{s}} \neq 0 \Longrightarrow \text{``superparticle''}\;.$$

> **Box 3.19** (Important formulae: electron pair/superparticle, interactions).

$\bar{\psi}$ = adjoint of wavefunction ψ, $\boldsymbol{A}^{\ni}$ = adjoint of general vectorial field function $\boldsymbol{A}_{\ni}$.

Preparation of self-interactions according to (7) and (9)
of the wave-related *and* particle-related energy balance equation (5.42):

$$-\frac{1}{4}\Big[\big(1-\bar{\psi}(\boldsymbol{x},t)\big)\nabla\psi(\boldsymbol{x},t)\Big]^{\mathrm{T}}\nabla\psi(\boldsymbol{x},t) - \frac{1}{2}\Big[\big(\boldsymbol{A}^{\ni}\cdot\boldsymbol{A}^{\ni}\big)\nabla\psi(\boldsymbol{x},t)\Big]^{\mathrm{T}}\nabla\psi(\boldsymbol{x},t) =$$

$$= \frac{1}{4}\big(1-\bar{\psi}(\boldsymbol{x},t)\big)\sum_{i=1}^{3}\frac{\partial\psi(\boldsymbol{x},t)}{\partial x^i}\frac{\partial\psi(\boldsymbol{x},t)}{\partial x^i} - \frac{1}{2}\sum_{i,j=1}^{3}\gamma^{j0}\gamma^{i0}\frac{\partial\psi(\boldsymbol{x},t)}{\partial x^i}\frac{\partial\psi(\boldsymbol{x},t)}{\partial x^j} = \quad (3.117)$$

$$= -\frac{1}{4}\big(1-\bar{\Psi}(\boldsymbol{x},t)\big)\nabla\psi(\boldsymbol{x},t)\nabla\psi(\boldsymbol{x},t)\,,$$

$$\bar{\Psi}(\boldsymbol{x},t) = \bar{\psi}(\boldsymbol{x},t)\,1 - 2\boldsymbol{A}^{\ni}\cdot\boldsymbol{A}^{\ni}\,,$$

$$\bar{\psi}(\boldsymbol{x},t) := \exp\left(-\mathrm{i}\omega t\right)\bar{\psi}(\boldsymbol{x})\,, \quad (3.118)$$
$$\big(1-\bar{\psi}(\boldsymbol{x},t)\big) := \big(1-\bar{\psi}(\boldsymbol{x})\big)\exp\left(-\mathrm{i}\omega t\right) + \mathrm{h.\,o.\,t.}\,,$$

$$\boldsymbol{A}^{\ni} := \exp\left(-\mathrm{i}\omega t\right)\boldsymbol{A}^{\ni}(\boldsymbol{x})\,, \quad (3.119)$$
$$\boldsymbol{A}^{\ni}\cdot\boldsymbol{A}^{\ni} := \boldsymbol{A}^{\ni}(\boldsymbol{x})\cdot\boldsymbol{A}^{\ni}(\boldsymbol{x})\exp\left(-\mathrm{i}\omega t\right) + \mathrm{h.\,o.\,t.}$$

$$\Longrightarrow$$

$$-\frac{1}{4}\big(1-\bar{\Psi}(\boldsymbol{x},t)\big)\nabla\psi(\boldsymbol{x},t)\nabla\psi(\boldsymbol{x},t) =$$
$$= -\frac{1}{4}\big(1-\bar{\Psi}(\boldsymbol{x})\big)\nabla\psi(\boldsymbol{x},t)\nabla\psi(\boldsymbol{x},t)\exp\left(-\mathrm{i}\omega t\right) + \mathrm{h.\,o.\,t.}\,, \quad (3.120)$$

$$\bar{\Psi}(\boldsymbol{x}) = \bar{\psi}(\boldsymbol{x})\,1 - 2\boldsymbol{A}^{\ni}(\boldsymbol{x})\cdot\boldsymbol{A}^{\ni}(\boldsymbol{x})\,. \quad (3.121)$$

Preparation of gradiental/rotational interactions according to (10) and (12)
of the wave-related *and* particle-related energy balance equation (5.42):

$$-\frac{1}{2}\Big[\big(\nabla\cdot\boldsymbol{A}_{\ni}\big)^{\mathrm{T}}\boldsymbol{A}^{\ni}\Big]^{\mathrm{T}}\nabla\psi(\boldsymbol{x},t) - \frac{1}{2}\big(\nabla\times\boldsymbol{A}_{\ni}\big)^{\mathrm{T}}\boldsymbol{A}^{\ni}\times\nabla\psi(\boldsymbol{x},t) =$$

$$= -\frac{1}{2}\sum_{i,j=1}^{3}\gamma^{j0}\frac{\partial\gamma_{i0}}{\partial x^j}\frac{\partial\psi(\boldsymbol{x},t)}{\partial x^i} - \frac{1}{2}\sum_{i,j=1}^{3}\gamma^{j0}\left(\frac{\partial\gamma_{i0}}{\partial x^j}-\frac{\partial\gamma_{j0}}{\partial x^i}\right)\frac{\partial\psi(\boldsymbol{x},t)}{\partial x^i}\,, \quad (3.122)$$

$$\boldsymbol{A}_{\ni} := \exp\left(\mathrm{i}\omega t\right)\boldsymbol{A}_{\ni}(\boldsymbol{x})\,,\quad \boldsymbol{A}^{\ni} := \exp\left(-\mathrm{i}\omega t\right)\boldsymbol{A}^{\ni}(\boldsymbol{x}) \quad (3.123)$$

$$\Longrightarrow$$

$$-\frac{1}{2}\Big[\big(\nabla\cdot\boldsymbol{A}_{\ni}\big)^{\mathrm{T}}\boldsymbol{A}^{\ni}\Big]^{\mathrm{T}}\nabla\psi(\boldsymbol{x},t) - \frac{1}{2}\big(\nabla\times\boldsymbol{A}_{\ni}\big)^{\mathrm{T}}\boldsymbol{A}^{\ni}\times\nabla\psi(\boldsymbol{x},t) =$$
$$= -\frac{1}{2}\Big[\big(\nabla\cdot\boldsymbol{A}_{\ni}(\boldsymbol{x})\big)^{\mathrm{T}}\boldsymbol{A}^{\ni}(\boldsymbol{x})\Big]^{\mathrm{T}}\nabla\psi(\boldsymbol{x},t) - \frac{1}{2}\big(\nabla\times\boldsymbol{A}_{\ni}(\boldsymbol{x})\big)^{\mathrm{T}}\boldsymbol{A}^{\ni}(\boldsymbol{x})\times\nabla\psi(\boldsymbol{x},t)\,. \quad (3.124)$$

Continuation of Box.

Integration of (3.120) and (3.124) into (3.111) and neglect of h. o. t. in order to obtain time-dependent complex functions, certainly evoking time-independent measurement quantities, but nevertheless incorporating circular frequencies ω:

$$F_{00}(\boldsymbol{x}, t) + E\frac{m}{\hbar^2}\psi(\boldsymbol{x}, t) =$$

$$= \frac{1}{2}\left[-\nabla^2\psi(\boldsymbol{x}, t) + 2\mathrm{i}\frac{\omega}{c}\exp(\mathrm{i}\omega t)\nabla\boldsymbol{A}_\ni(\boldsymbol{x}) - \mathrm{i}^2\frac{\omega^2}{c^2}\exp(\mathrm{i}\omega t)\operatorname{Tr}\boldsymbol{\theta}_\ni(\boldsymbol{x})\right] -$$

$$-\frac{1}{4}(1 - \bar{\psi}(\boldsymbol{x}))\nabla\psi(\boldsymbol{x}, t)\nabla\psi(\boldsymbol{x}, t)\exp(-\mathrm{i}\omega t) - \quad (3.125)$$

$$-\frac{1}{2}\left[(\nabla\cdot\boldsymbol{A}_\ni(\boldsymbol{x}))^{\mathrm{T}}\boldsymbol{A}^\ni(\boldsymbol{x})\right]^{\mathrm{T}}\nabla\psi(\boldsymbol{x}, t) - \frac{1}{2}(\nabla\times\boldsymbol{A}_\ni(\boldsymbol{x}))^{\mathrm{T}}\boldsymbol{A}^\ni(\boldsymbol{x})\times\nabla\psi(\boldsymbol{x}, t) +$$

$$+\mathrm{h. o. t.}$$

$$\stackrel{\mathrm{h. o. t.} \to 0}{\Longrightarrow}$$

$$\psi(\boldsymbol{x}, t) = \psi(\boldsymbol{x})\exp(\mathrm{i}\omega t) \;, \quad F_{00}(\boldsymbol{x}, t) = F_{00}(\boldsymbol{x})\exp(\mathrm{i}\omega t) \quad (3.126)$$

$$\Longrightarrow$$

$$F_{00}(\boldsymbol{x}) + E\frac{m}{\hbar^2}\psi(\boldsymbol{x}) =$$

$$= \frac{1}{2}\left[-\nabla^2\psi(\boldsymbol{x}) + 2\mathrm{i}\frac{\omega}{c}\nabla\boldsymbol{A}_\ni(\boldsymbol{x}) - \mathrm{i}^2\frac{\omega^2}{c^2}\operatorname{Tr}\boldsymbol{\theta}_\ni(\boldsymbol{x})\right] -$$

$$-\frac{1}{4}(1 - \bar{\psi}(\boldsymbol{x}))\nabla\psi(\boldsymbol{x})\nabla\psi(\boldsymbol{x}) - \quad (3.127)$$

$$-\frac{1}{2}\left[(\nabla\cdot\boldsymbol{A}_\ni(\boldsymbol{x}))^{\mathrm{T}}\boldsymbol{A}^\ni(\boldsymbol{x})\right]^{\mathrm{T}}\nabla\psi(\boldsymbol{x}) - \frac{1}{2}(\nabla\times\boldsymbol{A}_\ni(\boldsymbol{x}))^{\mathrm{T}}\boldsymbol{A}^\ni(\boldsymbol{x})\times\nabla\psi(\boldsymbol{x}) \;.$$

Implementation of
individual electrons/an electron pair:

$$\frac{\hbar^2}{m}F_{00}(\boldsymbol{x}) + E\psi(\boldsymbol{x}) =$$

$$= \frac{\hat{\boldsymbol{p}}^2}{2m}\psi(\boldsymbol{x}) - \hbar\omega\frac{\hat{\boldsymbol{p}}}{mc}\boldsymbol{A}_\ni(\boldsymbol{x}) + \hbar\omega\frac{\hbar\omega}{2mc^2}\operatorname{Tr}\boldsymbol{\theta}_\ni(\boldsymbol{x}) -$$

$$-\frac{\hbar^2}{4m}(1 - \bar{\psi}(\boldsymbol{x}))\nabla\psi(\boldsymbol{x})\nabla\psi(\boldsymbol{x}) - \quad (3.128)$$

$$-\frac{1}{2}\left[(\nabla\cdot\boldsymbol{A}_\ni(\boldsymbol{x}))^{\mathrm{T}}\boldsymbol{A}^\ni(\boldsymbol{x})\right]^{\mathrm{T}}\nabla\psi(\boldsymbol{x}) - \frac{1}{2}(\nabla\times\boldsymbol{A}_\ni(\boldsymbol{x}))^{\mathrm{T}}\boldsymbol{A}^\ni(\boldsymbol{x})\times\nabla\psi(\boldsymbol{x}) \;,$$

$$\boldsymbol{A}_\ni(\boldsymbol{x}), \boldsymbol{A}^\ni(\boldsymbol{x}), \operatorname{Tr}\boldsymbol{\theta}_\ni(\boldsymbol{x}) \neq 0 \Longrightarrow F_{00}(\boldsymbol{x}) \neq 0 \Longrightarrow \text{``individual electrons''} \;,$$

$$\boldsymbol{A}_\ni(\boldsymbol{x}), \boldsymbol{A}^\ni(\boldsymbol{x}), \operatorname{Tr}\boldsymbol{\theta}_\ni(\boldsymbol{x}) = 0 \Longrightarrow F_{00}(\boldsymbol{x}) = 0 \Longrightarrow \text{``electron pair''} \;.$$

Continuation of Box.

Implementation of
a superparticle:

$$A_{\ni}(x) := \frac{1}{2} P_s \psi(x) \,, \quad A^{\ni}(x) := \frac{1}{2} \bar{P}_s \bar{\psi}(x) \,, \tag{3.129}$$
$$1 \bullet \theta_{\ni}(x) = \mathrm{Tr}\, \theta_{\ni}(x) := P_s^2 \psi(x)$$

$$\Longrightarrow$$

$$E\psi(x) = \frac{\hat{p}_s^2}{2m}\psi(x) - \frac{\hbar^2}{4m}\big(1 - \bar{\psi}_s(x)\big)\nabla\psi(x)\nabla\psi(x) -$$
$$- \frac{\hbar^2}{8m}\big(\nabla \times P_s\psi(x)\big)\bar{P}_s\bar{\psi}(x) \times \nabla\psi(x) - \frac{\hbar^2}{8m}\Big[\big(\nabla \cdot P_s\psi(x)\big)\bar{P}_s\bar{\psi}(x)\Big]\nabla\psi(x)\,, \tag{3.130}$$
$$\hat{p}_s = -i\hbar\nabla - \hbar\frac{\omega}{c}P_s \,, \quad \bar{\psi}_s(x) = \bar{\psi}(x)\mathbf{1} - \frac{1}{2}\bar{P}_s \cdot \bar{P}_s \bar{\psi}(x)\bar{\psi}(x) \,,$$

at the phase transition point

$$P_s = 0 \,, \quad \bar{P}_s = 0 \tag{3.131}$$

$$\Longrightarrow$$

$$E\psi(x) = \frac{\hat{p}^2}{2m}\psi(x) - \frac{\hbar^2}{4m}\big(1 - \bar{\psi}(x)\big)\nabla\psi(x)\nabla\psi(x) \,,$$
$$\hat{p} = -i\hbar\nabla \tag{3.132}$$

("electron pair"),

below the phase transition point

$$P_s \neq 0 \,, \quad \bar{P}_s \neq 0 \tag{3.133}$$

$$\Longrightarrow$$

$$E\psi(x) = \frac{\hat{p}_s^2}{2m}\psi(x) - \frac{\hbar^2}{4m}\big(1 - \bar{\psi}_s(x)\big)\nabla\psi(x)\nabla\psi(x) -$$
$$- \frac{\hbar^2}{8m}\big(\nabla \times P_s\psi(x)\big)\bar{P}_s\bar{\psi}(x) \times \nabla\psi(x) - \frac{\hbar^2}{8m}\Big[\big(\nabla \cdot P_s\psi(x)\big)\bar{P}_s\bar{\psi}(x)\Big]\nabla\psi(x)\,,$$
$$\hat{p}_s = -i\hbar\nabla - \hbar\frac{\omega}{c}P_s \,, \quad \bar{\psi}_s(x) = \bar{\psi}(x)\mathbf{1} - \frac{1}{2}\bar{P}_s \cdot \bar{P}_s \bar{\psi}(x)\bar{\psi}(x) \,, \tag{3.134}$$
$$P_s \neq 0 \,, \quad \bar{P}_s \neq 0$$

("superparticle").

> **Continuation of Box.**
>
> Implementation of
> generation operators $(+)$, destruction operators $(-)$,
> and operators of partial aspects $(1,2)$
> (we compare with Figures 3.76 and 3.77):
>
> $$\hat{L}_{\mathrm{s}}^{+} = \boldsymbol{L}_{\mathrm{s},0} - \mathrm{i}\nabla - \frac{\omega}{c}\boldsymbol{P}_{\mathrm{s}}\,,\quad \hat{L}_{\mathrm{s}}^{-} = \boldsymbol{L}_{\mathrm{s},0} + \mathrm{i}\nabla + \frac{\omega}{c}\boldsymbol{P}_{\mathrm{s}}\,, \tag{3.135}$$
>
> $$\hat{G}^{1} = -\mathrm{i}\sqrt{\nabla\psi(\boldsymbol{x})}\,,\quad \hat{G}^{2} = +\mathrm{i}\sqrt{\nabla\psi(\boldsymbol{x})}\,, \tag{3.136}$$
>
> $$\hat{G}^{1}_{\bar{\psi}_{\mathrm{s}}} = -\mathrm{i}\sqrt{(1-\bar{\psi}_{\mathrm{s}}(\boldsymbol{x}))\,\nabla}\,,\quad \hat{G}^{2}_{\bar{\psi}_{\mathrm{s}}} = +\mathrm{i}\sqrt{(1-\bar{\psi}_{\mathrm{s}}(\boldsymbol{x}))\,\nabla}\,, \tag{3.137}$$
>
> $$\hat{G}^{1}_{\mathrm{grad},\boldsymbol{P}_{\mathrm{s}}} = -\mathrm{i}\sqrt{\nabla\cdot\boldsymbol{P}_{\mathrm{s}}\psi(\boldsymbol{x})}\,,\quad \hat{G}^{1}_{\mathrm{grad},\bar{\boldsymbol{P}}_{\mathrm{s}}} = -\mathrm{i}\sqrt{\bar{\boldsymbol{P}}_{\mathrm{s}}\bar{\psi}(\boldsymbol{x})\nabla}\,,$$
> $$\hat{G}^{2}_{\mathrm{grad},\boldsymbol{P}_{\mathrm{s}}} = +\mathrm{i}\sqrt{\nabla\cdot\boldsymbol{P}_{\mathrm{s}}\psi(\boldsymbol{x})}\,,\quad \hat{G}^{2}_{\mathrm{grad},\bar{\boldsymbol{P}}_{\mathrm{s}}} = +\mathrm{i}\sqrt{\bar{\boldsymbol{P}}_{\mathrm{s}}\bar{\psi}(\boldsymbol{x})\nabla}\,, \tag{3.138}$$
>
> $$\hat{G}^{1}_{\mathrm{rot},\boldsymbol{P}_{\mathrm{s}}} = -\mathrm{i}\sqrt{\nabla\times\boldsymbol{P}_{\mathrm{s}}\psi(\boldsymbol{x})}\,,\quad \hat{G}^{1}_{\mathrm{rot},\bar{\boldsymbol{P}}_{\mathrm{s}}} = -\mathrm{i}\sqrt{\bar{\boldsymbol{P}}_{\mathrm{s}}\bar{\psi}(\boldsymbol{x})}\times\nabla\,,$$
> $$\hat{G}^{2}_{\mathrm{rot},\boldsymbol{P}_{\mathrm{s}}} = +\mathrm{i}\sqrt{\nabla\times\boldsymbol{P}_{\mathrm{s}}\psi(\boldsymbol{x})}\,,\quad \hat{G}^{2}_{\mathrm{rot},\bar{\boldsymbol{P}}_{\mathrm{s}}} = +\mathrm{i}\sqrt{\bar{\boldsymbol{P}}_{\mathrm{s}}\bar{\psi}(\boldsymbol{x})}\times\nabla \tag{3.139}$$
>
> $$\Longrightarrow$$
>
> $$E\psi(\boldsymbol{x}) = \hat{\mathcal{H}}_{\mathrm{s}}\psi(\boldsymbol{x})\,, \tag{3.140}$$
>
> $$\hat{\mathcal{H}}_{\mathrm{s}} = \frac{\hat{\boldsymbol{p}}_{\mathrm{s}}^{2}}{2m} - \frac{\hbar^{2}}{4m}\nabla\psi(\boldsymbol{x})\bigl(1-\bar{\psi}_{\mathrm{s}}(\boldsymbol{x})\bigr)\nabla -$$
> $$-\frac{\hbar^{2}}{8m}\bigl[(\nabla\cdot\boldsymbol{P}_{\mathrm{s}}\psi(\boldsymbol{x}))\,\bar{\boldsymbol{P}}_{\mathrm{s}}\bar{\psi}(\boldsymbol{x})\bigr]\nabla - \frac{\hbar^{2}}{8m}(\nabla\times\boldsymbol{P}_{\mathrm{s}}\psi(\boldsymbol{x}))\,\bar{\boldsymbol{P}}_{\mathrm{s}}\bar{\psi}(\boldsymbol{x})\times\nabla =$$
> $$= -\frac{\hbar^{2}}{2m}\left(\hat{L}_{\mathrm{s}}^{+}\hat{L}_{\mathrm{s}}^{-} - \boldsymbol{L}_{\mathrm{s},0}^{2}\right) - \frac{\hbar^{2}}{4m}\hat{G}^{1}\hat{G}^{2}\hat{G}^{1}_{\bar{\psi}_{\mathrm{s}}}\hat{G}^{2}_{\bar{\psi}_{\mathrm{s}}} - \tag{3.141}$$
> $$-\frac{\hbar^{2}}{8m}\hat{G}^{1}_{\mathrm{grad},\boldsymbol{P}_{\mathrm{s}}}\hat{G}^{2}_{\mathrm{grad},\boldsymbol{P}_{\mathrm{s}}}\hat{G}^{1}_{\mathrm{grad},\bar{\boldsymbol{P}}_{\mathrm{s}}}\hat{G}^{2}_{\mathrm{grad},\bar{\boldsymbol{P}}_{\mathrm{s}}} -$$
> $$-\frac{\hbar^{2}}{8m}\hat{G}^{1}_{\mathrm{rot},\boldsymbol{P}_{\mathrm{s}}}\hat{G}^{2}_{\mathrm{rot},\boldsymbol{P}_{\mathrm{s}}}\hat{G}^{1}_{\mathrm{rot},\bar{\boldsymbol{P}}_{\mathrm{s}}}\hat{G}^{2}_{\mathrm{rot},\bar{\boldsymbol{P}}_{\mathrm{s}}}\,.$$

Inclusion of Magnetic Fields

Obviously, magnetic fields are not studied in the above GEFT model. Naturally, dealing with superconducting states, the magnetic field aspect plays an important role. Therefore, let us supplement the above GEFT model by the magnetic field aspect as it is shown in Box 3.20. Regarding the GEFT equations below, the remarks that follow should be helpful. (i) Firstly, in contrast to the order parameter relations (3.129) and the wave–particle energy balance equation (3.130), the order parameter relations (3.142) and the wave–particle energy balance equation (3.143) include magnetic fields such as interior magnetic fields of an electron pair current/a superparticle current or such as exterior magnetic fields applied to an electron pair current/a superparticle current. Based upon this extended formulation, the interplay between the vibrational potential $\boldsymbol{P}_\mathrm{s}$, defining constructive vibrations and compatible effects, the vector potential $\boldsymbol{A}$, defining interior magnetic fields and exterior magnetic fields, and the wavefunction ψ, which self-consistently adjusts itself to the special conditions that are implemented by the vibrational potential $\boldsymbol{P}_\mathrm{s}$ and the vector potential $\boldsymbol{A}$, can be modeled. (ii) Secondly, the momentum operator $\hat{\boldsymbol{p}}_\mathrm{s}$ now includes the vibrational potential $\boldsymbol{P}_\mathrm{s}$, in this way, implementing contributions to the superordinate state momentum that are evoked by constructive vibrations and compatible effects, and the vector potential $\boldsymbol{A}$, in this way, implementing contributions to the superordinate state momentum that are due to the remainders of the magnetic field. (iii) Thirdly, the field function $\bar{\psi}_\mathrm{s}$, which implements the action of the leading types of self-interaction, now includes the adjoint $\bar{\boldsymbol{P}}_\mathrm{s}$, in this way, implementing contributions to the action of the superordinate eigenfield that are evoked by constructive vibrations and compatible effects, and the adjoint $\bar{\boldsymbol{A}}$, in this way, implementing contributions to the action of the superordinate eigenfield that are due to the remainders of the magnetic field.

Say, the vibrational potential $\boldsymbol{P}_\mathrm{s}$ is given at a temperature that is associated with a superconducting state. Say, a critical magnetic field is applied that destroys the superconducting state. Neglecting that the vibrational potential $\boldsymbol{P}_\mathrm{s}$ can be altered by the critical magnetic field, without searching solutions of (3.143), already the following can be said. Increasing the exterior magnetic field up to the critical magnetic field, at first the wavefunction ψ self-consistently readjusts itself in such an extent that all parts, including the interior magnetic field that is associated with the superconducting state and the energy values E, form a physically/chemically possible entity, but reaching the critial magnetic field point, this is not possible anymore, i. e. at first solutions $\{\psi, \boldsymbol{A}, E\}$ that are compatible with boundary conditions and physical/chemical constraints can be derived, but reaching the critial magnetic field point, this is not possible anymore, defining the critial magnetic field point. The reason for this limit should be obvious. Considering a special substance/a special material as well as special electron classes building up the superconducting state, for example, electrons residing in special layers of the special substance/the special material populating special energy bands based upon special orbitals, the adaptability of solutions $\{\psi, \boldsymbol{A}, E\}$ is finite, sooner or later, reaching material-dependent limits. Congruent statements apply if we incorporate the vibrational potential $\boldsymbol{P}_\mathrm{s}$ in our considerations.

Box 3.20 (Important formulae: electron pair/superparticle, inclusion of magnetic fields).

$$\boldsymbol{B} = \text{magnetic field},$$
$$\boldsymbol{A} = \text{vector potential}, \bar{\boldsymbol{A}} = \text{adjoint of vector potential } \boldsymbol{A},$$
$$q = 2e = \text{charge of an electron pair/a superparticle}.$$

Inclusion of
magnetic fields $\boldsymbol{B} = \nabla \times \boldsymbol{A}$ via the vector potential $\boldsymbol{A}$:

$$\boldsymbol{A}_{\ni}(\boldsymbol{x}) := \frac{1}{2}\left(\boldsymbol{P}_{\text{s}} + \frac{cq}{\hbar\omega}\boldsymbol{A}\right)\psi(\boldsymbol{x}),$$

$$\boldsymbol{A}^{\ni}(\boldsymbol{x}) := \frac{1}{2}\left(\bar{\boldsymbol{P}}_{\text{s}} + \frac{cq}{\hbar\omega}\bar{\boldsymbol{A}}\right)\bar{\psi}(\boldsymbol{x}), \qquad (3.142)$$

$$1 \bullet \theta_{\ni}(\boldsymbol{x}) = \text{Tr}\,\theta_{\ni}(\boldsymbol{x}) := \left(\boldsymbol{P}_{\text{s}} + \frac{cq}{\hbar\omega}\boldsymbol{A}\right)^{2}\psi(\boldsymbol{x}),$$

$$\Longrightarrow$$

$$E\psi(\boldsymbol{x}) = \frac{\hat{\boldsymbol{p}}_{\text{s}}^{2}}{2m}\psi(\boldsymbol{x}) - \frac{\hbar^{2}}{4m}\left(1 - \bar{\Psi}_{\text{s}}(\boldsymbol{x})\right)\nabla\psi(\boldsymbol{x})\nabla\psi(\boldsymbol{x}) -$$

$$-\frac{\hbar^{2}}{8m}\left(\nabla \times \left(\boldsymbol{P}_{\text{s}} + \frac{cq}{\hbar\omega}\boldsymbol{A}\right)\psi(\boldsymbol{x})\right)\left(\bar{\boldsymbol{P}}_{\text{s}} + \frac{cq}{\hbar\omega}\bar{\boldsymbol{A}}\right)\bar{\psi}(\boldsymbol{x}) \times \nabla\psi(\boldsymbol{x}) -$$

$$-\frac{\hbar^{2}}{8m}\left[\left(\nabla \cdot \left(\boldsymbol{P}_{\text{s}} + \frac{cq}{\hbar\omega}\boldsymbol{A}\right)\psi(\boldsymbol{x})\right)\left(\bar{\boldsymbol{P}}_{\text{s}} + \frac{cq}{\hbar\omega}\bar{\boldsymbol{A}}\right)\bar{\psi}(\boldsymbol{x})\right]\nabla\psi(\boldsymbol{x}),$$

$$\hat{\boldsymbol{p}}_{\text{s}} = -i\hbar\nabla - \hbar\frac{\omega}{c}\left(\boldsymbol{P}_{\text{s}} + \frac{cq}{\hbar\omega}\boldsymbol{A}\right),$$

$$(3.143)$$

$$\bar{\Psi}_{\text{s}}(\boldsymbol{x}) = \bar{\psi}(\boldsymbol{x})\mathbf{1} - \frac{1}{2}\left(\bar{\boldsymbol{P}}_{\text{s}} + \frac{cq}{\hbar\omega}\bar{\boldsymbol{A}}\right) \cdot \left(\bar{\boldsymbol{P}}_{\text{s}} + \frac{cq}{\hbar\omega}\bar{\boldsymbol{A}}\right)\bar{\psi}(\boldsymbol{x})\bar{\psi}(\boldsymbol{x}),$$

$$\boldsymbol{A} = 0, \bar{\boldsymbol{A}} = 0, \boldsymbol{P}_{\text{s}} = 0, \bar{\boldsymbol{P}}_{\text{s}} = 0 \Longrightarrow \text{"electron pair"},$$

$$\boldsymbol{A} = 0, \bar{\boldsymbol{A}} = 0, \boldsymbol{P}_{\text{s}} \neq 0, \bar{\boldsymbol{P}}_{\text{s}} \neq 0 \Longrightarrow \text{"superparticle"},$$

$$\boldsymbol{A} \neq 0, \bar{\boldsymbol{A}} \neq 0, \boldsymbol{P}_{\text{s}} = 0, \bar{\boldsymbol{P}}_{\text{s}} = 0 \Longrightarrow \text{"electron pair current"},$$

$$\boldsymbol{A} \neq 0, \bar{\boldsymbol{A}} \neq 0, \boldsymbol{P}_{\text{s}} \neq 0, \bar{\boldsymbol{P}}_{\text{s}} \neq 0 \Longrightarrow \text{"superparticle current"}.$$

Inclusion of Spin Fields

Following [63], spin fields can be included by the 3×3 matrix $\boldsymbol{\theta}^{\ni}$, if specified suitably and then referred to as $\boldsymbol{\theta}^{\ni}_{\text{spin}}$, imprinting spin properties in the "wave–particle system". Following [63], a field of spins attached to a field of particles in GEFT models is recorded in the form $\boldsymbol{\theta}^{\ni}_{\text{spin}} \nabla \psi$ and $(\boldsymbol{\theta}^{\ni}_{\text{spin}} \nabla)^{\mathrm{T}} \nabla \psi$, reflecting that spins are carried along with particles. Concentrating on the kinetic energy terms (3.144) embracing spin fields and the gradiental/rotational interaction terms (3.145) embracing spin fields, we are led to (3.146), upgrading (3.143). Regarding the GEFT equations below, the remarks that follow should be helpful. (i) Firstly, we note that an electron pair/a superparticle, if no strong deformations are present, in accordance with Pauli's principle should not show spin properties so that $\boldsymbol{\theta}^{\ni}_{\text{spin}} = 0$. This possibility is included in (3.146). (ii) Secondly, we note that an electron pair/a superparticle, if strong deformations are present, in contrast to Pauli's principle could show spin properties so that $\boldsymbol{\theta}^{\ni}_{\text{spin}} \neq 0$, namely a spin separation could evolve, opening the way from antiparallelly arranged spins to "antiferromagnetically tasting" spins without leaving the superconducting state. This possibility is included in (3.146). (iii) Thirdly, we note that the first energy term of (3.144) should be interpreted as kinetic spin energy term in a generalized formulation, completing the kinetic energy term in a generalized formulation given by the first term of (3.107), while the result of the three energy terms (3.144), after application of the "order parameter relations" (3.142), should be interpreted as kinetic spin energy term of a superordinate unity concatenating "wave–particle parts" with interaction-field parts and spin parts in a generalized formulation, supplementing the kinetic energy term of a superordinate unity in a generalized formulation given by the r. h. s. terms of (3.113), having regard to the "order parameter relations" (3.142). In last consequence, the first r. h. s. term of (3.146) defining the kinetic energy term of a superordinate unity in an explicit manner then is supplemented by the second r. h. s. term of (3.146) defining the kinetic spin energy term of a superordinate unity in an explicit manner. (iv) Fourthly, we note that the first interaction energy term of (3.145) should be interpreted as an energy term that incorporates interactions of fields evoked by gradiental properties of $\boldsymbol{A}_{\ni}$ with spin properties (where $\boldsymbol{A}^{\ni}$ implements the action of these gradiental properties on the spin properties) in a generalized formulation, completing the energy term in a generalized formulation which is provided by the first energy term of (3.122), while the second interaction energy term of (3.145) should be interpreted as an energy term that incorporates interactions of fields evoked by rotational properties of $\boldsymbol{A}_{\ni}$ with spin properties (where $\boldsymbol{A}^{\ni} \times$ implements the action of these rotational properties on the spin properties) in a generalized formulation, completing the energy term in a generalized formulation which is provided by the second energy term of (3.122). In last consequence, we arrive at four interaction energy terms decomposable into two pairs of interaction energy terms each distinguished by a term showing $\boldsymbol{\theta}^{\ni}_{\text{spin}}$ and a term not showing $\boldsymbol{\theta}^{\ni}_{\text{spin}}$, we compare with (3.146), defining two central interaction classes that influence the particle motion ($\to \nabla \psi$) and the spin properties ($\to \boldsymbol{\theta}^{\ni}_{\text{spin}} \nabla \psi$). (v) Fifthly, we note that the momentum operator $\hat{\boldsymbol{p}}_{\mathrm{s}}$ and the field function $\bar{\boldsymbol{\psi}}_{\mathrm{s}}$, which implements the action of the leading types of self-interaction, do not depend on a field of spins attached to a field of particles.

> Box 3.21 (Important formulae: electron pair/superparticle, inclusion of spin fields).
>
> $\theta^{\ni}$ = adjoint of general tensorial function $\theta_{\ni}$,
> $\theta^{\ni}_{\text{spin}}$ = adjoint of general tensorial function $\theta_{\ni}$ if only spin properties are modeled.
>
> Inclusion of kinetic energy terms and gradiental/rotational interaction terms embracing spin fields according to (2), (4), (6), (11), and (13) of (5.42):
>
> $$-\frac{1}{2}\left(\theta^{\ni}_{\text{spin}}\nabla\right)^{\text{T}}\nabla\psi(\boldsymbol{x},t) + \frac{1}{c}\frac{\partial}{\partial t}(\theta^{\ni}_{\text{spin}}\nabla)^{\text{T}}\boldsymbol{A}_{\ni} - \frac{1}{2}\frac{1}{c^2}\left(\theta^{\ni}_{\text{spin}}\frac{\partial^2}{\partial t^2}\right)\bullet(\theta_{\ni}) , \quad (3.144)$$
>
> $$-\frac{1}{2}\left[(\nabla\cdot\boldsymbol{A}_{\ni})^{\text{T}}\boldsymbol{A}^{\ni}\right]^{\text{T}}\theta^{\ni}_{\text{spin}}\nabla\psi(\boldsymbol{x},t) - \frac{1}{2}(\nabla\times\boldsymbol{A}_{\ni})^{\text{T}}\boldsymbol{A}^{\ni}\times\theta^{\ni}_{\text{spin}}\nabla\psi(\boldsymbol{x},t) \quad (3.145)$$
>
> $$\Longrightarrow$$
>
> $$E\psi(\boldsymbol{x}) = \frac{\hat{\boldsymbol{p}}_s^2}{2m}\psi(\boldsymbol{x}) + \frac{(\theta^{\ni}_{\text{spin}}\hat{\boldsymbol{p}}_s)\hat{\boldsymbol{p}}_s}{2m}\psi(\boldsymbol{x}) - \frac{\hbar^2}{4m}(1-\bar{\psi}_s(\boldsymbol{x}))\nabla\psi(\boldsymbol{x})\nabla\psi(\boldsymbol{x}) -$$
> $$-\frac{\hbar^2}{8m}\left(\nabla\times\left(\boldsymbol{P}_s + \frac{cq}{\hbar\omega}\boldsymbol{A}\right)\psi(\boldsymbol{x})\right)\left(\bar{\boldsymbol{P}}_s + \frac{cq}{\hbar\omega}\bar{\boldsymbol{A}}\right)\bar{\psi}(\boldsymbol{x})\times\nabla\psi(\boldsymbol{x}) -$$
> $$-\frac{\hbar^2}{8m}\left[\left(\nabla\cdot\left(\boldsymbol{P}_s + \frac{cq}{\hbar\omega}\boldsymbol{A}\right)\psi(\boldsymbol{x})\right)\left(\bar{\boldsymbol{P}}_s + \frac{cq}{\hbar\omega}\bar{\boldsymbol{A}}\right)\bar{\psi}(\boldsymbol{x})\right]\nabla\psi(\boldsymbol{x}) -$$
> $$-\frac{\hbar^2}{8m}\left(\nabla\times\left(\boldsymbol{P}_s + \frac{cq}{\hbar\omega}\boldsymbol{A}\right)\psi(\boldsymbol{x})\right)\left(\bar{\boldsymbol{P}}_s + \frac{cq}{\hbar\omega}\bar{\boldsymbol{A}}\right)\bar{\psi}(\boldsymbol{x})\times\theta^{\ni}_{\text{spin}}\nabla\psi(\boldsymbol{x}) -$$
> $$-\frac{\hbar^2}{8m}\left[\left(\nabla\cdot\left(\boldsymbol{P}_s + \frac{cq}{\hbar\omega}\boldsymbol{A}\right)\psi(\boldsymbol{x})\right)\left(\bar{\boldsymbol{P}}_s + \frac{cq}{\hbar\omega}\bar{\boldsymbol{A}}\right)\bar{\psi}(\boldsymbol{x})\right]\theta^{\ni}_{\text{spin}}\nabla\psi(\boldsymbol{x}) ,$$
> $$\hat{\boldsymbol{p}}_s = -\mathrm{i}\hbar\nabla - \hbar\frac{\omega}{c}\left(\boldsymbol{P}_s + \frac{cq}{\hbar\omega}\boldsymbol{A}\right) ,$$
> $$\bar{\psi}_s(\boldsymbol{x}) = \bar{\psi}(\boldsymbol{x})\mathbf{1} - \frac{1}{2}\left(\bar{\boldsymbol{P}}_s + \frac{cq}{\hbar\omega}\bar{\boldsymbol{A}}\right)\cdot\left(\bar{\boldsymbol{P}}_s + \frac{cq}{\hbar\omega}\bar{\boldsymbol{A}}\right)\psi(\boldsymbol{x})\bar{\psi}(\boldsymbol{x}) ,$$
>
> $$(3.146)$$
>
> $\theta^{\ni}_{\text{spin}} = 0, \boldsymbol{A} = 0, \bar{\boldsymbol{A}} = 0, \boldsymbol{P}_s = 0, \bar{\boldsymbol{P}}_s = 0 \Longrightarrow$ "electron pair" ,
> $\theta^{\ni}_{\text{spin}} = 0, \boldsymbol{A} = 0, \bar{\boldsymbol{A}} = 0, \boldsymbol{P}_s \neq 0, \bar{\boldsymbol{P}}_s \neq 0 \Longrightarrow$ "superparticle" ,
> $\theta^{\ni}_{\text{spin}} = 0, \boldsymbol{A} \neq 0, \bar{\boldsymbol{A}} \neq 0, \boldsymbol{P}_s = 0, \bar{\boldsymbol{P}}_s = 0 \Longrightarrow$ "electron pair current" ,
> $\theta^{\ni}_{\text{spin}} = 0, \boldsymbol{A} \neq 0, \bar{\boldsymbol{A}} \neq 0, \boldsymbol{P}_s \neq 0, \bar{\boldsymbol{P}}_s \neq 0 \Longrightarrow$ "superparticle current" ,
> $\theta^{\ni}_{\text{spin}} \neq 0, \boldsymbol{A} = 0, \bar{\boldsymbol{A}} = 0, \boldsymbol{P}_s = 0, \bar{\boldsymbol{P}}_s = 0 \Longrightarrow$ "electron pair, spin separation" ,
> $\theta^{\ni}_{\text{spin}} \neq 0, \boldsymbol{A} = 0, \bar{\boldsymbol{A}} = 0, \boldsymbol{P}_s \neq 0, \bar{\boldsymbol{P}}_s \neq 0 \Longrightarrow$ "superparticle, spin separation" ,
> $\theta^{\ni}_{\text{spin}} \neq 0, \boldsymbol{A} \neq 0, \bar{\boldsymbol{A}} \neq 0, \boldsymbol{P}_s = 0, \bar{\boldsymbol{P}}_s = 0 \Longrightarrow$ "electron pair current, spin separation" ,
> $\theta^{\ni}_{\text{spin}} \neq 0, \boldsymbol{A} \neq 0, \bar{\boldsymbol{A}} \neq 0, \boldsymbol{P}_s \neq 0, \bar{\boldsymbol{P}}_s \neq 0 \Longrightarrow$ "superparticle current, spin separation" .

The Transition to the Ginzburg–Landau Theory of Superconductivity

We consider Box 3.22. The first segment shows formulae of classical electrodynamics, whereas the second segment deals with the London theory of superconductivity and the third segment deals with the Ginzburg–Landau theory of superconductivity. The deepening remarks that follow should be useful for the reader. (i) In the normal phase, following Ohm's law (3.147), a current density j is the answer to an electric field E. (ii) In the superconducting phase, following the London equation (3.153) or following the Ginzburg–Landau equation (3.158), a current density j, on the one hand, is a function of the magnetic field B implemented via the vector potential A, and on the other hand, is a function of wave properties implemented via the wavefunction ψ_{sc}. (iii) In the superconducting phase, following the London equations (3.151), on the one hand, changes of j in time $(\partial/\partial t)$ and E are mutually conditional, and on the other hand, rotational changes of j in space $(\nabla\times)$ and B are mutually conditional. (iv) Last but not least, in the superconducting phase, following the London law (3.155) or following the Ginzburg–Landau law (3.160), the penetration depth λ of B explicitly depends on the effective mass m of an electron pair, on the charge $q = 2e$ of an electron pair, and on the charge number density $n = |\psi_0|^2$.

The information content that is provided by the London equation (3.153) and the Ginzburg–Landau equation (3.157), on the one hand, is covered by our GEFT model, and on the other hand, is upgraded by our GEFT model. We here point at the following issues. Firstly, the GEFT model includes the notion of a kinetic energy, on the one hand, determined by a particle motion $(\to -i\hbar\nabla)$, and on the other hand, determined by an additional magnetic field $(\to q\boldsymbol{A})$. Secondly, the GEFT model includes the notion of an interaction energy that is distinguished by a mathematical structure of the type $\beta|\psi_{sc}|^2\psi_{sc}$ ($\to \bar{\psi}\nabla\psi\nabla\psi$ + further $\bar{\psi}\psi\psi$ parts $\Rightarrow \beta|\psi|^2\psi$). Thirdly, the GEFT model makes the vibrational potential $\boldsymbol{P}_s$ to a partner of the vector potential $\boldsymbol{A}$, upgrading the London equation (3.153) and the Ginzburg–Landau equation (3.157). Fourthly, we consider Box 3.23. (i) The Ginzburg–Landau free enthalpy is specified by interaction terms of fourth order in ψ and ψ^*, however, the GEFT model free enthalpy is specified by interaction terms of third order, fourth order, and fifth order in ψ and ψ^*, allowing a more complicated state topology. (ii) The Ginzburg–Landau current density works with the vector potential thus reflecting our conception "electron pair", however, the GEFT model current density additionally works with the vibrational potential thus reflecting our conception "electron pair" and our conception "superparticle", allowing a much more complicated state topology. (iii) We diagnose that structural properties including chemical bond, crystal structure, and microstructure of substances/materials, in comparison to the Ginzburg–Landau model, in the GEFT model can be implemented in subtle details. For example, the vibrational potential $\boldsymbol{P}_s$ depends on parameters that characterize the chemical bonds, the crystal structures, and the microstructures of substances/materials so that models for the influence of such structural characteristics step by step can be designed. In this context, the reader may think of the dependence of free enthalpy and current density of the chemical bonds, the crystal structures, and the microstructures of substances/materials.

Box 3.22 (Important formulae: Ohm, Maxwell, London, Ginzburg, Landau).

E = electric field, B = magnetic field, A = vector potential,
j = current density, j_L = London superconducting current density,
j_GL = Ginzburg–Landau superconducting current density,
σ = tensor of conductivity,
ϵ_0 = dielectric constant, μ_0 = magnetic constant,
ε_r = tensor of relative permittivity, ε = tensor of permittivity,
μ_r = tensor of relative permeability, μ = tensor of permeability,
e = elementary charge, q = charge, m = effective mass,
ρ = charge density, n = charge number density,
ψ_sc = superconducting wavefunction,
ψ_0 = equilibrium value of the superconducting wavefunction,
$S(\boldsymbol{x})$ = phase of the superconducting wavefunction,
$\boldsymbol{x}$ = position vector, x = penetration direction,
$\hat{\boldsymbol{p}}_A$ = momentum operator in the presence of a magnetic field,
$\hbar$ = Planck's constant,
$\boldsymbol{v}$ = vector of velocity,
G_GL = Ginzburg–Landau free enthalpy, G_0 = free enthalpy in the normal phase,
α, β = phenomenological parameters,
Re = real part.

Classical electrodynamics.

Ohm's law (valid: "normal phase"):

$$j = \sigma E \ . \tag{3.147}$$

Maxwell equations (valid: "normal phase", "superconducting phase"):

$$\mathrm{curl}\, \boldsymbol{H} = \nabla \times \boldsymbol{H} = \boldsymbol{j} + \frac{\partial \boldsymbol{D}}{\partial t} \ , \quad \mathrm{curl}\, \boldsymbol{E} = \nabla \times \boldsymbol{E} = -\frac{\partial \boldsymbol{B}}{\partial t} \ , \tag{3.148}$$

$$\mathrm{div}\, \boldsymbol{D} = \nabla \boldsymbol{D} = \rho \ , \quad \mathrm{div}\, \boldsymbol{B} = \nabla \boldsymbol{B} = 0 \ ,$$

$$\boldsymbol{D} = \varepsilon_\mathrm{r} \epsilon_0 \boldsymbol{E} = \varepsilon \boldsymbol{E} \ , \quad \boldsymbol{B} = \mu_\mathrm{r} \mu_0 \boldsymbol{H} = \mu \boldsymbol{H} \ , \tag{3.149}$$

$$\boldsymbol{B} = \mathrm{curl}\, \boldsymbol{A} = \nabla \times \boldsymbol{A} \ .$$

Light velocity in the vacuum:

$$c = \sqrt{\frac{1}{\epsilon_0 \mu_0}} \ . \tag{3.150}$$

Continuation of Box.

London theory of superconductivity.

London equations ("London equation" + "first and second London equation"):

$$j_L = \frac{nq\hbar}{m}\nabla S - \frac{nq^2}{m}A\,, \quad \frac{\partial j_L}{\partial t} = \frac{nq^2}{m}E\,, \quad \nabla \times j_L = -\frac{nq^2}{m}B\,. \quad (3.151)$$

Derivation of the London equation:

$$\hat{p}_A = -i\hbar\nabla - qA\,, \quad \psi_{sc}(x) = \psi_0\exp[iS(x)]\,, \quad \hat{p}_A\psi_{sc}(x) = mv\psi_{sc}(x)$$
$$\Downarrow \quad\quad\quad\quad\quad\quad\quad\quad\quad\quad\quad\quad\quad (3.152)$$
$$v = \frac{\hbar}{m}\nabla S(x) - \frac{q}{m}A\,,$$

$$j_L = qnv \Rightarrow j_L = \frac{nq\hbar}{m}\nabla S(x) - \frac{nq^2}{m}A\,. \quad (3.153)$$

Derivation of the first and the second London equation
(using the second Maxwell equation, i.e. the law of induction):

$$j_L = \frac{nq\hbar}{m}\nabla S - \frac{nq^2}{m}A$$
$$\Downarrow$$
$$\frac{\partial j_L}{\partial t} = \frac{nq^2}{m}E$$
$$\Downarrow$$
$$\frac{\partial j_L}{\partial t} = \frac{nq^2}{m}E \to E = \frac{m}{nq^2}\frac{\partial j_L}{\partial t} \to \nabla\times E = \frac{m}{nq^2}\nabla\times\frac{\partial j_L}{\partial t} \quad (3.154)$$
$$\to -\frac{\partial B}{\partial t} = \frac{m}{nq^2}\nabla\times\frac{\partial j_L}{\partial t} \to \frac{\partial}{\partial t}\left(\nabla\times j_L + \frac{nq^2}{m}B\right) = 0$$
$$\Downarrow$$
$$\nabla\times j_L = -\frac{nq^2}{m}B\,.$$

Calculation of the penetration depth λ
(using the first Maxwell equation, i.e. the law of magnetic flux):

$$\nabla\times j_L = -\frac{nq^2}{m}B \Rightarrow \nabla^2 B = \frac{1}{\lambda^2}B$$
$$\Downarrow \quad\quad\quad\quad\quad\quad\quad\quad\quad (3.155)$$
$$B = B_0\exp[-x/\lambda]\,, \quad \lambda = \sqrt{\frac{m}{\mu_0 q^2 n}}\,.$$

Continuation of Box.

Ginzburg–Landau theory of superconductivity.

The free enthalpy of a superconductor near the superconducting phase transition according to the Ginzburg–Landau theory of superconductivity:

$$G_{\text{GL}} = G_0 + \alpha \left|\psi_{\text{sc}}\right|^2 + \frac{\beta}{2} \left|\psi_{\text{sc}}\right|^4 + \frac{1}{2m} \left|(-i\hbar\nabla - 2e\boldsymbol{A})\psi_{\text{sc}}\right|^2 + \frac{1}{2\mu_0} \left|\boldsymbol{B}\right|^2 \,. \quad (3.156)$$

The minimum of the free enthalpy of a superconductor
near the superconducting phase transition
according to the Ginzburg–Landau theory of superconductivity
is defined by the Ginzburg–Landau equations:

$$-\alpha\psi_{\text{sc}} - \beta\left|\psi_{\text{sc}}\right|^2 \psi_{\text{sc}} = \frac{1}{2m}\left(-i\hbar\nabla - 2e\boldsymbol{A}\right)^2 \psi_{\text{sc}} \,, \quad (3.157)$$

$$\boldsymbol{j}_{\text{GL}} = \frac{2e}{m}\text{Re}\left[\psi_{\text{sc}}^{*}\left(-i\hbar\nabla - 2e\boldsymbol{A}\right)\psi_{\text{sc}}\right] \,. \quad (3.158)$$

The coherence length ξ
according to the Ginzburg–Landau theory of superconductivity:

$$\xi = \sqrt{\frac{\hbar^2}{2m\left|\alpha\right|}} \,. \quad (3.159)$$

The penetration depth λ
according to the Ginzburg–Landau theory of superconductivity:

$$\lambda = \sqrt{\frac{m}{4\mu_0 e^2 \left|\psi_0\right|^2}} \,. \quad (3.160)$$

Notes:
We note that the Ginzburg–Landau parameter $\kappa = \lambda/\xi$, concatenating the coherence length ξ describing the size of electron pairs and the penetration depth λ describing the penetration depth of a magnetic field, separates superconductors of type I from superconductors of type II according to the relations type I $\Leftrightarrow \kappa < 1/\sqrt{2}$ and type II $\Leftrightarrow \kappa > 1/\sqrt{2}$. We note that the penetration depth in the case of the London theory and the penetration depth in the case of the Ginzburg–Landau theory are identical. Setting $q = 2e$, in this way, implementing electron pairs, and setting $n = \left|\psi_0\right|^2$, in this way, switching from classical notions to wave–particle notions, this becomes obvious.

Box 3.23 (Important formulae: the way from GEFT-type to GL-type relations).

GEFT model free enthalpy versus Ginzburg–Landau (GL) free enthalpy:

$$G_{\mathrm{GL}} = G_0 + \alpha \left|\psi_{\mathrm{sc}}\right|^2 + \frac{1}{2m}\left|(-\mathrm{i}\hbar\nabla - 2e\boldsymbol{A})\psi_{\mathrm{sc}}\right|^2 + \frac{1}{2\mu_0}\left|\boldsymbol{B}\right|^2 +$$

$$+ \frac{\beta}{2}\left|\psi_{\mathrm{sc}}\right|^4$$

$$\Updownarrow \qquad (3.161)$$

$$G_{\mathrm{s}} = G_0 - E\left|\psi\right|^2 + \frac{1}{2m}\left|\left[-\mathrm{i}\hbar\nabla - \hbar\frac{\omega}{c}\left(\boldsymbol{P}_{\mathrm{s}} + \frac{cq}{\hbar\omega}\boldsymbol{A}\right)\right]\psi\right|^2 -$$

$$- \frac{\hbar^2}{4m}V_1\left|\psi\right|^2\psi + \frac{\hbar^2}{8m}V_2\left|\psi\right|^4 + \cdots ,$$

with

$$-\frac{\hbar^2}{4m}\nabla\psi\nabla\psi = -\frac{\hbar^2}{4m}V_1\psi\psi \Rightarrow -\frac{\hbar^2}{4m}\psi^*\nabla\psi\nabla\psi = -\frac{\hbar^2}{4m}V_1\left|\psi\right|^2\psi ,$$

$$+\frac{\hbar^2}{4m}\bar{\psi}\nabla\psi\nabla\psi = +\frac{\hbar^2}{4m}V_2\left|\psi\right|^2\psi \Rightarrow +\frac{\hbar^2}{8m}\psi^*\bar{\psi}\nabla\psi\nabla\psi + \frac{\hbar^2}{8m}V_2\left|\psi\right|^4 , \qquad (3.162)$$

$$\cdots .$$

GEFT model current density versus Ginzburg–Landau (GL) current density:

$$\boldsymbol{j}_{\mathrm{GL}} = \frac{2e}{m}\mathrm{Re}\left[\psi_{\mathrm{sc}}^*\left(-\mathrm{i}\hbar\nabla - 2e\boldsymbol{A}\right)\psi_{\mathrm{sc}}\right]$$

$$\Updownarrow \qquad (3.163)$$

$$\boldsymbol{j}_{\mathrm{s}} = \frac{q}{m}\mathrm{Re}\left(\psi^*\left[-\mathrm{i}\hbar\nabla - \hbar\frac{\omega}{c}\left(\boldsymbol{P}_{\mathrm{s}} + \frac{cq}{\hbar\omega}\boldsymbol{A}\right)\right]\psi\right) .$$

GEFT model penetration depth versus Ginzburg–Landau (GL) penetration depth:

$$\lambda_{\mathrm{GL}} = \sqrt{\frac{m}{4\mu_0 e^2\left|\psi_0\right|^2}}$$

$$= \qquad (3.164)$$

$$\lambda_{\mathrm{s}} = \sqrt{\frac{m}{\mu_0 q^2\left|\psi_0\right|^2}} ,$$

Continuation of Box.

which follows from

$$j_s = \frac{q}{m} \operatorname{Re}\left(\psi^* \left[-i\hbar\nabla - \hbar\frac{\omega}{c}\left(P_s + \frac{cq}{\hbar\omega}A\right)\right]\psi\right) \tag{3.165}$$

$$\Longrightarrow$$

$$\nabla \times j_s = \nabla \times \frac{q}{m} \operatorname{Re}\left(\psi^* \left[-i\hbar\nabla - \hbar\frac{\omega}{c}\left(P_s + \frac{cq}{\hbar\omega}A\right)\right]\psi\right), \tag{3.166}$$

$$\psi = \psi_0 \exp\left[iS(\boldsymbol{x})\right] \tag{3.167}$$

$$\Longrightarrow$$

$$\nabla \times j_s = \frac{q}{m} \operatorname{Re}\left[-i\hbar\nabla \times \left[\nabla S(\boldsymbol{x})\right] - \hbar\frac{\omega}{c}\left(\nabla \times P_s + \frac{cq}{\hbar\omega}\nabla \times A\right)\right]|\psi_0|^2, \tag{3.168}$$

$$\nabla \times \left[\nabla S(\boldsymbol{x})\right] = \operatorname{curl grad} S(\boldsymbol{x}) = 0,$$

$$\nabla \times P_s = \operatorname{curl} P_s := 0 \tag{3.169}$$

(no rotational properties of the vibrational potential)

$$\Longrightarrow$$

$$\nabla \times j_s = -\frac{q^2}{m} \operatorname{Re}\left[\nabla \times A\right]|\psi_0|^2 = -\frac{q^2|\psi_0|^2}{m}\nabla \times A = -\frac{q^2|\psi_0|^2}{m}B, \tag{3.170}$$

$$\nabla \times j_s = \nabla \times \nabla \times H = -\nabla^2 H = -\frac{1}{\mu_0}\nabla^2 B \tag{3.171}$$

(using the first Maxwell equation, i.e. the law of magnetic flux)

$$\Longrightarrow$$

$$\nabla^2 B = \frac{1}{\lambda_s^2} B \tag{3.172}$$

$$\Longrightarrow$$

$$B = B_0 \exp\left[-x/\lambda_s\right], \quad \lambda_s = \sqrt{\frac{m}{\mu_0 q^2 |\psi_0|^2}}. \tag{3.173}$$

Notes:
We utilize that the application of differential operators to wavefunctions ψ leads to results that can be expressed as products of interaction energy terms V_i and wavefunctions ψ, and we utilize that the adjoint $\bar{\psi}$ of a wavefunction ψ primarily is the conjugate complex ψ^* of a wavefunction ψ.

The Transition to the BCS Theory of Superconductivity

In Box 3.24, we present a collection of typical BCS-type Hamiltonians. The following remarks should be useful. (i) The effective BCS model Hamiltonian $\hat{H}_{\text{BCS}}$ that is defined by (3.174) assumes one single energy band populated by electrons that are defined by wave vectors $\boldsymbol{k}$ and spin orientations $\sigma = \{\uparrow, \downarrow\}$. Applied to a suitable state vector, the first term of (3.174) sets up the kinetic energies $E_{\boldsymbol{k}}$ of the electrons in the energy band relative to the chemical potential μ of the energy band, whereas the second term of (3.174) sets up the interaction energies $V_{\boldsymbol{k},\boldsymbol{k}'}$ in each case resulting from two electrons that interact with each other in such a way that electron states $\boldsymbol{k}' \uparrow$ and $-\boldsymbol{k}' \downarrow$ are destroyed and electron states $-\boldsymbol{k} \downarrow$ and $\boldsymbol{k} \uparrow$ are generated. (ii) A modification of the effective BCS model Hamiltonian $\hat{H}_{\text{BCS}}$ that assumes two overlapping energy bands in the vicinity of the Fermi surface is defined by (3.175). Applied to a suitable state vector, $\hat{H}_1$ sets up the kinetic energies $E_{\boldsymbol{k}}^a$ and $E_{\boldsymbol{k}}^b$ of the electrons in band a and band b relative to the chemical potentials μ^a and μ^b of band a and band b. Furthermore, $\hat{H}_2$ sets up the effective attractive intraband interaction energies V^a and V^b. Moreover, $\hat{H}_3$ sets up the effective attractive interband interaction energy V^{ab}. Naturally, the interaction energies V^a, V^b, and V^{ab} must reflect the impact of constructive vibrations and compatible effects in central parts. (iii) The two-band model (3.175), together with a thermodynamic model allowing us to incorporate the notion temperature T, implies a two-band-gap structure which explicitly depends on the temperature T. The two-band-gap equations are given by (3.176). Considering relatively small band gaps Δ^a, Δ^b, i.e. $\Delta^a, \Delta^b \to 0$, the temperature T approaches the critical temperature T_{c}. The two-band-gap equations then are given by (3.178). SrTiO$_3$ doped with Nb is an example for a ceramic material that shows a two-band-gap structure. (iv) Regarding the thermodynamic model allowing us to incorporate the notion temperature T, let us here collect the theoretical details that follow. (1) In a first step, one applies the Bogoljubov transformation [9] to the Fermi operators ($\hat{a}^{\pm}$ and $\hat{b}^{\pm}$) of (3.175), in last consequence, introducing a transformed Hamiltonian containing unknown functions that can be determined during the calculations that follow. (2) In a second step, one utilizes that connections of thermodynamic potentials and Hamiltonians are established via mean values, in last consequence, introducing the absolute temperature T via the Boltzmann term $k_{\text{B}}T$. Following [8], one specifies such a thermodynamic potential by applying the transformed Hamiltonian, leading to contributions of various order to the thermodynamic potential. (3) In a third step, one applies additional conditions to the contributions of various order to the thermodynamic potential, in last consequence, evoking relations that fix the unknown functions. Following [29], one easily derives (3.176) and (3.177), finally leading to (3.178). (4) In a fourth step, one has to fix the interaction energies V^a, V^b, and V^{ab} as well as the kinetic energies $E_{\boldsymbol{k}}^a$ and $E_{\boldsymbol{k}}^b$ and the chemical potentials μ^a and μ^b. On the one hand, this can be systematically done by adapting (3.176), (3.177), and (3.178) to measurement results. On the other hand, this can be systematically done by developing interaction models of electrons populating given energy bands, in last consequence, by starting from classical models. (v) We here additionally note that a lot of refinements of this method are known today. In this context, however, we want to refer to [37].

Box 3.24 (Important formulae: Bardeen, Cooper, Schrieffer).

The situation assumed by the BCS theory in terms of generation operators $\hat{c}^+$ and destruction operators $\hat{c}^-$ is described by the effective BCS model Hamiltonian $\hat{H}_{\text{BCS}}$:

$$\hat{H}_{\text{BCS}} = \sum_{k,\sigma} (E_k - \mu)\hat{c}^+_{k,\sigma}\hat{c}^-_{k,\sigma} + \sum_{k,k'} V_{k,k'}\hat{c}^+_{k\uparrow}\hat{c}^+_{-k\downarrow}\hat{c}^-_{-k'\downarrow}\hat{c}^-_{k'\uparrow} . \quad (3.174)$$

A modification of the effective BCS model Hamiltonian $\hat{H}_{\text{BCS}}$ according to [37, 31, 54]:

$$\hat{H}_{\text{SM}} = \hat{H}_1 + \hat{H}_2 + \hat{H}_3 ,$$

$$\hat{H}_1 = \sum_{k,\sigma} (E^a_k - \mu^a)\hat{a}^+_{k,\sigma}\hat{a}^-_{k,\sigma} + \sum_{k,\sigma} (E^b_k - \mu^b)\hat{b}^+_{k,\sigma}\hat{b}^-_{k,\sigma} ,$$

$$\hat{H}_2 = -\sum_{k \neq g} V^a \hat{a}^+_{k,\uparrow}\hat{a}^+_{-k,\downarrow}\hat{a}^-_{-g,\downarrow}\hat{a}^-_{g,\uparrow} + \sum_{k \neq g} V^b \hat{b}^+_{k,\uparrow}\hat{b}^+_{-k,\downarrow}\hat{b}^-_{-g,\downarrow}\hat{b}^-_{g,\uparrow} , \quad (3.175)$$

$$\hat{H}_3 = -\sum_{k \neq g} V^{ab} \left(\hat{a}^+_{k,\uparrow}\hat{a}^+_{-k,\downarrow}\hat{b}^-_{-g,\downarrow}\hat{b}^-_{g,\uparrow} + \hat{b}^+_{k,\uparrow}\hat{b}^+_{-k,\downarrow}\hat{a}^-_{-g,\downarrow}\hat{a}^-_{g,\uparrow} \right) .$$

The two-band-gap equations of the two-band model (3.175) according to [37, 31, 54]:

$$\Delta^a = \frac{V^a}{2} \sum_k \frac{\Delta^a}{\Omega^a(k)} \tanh\left[\frac{\Omega^a(k)}{2k_B T}\right] + \frac{V^{ab}}{2} \sum_k \frac{\Delta^b}{\Omega^b(k)} \tanh\left[\frac{\Omega^b(k)}{2k_B T}\right] ,$$

$$\Delta^b = \frac{V^b}{2} \sum_k \frac{\Delta^b}{\Omega^b(k)} \tanh\left[\frac{\Omega^b(k)}{2k_B T}\right] + \frac{V^{ab}}{2} \sum_k \frac{\Delta^a}{\Omega^a(k)} \tanh\left[\frac{\Omega^a(k)}{2k_B T}\right] , \quad (3.176)$$

$$\Omega^a(k) = \sqrt{\epsilon^a(k)^2 + \Delta^{a\,2}} , \quad \Omega^b(k) = \sqrt{\epsilon^b(k)^2 + \Delta^{b\,2}} ,$$

$$\epsilon^a(k) = E^a_k - \mu^a , \quad \epsilon^b(k) = E^b_k - \mu^b . \quad (3.177)$$

$\Delta^a, \Delta^b \to 0$:

$$\Delta^a = \frac{V^a}{2} \sum_k \frac{\Delta^a}{\epsilon^a(k)} \tanh\left[\frac{\epsilon^a(k)}{2k_B T_c}\right] + \frac{V^{ab}}{2} \sum_k \frac{\Delta^b}{\epsilon^b(k)} \tanh\left[\frac{\epsilon^b(k)}{2k_B T_c}\right] ,$$

$$\Delta^b = \frac{V^b}{2} \sum_k \frac{\Delta^b}{\epsilon^b(k)} \tanh\left[\frac{\epsilon^b(k)}{2k_B T_c}\right] + \frac{V^{ab}}{2} \sum_k \frac{\Delta^a}{\epsilon^a(k)} \tanh\left[\frac{\epsilon^a(k)}{2k_B T_c}\right] . \quad (3.178)$$

The way from GEFT-type to BCS-type relations is shown in Box 3.25. The remarks that follow should be useful. (i) The GEFT-type model (3.179)–(3.182) eventually resorts to the GEFT-type Hamiltonian (3.141), according to (3.142), completed by the vector potential $\boldsymbol{A}$ allowing the implementation of magnetic fields $\boldsymbol{B} = \nabla \times \boldsymbol{A}$. The GEFT-type Hamiltonian (3.141) offers us a parameter $\boldsymbol{L}_{s,0}$, allowing us to target relative total energies $E - \hbar^2 \boldsymbol{L}_{s,0}^2/2m = E - E_0 = \Delta E$ and relative kinetic energies $T - \hbar^2 \boldsymbol{L}_{s,0}^2/2m = T - E_0 = \Delta T$. This GEFT-type Hamiltonian, on the one hand, covers the two-particle states "electron pair"/"superparticle", and on the other hand, covers the many-particle states "coherent electron pairs"/"coherent superparticles", and $\boldsymbol{P}_s, \bar{\boldsymbol{P}}_s = 0$ evokes electron pair states, $\boldsymbol{P}_s, \bar{\boldsymbol{P}}_s \neq 0$ evokes superparticle states, and the field quantities $\psi, \bar{\psi}, \boldsymbol{A}, \bar{\boldsymbol{A}}$ etc. specify the diverse electron pair states and the diverse superparticle states. (ii) The BCS-type model (3.183)–(3.185) is readily obtained following Box 3.25. This BCS-type model contains position information and energy information ("position representation"). This BCS-type model can be used as starting point for the introduction of alternative representations such as informationally reduced BCS-type models of the kind (3.174) exclusively containing energy information ("energy representation") by utilizing commutator relations and eigenvalue equations and by applying methods of group theory. (iii) A special feature of BCS-type operators is the "normalization" feature, i. e. BCS-type operators, acting on state vectors such as wavefunctions, do not generate energy/momentum eigenvalues/functions etc. in all their complexity, but do generate quantum numbers counting fundamental elements of energy/momentum eigenvalues/functions etc., and these must be preset. In order to implement the "normalization" feature, we proceed following the part (3.190)–(3.192). Another special feature of BCS-type operators is the "individualization" feature, i. e. BCS-type operators are raising operators and lowering operators, in a formal manner, filling/emptying possible energy states with/of "particles" and/or "particle aspects" like electrons and/or phonons. In order to implement the "individualization" feature, we proceed following the part (3.193). Another special feature of BCS-type operators is the "compaction" feature, i. e. BCS-type operators operate on state vectors such as wavefunctions that incorporate all "particles" and/or "particle aspects" like electrons and/or phonons. In order to implement the "compaction" feature, we proceed following the part (3.194). (iv) The BCS-type model (3.183)–(3.185) that comes into being in this way replaces the total field quantities $\psi, \bar{\psi}, \boldsymbol{A}, \bar{\boldsymbol{A}}$ etc., describing total systems of coherent electron pairs/coherent superparticles, by system-specific sub-quantities $\psi_i(\boldsymbol{x}_i), \bar{\psi}_i(\boldsymbol{x}_i), \boldsymbol{A}(\boldsymbol{x}_i/\boldsymbol{x}_j), \bar{\boldsymbol{A}}(\boldsymbol{x}_i/\boldsymbol{x}_j)$ etc., depending on their specification, describing system-specific sub-systems i such as electron-like sub-systems completed by remains or electron-pair-like/superparticle-like sub-systems completed by remains showing us that superordinate unities are present that are not equivalent to interacting particles. Hence, the neglection of the system-specific sub-quantities representing the remains finally launches the notion of particles such as electrons that form total systems of coherent electron pairs/coherent superparticles or such as electron pairs/superparticles that form total systems of coherent electron pairs/coherent superparticles, and this is true for sum-type decompositions and product-type decompositions. Naturally, since wavefunctions and vibrational functions/compatible functions have been merged by "order parameter relations", no phonon parts/compatible parts occur in (3.185).

The way from GEFT-type to BCS-type interaction terms is shown in Box 3.25 by using the first interaction term of (3.140) + (3.141). The remarks that follow should be useful. (i) As it is illustrated in Figure 3.76 by considering the first interaction term of (3.140) + (3.141), the operative expressions of GEFT-type interaction terms evoke partial aspects of the interaction, approaching the notion of particle states that are destroyed and particle states that are generated. For instance, in a first step, $\hat{G}^1_{\bar{\psi}_s}\hat{G}^2_{\bar{\psi}_s}$ gives birth to the interaction field $\bar{\psi}_s$, and in a second step, $\hat{G}^1\hat{G}^2$ gives birth to a final state ψ' the deviation from the initial state ψ of which is given by $V_{\bar{\psi}_s}$, approaching the notion of an initial particle state ψ that is destroyed and a final particle state $\psi' = V_{\bar{\psi}_s}\psi$ that is generated. (ii) As it is illustrated in Figure 3.77 by considering the first interaction term of (3.140) + (3.141), the "individualization" feature that converts GEFT-type interaction terms into BCS-type interaction terms finally leads to a central feature of BCS-type interaction terms, namely it leads from partial aspects of the interaction to particle aspects of the interaction. We note that the $\hat{W}^+_i\hat{W}^-_i$ shown in Box 3.25 and Figure 3.77 evoke the destruction of an initial state of particle i and the generation of a final state of particle i. We also note that the uprightly placed operative expressions of Figure 3.77 evoke self-interaction contributions to the $\hat{W}^+_i\hat{W}^-_i$. We further note that this takes for granted that the $\hat{W}^+_j\hat{W}^-_j$ that are associated with remains describing deviations from the notion of individual particles are neglected. (iii) For example, $\psi = \psi_{e^-_1} + \psi_{e^-_2} + \cdots$ characterizes an entity that in some approximation can be considered as composed of two electrons. Neglecting the remains $\cdots$ describing deviations from the notion of two electrons, we arrive at $\psi = \psi_{e^-_1} + \psi_{e^-_2}$ and thus at $\hat{V}^+_{\bar{\psi}_s,2}\hat{V}^+_{\bar{\psi}_s,1}\hat{V}^-_{\bar{\psi}_s,2}\hat{V}^-_{\bar{\psi}_s,1} = \hat{W}^+_1\hat{W}^-_1 + \hat{W}^+_2\hat{W}^-_2$, in last consequence, establishing the notion of two electrons the interaction of which is described by generation operators generating particle states and destruction operators destroying particle states.

What can we learn following these threads? Well, especially we become aware that generation operators and destruction operators relating to kinetic energy aspects, in the case of superconducting states, firstly are dealing with the kinetic energy of electrons ($\rightarrow \pm i\nabla$ in $\hat{L}^\mp_s$), secondly are dealing with kinetic energy contributions that are evoked by field contributions to the electrons that are due to constructive vibrations and compatible effects ($\rightarrow \pm\omega\boldsymbol{P}_s/c$ in $\hat{L}^\mp_s$), and thirdly are dealing with kinetic energy contributions that are evoked by field contributions to the electrons that are due to the remainders of the magnetic field ($\rightarrow \pm q\boldsymbol{A}/\hbar$ in $\hat{L}^\mp_s$). Well, going beyond this aspect, we become aware that four-party operators such as $\hat{V}^+_{\bar{\psi}_s,2}\hat{V}^+_{\bar{\psi}_s,1}\hat{V}^-_{\bar{\psi}_s,2}\hat{V}^-_{\bar{\psi}_s,1}$ as well as two-party operators such as $\hat{W}^+_1\hat{W}^-_1$ and $\hat{W}^+_2\hat{W}^-_2$ are eventually capable of breeding generation aspects and destruction aspects which are connected with interactions of two electrons, provided quantum correlations [33, 36] are taken into account dealing with two-party operators ($\rightarrow \hat{G}^j_i, \hat{G}^i_{\bar{\psi}_s,i}$). Well, going further beyond this aspect, we become aware that four-party operators as well as two-party operators are eventually capable of resolving generation aspects and destruction aspects in space ($\rightarrow \hat{G}^j_i, \hat{G}^i_{\bar{\psi}_s,i}$) – and also in time if time dependencies are not neglected, as it is done in this study for the sake of simplicity. Following these hints, further classes of upgraded BCS models can be immediately established.

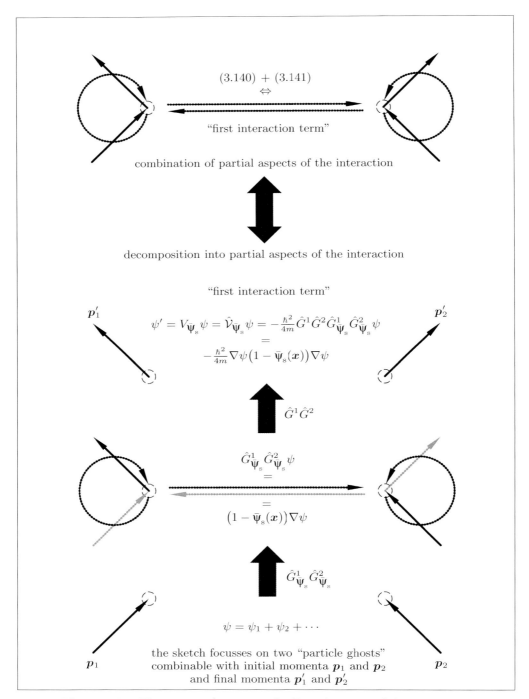

Figure 3.76. Electron pair/superparticle. Partial aspects of the interaction.

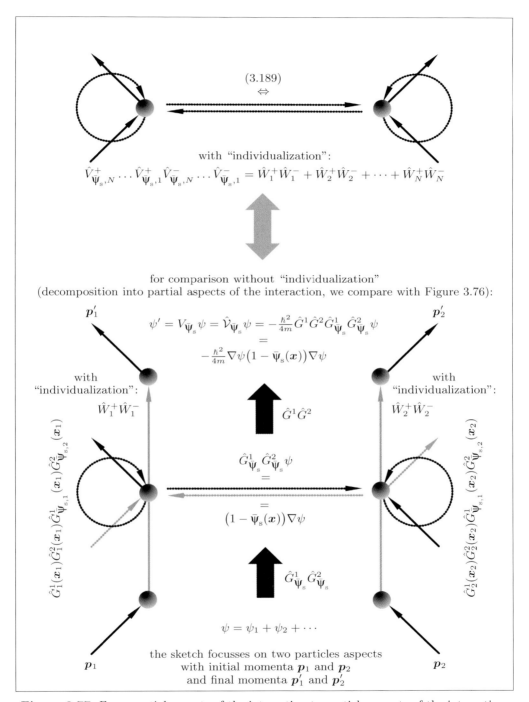

Figure 3.77. From partial aspects of the interaction to particle aspects of the interaction.

> **Box 3.25** (Important formulae: the way from GEFT-type to BCS-type relations).
>
> $\hat{\mathcal{H}}_s$ = Hamiltonian (superconducting state),
> $\hat{\mathcal{T}}_s$ = kinetic energy operator (superconducting state).
>
> GEFT-type model:
>
> $$E\psi(\boldsymbol{x}) = \hat{\mathcal{H}}_s \psi(\boldsymbol{x}) \ , \ \hat{\mathcal{H}}_s := \hat{\mathcal{T}}_s + \cdots \ , \tag{3.179}$$
>
> or
>
> $$\left(E - \frac{\hbar^2}{2m}\boldsymbol{L}_{s,0}^2\right)\psi(\boldsymbol{x}) = \hat{\mathcal{H}}'_s \psi(\boldsymbol{x}) \ , \ \hat{\mathcal{H}}'_s := -\frac{\hbar^2}{2m}\hat{L}_s^+ \hat{L}_s^- + \cdots \ , \tag{3.180}$$
>
> $$\hat{\mathcal{T}}_s = -\frac{\hbar^2}{2m}\left(\hat{L}_s^+ \hat{L}_s^- - \boldsymbol{L}_{s,0}^2\right) \ , \tag{3.181}$$
>
> $$\begin{aligned}\hat{L}_s^+ &= \boldsymbol{L}_{s,0} - \mathrm{i}\nabla - \frac{\omega}{c}\left(\boldsymbol{P}_s + \frac{cq}{\hbar\omega}\boldsymbol{A}\right) \ , \\ \hat{L}_s^- &= \boldsymbol{L}_{s,0} + \mathrm{i}\nabla + \frac{\omega}{c}\left(\boldsymbol{P}_s + \frac{cq}{\hbar\omega}\boldsymbol{A}\right) \ .\end{aligned} \tag{3.182}$$
>
> BCS-type model:
>
> $$\begin{aligned}\left(E - \frac{\hbar^2}{2m}\boldsymbol{L}_{s,0}^2\right)\psi(\boldsymbol{x}_1, \boldsymbol{x}_2, \ldots, \boldsymbol{x}_N) &= \hat{H}'_s \psi(\boldsymbol{x}_1, \boldsymbol{x}_2, \ldots, \boldsymbol{x}_N) \ , \\ \hat{H}'_s &= \sum_{i=1}^N \left(T_i - \frac{\hbar^2}{2m}\boldsymbol{L}_{s,0}^2\right)\hat{L}_{s,i}^+(\boldsymbol{x}_i)\hat{L}_{s,i}^-(\boldsymbol{x}_i) + \cdots \ ,\end{aligned} \tag{3.183}$$
>
> $$\hat{L}_{s,i}^+(\boldsymbol{x}_i) = \frac{\mathrm{i}\hbar \hat{L}_s^+(\boldsymbol{x}_i)}{\sqrt{2m\left(T_i - \frac{\hbar^2}{2m}\boldsymbol{L}_{s,0}^2\right)}} \ , \ \hat{L}_{s,i}^-(\boldsymbol{x}_i) = \frac{\mathrm{i}\hbar \hat{L}_s^-(\boldsymbol{x}_i)}{\sqrt{2m\left(T_i - \frac{\hbar^2}{2m}\boldsymbol{L}_{s,0}^2\right)}} \ ,$$
>
> $$\begin{aligned}\hat{L}_s^+(\boldsymbol{x}_i) &= \boldsymbol{L}_{s,0} - \mathrm{i}\nabla_{\boldsymbol{x}_i} - \frac{\omega}{c}\left(\boldsymbol{P}_s(\boldsymbol{x}_i/\boldsymbol{x}_j) + \frac{cq}{\hbar\omega}\boldsymbol{A}(\boldsymbol{x}_i/\boldsymbol{x}_j)\right) \ , \\ \hat{L}_s^-(\boldsymbol{x}_i) &= \boldsymbol{L}_{s,0} + \mathrm{i}\nabla_{\boldsymbol{x}_i} + \frac{\omega}{c}\left(\boldsymbol{P}_s(\boldsymbol{x}_i/\boldsymbol{x}_j) + \frac{cq}{\hbar\omega}\boldsymbol{A}(\boldsymbol{x}_i/\boldsymbol{x}_j)\right) \ ,\end{aligned} \tag{3.184}$$
>
> $$\boldsymbol{P}_s(\boldsymbol{x}_i/\boldsymbol{x}_j) = \begin{cases}\boldsymbol{P}_s(\boldsymbol{x}_i) & \text{if applied to } \psi_i(\boldsymbol{x}_i) \\ 0 & \text{if applied to } \psi_j(\boldsymbol{x}_j)\end{cases} \ ,$$
>
> $$\boldsymbol{A}(\boldsymbol{x}_i/\boldsymbol{x}_j) = \begin{cases}\boldsymbol{A}(\boldsymbol{x}_i) & \text{if applied to } \psi_i(\boldsymbol{x}_i) \\ 0 & \text{if applied to } \psi_j(\boldsymbol{x}_j)\end{cases}$$
>
> $$\psi(\boldsymbol{x}_1, \boldsymbol{x}_2, \ldots, \boldsymbol{x}_N) = \sum_{i=1}^N \psi_i(\boldsymbol{x}_i) \ \text{ or } \ \psi(\boldsymbol{x}_1, \boldsymbol{x}_2, \ldots, \boldsymbol{x}_N) = \prod_{i=1}^N \psi_i(\boldsymbol{x}_i) \ . \tag{3.185}$$

> **Continuation of Box.**

GEFT-type interaction model:

$$V_{\bar{\psi}_s}\psi(x) = \hat{\mathcal{V}}_{\bar{\psi}_s}\psi(x) = -\frac{\hbar^2}{4m}\hat{G}^1\hat{G}^2\hat{G}^1_{\bar{\psi}_s}\hat{G}^2_{\bar{\psi}_s}\psi(x) , \quad (3.186)$$

$$\hat{\mathcal{V}}_{\bar{\psi}_s} = -\frac{\hbar^2}{4m}\hat{G}^1\hat{G}^2\hat{G}^1_{\bar{\psi}_s}\hat{G}^2_{\bar{\psi}_s} = -\frac{\hbar^2}{4m}\nabla\psi(x)\big(1-\bar{\psi}_s(x)\big)\nabla , \quad (3.187)$$

$$\bar{\psi}_s(x) = \bar{\psi}(x)\,1 - \frac{1}{2}\bar{P}_s\cdot\bar{P}_s\,\bar{\psi}(x)\bar{\psi}(x) . \quad (3.188)$$

BCS-type interaction model:

$$\hat{V}_{\bar{\bar{\psi}}_s}\psi(x_1,x_2,\ldots,x_N)$$
$$=$$
$$V_{\bar{\bar{\psi}}_s}\big(\hat{V}^+_{\bar{\psi}_s,N}\cdots\hat{V}^+_{\bar{\psi}_s,1}\hat{V}^-_{\bar{\psi}_s,N}\cdots\hat{V}^-_{\bar{\psi}_s,1}\big)\psi(x_1,x_2,\ldots,x_N) ,$$

$$\hat{V}^+_{\bar{\psi}_s,N}\cdots\hat{V}^+_{\bar{\psi}_s,1}\hat{V}^-_{\bar{\psi}_s,N}\cdots\hat{V}^-_{\bar{\psi}_s,1} = \hat{W}^+_1\hat{W}^-_1 + \hat{W}^+_2\hat{W}^-_2 + \cdots + \hat{W}^+_N\hat{W}^-_N ,$$

$$\hat{W}^+_i\hat{W}^-_i = \sum_{j=1}^N \hat{G}^1_j(x_j)\hat{G}^2_j(x_j)\hat{G}^1_{\bar{\psi}_s,1}(x_i)\hat{G}^2_{\bar{\psi}_s,2}(x_i) ,$$

$$\hat{G}^1_1(x_1) = \big[-\mathrm{i}\sqrt{\nabla_{x_1}\psi_1(x_1)}\big] , \quad \hat{G}^2_1(x_1) = \big[\mathrm{i}\sqrt{\nabla_{x_1}\psi_1(x_1)}\big] , \ldots$$
$$(3.189)$$
$$\hat{G}^1_2(x_2) = \big[-\mathrm{i}\sqrt{\nabla_{x_2}\psi_2(x_2)}\big] , \quad \hat{G}^2_2(x_2) = \big[\mathrm{i}\sqrt{\nabla_{x_2}\psi_2(x_2)}\big] , \ldots$$

$$\hat{G}^1_{\bar{\psi}_s,1}(x_i) = \frac{\sqrt{-\frac{\hbar^2}{4m}}\big[-\mathrm{i}\sqrt{(1-\bar{\psi}_s(x_i))}\,\nabla_{x_i}\big]}{\sqrt{V_{\bar{\psi}_s}}} ,$$

$$\hat{G}^2_{\bar{\psi}_s,2}(x_i) = \frac{\sqrt{-\frac{\hbar^2}{4m}}\big[\mathrm{i}\sqrt{(1-\bar{\psi}_s(x_i))}\,\nabla_{x_i}\big]}{\sqrt{V_{\bar{\psi}_s}}} ,$$

$$\psi(x_1,x_2,\ldots,x_N) = \sum_{i=1}^N \psi_i(x_i) \quad\text{or}\quad \psi(x_1,x_2,\ldots,x_N) = \prod_{i=1}^N \psi_i(x_i) .$$

> **Continuation of Box.**

The way from the GEFT-type model to the BCS-type model:

$$\psi = \psi_1 + \psi_2 + \cdots ,$$

$$\left(E - \frac{\hbar^2}{2m}\boldsymbol{L}_{s,0}^2\right)(\psi_1 + \psi_2 + \cdots) = \left(-\frac{\hbar^2}{2m}\hat{L}_s^+\hat{L}_s^- + \cdots\right)(\psi_1 + \psi_2 + \cdots) \quad (3.190)$$

$$\Longrightarrow$$

$$\left(E - \frac{\hbar^2}{2m}\boldsymbol{L}_{s,0}^2\right)(\psi_1 + \psi_2 + \cdots) = -\frac{\hbar^2}{2m}\hat{L}_s^+\hat{L}_s^-\psi_1 - \frac{\hbar^2}{2m}\hat{L}_s^+\hat{L}_s^-\psi_2 + \cdots \quad (3.191)$$

$$\Longrightarrow$$

$$\left(E - \frac{\hbar^2}{2m}\boldsymbol{L}_{s,0}^2\right)(\psi_1(\boldsymbol{x}) + \psi_2(\boldsymbol{x}) + \cdots) =$$

$$= \left(T_1 - \frac{\hbar^2}{2m}\boldsymbol{L}_{s,0}^2\right)\hat{L}_{s,1}^+\hat{L}_{s,1}^-\psi_1(\boldsymbol{x}) + \left(T_2 - \frac{\hbar^2}{2m}\boldsymbol{L}_{s,0}^2\right)\hat{L}_{s,2}^+\hat{L}_{s,2}^-\psi_2(\boldsymbol{x}) + \cdots ,$$

$$\hat{L}_{s,i}^+ = \frac{\mathrm{i}\hbar\hat{L}_s^+}{\sqrt{2m\left(T_i - \frac{\hbar^2}{2m}\boldsymbol{L}_{s,0}^2\right)}} , \quad \hat{L}_{s,i}^- = \frac{\mathrm{i}\hbar\hat{L}_s^-}{\sqrt{2m\left(T_i - \frac{\hbar^2}{2m}\boldsymbol{L}_{s,0}^2\right)}}$$
$$(3.192)$$

("normalization")

$$\Longrightarrow$$

$$\left(E - \frac{\hbar^2}{2m}\boldsymbol{L}_{s,0}^2\right)(\psi_1(\boldsymbol{x}_1) + \psi_2(\boldsymbol{x}_2) + \cdots) =$$

$$= \left(T_1 - \frac{\hbar^2}{2m}\boldsymbol{L}_{s,0}^2\right)\hat{L}_{s,1}^+(\boldsymbol{x}_1)\hat{L}_{s,1}^-(\boldsymbol{x}_1)\psi_1(\boldsymbol{x}_1) + \left(T_2 - \frac{\hbar^2}{2m}\boldsymbol{L}_{s,0}^2\right)\hat{L}_{s,2}^+(\boldsymbol{x}_2)\hat{L}_{s,2}^-(\boldsymbol{x}_2)\psi_2(\boldsymbol{x}_2)$$

$$+ \cdots ,$$

$$\hat{L}_{s,i}^+(\boldsymbol{x}_i) = \frac{\mathrm{i}\hbar\hat{L}_s^+(\boldsymbol{x}_i)}{\sqrt{2m\left(T_i - \frac{\hbar^2}{2m}\boldsymbol{L}_{s,0}^2\right)}} , \quad \hat{L}_{s,i}^-(\boldsymbol{x}_i) = \frac{\mathrm{i}\hbar\hat{L}_s^-(\boldsymbol{x}_i)}{\sqrt{2m\left(T_i - \frac{\hbar^2}{2m}\boldsymbol{L}_{s,0}^2\right)}} ,$$

$$\hat{L}_s^+(\boldsymbol{x}_i) = \boldsymbol{L}_{s,0} - \mathrm{i}\nabla_{\boldsymbol{x}_i} - \frac{\omega}{c}\left(\boldsymbol{P}_s(\boldsymbol{x}_i/\boldsymbol{x}_j) + \frac{cq}{\hbar\omega}\boldsymbol{A}(\boldsymbol{x}_i/\boldsymbol{x}_j)\right) ,$$
$$(3.193)$$

$$\hat{L}_s^-(\boldsymbol{x}_i) = \boldsymbol{L}_{s,0} + \mathrm{i}\nabla_{\boldsymbol{x}_i} + \frac{\omega}{c}\left(\boldsymbol{P}_s(\boldsymbol{x}_i/\boldsymbol{x}_j) + \frac{cq}{\hbar\omega}\boldsymbol{A}(\boldsymbol{x}_i/\boldsymbol{x}_j)\right) ,$$

$$\boldsymbol{P}_s(\boldsymbol{x}_i/\boldsymbol{x}_j) = \begin{cases} \boldsymbol{P}_s(\boldsymbol{x}_i) & \text{if applied to } \psi_i(\boldsymbol{x}_i) \\ 0 & \text{if applied to } \psi_j(\boldsymbol{x}_j) \end{cases} ,$$

$$\boldsymbol{A}(\boldsymbol{x}_i/\boldsymbol{x}_j) = \begin{cases} \boldsymbol{A}(\boldsymbol{x}_i) & \text{if applied to } \psi_i(\boldsymbol{x}_i) \\ 0 & \text{if applied to } \psi_j(\boldsymbol{x}_j) \end{cases}$$

("individualization")

> **Continuation of Box.**
>
> $$\left(E - \frac{\hbar^2}{2m}\boldsymbol{L}_{s,0}^2\right)\psi(\boldsymbol{x}_1, \boldsymbol{x}_2, \ldots, \boldsymbol{x}_N) = \hat{H}'_s\psi(\boldsymbol{x}_1, \boldsymbol{x}_2, \ldots, \boldsymbol{x}_N) \;,$$
>
> $$\hat{H}'_s = \sum_{i=1}^{N}\left(T_i - \frac{\hbar^2}{2m}\boldsymbol{L}_{s,0}^2\right)\hat{L}^+_{s,i}(\boldsymbol{x}_i)\hat{L}^-_{s,i}(\boldsymbol{x}_i) + \cdots \;,$$
>
> $$\hat{L}^+_{s,i}(\boldsymbol{x}_i) = \frac{i\hbar \hat{L}^+_s(\boldsymbol{x}_i)}{\sqrt{2m\left(T_i - \frac{\hbar^2}{2m}\boldsymbol{L}_{s,0}^2\right)}} \;,\quad \hat{L}^-_{s,i}(\boldsymbol{x}_i) = \frac{i\hbar \hat{L}^-_s(\boldsymbol{x}_i)}{\sqrt{2m\left(T_i - \frac{\hbar^2}{2m}\boldsymbol{L}_{s,0}^2\right)}} \;,$$
>
> $$\hat{L}^+_s(\boldsymbol{x}_i) = \boldsymbol{L}_{s,0} - i\nabla_{\boldsymbol{x}_i} - \frac{\omega}{c}\left(\boldsymbol{P}_s(\boldsymbol{x}_i/\boldsymbol{x}_j) + \frac{cq}{\hbar\omega}\boldsymbol{A}(\boldsymbol{x}_i/\boldsymbol{x}_j)\right) \;,$$
> $$\hat{L}^-_s(\boldsymbol{x}_i) = \boldsymbol{L}_{s,0} + i\nabla_{\boldsymbol{x}_i} + \frac{\omega}{c}\left(\boldsymbol{P}_s(\boldsymbol{x}_i/\boldsymbol{x}_j) + \frac{cq}{\hbar\omega}\boldsymbol{A}(\boldsymbol{x}_i/\boldsymbol{x}_j)\right) \;,$$
> (3.194)
>
> $$\boldsymbol{P}_s(\boldsymbol{x}_i/\boldsymbol{x}_j) = \begin{cases}\boldsymbol{P}_s(\boldsymbol{x}_i) & \text{if applied to } \psi_i(\boldsymbol{x}_i)\\ 0 & \text{if applied to } \psi_j(\boldsymbol{x}_j)\end{cases} \;,$$
>
> $$\boldsymbol{A}(\boldsymbol{x}_i/\boldsymbol{x}_j) = \begin{cases}\boldsymbol{A}(\boldsymbol{x}_i) & \text{if applied to } \psi_i(\boldsymbol{x}_i)\\ 0 & \text{if applied to } \psi_j(\boldsymbol{x}_j)\end{cases}$$
>
> $$\psi(\boldsymbol{x}_1, \boldsymbol{x}_2, \ldots, \boldsymbol{x}_N) = \sum_{i=1}^{N}\psi_i(\boldsymbol{x}_i)$$
>
> ("compaction").
>
> Product-type decompositions $\psi = \psi_1\psi_2\cdots$ proceed analogously to sum-type decompositions $\psi = \psi_1 + \psi_2 + \cdots$, i.e.
>
> $$\psi(\boldsymbol{x}_1, \boldsymbol{x}_2, \ldots, \boldsymbol{x}_N) = \sum_{i=1}^{N}\psi_i(\boldsymbol{x}_i) \quad \text{or} \quad \psi(\boldsymbol{x}_1, \boldsymbol{x}_2, \ldots, \boldsymbol{x}_N) = \prod_{i=1}^{N}\psi_i(\boldsymbol{x}_i) \;. \qquad (3.195)$$
>
> Commutator relations and eigenvalue equations define structure equivalents of generation operators, destruction operators, and wavefunctions, in combination with the methods of group theory, allowing the transition to alternative representations, including the transition to an "energy representation", certainly not containing position information, but yet containing energy information, i.e.
>
> $$\psi_i(\boldsymbol{x}_i), \hat{L}^\pm_s(\boldsymbol{x}_i), \ldots \quad \text{or} \quad |\psi_i\rangle, \hat{C}^\pm_s, \ldots \;, \qquad (3.196)$$
>
> $$\left[\hat{L}^+_s(\boldsymbol{x}_i), \hat{L}^-_s(\boldsymbol{x}_i)\right]_\mp = f_\mp\left(\hat{L}^+_s(\boldsymbol{x}_i), \hat{L}^-_s(\boldsymbol{x}_i)\right) \;,\quad \hat{L}^\pm_s(\boldsymbol{x}_i)\psi_i(\boldsymbol{x}_i) = L^\pm_s\psi_i(\boldsymbol{x}_i) \ldots$$
> $$\Downarrow \qquad\qquad (3.197)$$
> $$\left[\hat{C}^+_s, \hat{C}^-_s\right]_\mp = f_\mp\left(\hat{C}^+_s, \hat{C}^-_s\right) \;,\quad \hat{C}^\pm_s|\psi_i\rangle = L^\pm_s|\psi_i\rangle \ldots \;.$$

> **Continuation of Box.**
>
> The way from the GEFT-type interaction model to the BCS-type interaction model:
>
> $$\psi(\boldsymbol{x}) = \psi_1(\boldsymbol{x}) + \psi_2(\boldsymbol{x}) + \cdots,$$
>
> $$-\frac{\hbar^2}{4m} \hat{G}^1 \hat{G}^2 \hat{G}^1_{\overline{\psi}_s} \hat{G}^2_{\overline{\psi}_s} \left(\psi_1(\boldsymbol{x}) + \psi_2(\boldsymbol{x}) + \cdots \right) \quad (3.198)$$
> $$=$$
> $$-\frac{\hbar^2}{4m} \left[-\mathrm{i}\sqrt{\nabla} \right] \left[\mathrm{i}\sqrt{\nabla} \right] \left(\psi_1(\boldsymbol{x}) + \psi_2(\boldsymbol{x}) + \cdots \right) \hat{G}^1_{\overline{\psi}_s} \hat{G}^2_{\overline{\psi}_s} \left(\psi_1(\boldsymbol{x}) + \psi_2(\boldsymbol{x}) + \cdots \right)$$
> $$\Longrightarrow$$
> $$-\frac{\hbar^2}{4m} \left[-\mathrm{i}\sqrt{\nabla} \right] \left[\mathrm{i}\sqrt{\nabla} \right] \left(\psi_1(\boldsymbol{x}) + \psi_2(\boldsymbol{x}) + \cdots \right) \hat{G}^1_{\overline{\psi}_s} \hat{G}^2_{\overline{\psi}_s} \left(\psi_1(\boldsymbol{x}) + \psi_2(\boldsymbol{x}) + \cdots \right)$$
> $$= \quad (3.199)$$
> $$-\frac{\hbar^2}{4m} \left(\hat{G}^1_1 \hat{G}^2_1 + \hat{G}^1_2 \hat{G}^2_2 + \cdots \right) \hat{G}^1_{\overline{\psi}_s} \hat{G}^2_{\overline{\psi}_s} \left(\psi_1(\boldsymbol{x}) + \psi_2(\boldsymbol{x}) + \cdots \right)$$
> $$\Longrightarrow$$
> $$-\frac{\hbar^2}{4m} \left(\hat{G}^1_1 \hat{G}^2_1 + \hat{G}^1_2 \hat{G}^2_2 + \cdots \right) \hat{G}^1_{\overline{\psi}_s} \hat{G}^2_{\overline{\psi}_s} \left(\psi_1(\boldsymbol{x}) + \psi_2(\boldsymbol{x}) + \cdots \right)$$
> $$= \quad (3.200)$$
> $$-\frac{\hbar^2}{4m} \left(\hat{G}^1_1 \hat{G}^2_1 + \hat{G}^1_2 \hat{G}^2_2 + \cdots \right) \left(\hat{G}^1_{\overline{\psi}_s} \hat{G}^2_{\overline{\psi}_s} \psi_1(\boldsymbol{x}) + \hat{G}^1_{\overline{\psi}_s} \hat{G}^2_{\overline{\psi}_s} \psi_2(\boldsymbol{x}) + \cdots \right)$$
> $$\Longrightarrow$$
> $$-\frac{\hbar^2}{4m} \left(\hat{G}^1_1 \hat{G}^2_1 + \hat{G}^1_2 \hat{G}^2_2 + \cdots \right) \left(\hat{G}^1_{\overline{\psi}_s} \hat{G}^2_{\overline{\psi}_s} \psi_1(\boldsymbol{x}) + \hat{G}^1_{\overline{\psi}_s} \hat{G}^2_{\overline{\psi}_s} \psi_2(\boldsymbol{x}) + \cdots \right)$$
> $$=$$
> $$V_{\overline{\psi}_s} \left(\hat{G}^1_1 \hat{G}^2_1 + \hat{G}^1_2 \hat{G}^2_2 + \cdots \right) \left(\hat{G}^1_{\overline{\psi}_s,1} \hat{G}^2_{\overline{\psi}_s,2} \psi_1(\boldsymbol{x}) + \hat{G}^1_{\overline{\psi}_s,1} \hat{G}^2_{\overline{\psi}_s,2} \psi_2(\boldsymbol{x}) + \cdots \right),$$
>
> $$\hat{G}^1_1 = \left[-\mathrm{i}\sqrt{\nabla \psi_1(\boldsymbol{x})} \right], \; \hat{G}^2_1 = \left[\mathrm{i}\sqrt{\nabla \psi_1(\boldsymbol{x})} \right], \; \ldots$$
> $$\quad (3.201)$$
> $$\hat{G}^1_2 = \left[-\mathrm{i}\sqrt{\nabla \psi_2(\boldsymbol{x})} \right], \; \hat{G}^2_2 = \left[\mathrm{i}\sqrt{\nabla \psi_2(\boldsymbol{x})} \right], \; \ldots$$
>
> $$\hat{G}^1_{\overline{\psi}_s,1} = \frac{\sqrt{-\frac{\hbar^2}{4m}} \left[-\mathrm{i}\sqrt{(1 - \overline{\psi}_s(\boldsymbol{x}))} \nabla \right]}{\sqrt{V_{\overline{\psi}_s}}}, \quad \hat{G}^2_{\overline{\psi}_s,2} = \frac{\sqrt{-\frac{\hbar^2}{4m}} \left[\mathrm{i}\sqrt{(1 - \overline{\psi}_s(\boldsymbol{x}))} \nabla \right]}{\sqrt{V_{\overline{\psi}_s}}}$$
>
> ("normalization")

Continuation of Box.

$$\Longrightarrow$$

$$V_{\bar{\Psi}_s}\left(\hat{G}_1^1(\boldsymbol{x}_1)\hat{G}_1^2(\boldsymbol{x}_1)+\hat{G}_2^1(\boldsymbol{x}_2)\hat{G}_2^2(\boldsymbol{x}_2)+\cdots\right)$$

$$\cdot$$

$$\left(\hat{G}_{\bar{\Psi}_{s,1}}^1(\boldsymbol{x}_1)\hat{G}_{\bar{\Psi}_{s,2}}^2(\boldsymbol{x}_1)\psi_1(\boldsymbol{x}_1)+\hat{G}_{\bar{\Psi}_{s,1}}^1(\boldsymbol{x}_2)\hat{G}_{\bar{\Psi}_{s,2}}^2(\boldsymbol{x}_2)\psi_2(\boldsymbol{x}_2)+\cdots\right)$$

$$=$$

$$V_{\bar{\Psi}_s}\hat{G}_1^1(\boldsymbol{x}_1)\hat{G}_1^2(\boldsymbol{x}_1)\hat{G}_{\bar{\Psi}_{s,1}}^1(\boldsymbol{x}_2)\hat{G}_{\bar{\Psi}_{s,2}}^2(\boldsymbol{x}_2)\psi_2(\boldsymbol{x}_2)$$

$$+$$

$$V_{\bar{\Psi}_s}\hat{G}_2^1(\boldsymbol{x}_2)\hat{G}_2^2(\boldsymbol{x}_2)\hat{G}_{\bar{\Psi}_{s,1}}^1(\boldsymbol{x}_1)\hat{G}_{\bar{\Psi}_{s,2}}^2(\boldsymbol{x}_1)\psi_1(\boldsymbol{x}_1)$$

$$+\cdots$$

$$=$$

$$V_{\bar{\Psi}_s}\hat{G}_1^1(\boldsymbol{x}_1)\hat{G}_1^2(\boldsymbol{x}_1)\hat{G}_{\bar{\Psi}_{s,1}}^1(\boldsymbol{x}_2)\hat{G}_{\bar{\Psi}_{s,2}}^2(\boldsymbol{x}_2)\big(\psi_1(\boldsymbol{x}_1)+\psi_2(\boldsymbol{x}_2)+\cdots\big)$$

$$+$$

$$V_{\bar{\Psi}_s}\hat{G}_2^1(\boldsymbol{x}_2)\hat{G}_2^2(\boldsymbol{x}_2)\hat{G}_{\bar{\Psi}_{s,1}}^1(\boldsymbol{x}_1)\hat{G}_{\bar{\Psi}_{s,2}}^2(\boldsymbol{x}_1)\big(\psi_1(\boldsymbol{x}_1)+\psi_2(\boldsymbol{x}_2)+\cdots\big)$$

$$+\cdots$$

$$=$$ (3.202)

$$V_{\bar{\Psi}_s}\left(\hat{G}_1^1(\boldsymbol{x}_1)\hat{G}_1^2(\boldsymbol{x}_1)\hat{G}_{\bar{\Psi}_{s,1}}^1(\boldsymbol{x}_2)\hat{G}_{\bar{\Psi}_{s,2}}^2(\boldsymbol{x}_2)+\hat{G}_2^1(\boldsymbol{x}_2)\hat{G}_2^2(\boldsymbol{x}_2)\hat{G}_{\bar{\Psi}_{s,1}}^1(\boldsymbol{x}_1)\hat{G}_{\bar{\Psi}_{s,2}}^2(\boldsymbol{x}_1)+\cdots\right)$$

$$\cdot$$

$$\big(\psi_1(\boldsymbol{x}_1)+\psi_2(\boldsymbol{x}_2)+\cdots\big),$$

$$\hat{G}_1^1(\boldsymbol{x}_1)=\left[-\mathrm{i}\sqrt{\nabla_{\boldsymbol{x}_1}\psi_1(\boldsymbol{x}_1)}\right],\quad \hat{G}_1^2(\boldsymbol{x}_1)=\left[\mathrm{i}\sqrt{\nabla_{\boldsymbol{x}_1}\psi_1(\boldsymbol{x}_1)}\right],\ \ldots$$

$$\hat{G}_2^1(\boldsymbol{x}_2)=\left[-\mathrm{i}\sqrt{\nabla_{\boldsymbol{x}_2}\psi_2(\boldsymbol{x}_2)}\right],\quad \hat{G}_2^2(\boldsymbol{x}_2)=\left[\mathrm{i}\sqrt{\nabla_{\boldsymbol{x}_2}\psi_2(\boldsymbol{x}_2)}\right],\ \ldots$$

$$\hat{G}_{\bar{\Psi}_s,1}^1(\boldsymbol{x}_i)=\frac{\sqrt{-\frac{\hbar^2}{4m}}\left[-\mathrm{i}\sqrt{(1-\bar{\Psi}_s(\boldsymbol{x}_i))}\,\nabla_{\boldsymbol{x}_i}\right]}{\sqrt{V_{\bar{\Psi}_s}}},$$

$$\hat{G}_{\bar{\Psi}_s,2}^2(\boldsymbol{x}_i)=\frac{\sqrt{-\frac{\hbar^2}{4m}}\left[\mathrm{i}\sqrt{(1-\bar{\Psi}_s(\boldsymbol{x}_i))}\,\nabla_{\boldsymbol{x}_i}\right]}{\sqrt{V_{\bar{\Psi}_s}}}$$

("individualization")

> **Continuation of Box.**
>
> $$\Longrightarrow$$
>
> $$\hat{V}_{\bar{\Psi}_s}\psi(\boldsymbol{x}_1,\boldsymbol{x}_2,\ldots,\boldsymbol{x}_N)$$
>
> $$=$$
>
> $$V_{\bar{\Psi}_s}\hat{V}^+_{\bar{\Psi}_s,N}\cdots\hat{V}^+_{\bar{\Psi}_s,1}\hat{V}^-_{\bar{\Psi}_s,N}\cdots\hat{V}^-_{\bar{\Psi}_s,1}\psi(\boldsymbol{x}_1,\boldsymbol{x}_2,\ldots,\boldsymbol{x}_N)\,,$$
>
> $$\hat{V}^+_{\bar{\Psi}_s,N}\cdots\hat{V}^+_{\bar{\Psi}_s,1}\hat{V}^-_{\bar{\Psi}_s,N}\cdots\hat{V}^-_{\bar{\Psi}_s,1}=\hat{W}^+_1\hat{W}^-_1+\hat{W}^+_2\hat{W}^-_2+\cdots+\hat{W}^+_N\hat{W}^-_N\,,$$
>
> $$\hat{W}^+_i\hat{W}^-_i=\sum_{j=1}^{N}\hat{G}^1_j(\boldsymbol{x}_j)\hat{G}^2_j(\boldsymbol{x}_j)\hat{G}^1_{\bar{\Psi}_s,1}(\boldsymbol{x}_i)\hat{G}^2_{\bar{\Psi}_s,2}(\boldsymbol{x}_i)\,,$$
>
> $$\hat{G}^1_1(\boldsymbol{x}_1)=\left[-\mathrm{i}\sqrt{\nabla_{\boldsymbol{x}_1}\psi_1(\boldsymbol{x}_1)}\right]\,,\quad \hat{G}^2_1(\boldsymbol{x}_1)=\left[\mathrm{i}\sqrt{\nabla_{\boldsymbol{x}_1}\psi_1(\boldsymbol{x}_1)}\right]\,,\ \ldots$$
>
> $$\hat{G}^1_2(\boldsymbol{x}_2)=\left[-\mathrm{i}\sqrt{\nabla_{\boldsymbol{x}_2}\psi_2(\boldsymbol{x}_2)}\right]\,,\quad \hat{G}^2_2(\boldsymbol{x}_2)=\left[\mathrm{i}\sqrt{\nabla_{\boldsymbol{x}_2}\psi_2(\boldsymbol{x}_2)}\right]\,,\ \ldots \qquad (3.203)$$
>
> $$\hat{G}^1_{\bar{\Psi}_s,1}(\boldsymbol{x}_i)=\frac{\sqrt{-\frac{\hbar^2}{4m}}\left[-\mathrm{i}\sqrt{(1-\bar{\Psi}_s(\boldsymbol{x}_i))\,\nabla_{\boldsymbol{x}_i}}\right]}{\sqrt{V_{\bar{\Psi}_s}}}\,,$$
>
> $$\hat{G}^2_{\bar{\Psi}_s,2}(\boldsymbol{x}_i)=\frac{\sqrt{-\frac{\hbar^2}{4m}}\left[\mathrm{i}\sqrt{(1-\bar{\Psi}_s(\boldsymbol{x}_i))\,\nabla_{\boldsymbol{x}_i}}\right]}{\sqrt{V_{\bar{\Psi}_s}}}\,,$$
>
> $$\psi(\boldsymbol{x}_1,\boldsymbol{x}_2,\ldots,\boldsymbol{x}_N)=\sum_{i=1}^{N}\psi_i(\boldsymbol{x}_i)$$
>
> ("compaction").
>
> As before,
> product-type decompositions $\psi=\psi_1\psi_2\cdots$ proceed analogeously
> to sum-type decompositions $\psi=\psi_1+\psi_2+\cdots$.
>
>
> As before,
> commutator relations and eigenvalue equations define structure equivalents
> of generation operators, destruction operators, and wavefunctions,
> in combination with the methods of group theory,
> allowing the transition to alternative representations, including
> the transition to an "energy representation", certainly not containing
> position information, but yet containing energy information.

Working with GEFT-Type Hamiltonians

How can we work with GEFT-type Hamiltonians? (i) Firstly, follwing the above scheme, we can derive BCS-type Hamiltonians in the position representation. Using these within thermodynamic models allowing us to incorporate the notion temperature T, or going over to BCS-type Hamiltonians in the energy representation and using these within thermodynamic models allowing us to incorporate the notion temperature T, a basic band gap analysis is possible [37, 31, 54]. In practice, one establishes a special model by determining the energy band structure of the crystal, eventually using special electrons residing in special energy bands. Unknown energy terms, for example, the energy term $V_{\bar{\psi}_s}$ could be unknown, on the one hand, can be fixed via suitable energy models, and on the other hand, can be fixed by adapting these to suitable measurement results. (ii) Secondly, the "wave–particle energy balance equations" to be considered here are completed by "wave–particle momentum balance equations" [63], in combination with intial conditions, boundary conditions, and additional constraints defining the crystal, allowing a systematic analytical access as well as a systematic numerical access to superconducting states. However, since this proceeding is rather lengthy, in practice, one mostly will focus on the "wave–particle energy balance equations", will preset the wavefunctions according to the energy band structure of the crystal, and will develop models for unknown energy terms.

A more detailed insight into the handling of GEFT-type Hamiltonians provide the mathematical post-its that are attached below. (i) (3.210) is the kinetic energy and (3.211) is the momentum of an electron pair/a superparticle with the characteristics (3.208) provided m is the "effective mass" of the electron pair/superparticle. We read off from (3.211) that the momentum of the electron pair/superparticle then is defined by a wave vector $\boldsymbol{k}$ reflecting the existence of a particle, by a wave vector $\omega \boldsymbol{P}_s/c$ making constructive vibrations and compatible effects to an inherent part of the superior unity, and by a wave vector $q\boldsymbol{A}/\hbar$ additionally making the remainders of the magnetic field to an inherent part of the superior unity. However, since the vector potential $\boldsymbol{A}$ surely depends on the position coordinates, E in combination with ψ describes a flow profile that surely depends on the position coordinates. (ii) (3.219) traces the vibrational potential $\boldsymbol{P}_s$ back to the energy contributions $E_x(\boldsymbol{x})$, $E_y(\boldsymbol{x})$, and $E_z(\boldsymbol{x})$ collecting the vibrational energy of vectorial contributions of constructive vibrations and compatible effects in the three spatial directions x, y, and z. We read off from (3.219) that on the one hand the knowledge of the energy contributions $E_x(\boldsymbol{x})$, $E_y(\boldsymbol{x})$, and $E_z(\boldsymbol{x})$ and on the other hand the knowledge of the wavefunction $\psi(\boldsymbol{x})$ is needed to obtain a closed expression for the vibrational potential $\boldsymbol{P}_s$. (iii) Reflecting on these examples, we realize that this top-down approach to superconductivity breeds the wavefunctions, energies, and energy gaps, and thus the Fermi surface including Fermi surface gaps, by solving partial differential equations attached to the partial differential equations of quantum mechanics and quantum field theory. Reflecting on these examples, we realize that substances or domains of states without superconductivity reveal oneself via partial differential equations certainly containing the conditions for superconductivity, but having no solutions compatible with physical/chemical constraints, which always define an inherent part of such methods.

Mathematical post-it. Electron pair/superparticle momentum $\boldsymbol{p}_\mathrm{s} = \hbar \boldsymbol{k}_\mathrm{s}$.

$$\left(E - \frac{\hbar^2}{2m}\boldsymbol{L}_{\mathrm{s},0}^2\right)\psi(\boldsymbol{x}) = \hat{\mathcal{H}}_\mathrm{s}'\psi(\boldsymbol{x}) , \qquad (3.204)$$

$$\hat{\mathcal{H}}_\mathrm{s}' := -\frac{\hbar^2}{2m}\hat{L}_\mathrm{s}^+\hat{L}_\mathrm{s}^- , \qquad (3.205)$$

$$\hat{L}_\mathrm{s}^+ = \boldsymbol{L}_{\mathrm{s},0} - \mathrm{i}\nabla - \frac{\omega}{c}\left(\boldsymbol{P}_\mathrm{s} + \frac{cq}{\hbar\omega}\boldsymbol{A}\right) ,$$
$$\hat{L}_\mathrm{s}^- = \boldsymbol{L}_{\mathrm{s},0} + \mathrm{i}\nabla + \frac{\omega}{c}\left(\boldsymbol{P}_\mathrm{s} + \frac{cq}{\hbar\omega}\boldsymbol{A}\right) \qquad (3.206)$$

$$\Longrightarrow$$

$$E\psi(\boldsymbol{x}) =$$
$$= -\frac{\hbar^2}{2m}\triangle\psi(\boldsymbol{x}) +$$
$$+ \mathrm{i}\frac{\hbar^2}{2m}\frac{\omega}{c}\left[\nabla\left(\boldsymbol{P}_\mathrm{s} + \frac{cq}{\hbar\omega}\boldsymbol{A}\right) + \left(\boldsymbol{P}_\mathrm{s} + \frac{cq}{\hbar\omega}\boldsymbol{A}\right)\nabla\right]\psi(\boldsymbol{x}) +$$
$$+ \hbar\omega\frac{\hbar\omega}{2mc^2}\left(\boldsymbol{P}_\mathrm{s} + \frac{cq}{\hbar\omega}\boldsymbol{A}\right)^2\psi(\boldsymbol{x}) , \qquad (3.207)$$

$$\psi(\boldsymbol{x}) := \psi_0 \exp(\mathrm{i}\boldsymbol{k}\boldsymbol{x}) , \quad \nabla\left(\boldsymbol{P}_\mathrm{s} + \frac{cq}{\hbar\omega}\boldsymbol{A}\right) := 0 \qquad (3.208)$$

$$\Longrightarrow$$

$$E\psi(\boldsymbol{x}) =$$
$$= \frac{\hbar^2}{2m}\left[\boldsymbol{k}^2 - \boldsymbol{k}\frac{\omega}{c}\left(\boldsymbol{P}_\mathrm{s} + \frac{cq}{\hbar\omega}\boldsymbol{A}\right) + \frac{\omega^2}{c^2}\left(\boldsymbol{P}_\mathrm{s} + \frac{cq}{\hbar\omega}\boldsymbol{A}\right)^2\right]\psi(\boldsymbol{x}) \qquad (3.209)$$

$$\Longrightarrow$$

$$E = \frac{\boldsymbol{p}_\mathrm{s}^2}{2m} = \frac{\hbar^2\boldsymbol{k}_\mathrm{s}^2}{2m} , \qquad (3.210)$$

$$\boldsymbol{p}_\mathrm{s} = \hbar\boldsymbol{k}_\mathrm{s} , \quad \boldsymbol{k}_\mathrm{s} = \sqrt{\boldsymbol{k}^2 - \boldsymbol{k}\frac{\omega}{c}\left(\boldsymbol{P}_\mathrm{s} + \frac{cq}{\hbar\omega}\boldsymbol{A}\right) + \frac{\omega^2}{c^2}\left(\boldsymbol{P}_\mathrm{s} + \frac{cq}{\hbar\omega}\boldsymbol{A}\right)^2} . \qquad (3.211)$$

Mathematical post-it. Vibrational potential $\boldsymbol{P}_\text{s}$.

$$\boldsymbol{A}_\ni(\boldsymbol{x},t) := \exp\left(\mathrm{i}\omega t\right) \boldsymbol{A}_\ni(\boldsymbol{x}) \tag{3.212}$$

(compare with (3.110))

$$\Longrightarrow$$

$$+\frac{1}{c}\frac{\partial}{\partial t}(\nabla)^\mathrm{T}\boldsymbol{A}_\ni := \mathrm{i}\frac{\omega}{c}\exp\left(\mathrm{i}\omega t\right)\nabla \boldsymbol{A}_\ni(\boldsymbol{x}) \tag{3.213}$$

(compare with (3.110))

$$\Longrightarrow$$

$$+\frac{1}{c}\frac{\partial}{\partial t}(\nabla)^\mathrm{T}\boldsymbol{A}_\ni := \mathrm{i}\frac{1}{2}\frac{\omega}{c}\exp\left(\mathrm{i}\omega t\right)\nabla \boldsymbol{P}_\text{s}\psi(\boldsymbol{x}) \tag{3.214}$$

(superconducting state, wave-vector-square dimension)

$$\Longrightarrow$$

$$+\frac{\hbar^2}{m}\frac{1}{c}\frac{\partial}{\partial t}(\nabla)^\mathrm{T}\boldsymbol{A}_\ni = \mathrm{i}\frac{1}{2}\frac{\hbar^2}{m}\frac{\omega}{c}\exp\left(\mathrm{i}\omega t\right)\nabla \boldsymbol{P}_\text{s}\psi(\boldsymbol{x}) \tag{3.215}$$

(superconducting state, energy dimension)

$$\Longrightarrow$$

$$[E_x(\boldsymbol{x},t) + E_y(\boldsymbol{x},t) + E_z(\boldsymbol{x},t)]\psi(\boldsymbol{x}) = \mathrm{i}\frac{1}{2}\frac{\hbar^2}{m}\frac{\omega}{c}\exp\left(\mathrm{i}\omega t\right)\nabla \boldsymbol{P}_\text{s}\psi(\boldsymbol{x}) \tag{3.216}$$

$$\Longrightarrow$$

$$\nabla \boldsymbol{P}_\text{s}\psi(\boldsymbol{x}) =$$
$$= \left(\mathrm{i}\frac{1}{2}\frac{\hbar^2}{m}\frac{\omega}{c}\right)^{-1}[E_x(\boldsymbol{x},t) + E_y(\boldsymbol{x},t) + E_z(\boldsymbol{x},t)]\exp\left(-\mathrm{i}\omega t\right)\psi(\boldsymbol{x}) \tag{3.217}$$

$$\Longrightarrow$$

$$\nabla \boldsymbol{P}_\text{s}\psi(\boldsymbol{x}) = \left(\mathrm{i}\frac{1}{2}\frac{\hbar^2}{m}\frac{\omega}{c}\right)^{-1}[E_x(\boldsymbol{x}) + E_y(\boldsymbol{x}) + E_z(\boldsymbol{x})]\psi(\boldsymbol{x}) \tag{3.218}$$

$$\Longrightarrow$$

$$\begin{aligned}
P_{\text{s}x} &= \frac{1}{\psi(\boldsymbol{x})}\int \left(\mathrm{i}\frac{1}{2}\frac{\hbar^2}{m}\frac{\omega}{c}\right)^{-1} E_x(\boldsymbol{x})\psi(\boldsymbol{x})\,\mathrm{d}x + C_x\,, \\
P_{\text{s}y} &= \frac{1}{\psi(\boldsymbol{x})}\int \left(\mathrm{i}\frac{1}{2}\frac{\hbar^2}{m}\frac{\omega}{c}\right)^{-1} E_y(\boldsymbol{x})\psi(\boldsymbol{x})\,\mathrm{d}y + C_y\,, \\
P_{\text{s}z} &= \frac{1}{\psi(\boldsymbol{x})}\int \left(\mathrm{i}\frac{1}{2}\frac{\hbar^2}{m}\frac{\omega}{c}\right)^{-1} E_z(\boldsymbol{x})\psi(\boldsymbol{x})\,\mathrm{d}z + C_z\,.
\end{aligned} \tag{3.219}$$

Let us finish this section with hints regarding the embedding of the GEFT model for superconducting electrons into a GEFT model for the crystal that gives shelter to the superconducting electrons. Firstly, we want to advise the reader of the fact that the situation of electrons embedded into a crystal can be compared with the situation of electrons embedded into molecules. In particular, applying separations of the Born–Oppenheimer type, in both cases one can model the statics and the dynamics of the electrons without considering the complementary statics and the complementary dynamics of the molecular background or the crystal background, respectively, i. e. the molecular background or the crystal background, respectively, is simply mapped into the electron system via conjoint quantities, i. e. in the first case via the distance of the centers of the nuclei, i. e. in the second case via the conjoint potential $\boldsymbol{P}_\text{s}$. Secondly, we want to advise the reader of the fact that the situation of the crystal then can be subsequently modeled, i. e. the states of the electron system are described by energies and wavefunctions fixed by a differential equation for the electron system, and the energies and wavefunctions enter a differential equation for the crystal so that the crystal then can be subsequently modeled. Exactly this is the central prerequisite the GEFT model for superconducting electrons presented above resorts to!

3.4.3 Bond Sensitivity

Meanwhile, experimental studies and theoretical studies reveal that even subtleties of a chemical bond in a solid substance can alter physical properties of the solid substance. Observing Figure 3.78, this becomes clear. Firstly, it shows us that atomic orbitals such as $(1,2,3\dots)$ s orbitals and $(1,2,3\dots)$ p_z orbitals can evoke "pure molecular bonds" such as $p_z \cdots p_z$ (π) bonds and "mixed molecular bonds" such as sp^2 bonds. Secondly, it shows us that atomic orbitals such as $(1,2,3\dots)$ s orbitals and $(1,2,3\dots)$ p_z orbitals can evoke bonding molecular orbitals and antibonding molecular orbitals. Thirdly, it shows us that different molecular orbitals evoke different classes of crystal orbitals associated with energy bands of different shape. Reflecting about it, we become aware that even fine variations of a crystal structure (for example, B atom replaces A atom) can alter subtleties of the chemical bond of the crystal structure (for example, 2s orbital replaces 1s orbital), thereby can alter energy bands of the crystal structure, and thereby can alter physical properties of the crystal structure.

In Figure 3.79, we show an interesting example of bond sensitivity, namely we show ABO_3 perovskite-type structures the electrical conductivity of which is directed by subtleties of the chemical bond of the ABO_3 perovskite-type structures. Certainly, one could expect that the different size of the atoms is responsible for the transition from the metallic character to the insulating character. Certainly, one could expect that the different crystal fields are responsible for the transition from the metallic character to the insulating character. However, as experimental studies and theoretical studies reveal, the tilt and the rotation of the BO_3 octahedra increase due to altering subtleties of the chemical bond, leading to an increasing localization of the B $3d_{t_{2g}}$ electron, finally paving the way from metallic behavior (metallic band structure) to insulating behavior (insulating band structure).

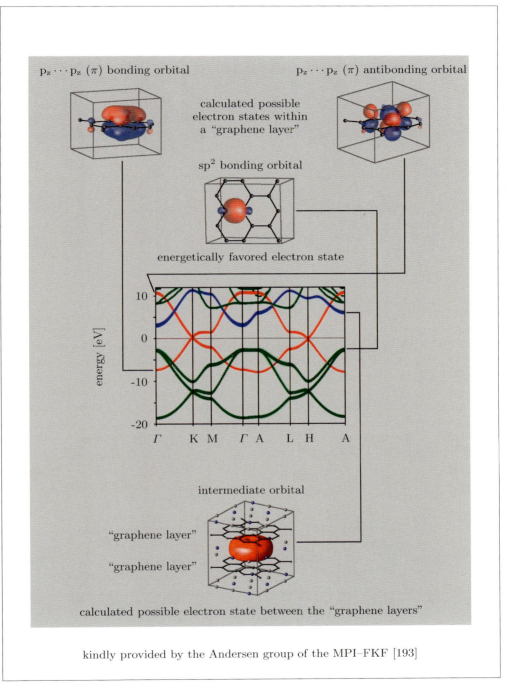

Figure 3.78. Molecular orbitals and energy bands of graphite.

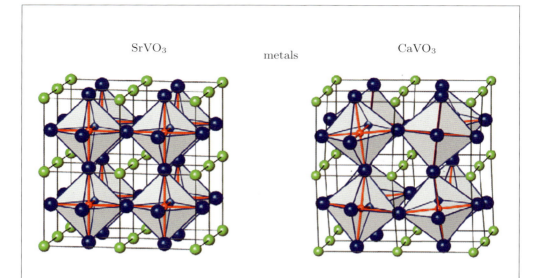

ABO$_3$ perovskite-type structures,
the tilt and the rotation of the BO$_3$ octahedra increase
caused by slightly different chemical bonds,
and so does the localization of the B 3d$_{t_{2g}}$ electron,
finally paving the way from metallic behavior (metallic band structure)
to insulating behavior (insulating band structure)

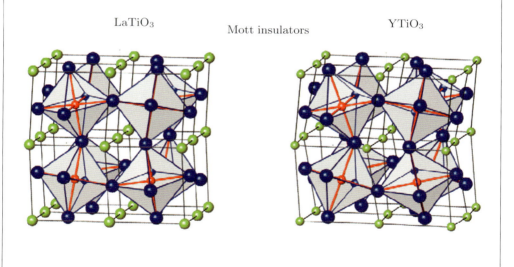

kindly provided by the Andersen group of the MPI–FKF [193]

Figure 3.79. Bond sensitivity: electrical conductivity.

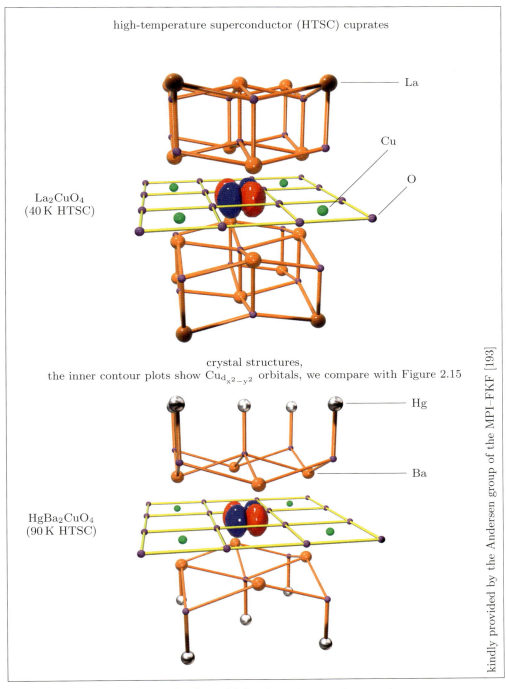

Figure 3.80. Bond sensitivity: jump temperature, part one.

330   3. Properties

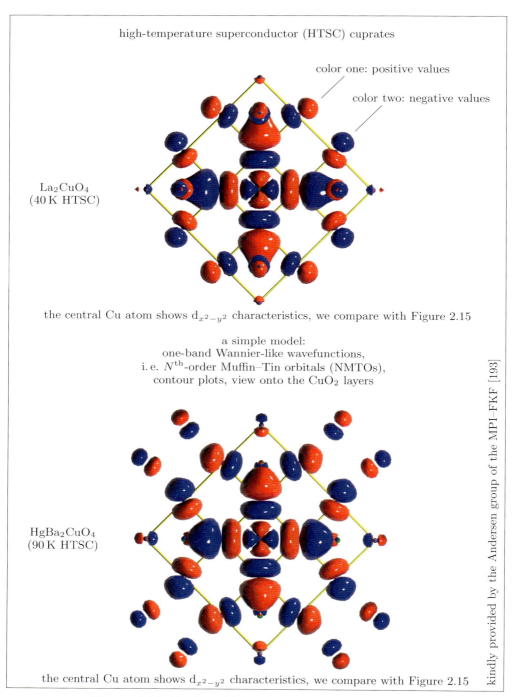

Figure 3.81. Bond sensitivity: jump temperature, part two.

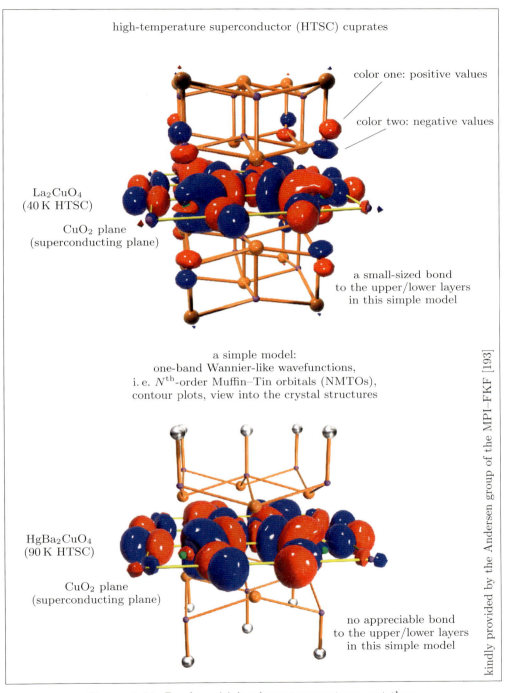

Figure 3.82. Bond sensitivity: jump temperature, part three.

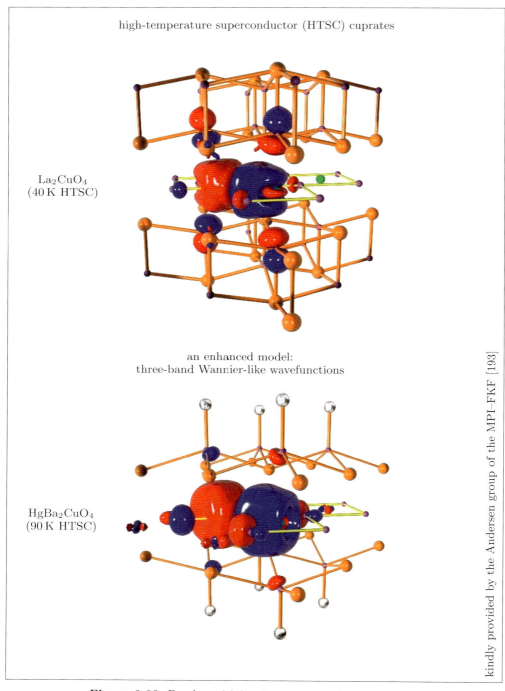

Figure 3.83. Bond sensitivity: jump temperature, part four.

In Figures 3.80–3.83, we show a further interesting example of bond sensitivity, namely we show HTSC cuprates the jump temperature of which is directed by subtleties of the chemical bond of the HTSC cuprates. Figure 3.80 presents the crystal structure of the 40 K HTSC La_2CuO_4 and the 90 K HTSC $HgBa_2CuO_4$. Figure 3.81 tells us that the CuO_2 layer of the 40 K HTSC and the CuO_2 layer of the 90 K HTSC tolerate different classes of crystal orbitals. Figure 3.82, on the one hand, tells us that the chemical bond perpendicular to the CuO_2 layers is responsible for the different classes of crystal orbitals, and on the other hand, tells us that the difference in character and strength of the chemical bond perpendicular to the CuO_2 layers is relatively small, and this is also valid if less simple models such as this one shown in Figure 3.83 are constructed. Figure 3.81–3.83 tell us that subtle differences of the chemical bond perpendicular to the CuO_2 layers generate slightly different quantum oscillators with slightly different CuO_2-wavefunction-part and slightly different lattice-vibration-part associated with relatively widely diverging jump temperatures.

Recalling the GEFT model for superconducting electrons presented above, we easily understand the alteration of the jump temperature that leads from normal conductivity to superconductivity as follows. Superconductivity evolves when the wavefunction of the superconducting electrons and the lattice vibrations or compatible effects obey special conditions, in the GEFT model for superconducting electrons presented above, expressed as order parameter relations. Say, we consider a quantum oscillator with CuO_2-wavefunction-part and lattice-vibration-part in a superconducting state. Say, we detune the quantum oscillator by implementation of slightly different chemical bonds. Question: What can we do to obtain again a superconducting state? Answer: We have to readjust the quantum oscillator via accessible parameters, eventually re-establishing the order parameter relations! For example, we have to adapt the temperature, eventually re-establishing the order parameter relations!

Manufacturing Ceramics

4. Technologies

It has been a long way from the first trials of the people of the Stone Age to burn clay to the technological standards of the manufacturing plants of the ceramic industry of nowadays. We should appreciate that at the beginning there were soil and fire, not any manufacturing plants of the type shown in Figures 4.1–4.4. We should appreciate that at the beginning there were glazed pebbles (12 000 B. C.), molded glassware (7 000 B. C.), and clay bricks (3 000 B. C.), not any dental prostheses or hip joint endoprostheses. We should appreciate that a steady development process took place throughout the centuries. Already 1 000 AD, magnifying glasses were manufactured in China. At the beginning of the 18th century, the European hard porcelain was invented. In the middle of the 19th century, the cement was reinvented. Important milestones are provided by the Acheson process (1893) for the synthesis of SiC, the development of the Al_2O_3 ball bearing (1913), the development of the Al_2O_3 ignition plug (1931), and the development of the TiO_2 condenser (1933). Further important milestones are the discovery of the fiber optic cable (1964) and the discovery of the superconducting ceramic material lanthanum barium copper oxide ($La_{1.85}Ba_{0.15}CuO_4$) [6]. Stabilized cubic ZrO_2 is known since 1956. Reaction-sintered Si_3N_4 is known since 1955 and liquid-phase-sintered Si_3N_4 is known since 1961. Y_2O_3 is used as sinter additive for AlN since 1971. B and C are used as sinter additives for SiC since 1974. Furthermore, isn't it remarkable that already the Egyptians used the potter's wheel? Moreover, isn't it remarkable that already the Romans knew the hydraulic properties of clayey lime?

The classical production methods of ceramics are powder-based methods. These are precisely described by the keywords *powder*, *mass*, *forming*, *heat*. These resort to natural starting materials. The non-classical production methods of ceramics can be subdivided into powder-based methods and powder-free methods and nowadays also include bio-inspired techniques. In any case, those resort to synthetically fabricated starting materials. By way of example, here casting a glance at Figures 4.1–4.4, we anticipate that those require well-manageable technical equipment that is placed within cleanrooms. By way of example, here casting a glance at Figures 4.3–4.5, we anticipate that those allow us the creation of ceramic materials and ceramic products that meet the highest demands. Indeed, the ceramic heads and the ceramic inserts of hip joint endoprostheses collected in Figure 4.5 endure extraordinary loadings! We compare with the scratched metal head shown in Figure 1.10! This metallic material surely has not the ability to fulfil its task inside of the human body!

336 4. Technologies

images from [71]

Figure 4.1. An example of a manufacturing plant of the ceramic industry in Germany (CeramTec, Marktredwitz, Germany).

4. Technologies   337

images from [71]

Figure 4.2. A view into the CeramTec factory.

image from [71]

Figure 4.3. Technological details. HIP.

4. Technologies 339

image from [71]

Figure 4.4. Technological details. HIP.

images from [71]

Figure 4.5. Product details. Ceramic heads and inserts.

4.1 Powder-Based Technologies

The principle of powder-based production methods of ceramics is illustrated by the flow chart shown in Figure 4.6. We read off from the flow chart of Figure 4.6 that in a first principal step the base product, the so-called *powder*, is fabricated, that in a second principal step the shapeable product, the so-called *mass*, is fabricated, that in a third principal step the mass is shaped, evoking a so-called *green body*, that in a fourth principal step the auxiliaries are burned out from the green body, evoking a solidified green body, that in a fifth principal step the solidified green body is sintered, evoking a raw ceramic, and that in a sixth principal step the raw ceramic finally is hard machined, evoking the component. Mind you, the flow chart shown in Figure 4.6 only imparts knowledge about *the principle* of powder-based production methods of ceramics. Mind you, the flow chart shown in Figure 4.6 can be refined in many subtle details. The reader, on the one hand, should compare with Figure 4.7, supplying us with a refinement for the core steps "forming and sintering", and on the other hand, should retrace the following explanations, supplying us with a collection of details concerning the standard sub-processes *powder fabrication*, *mass fabrication*, *forming*, *burnout*, *sintering*, and *hard machining* including such refinements.

4.1.1 Basic Procedures

Certainly, in the sections that follow, in the context of formal technical schemes and technologies, we refer to ceramic materials. However, the reader should appreciate that many of the formal technical schemes and technologies to be discussed in the sections that follow also play an important role in the framework of the powder-metallurgical production of metallic materials. To keep this soever in mind, let us here and there include illustrating matter referring to metallic materials.

4.1.1.1 Powder Fabrication Procedures

Taking up the modern position, a basic requirement for the production of ceramics are synthetically fabricated powders with high-level sintering activity, high-level purity, and particle sizes in the micron or even sub-micron range, and in order to achieve quality standards such as well-defined distributions of particle sizes, additional steps of powder processing such as fine milling are necessary. As it is shown in Box 4.1, for the synthesis of silicon carbide (SiC) powder, in particular, the Acheson process is applied, and for the synthesis of aluminum oxide (Al_2O_3) powder, in particular, the Bayer process is applied. Processes for the synthesis of zirconium oxide (ZrO_2) powder, silicon nitride (Si_3N_4) powder, and aluminum nitride (AlN) powder are provided by the hydrothermal oxidation (for ZrO_2), the chemical vapor deposition (for ZrO_2), the diimide process (for Si_3N_4), the direct nitridation (for Si_3N_4 and for AlN), and the carbon-thermal nitridation (for AlN). We compare with Figures 4.8 and 4.9. Due to the technical importance of the Bayer process and the Acheson process, some details are given in the next sections.

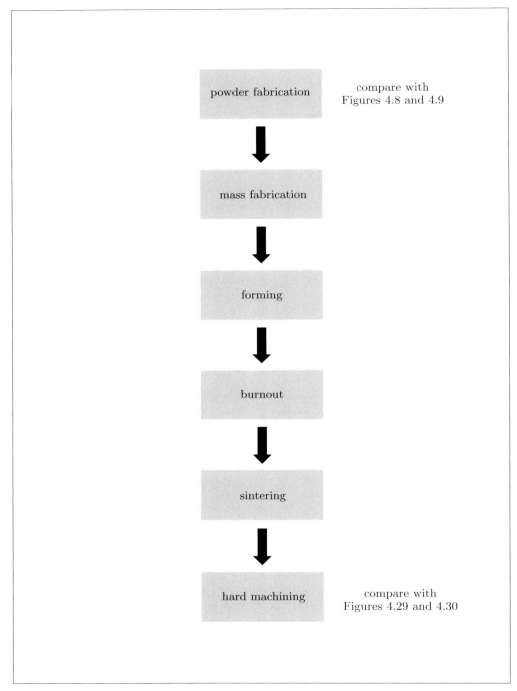

Figure 4.6. The principle of powder-based production methods of ceramics.

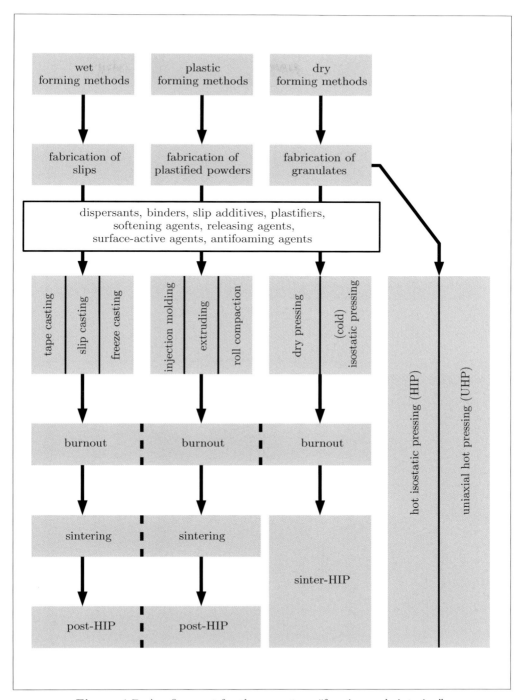

Figure 4.7. A refinement for the core steps "forming and sintering".

Figure 4.8. Ceramic powders. Si_3N_4 powder.

W powder ("metallic")

Fe–Cu powder ("metallic")

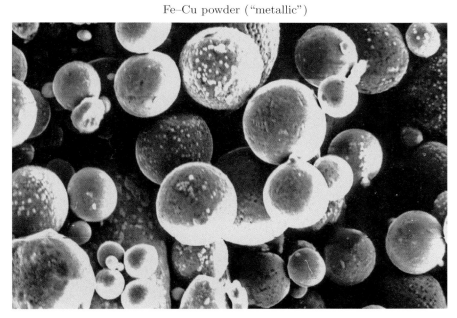

taken from the PML picture archive

Figure 4.9. Metallic powders. W powder and Fe–Cu powder.

Box 4.1 (Important data: important processes for the synthesis of powders).

Acheson process (SiC):

$$3C + SiO_2 \xrightarrow{2000°C} \alpha\text{-SiC} + 2CO \;.$$

Bayer process (Al_2O_3):

$$\text{bauxite } (Al_2O_3, SiO_2, Fe_2O_3) + 2NaOH + 3H_2O \xrightarrow{170-180°C,\, 5-7\,atm} 2NaAl(OH)_4 \;,$$

$$2NaAl(OH)_4 \longrightarrow 2Al(OH)_3 + 2NaOH \;,$$

$$2Al(OH)_3 \xrightarrow{1100-1300°C} \alpha\text{-}Al_2O_3 + 3H_2O \;.$$

Hydrothermal oxidation (ZrO_2):

$$Zr + 2H_2O \xrightarrow{250-700°C,\, 1000\,atm} ZrO_2 + 2H_2 \;,$$

$$Zr + H_2 \xrightarrow{250-700°C,\, 1000\,atm} ZrH_2 \;,$$

$$ZrH_2 + 2H_2O \xrightarrow{250-700°C,\, 1000\,atm} ZrO_2 + 3H_2 \;.$$

Chemical vapor deposition (ZrO_2):

$$ZrCl_4 + O_2 \xrightarrow{900°C,\, +O_2/H_2O} ZrO_2 + 2Cl_2 \;,$$

$$ZrCl_4 + 2H_2O \xrightarrow{900°C,\, +O_2/H_2O} ZrO_2 + 4HCl \;.$$

Diimide process (Si_3N_4):

$$SiCl_4 + 6NH_3 \longrightarrow Si(NH)_2 + 4NH_4Cl \;,$$

$$3Si(NH)_2 \longrightarrow \alpha\text{-}Si_3N_4 + 2NH_3 \;.$$

Direct nitridation (Si_3N_4, AlN):

$$3Si + 2N_2 \longrightarrow Si_3N_4 \;, \quad 2Al + N_2 \longrightarrow 2AlN \;.$$

Carbon-thermal nitridation (AlN):

$$2AlOOH \xrightarrow{1000-1500°C} \alpha\text{-}Al_2O_3 + H_2O \;,$$

$$\alpha\text{-}Al_2O_3 + 3C \xrightarrow{1500-1700°C} 2AlN + 3CO \;.$$

Cleaning, combustion of C:

$$AlN + C + \tfrac{1}{2}O_2 \xrightarrow{500-800°C} AlN + CO \;.$$

Bayer Process

When we speak of the Bayer process, however, we mean the alkaline pulping of bauxite. In the Bayer process, bauxite is washed with a hot solution of sodium hydroxide (NaOH) at 170–180°C and 5–7 atm, evoking a hot solution of aluminum hydroxide (Al(OH)$_3$). Next, the remaining contributions to bauxite, in particular, silica and iron oxide, are filtered out as solid impurities ("red mud"). Next, cooling down the resulting solution, the dissolved aluminum hydroxide (Al(OH)$_3$) is precipitated out. Finally, heating up the dissolved aluminum hydroxide (Al(OH)$_3$) to a critical temperature, referred to as *calcination*, alumina (α–Al$_2$O$_3$) comes into being, releasing water vapor. We annotate that the solution of raw materials, solute purification, precipitation, and calcination are the major steps of the Bayer Process, and these steps are typical for the production of a lot of other oxidic ceramics.

Acheson Process

The technical implementation of the Acheson process is shown in Figures 4.10 and 4.11. In Figure 4.10, we present the flow chart that collects the various steps leading from the starting materials, namely silica sand (SiO$_2$) and coke (C) supplemented firstly by sawdust and secondly by common salt, to the SiC powder. In Figure 4.11, we outline the situation after the reaction. We here note that the mixture of the reactants is placed around the carbon core (graphite and coke). We here also note that the carbon core is heated by means of direct current, finally heating up the mixture of the reactants to a temperature of 2700°C, after which the temperature is gradually lowered. The reader should know that near the carbon core, where the highest temperatures are reached, in a first step, SiC is generated after crossing a first critical temperature, and in a second step, SiC is decomposed into Si parts and C parts after crossing a second critical temperature. The reader should know, too, that the C parts finally form the graphite layer around the carbon core, that the Si parts in colder areas of the oven react with C to form the SiC layer, and that the SiC crystals (mostly α-SiC) after reaction and cooling are separated from the unreacted material. We here just want to scrible down that the sawdust makes the mixture of the reactants porous so that CO parts may escape and that the common salt acts as a purifier, namely the Cl parts react with impurities, forming chlorides which escape.

4.1.1.2 Mass Fabrication Procedures

Depending on the forming method, the powder must be processed further, leading to the so-called "mass". (i) Regarding the wet forming methods, powder suspensions with adjusted viscosity, also called "the slips", are a prerequisite. (ii) Regarding the dry forming methods, free flowing powder agglomerates are a prerequisite. (iii) Regarding the plastic forming methods, plastified powders are a prerequisite.

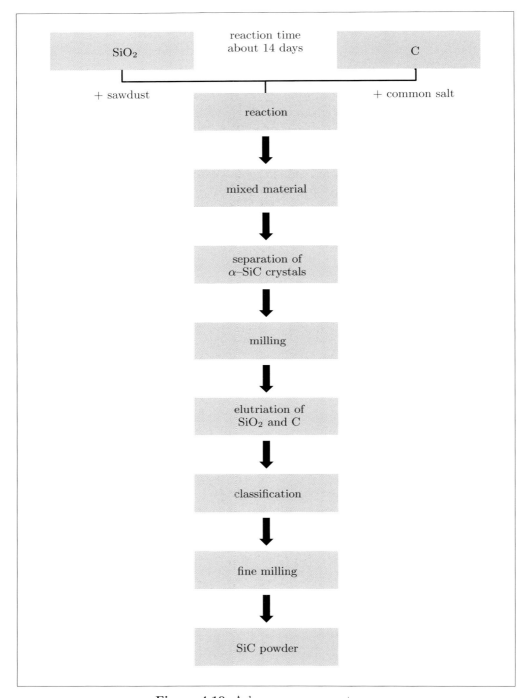

Figure 4.10. Acheson process, part one.

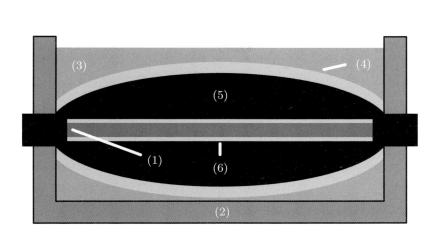

(1) electrodes with carbon core (graphite and coke); (2) furnace bed; (3) not reacted material; (4) not reacted material + β–SiC layer; (5) α–SiC layer; (6) graphite layer

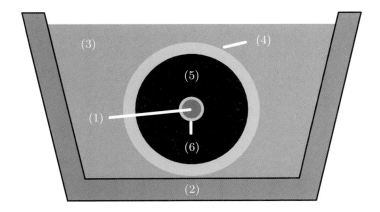

Figure 4.11. Acheson process, part two.

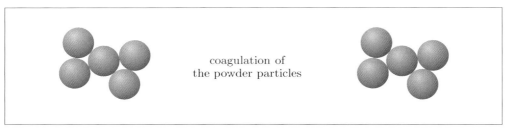

Figure 4.12. Without sterical stabilisation.

The deepening remarks that follow should be useful for the reader. (i) Slips are colloidal solutions. The stability of such colloidal solutions is evoked by sterical effects or electrostatic effects. As it is shown in Figures 4.12 and 4.13, the individuality of the powder particles in the first case is maintained by adsorbed chain molecules, keeping the powder particles at distance. As it is shown in Figure 4.14, the individuality of the powder particles in the second case is maintained by a bilayer, which is formed in polar solvents around each powder particle, and which compels a loose arrangement of the powder particles. The reader should know that this electrostatic stabilization is modelled by the DLVO (Derjaguin, Landau, Vervey, Overbeek) theory, in principle, leading back the electrostatic stabilization effect to an overlay of a bilayer repulsion and a van-der-Waals attraction as it is outlined in Figure 4.14. The reader should also know that the viscosity of such colloidal solutions depends on the pH-value, the ionic strength, and the properties of the solvents and the dispersants. (ii) We here note that free flowing powder agglomerates are fabricated by spray drying of powder suspensions. We here also note that free flowing powder agglomerates contain a plastification mass (wax, stearin acid, polyvinyl alcohol etc.) of about 1–2% of the total mass, and the plastification mass not only acts as a binder, but also as a slip additive reducing the friction between the powder particles as well as between the mass and the machine components during the powder compaction. (iii) We here note that plastified powders are fabricated by making use of binders. Here also auxiliaries are added to reduce the friction between the powder particles as well as between the mass and the machine components needed for compounding steps such as kneading and molding.

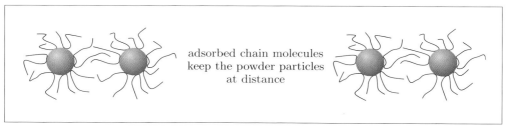

Figure 4.13. With sterical stabilisation.

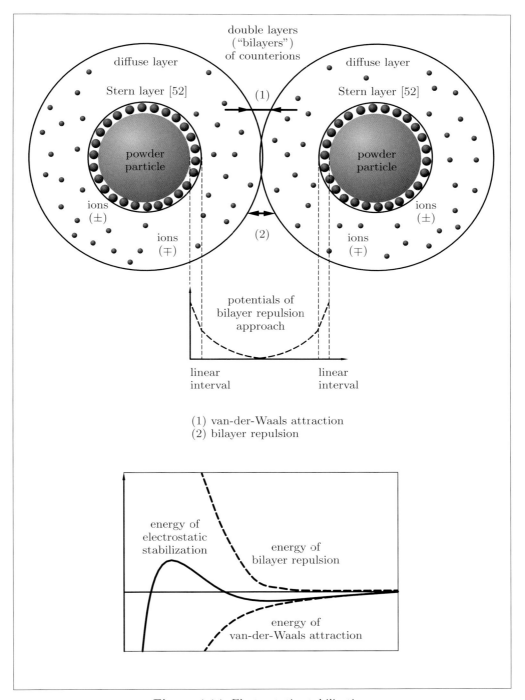

Figure 4.14. Electrostatic stabilisation.

4.1.1.3 Forming Procedures

The forming of the outcome of the powder preparation, i. e. the subsequent forming of the mass, leads to the green body. Capitalising on the advantages of each technology, the various branches of the ceramic industry produce different ceramic products with different forming methods. Figures 4.15–4.17 outline the principal elements of the most important wet forming methods, dry forming methods, and plastic forming methods. In combination with Box 4.2, Figures 1.3–1.10 presented in the first chapter link the different forming methods to special ceramic products.

Injection molding, extruding, and *roll compaction* are basic plastic forming methods. In all of these cases, one makes use of the fact that special types of molding masses under high pressure become plastic. Firstly, we note that injection molding means that the molding mass is compressed into a tool of any shape, and this enables us to produce green bodies that can be instantly removed and processed, guaranteeing low clock cycles and large numbers of pieces. Secondly, we note that extruding means that the molding mass is pressed through nozzles, and this enables us to produce endless strings with nearly any cross-sectional profile. Thirdly, we note that roll compaction means that plates are thinned down between pairs of rolls.

Tape casting, slip casting, and *freeze casting* are the major wet forming methods. Firstly, we note that tape casting means that slip is cast onto an endless carrier band. Passing through a drying line, a leather-like foil of about 0.1–1.0 mm thickness results. Secondly, we note that slip casting means that slip is cast into porous molds such as porous plaster molds. Absorbing the solvent, the mold walls enforce the formation of bodies, which after pouring out the remaining slip and after a final preparation can be taken from the mold, we compare with Figure 4.18. Thirdly, we note that freeze casting means that a suspension, which consists of the mass and the solvent, for example, water, is cast into a mold, for example, a copper or a silicon rubber mold, and frozen. Eliminating the water by freeze drying, a green body that is ready for burnout and sintering comes into being, we compare with Figure 4.19.

Dry pressing and *(cold) isostatic pressing* are the major dry forming methods. Firstly, regarding dry pressing, we note that especially the uniaxial pressing of powders is applied. We here take notice that the powders should contain just as much binder or slip additives that the agglomerates are free-flowing, easily compressible, and easily manageable after the dry pressing. We here take notice, too, that the slip additives can be burned out during heating the parts in the sintering furnace. In the case of the uniaxial pressing of powders, due to density gradients, the height-to-diameter ratio of the green bodies is restricted to about 2 : 1. In the case of the uniaxial pressing of powders, subdivided tools are applied to achieve a preferably uniform density of the green bodies. In any case, dry pressing needs precise dies which are costly. In any case, dry pressing is simple and suitable for low clock cycles and large numbers of pieces (100 parts and more per minute). Secondly, regarding the (cold) isostatic pressing, we note that especially the wet matrix technique and the dry matrix technique are applied. Naturally, these techniques are expensive and the clock cycle is long in comparison to the clock cycle of dry pressing. However, these techniques enable us to manufacture green bodies with almost every shape and almost every dimension.

Let us here collect some deepening remarks concerning slip casting. The principle of slip casting is simple. Firstly, slip is cast into porous molds such as plaster molds. Secondly, absorbing the solvent, the mold walls evoke the formation of bodies. Thirdly, the bodies are taken from the mold after pouring out the remaining slip. As it is shown in Figure 4.18, the application of additional pressure usually is necessary. As it is also shown in Figure 4.18, a final preparation usually is necessary. We here note that the forming is achieved by dewatering and flocculation of the suspension. We here also note that slip casting is a method of powder compaction that reduces the defect sizes to the level of powder particles sizes. Certainly, it is associated with low tool costs. However, it requires a large stacking ground and time for the drying of the bodies. Beyond that, due to the low mechanical strength of porous plaster, the wear of the mold walls is high, leading to a fast reduction of the surface quality and the dimensional accuracy of the green bodies, and thereby of the ceramic products.

Table 4.1. Alumina powders from ALCOA World Chemicals.

powder name	d_{50} [μm]	BET [m^2/g]	distribution
CT1200 SG	1.2	3.3	unimodal
A16 SG	0.4	9.5	unimodal
A17 NE	2.5	2.9	bimodal

Let us here collect some deepening remarks concerning freeze casting. Freeze casting has several advantages in comparison to other forming methods. Firstly, freeze casting makes it possible to produce agglomerate-free green bodies. Secondly, without needing complicated technical equipment, freeze casting makes it possible to form green bodies with complex surface geometries and fine surface structures, for example, relatively simple molding techniques such as silicon rubber molding techniques are sufficient to form green bodies with complex surface geometries and fine surface structures. Thirdly, as a consequence, high-quality ceramic products can be produced at low costs. We compare with Figure 4.19, which illustrates the typical production chain, which consists of eight steps, by example. In a first step, the mixing step, we prepare a suspension out of water, 20% glycerol, and alumina powder, and we use triammonium citrate as dispersant. In a second step, the degasing step, we prepare a gas-free suspension. In a third step, the casting step, we fill a silicon rubber mold with the gas-free suspension. The freezing step and the demolding step evoke the green body. The freeze drying, the burnout at 500°C, and the sintering, which needs 1650°C and takes 5 hours, evoke the ceramic product. We here note that also freeze casting supplies us with a method of powder compaction that reduces the defect sizes to the level of powder particles sizes. We here also note that freeze casting was successfully applied for the manufacturing of alumina ceramics, aluminum nitride ceramics, and zircon silicate ceramics, we compare with Table 4.1, collecting corresponding alumina powders. We here further note that silicon rubber molds are cheaply designed by rapid prototyping methods.

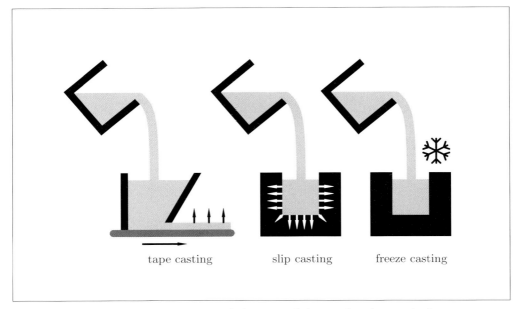

Figure 4.15. The principal elements of the wet forming methods.

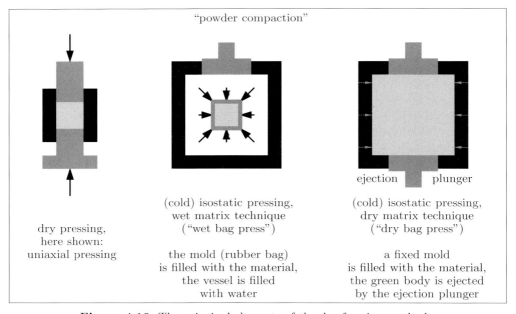

Figure 4.16. The principal elements of the dry forming methods.

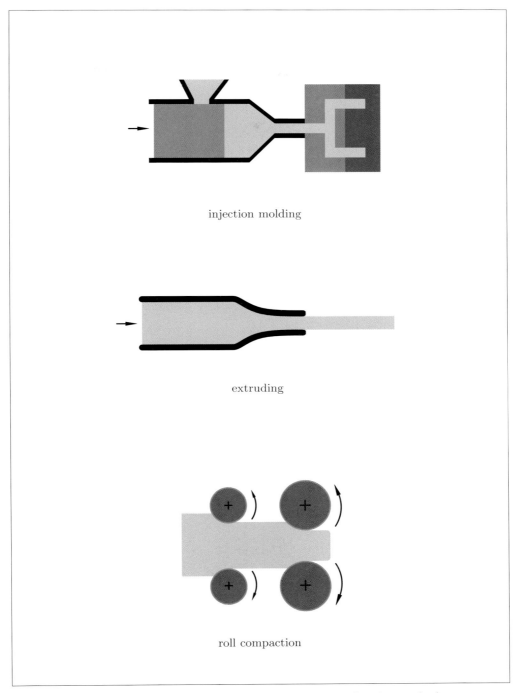

Figure 4.17. The principal elements of the plastic forming methods.

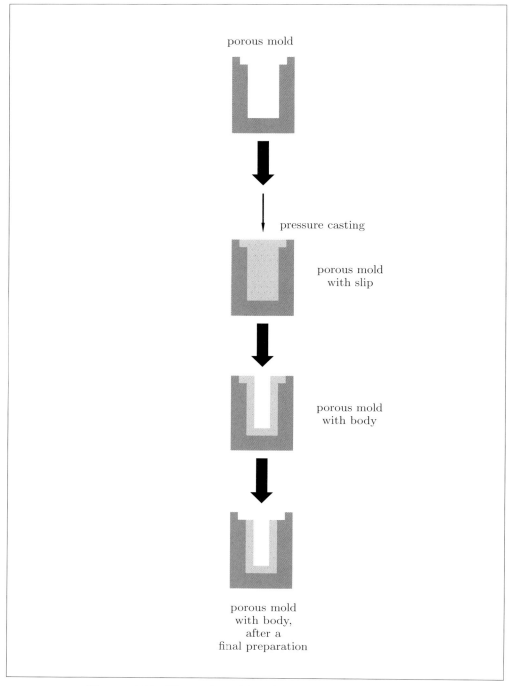

Figure 4.18. Slip casting within the typical production chain.

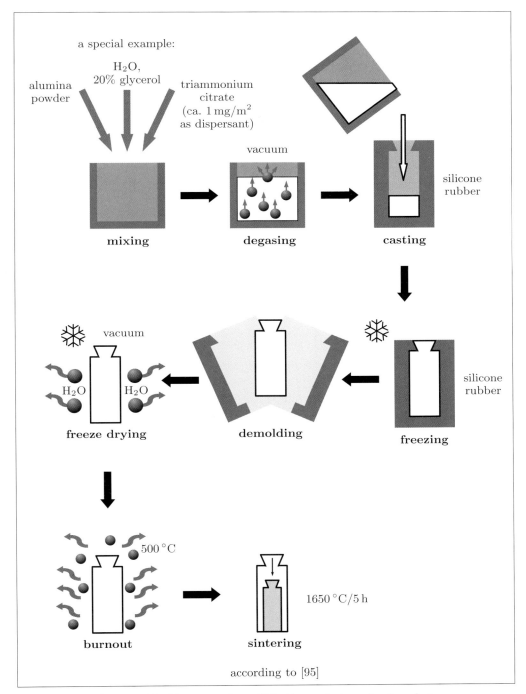

Figure 4.19. Freeze casting within the typical production chain.

> **Box 4.2** (Important data: forming, selected ceramic products).
>
> Wet forming methods.
>
> Tape casting:
>
> substrates, electrical circuits, and power electronics (Al_2O_3), integrated circuit packages, multi-chip packages, fuel cells (ZrO_2), heat exchangers (SiC).
>
> Slip casting:
>
> blower pipes (SiC), portliners (Al_2TiO_5).
>
> Dry forming methods.
>
> Dry pressing:
>
> overload protection switch (Al_2O_3), hip joint endoprotheses (Al_2O_3), sealing washers (Al_2O_3), sound generators (PZT).
>
> (Cold) isostatic pressing:
>
> glide ring seals (SiSiC), ignition plug insulators (Al_2O_3).
>
> Powder rolling:
>
> substrates (Al_2O_3).
>
> Plastic forming methods.
>
> Injection molding:
>
> pre-combustion chambers (Si_3N_4), turbo charger rotors (Si_3N_4), valves (Si_3N_4), valve guides (Si_3N_4).
>
> Extruding:
>
> catalytic converter carriers (cordierite), diesel soot filter, heat exchanger pipes (SiSiC).
>
> Roll compaction:
>
> substrates.
>
> Compare with Figures 4.15–4.17 and take a look at Figures 1.3–1.10.

Going a little bit beyond these explanations, let us here point out that there are forming sub-processes which concatenate forming with heat. Let us here point at the *hot isostatic pressing (HIP)* and the *uniaxial hot pressing (UHP)*. (i) HIP outlines the concatenation of high temperature with pressurizing gases or pressurizing liquids. HIP means that the powder is filled into a metal container or glass container that is evacuated, sealed, and finally placed into a high pressure storage vessel, which is heated, on the one hand, evoking the necessary temperature, and on the other hand, causing a basis pressure that is elevated up to the necessary pressure by gas pumping. HIP is also an effective method to reduce the microporosity that usually remains after normal sintering. HIP evokes plastic deformations including creep deformations evoked by diffusion processes, and these are responsible for the reduction of the microporosity. In the case that the pieces to be pressed certainly reveal closed pores, but no porosity open to the outside, the process needs no containers and no further treatments of the pieces. In any case, the process is rather costly and only applied if no cheaper process is available. (ii) UHP outlines the concatenation of high temperature with high mechanical pressure. UHP means that the powder is filled into a die of graphite, and since only small numbers of pieces can be produced with a single die, the economy of UHP is restricted. UHP facilitates the compacting of all kinds of oxides, nitrides, borides, carbides, and sulfides to near theoretical densities. (iii) Following the scheme of Figure 4.7, on the one hand, we also realize that HIP facilitates the redensification of sintered materials ("post-HIP"), and on the other hand, we also realize that HIP facilitates the merging of sintering and post-HIP ("sinter-HIP"). (iv) Going beyond the scheme of Figure 4.7, we keep in mind that HIP and UHP are also widely used for the creation of metal matrix composite (MMC) microstructures.

There is a variation of uniaxial hot pressing (UHP), the so-called *sinter forging*, we consult Figures 4.21–4.24. This variation of uniaxial hot pressing (UHP) was applied to produce nanograined SiC. (i) SiC, on the one hand, is one of the hardest materials, and on the other hand, exhibits a remarkable thermomechanical stability as well as a remarkable thermochemical stability. It is therefore one of the most important materials of technological relevance. (ii) Naturally, the properties of this material are dictated by the microstructure of this material, and thus depend on the manufacturing process including temperatures used, pressures used, and additives used. Naturally, it thus is important to study different manufacturing processes just like the results of the different manufacturing processes. (iii) The following process scheme was taken as the basis. 1. SiC powder (50 nm) was washed with hydrofluoric acid (HF). 2. SiC powder was mixed with the additives Y_2O_3 (20–70 nm) and Al_2O_3 (20 nm). 3. The green body was manufactured by (cold) isostatic pressing. 4. The green body was compacted by sinter forging at different temperatures and different pressures. (iv) The results that follow are of technological importance. Firstly, increasing temperature or pressure, we achieve a compacting to near theoretical densities. Secondly, in comparison to the gas-pressure-sintered sample, lower temperatures are sufficient. Thirdly, in comparison to the uniaxial hot pressed sample, if at all only finer pores are observable. Fourthly, in comparison to other processes, very fine grain sizes are achievable (≈ 55 nm) which reveal superplastic deformation. (v) The additives are necessary because the strong covalent contributions of SiC bonds do not allow diffusion processes.

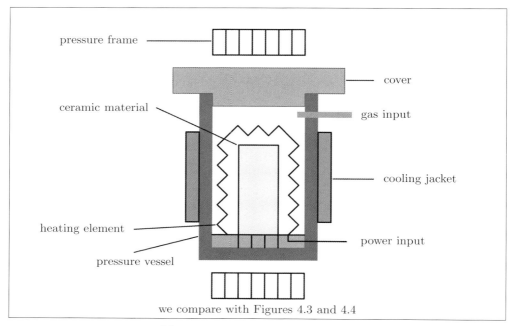

Figure 4.20. The principle of HIP.

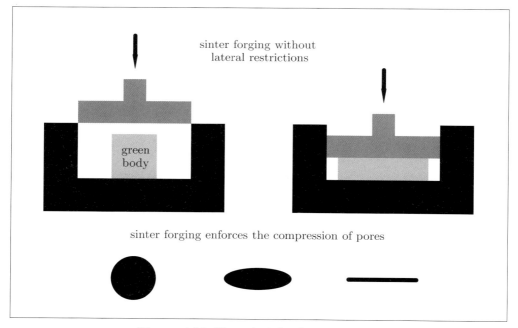

Figure 4.21. The principle of sinter forging.

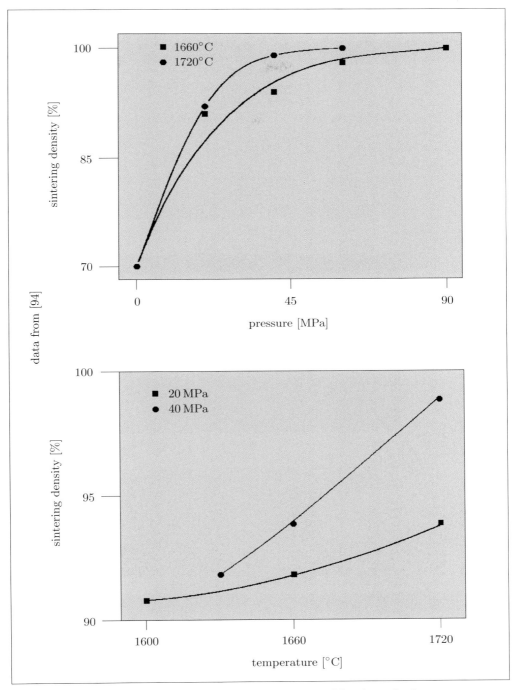

Figure 4.22. SiC sintering densities obtained by sinter forging.

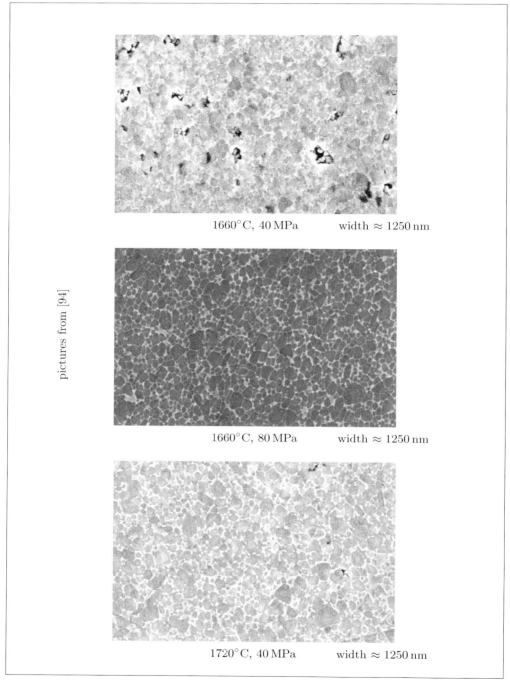

Figure 4.23. SiC microstructures obtained by sinter forging.

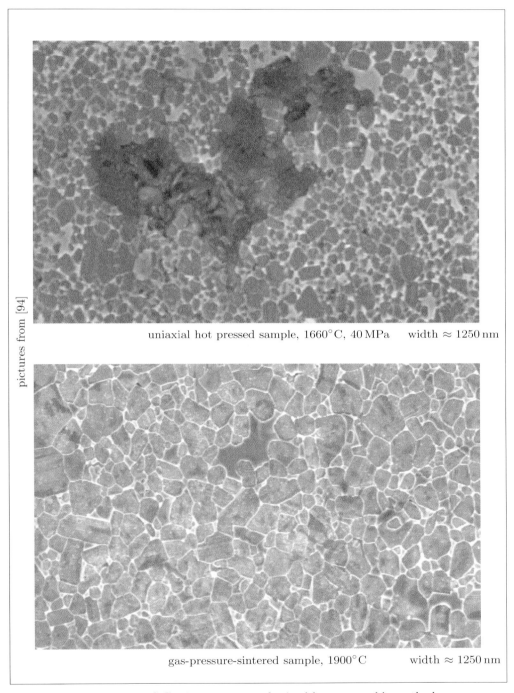

Figure 4.24. SiC microstructures obtained by comparable methods.

4.1.1.4 Burnout Procedures

The burnout removes the organic additives from the green body by evaporation, pyrolysis, and oxidation so that the green body during burnout gradually loses weight. In a first step, high-molecular supplies disintegrate and low-molecular supplies such as solvents evaporate, and in a second step, the high-molecular binders burn up. The reader should appreciate that the burnout period is a highly critical period because the local volume increasing caused by the melting of the binders can be huge, eventually leading to crack formation. Certainly, with high- and low-molecular polymers such as polyethylene, in combination with slip additives such as waxes, green bodies showing wall thicknesses $< 10\,\mathrm{mm}$ can be burned out. However, for thicker parts, the burnout may need several weeks, which makes the process uneconomical. Fortunately, starting from the surface of the green body, binders such as polyoxymethylene disintegrate in acidic gases such as HCl and BF_3 during the heating, evoking a reaction mechanism that can be easily utilized to reduce the expenditure of time, which makes the process economical. We here yet want to annotate that alternative methods to get rid of the unwanted remains are the extraction of the supplies by solvents or supercritical gases, in the latter case, evoking degasing channels that facilitate the disintegration of the unwanted remains.

4.1.1.5 Sintering Procedures

The sintering restructures the microstructure of the green body. At temperatures much higher than those applied during the burnout, extensive material transport mechanisms are excited, evoking the compacting and the hardening of the green body. In particular, one distinguishes between "solid-state sintering" with or without additives enhancing the sintering process and "liquid-phase sintering" with additives that become fluid during the heating, leading to a glass phase. In particular, one distinguishes between sintering processes without additionally applied pressure and sintering processes with additionally applied pressure. (i) The reader should take a look at Figure 4.25, where a shrinkage curve typical for the shrinking of a green body during solid-state sintering is depicted. There are typical evolution stages. With respect to structural parameters *and* processing parameters, one distinguishes between initial stage, intermediate stage, and final stage. With respect to structural parameters alone, one distinguishes between *contact evolution stage*, *shrinkage stage*, and *after-shrinkage stage*. (ii) The reader also should take a look at Figure 4.26, where the neck evolution of powder particles during solid-state sintering is illustrated by means of a basic two-particle model. There are diffusion paths that lead through the particle volumes called "volume diffusion" and "grain boundary diffusion", paths that lead along the particle surfaces, and paths that lead through the vapor phase. (iii) The reader also should take a look at Figure 4.28, where neck evolution and grain growth of powder particles during liquid-phase sintering is illustrated. The difference to solid-state sintering is that the liquid phase evoked by the additives leads to particle rearrangement and boundary condition alteration affecting the diffusion processes. The solution processes and redeposition processes that come into being during liquid-phase sintering depend on the liquid pressure and the surface curvature.

We here point out that shrinkage only occurs when the particle centers approach, implying material transport from the grain boundary between the particles to the neck, namely grain boundary diffusion 3. and volume diffusion 5. and 6., whereas the other transport mechanisms create neck formation, but no shrinkage. Nevertheless, the other transport mechanisms are important for the collective sintering kinetics. For example, at given temperature, surface diffusion 1. exceeds grain boundary diffusion 3. Hence, additional grain boundary diffusion paths can come into being. For example, at given temperature, grain boundary diffusion 3. exceeds volume diffusion 4., 5., and 6. Hence, the grain growth during the sintering process may decrease and the shrinkage evoked by grain boundary diffusion 3. and volume diffusion 5. and 6. may stop, but pores that are enclosed in the grains that emerge can be further reduced by volume diffusion 4, however, which is slow.

4.1.1.6 Hard Machining Procedures

In order to meet technical demands, in most practical cases, on the one hand, an advanced surface finishing of the ceramic object is necessary, and on the other hand, an advanced mechanical processing of the ceramic object is necessary. For example, we here compare with Figure 4.29, which shows stamped substrates, and Figure 4.30, which shows lasered substrates. Certainly, in order to reduce costs, the process chains nowadays are designed to produce so-called "near-net-shape ceramic parts", namely ceramic parts the skins of which almost have the surface qualities of the desired final ceramic products and the contours of which almost have the proportions of the desired final ceramic products. However, the experience shows us that a postprocessing in most practical cases rarely is avoidable.

The reader probably should know that *polishing* of a ceramic object means that the surface quality of the ceramic object is improved by applying rubbing techniques or by applying chemical techniques. The reader probably should know that *lapping* of a ceramic object means that the surface of the ceramic object and the surface of another object are rubbed together with an abrasive in between. The reader probably should know that one speaks of *grinding* if a brittle material such as a ceramic material is rubbed against a hard material such as iron including abrasives such as diamond, aluminum oxide, or silicon carbide. We here note in passing that polishing also is a very good means for the preparation of ceramic objects the microstructure of which shall be pictured by (light-optical/electron) microscopes.

The reader probably should know that for the production of ceramic objects with well-defined dimensions, holes, slots, and edges, in particular, for the production of substrates with well-defined dimensions, holes, slots, and edges, *stamping techniques* and *lasering techniques* are applied. We here note in passing that lasering techniques also facilitate the "physical cleaning" of surfaces, for example, lasering techniques also facilitate the stripping of bonded hydrogen atoms or bonded carbon hydrogens from surfaces. We here also note in passing that the stripping of hydrogen atoms from surfaces is selective, for example, irradiating a silicon surface, on the one hand, showing hydrogen atoms, and on the other hand, showing deuterium atoms, hydrogen atoms can be removed without detaching a large number of deuterium atoms.

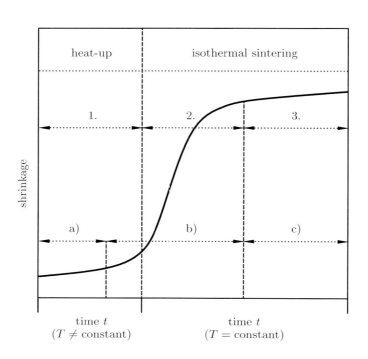

1. initial stage
2. intermediate stage
3. final stage

a) contact evolution stage
b) shrinkage stage
c) after-shrinkage stage

During stage a), caused by beginning transport processes evoked by the increasing temperature, the powder particles begin to grow together, in particular, necks begin to develop ("contact evolution stage"). During stage b), the high temperature generates a high diffusion activity, necks develop strongly, grain boundaries emerge, pore channels enabling gas transport emerge ("shrinkage stage"). During stage c), grains develop further, porosity reduces further ("after-shrinkage stage").

We compare with Figures 4.26–4.28.

Figure 4.25. Solid-state sintering. Shrinkage curve.

Neck evolution (two-particle model):

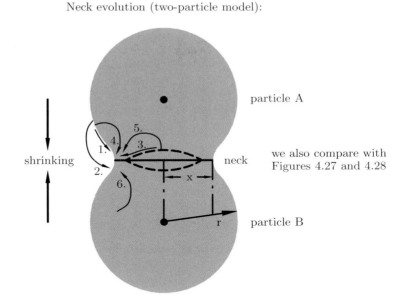

Transport mechanisms:

1. diffusion at the surface   (surface → neck)
2. vaporisation and resublimation   (surface → neck)
3. diffusion along the grain boundary (boundary → neck)
4. diffusion through the volume   (surface → neck)
5. diffusion through the volume   (boundary → neck)
6. diffusion through the volume   (volume → neck)

Surface diffusion processes are faster than
grain boundary diffusion processes or volume diffusion processes.

Responsible for the shrinking:

3., 5., 6.

Radius of neck (sintering kinetics):
$$x^n = r^m f(T)\, t$$
(x = radius of neck, r = radius of particle, T = temperature, t = sintering time).

Thermodynamic modeling by the "sintering equation" (2.76).
On the one hand, (2.76) tells us how strong the free enthalpy G decreases
if the volume of the ceramic material decreases.
On the other hand, (2.76) tells us how strong the free enthalpy G decreases
if the surfaces of the ceramic material decrease.

Figure 4.26. Solid-state sintering. Neck evolution.

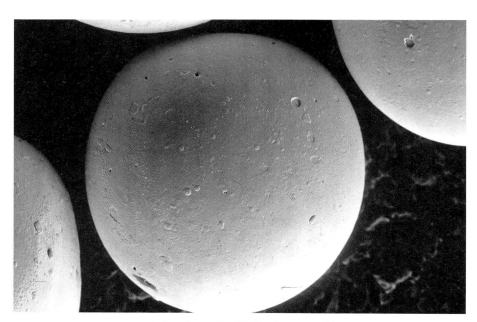

taken from the PML picture archive

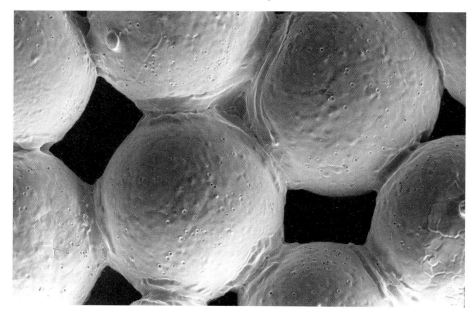

Figure 4.27. Neck evolution of a Ni–Cu system during solid-state sintering.

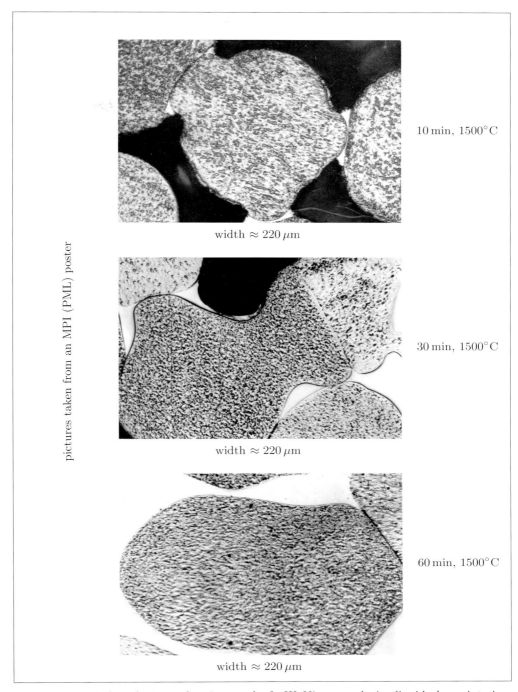

Figure 4.28. Neck evolution and grain growth of a W–Ni system during liquid-phase sintering.

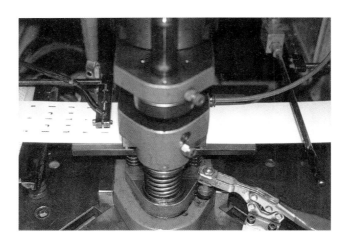

images from [71]

stamped substrates

stamped hole   stamped edge

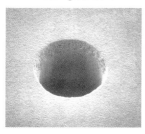   

Figure 4.29. Hard machining: stamping.

lasered substrates

images from [71]

lasered substrates

lasered edge

Figure 4.30. Hard machining: lasering.

4.1.2 Advanced Procedures

An advanced procedure within the familiy of powder-based procedures is the so-called *reaction sintering*, also called *reaction bonding*. The reader should particularly know that reaction sintering means that metallic powders such as silicon (Si) powders are formed by cold compression methods such as (cold) isostatic pressing and subsequently are reacted in reactive atmospheres such as nitrogen/hydrogen at high temperatures, following the reaction scheme (4.1), evoking very dense ceramic materials such as RBSN (reaction bonded silicon nitride). The reader should take notice that ceramic materials produced by reaction sintering conserve their strength up to temperatures $> 1400°C$, while ceramic materials produced by liquid-phase sintering only conserve their strength up to temperatures $\leq 1200°C$ due to the softening of the glass phase. The reader should also take notice that there are modifications of reaction sintering. Regarding Box 4.3, on the one hand, we point at the Lanxide process, which departs from a suitable mixture of a metal powder and a ceramic powder, namely Al and Al_2O_2, and proceeds at temperatures $> 600°C$, and on the other hand, we point at the thermite processes, which are applied as initiating processes in pyrotechnics (fireworks).

Another advanced procedure within the familiy of powder-based procedures to be mentioned here is the so-called *powder spraying*. We here should especially note that one distinguishes between wet powder spraying procedures, where green films are manufactured by the spraying of powder suspensions with the "paint spraying gun" and are sintered in a further step, and thermal powder spraying procedures such as flame spraying, where a mixture of powder and carrier gas is injected into a flame fed by fuel gases, or plasma spraying, where a mixture of powder and carrier gas is injected into an accelerated hot gas, in both cases, leading to a ceramic film on a substrate, at which plasma spraying is a good choice if we want to produce dense thick films or bulk samples. In Figure 4.31, the direct current version of plasma spraying is presented. In Figure 4.33, a plasma-sprayed film of the solid electrolyte $La_{0.85}Sr_{0.15}Ga_{0.9}Mg_{0.1}O_{3-\delta}$, which is a member of the familiy of LSGM ceramics, is presented. We immediately diagnose that the plasma-sprayed film of the solid electrolyte $La_{0.85}Sr_{0.15}Ga_{0.9}Mg_{0.1}O_{3-\delta}$, apart from some chemical inhomogeneities, exhibits a crystalline, on the one hand, very dense, and on the other hand, very uniform structure.

Another advanced procedure within the familiy of powder-based procedures to be mentioned here is the so-called *porous printing*. We here should especially note that a viscous paste of powdered material is pressed with a doctor blade through a metal net onto a substrate, together with an additional mask, facillitating a lateral structuring in the $100\,\mu m$ range. Porous printing plays an important role in the production of printed circuit boards. In a first step, a lacquer imprint of the conducting path scheme is applied by porous printing to a copper-coated substrate, evoking a partially lacquered copper-coated substrate, reflecting the conducting path scheme. In a second step, the partially lacquered copper-coated substrate is etched, removing the bare copper parts, eventually evoking the desired printed circuit board.

> **Box 4.3** (Important formulae: reaction synthesis).
>
> Reaction sintering.
>
> Example:
>
> $$\text{Si powder preform} \xrightarrow{N_2, 1\,\text{bar}, \approx 1400^\circ C} Si_3N_4 \qquad (4.1)$$
>
> (RBSN = reaction bonded silicon nitride).
>
> Variations of reaction sintering.
>
> Examples:
>
> $$\text{Si + C powder preform} \xrightarrow{Si_{liquid}, \approx 1450-1700^\circ C} SiC \qquad (4.2)$$
>
> (RBSC = reaction bonded silicon carbide),
>
> $$Al_2O_3 \text{ powder preform} + Al \xrightarrow{O_2, \text{temperature}} Al_2O_3/Al \text{ composite} \qquad (4.3)$$
>
> (Lanxide process),
>
> $$Fe_2O_3 + 2Al \longrightarrow 2Fe + Al_2O_3 + \text{heat}\,(\Delta H) \qquad (4.4)$$
>
> (iron thermite process),
>
> $$3CuO + 2Al \longrightarrow 3Cu + Al_2O_3 + \text{heat}\,(\Delta H) \qquad (4.5)$$
>
> (copper thermite process).
>
> Notes:
> We note that thermite processes define a sub-class of a class of high-temperature processes summarized as *self-propagating high-temperature synthesis* (SHS) and characterized by the release of a relatively big amount of heat energy (exothermic processes).
> Furthermore, we note that metals reacting with metalloids such as B, Si, As, Te, Ge, and Sb define another SHS sub-class.

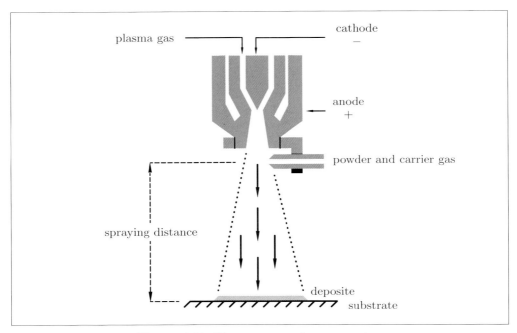

Figure 4.31. Direct current plasma spraying.

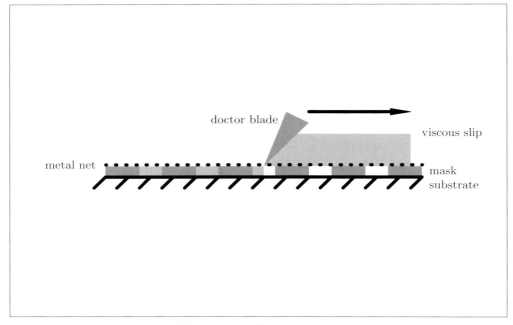

Figure 4.32. Porous printing.

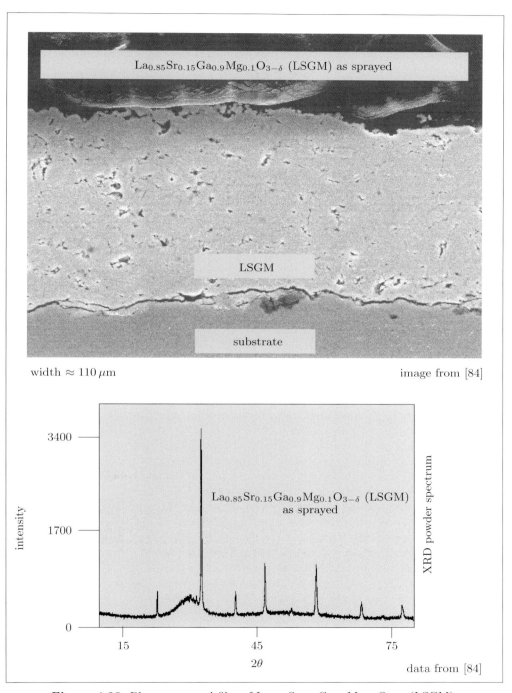

Figure 4.33. Plasma-sprayed film of $La_{0.85}Sr_{0.15}Ga_{0.9}Mg_{0.1}O_{3-\delta}$ (LSGM).

For completion, let us here talk about an alternative gateway to LSGM ceramics, we compare with Figure 4.34. LaGaO$_3$-type (lanthanum-gallate-type) ceramics doped with Sr and Mg show excellent oxygen ion conducting properties and thereby show excellent electrode performance, for example, desirable for solid oxide fuel cells (SOFCs). For completion, let us here record the following well-manageable production method of LSGM ceramics [57, 92, 107, 108]. (i) The first step is the powder synthesis according to the Pechini method [43]. La-, Ga-, Sr-, and Mg-nitrate aqueous solutions, together with citric acid and ethylene glycol, are used to form polymeric gels. The citric acid forms polybasic acid chelates with the metal cations, and the polybasic acid chelates undergo polyesterification, when heated with ethylene glycol at a temperature of about 150°C, forming polymeric precursor resin. The subsequent heating of the polymeric precursor resin at temperatures > 400°C, on the one hand, removes the organics, and on the other hand, generates chars with controlled cation stoichiometry and little cation segregation or no cation segregation. The subsequent heating of the chars to higher temperatures in air evokes the oxidation, in last consequence, generating a sinterable LSGM powder, however, which may contain significant amounts of C, depending on the temperature of calcination, i.e. a lower temperature of calcination leads to a bigger amount of C. (ii) The second step is the sintering. It is sufficient to remark that the LSGM powder even as loose LSGM powder compact can be sintered, we compare with Figure 4.35, showing LSGM material of the type $La_{0.8}Sr_{0.2}Ga_{0.8}Mg_{0.2}O_{2.8}$.

For completion, let us here talk about a comparable gateway to LSGZ ceramics, we also compare with Figure 4.34. LaGaO$_3$-type (lanthanum-gallate-type) ceramics doped with Sr and Zn show excellent oxygen ion conducting properties, too, and thereby show excellent electrode performance, too. For completion, let us here record the following well-manageable production method of LSGZ ceramics [58]. (i) The first step is the powder synthesis by homogeneous precipitation [56]. La-, Ga-, Sr-, and Zn-nitrate aqueous solutions, together with urea (with or without enzyme urease), are used to form starting clear solutions, when heated at a temperature of about 90°C, leading to an homogeneous precipitation, forming hydroxycarbonate precursors. The subsequent heating of the hydroxycarbonate precursors to higher temperatures in air evokes the oxidation, in last consequence, generating a sinterable LSGZ powder. On the one hand, it is remarkable that during homogeneous precipitation the conditions of formation of the hydroxycarbonate precursors are especially governed by the controlled generation of hydroxide ions and carbonate ions through the decomposition of urea. On the other hand, it is remarkable that enzyme urease enhances the rate of decomposition of urea. (ii) The second step is the sintering. It is sufficient to remark that the LSGZ powder even as loose LSGZ powder compact can be sintered, we compare with Figure 4.35, showing LSGZ material of the type $La_{0.8}Sr_{0.2}Ga_{0.8}Zn_{0.2}O_{2.8}$.

By the way, the method of homogeneous precipitation in the presence of urea (with or without enzyme urease) is also applied to produce nanopowders of phase pure gallium oxide hydroxide GaOOH, and gallium oxide hydroxide GaOOH is immediately converted into Ga_2O_3 upon a light calcination in air, which after doping with Mn/Eu is applied in electroluminescent display devices, we compare with Figure 4.36, showing precipitated Zeppelin-shaped single GaOOH crystals [59].

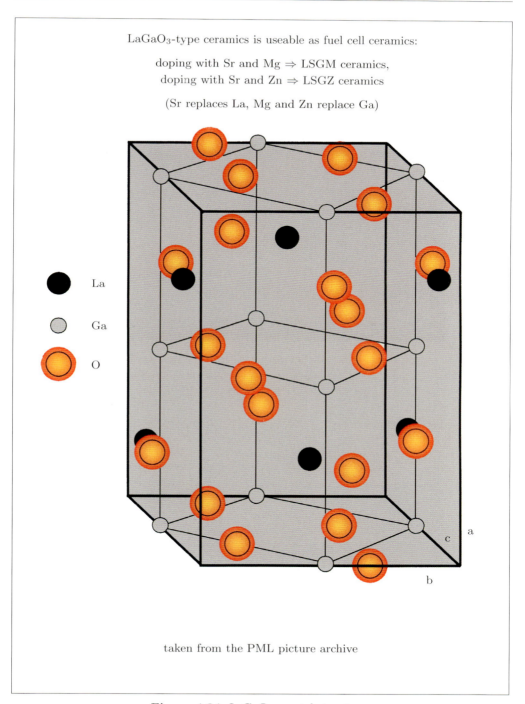

Figure 4.34. LaGaO$_3$ crystal structure.

LSGM material obtained by heating of a loose LSGM powder compact at 1300°C for 6 h

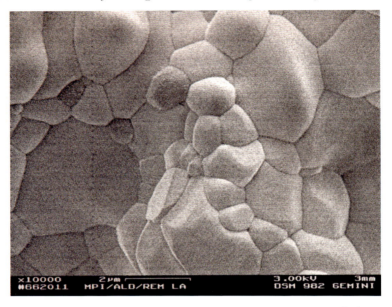

SEM micrographs, taken from the PML picture archive [92]

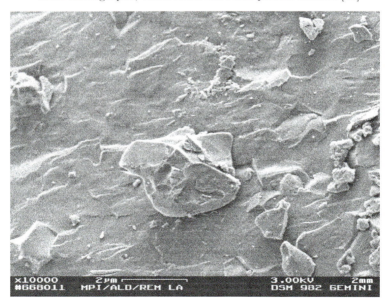

LSGZ material obtained by heating of a loose LSGM powder compact at 1300°C for 6 h

Figure 4.35. LSGM material/LSGZ material.

Zeppelin-shaped single crystals (2 μm scale)

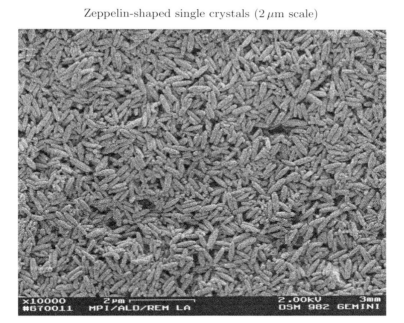

SEM micrographs, taken from the PML picture archive [92]

Zeppelin-shaped single crystals (500 nm scale)

Figure 4.36. Precipitated Zeppelin-shaped single GaOOH crystals.

4.2 Powder-Free Technologies

Certainly, for the production of bulk components of ceramics, powder-free technologies in most cases are out of choice. However, following the calls for advanced technologies consuming less energy or/and allowing the production of microparts exhibiting low cross-sections or/and thin-layered patterns, a vast of powder-free technologies have been developed. In the sections that follow, let us focus on chemical procedures such as the *sol–gel processes* and the *precursor thermolysis*, on vapor deposition procedures such as the *chemical vapor deposition* and the *physical vapor deposition*, and on novel bio-inspired mineralization procedures such as the *chemical liquid deposition*.

Beforehand, however, some annotations about a familiy of powder-free technologies circumscribed as *melting and casting* should here be appended for the convenience of the reader. Melting and casting, for example, is applied to process zirconium oxide to obtain grinding balls, or for example, is applied to process ceramic substances to obtain superconducting devices. Nevertheless, since many ceramic substances are not fusible (for example, Si_3N_4 disintegrates at $\approx 1841°C$) or melt only at very high temperatures (for example, SiC melts at $\approx 2837°C$ and B_4C melts above $3000°C$), and since many ceramic substances always develop coarse-grained structures and/or porous structures when these are melted and casted (for example, aluminum oxide always develops such coarse-grained structures), this technique has no large-scale importance. (Regarding the latter issue, we observe that the brittleness of ceramic materials increases with increasing grain size so that coarse-grained structures are mechanically weak.)

4.2.1 Chemical Procedures

Sol–gel processes enable the production of materials for further processing, so-called "precursors". As it is shown in Figure 4.37, firstly the starting material is dissolved via polycondensation evoking a dispersion of colloidal particles ("sol"), secondly the sol is dehydrated via polymerisation evoking a gel, and thirdly and fourthly the gel is dried and calcined, evoking the "precursor". Two important sol–gel processes, namely the *alkoxide process* and the *silica gel process*, are illustrated in Figures 4.39 and 4.40. In both cases, one fabricates a solution containing the hydroxide of the metal component of the starting material. Firstly, colloidal particles emerge via a polycondensation process, leading to a sol. Secondly, the elimination of water evokes a polymerisation process, leading to a gel, namely to a three-dimensional polymeric network which is linked by M–O–M bridge bonds, within superfine pores, still containg a considerable amount of organic solvents. Thirdly, a sequence of heat treatments removes waste water and the organic solvents. Comparing with Figures 4.41 and 4.42, which show the genesis and the subsequent treatment of boron-modified polysilylcarbodiimide in the context of a sol–gel process, we realize that a sol–gel process can be carried out in a test tube, so to speak, "in vitro". Comparing with literature, we realize that sol–gel processes are frequently used for the coating of substrates. Applying a sol by dipping a substrate into the sol ("dip coating") or applying a sol by spraying a sol onto a rotaing substrate ("spin coating"), we can proceed according to the recipe presented in Figure 4.37.

Precursor thermolysis enables us to produce "precursor-derived ceramics", namely crystalline ceramics manufactured from "precursors" (pre-ceramic polymers), which on their part are manufactured from "molecular precursors" (monomers), or which are the result of diverging procedures such as sol–gel processes, which we do not want to consider in this context, though. As it is shown in Figure 4.38, the starting material consists of organometallic monomers from which organometallic polymers are directly manufactured by synthesis, followed by a sequence of heating steps, in a first step, evoking a pre-ceramic network, in a second step, evoking an amorphous ceramic, and in a third step, evoking a crystalline ceramic. We here point at the advantages of the precursor thermolysis. Firstly, the temperatures needed for the precursor thermolysis are relatively low. Secondly, organometallic polymers obtained in the framework of the precursor thermolysis facilitate a versatile shaping. Thirdly, ceramic substances such as silicon nitride, which otherwise only can be sintered by using additives, can be processed without additives. We here also point at the application of the precursor thermolysis. Firstly, we here note that precursor thermolysis allows the making of ceramic powders, ceramic coatings, ceramic fibers, infiltrated media, and much more. Secondly, we here note that precursor thermolysis is used to fabricate high-temperature ceramics such as Si–C–N ceramics and Si–B–C–N ceramics.

For the convenience of the reader, we present a case study in Figures 4.43–4.52. The following remarks thereto. (i) The thermal degradations of the two variants VT50 ($Si_{2.54}N_{3.38}C_{4.07}$) and NCP200 ($Si_{3.82}N_{3.89}C_{2.29}$) of Si–C–N ceramics turned out to be much greater than those of the variants T2-1, MW33, and MW36 of Si–B–C–N ceramics provided we go beyond a temperature of $\approx 1550°C$, i. e. the incorporation of boron (B) leads to a much higher thermal stability. We compare with Figure 4.46, which shows the mass loss as a function of temperature. (ii) T2-1 ($Si_{2.41}B_{0.83}C_{4.49}N_{2.26}$) turned out to be the most stable variant. At 2200°C, the composition of T2-1 can be characterized as a mixture of β–Si_3N_4 domains and β–SiC domains encapsulated by a BNC_x matrix phase composed of BN and C in varying compositions. We compare with Figure 4.44, which shows a microstructure reflecting the composition at 2200°C. (iii) Firstly, we note that the thermal stability of Si–C–N ceramics is limited to $\approx 1550°C$ because then the endothermic phase reaction $Si_3N_4 + 3C \leftrightarrow 3SiC + 2N_2$ comes into play. Secondly, we note that at higher temperatures the endothermic phase reaction $Si_3N_4 \leftrightarrow 3Si + 2N_2$ comes into play, restricting the thermal stability of Si–C–N ceramics further. We read off from Figure 4.45 that the VT50 variant at relatively low temperatures exhibits Si_3N_4 and graphite parts, whereas the NCP200 variant at relatively low temperatures exhibits Si_3N_4, SiC, and graphite parts. We also read off from Figure 4.45 that in both cases at higher temperatures gas species such as Si, N_2, N_3 etc. come into being, strongly contributing to the mass loss. (iv) For the convenience of the reader, we here also show the calculated phase diagram of the quarternary system Si–B–C–N including the phase diagram of the ternary sub-system Si–B–C and the phase diagrams of the binary sub-systems in Figures 4.47–4.50. For the convenience of the reader, we here also show elemental maps obtained by energy filtering transmission electron microscopy (EFTEM) of a Si–B–C–N ceramic in Figures 4.50 and 4.51.

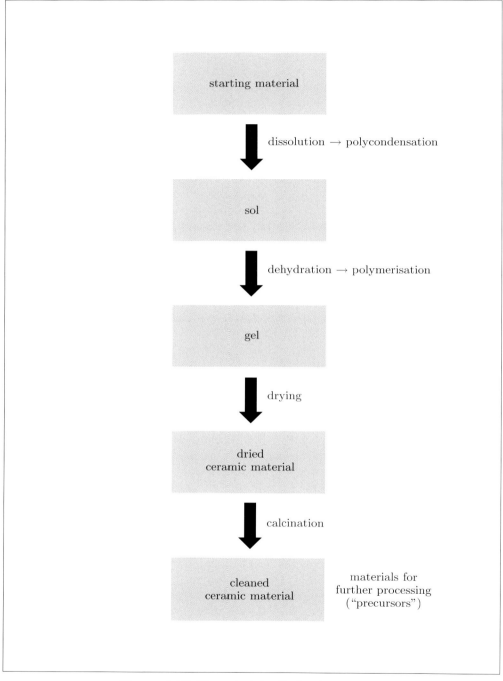

Figure 4.37. The principle of sol–gel processes.

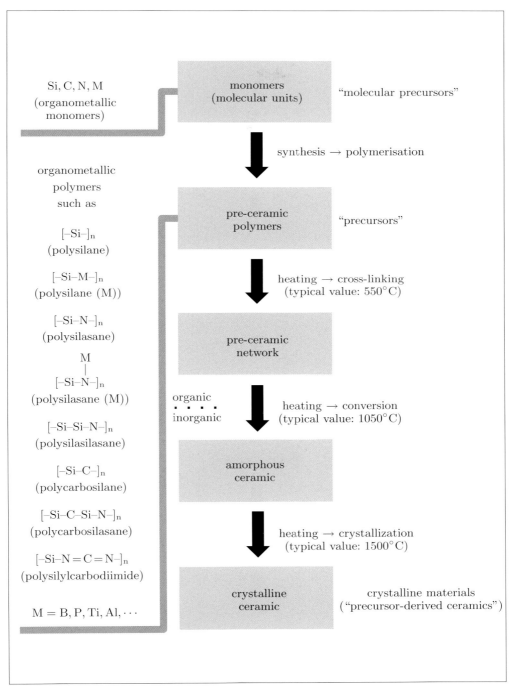

Figure 4.38. The principle of precursor thermolysis.

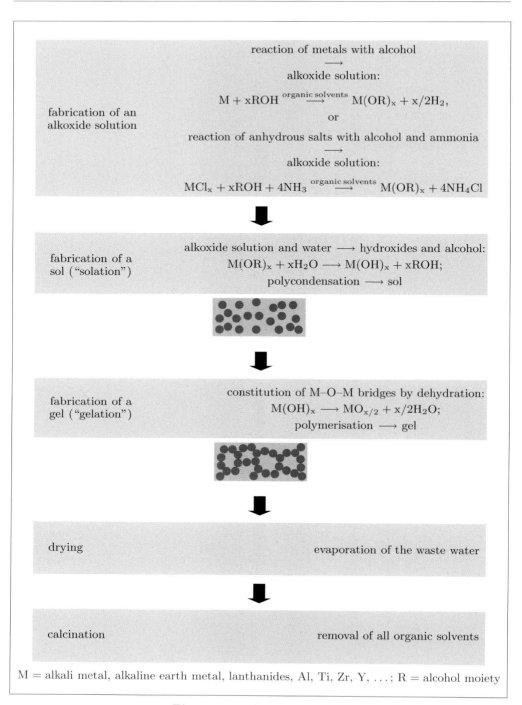

Figure 4.39. Alkoxide process.

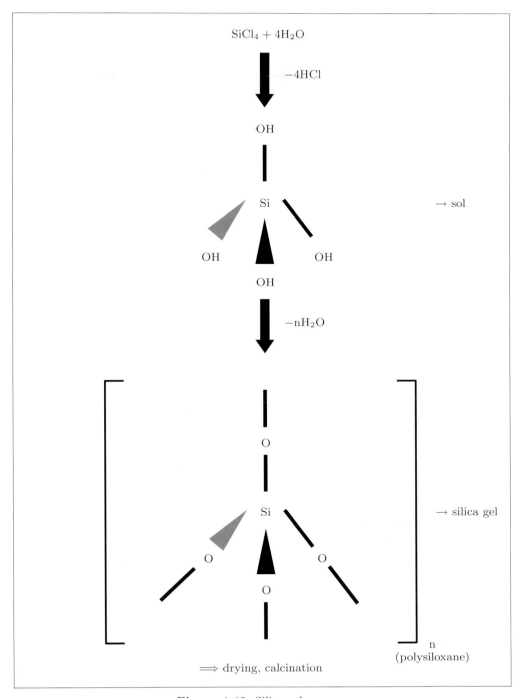

Figure 4.40. Silica gel process.

sol in test tube

gel in test tube

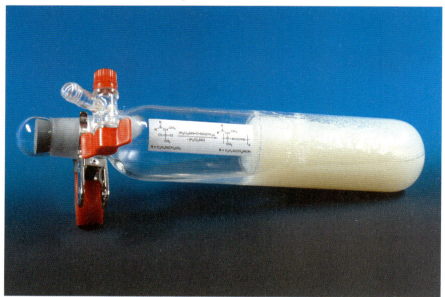

images from [93]

Figure 4.41. On the sol–gel process, part one.

dried and cleaned ceramic material

pulverized ceramic material

images from [93]

Figure 4.42. On the sol–gel process, part two.

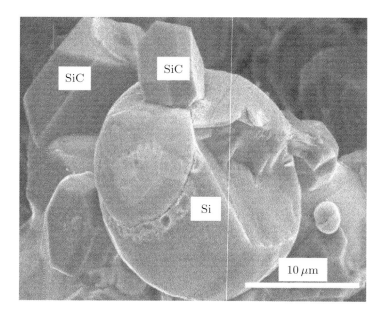

Figure 4.43. Precursor-derived Si–C–N ceramics.

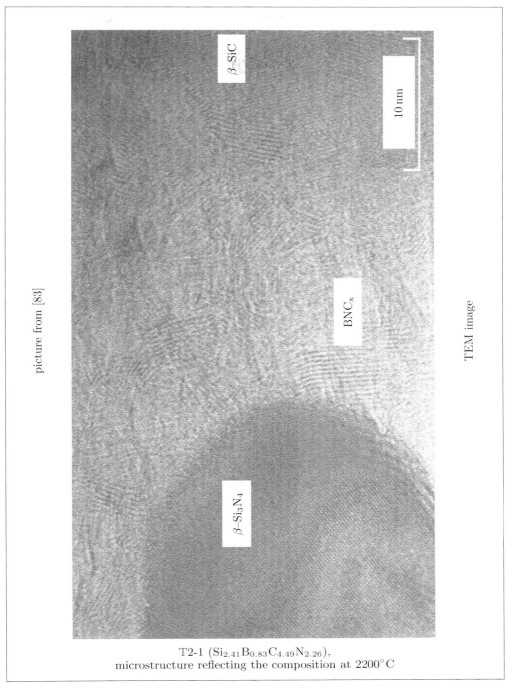

Figure 4.44. Precursor-derived Si–B–C–N ceramics.

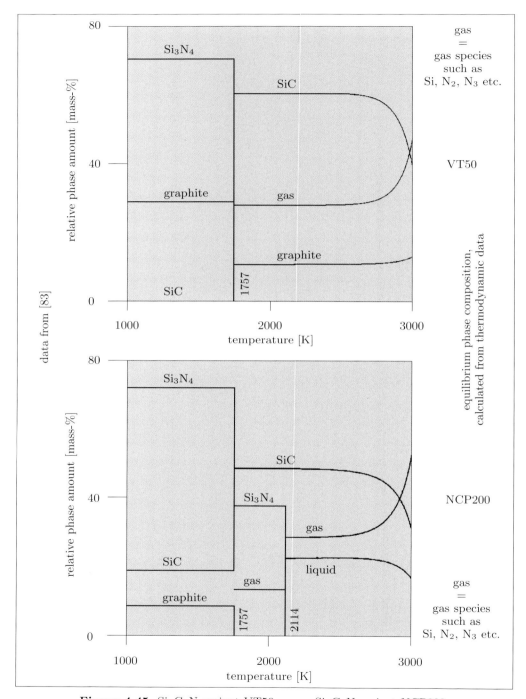

Figure 4.45. Si–C–N variant VT50 versus Si–C–N variant NCP200.

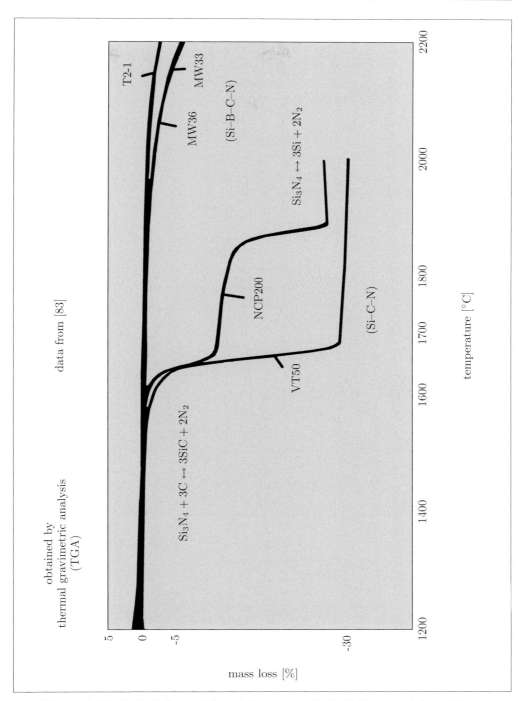

Figure 4.46. Si–C–N thermal degradation versus Si–B–C–N thermal degradation.

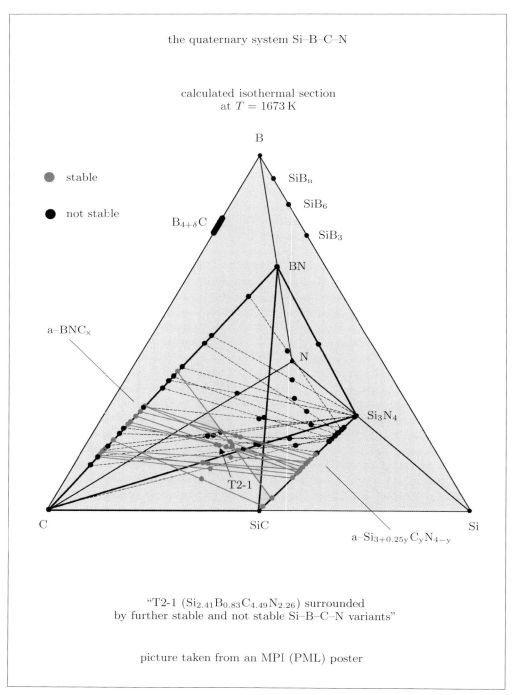

Figure 4.47. Si–B–C–N phase diagram.

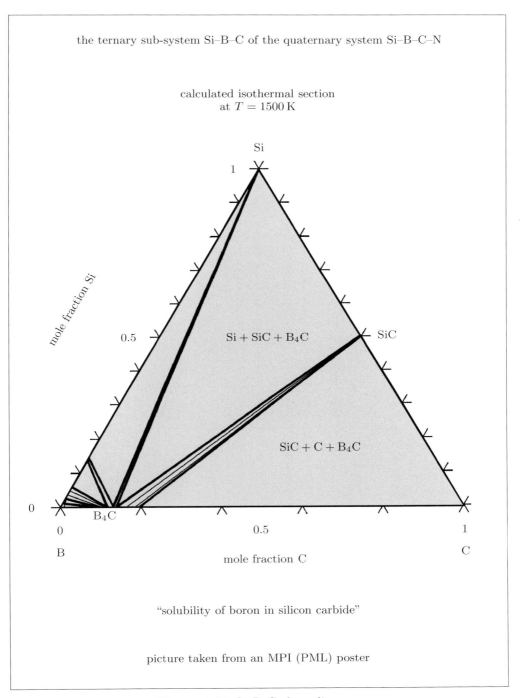

Figure 4.48. Si–B–C phase diagram.

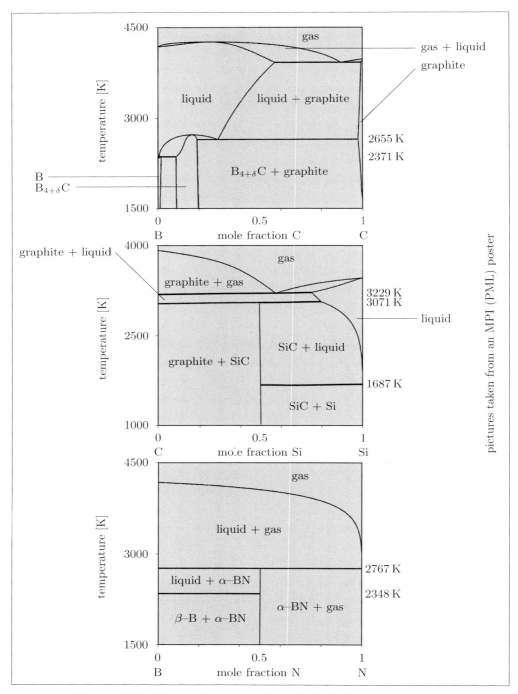

Figure 4.49. Binary sub-systems of the Si–B–C–N system, part one.

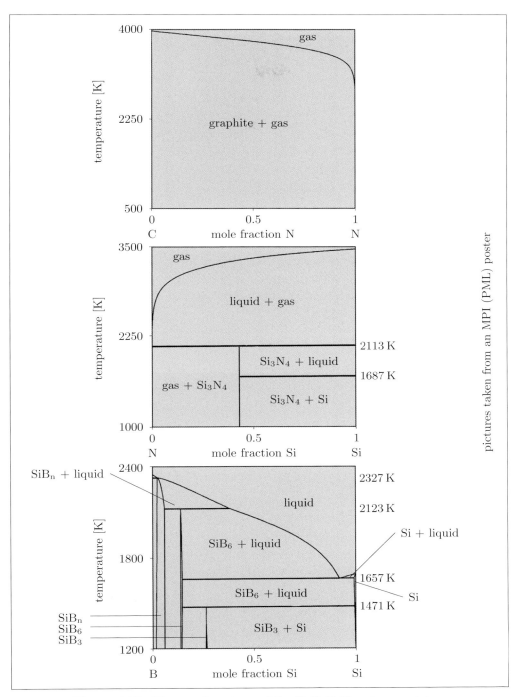

Figure 4.50. Binary sub-systems of the Si–B–C–N system, part two.

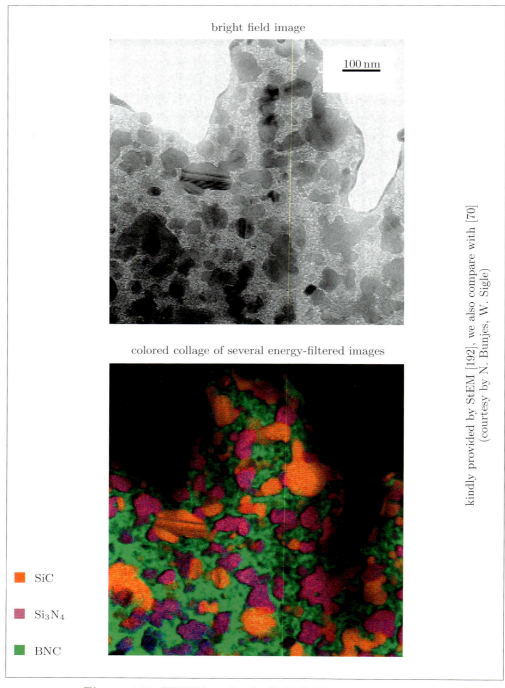

Figure 4.51. EFTEM study of a Si–B–C–N ceramic, part one.

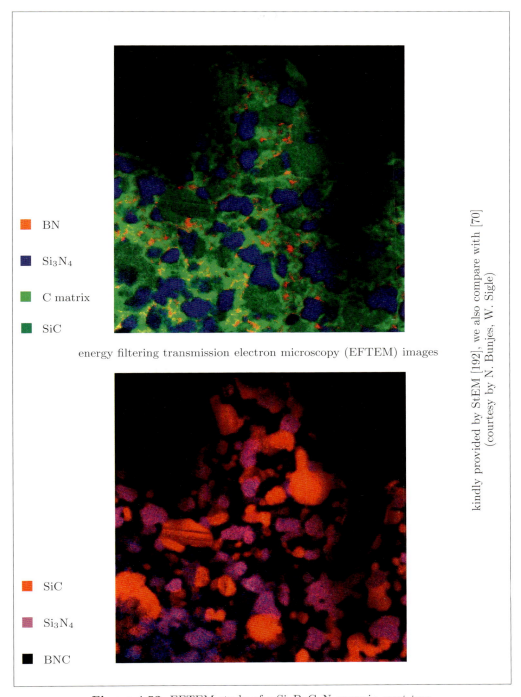

Figure 4.52. EFTEM study of a Si–B–C–N ceramic, part two.

4.2.2 Vapour Deposition Procedures

Epitaxy means the regular growth of a crystalline layer on a monocrystalline substrate imposing its molecular order on the growing crystalline layer. We here note that if the monocrystalline substrate and the crystalline layer consist of the same material, we speak of *homoepitaxy*. We here also note that if the monocrystalline substrate and the crystalline layer consist of different materials, we speak of *heteroepitaxy*. We remark that one distinguishes between liquid-phase epitaxy (LPE), molecular beam epitaxy (MBE), ion beam aided deposition (IBAD), organometallic vapor phase epitaxy (MOVPE), *chemical vapor deposition (CVD)*, and *physical vapor deposition (PVD)*. We further remark that the latter two procedures, which we want to consider in more detail in the sections that follow, are not only a good choice if we want to initiate epitaxial growth, but likewise if we want to create amorphous or crystalline layers with well-defined chemical composition. We further remark that MBE, IBAD, and MOVPE are also considered as sub-classes of the class of vapor deposition procedures, what we here do not want to do, though. We additionally remark that these techniques are suitable to produce almost any kind of compound and almost any kind of composition of materials regardless of their thermodynamic stability.

Chemical vapor deposition is an umbrella term for deposition processes where educts (reactants) are pyrolysed in the vapor phase and/or on a hot substrate and/or chemical reactions of educts (reactants) take place in the vapor phase and/or on a hot substrate, leading to a crystalline layer on the substrate. The principle of CVD is illustrated by Figure 4.53, completed by Figure 4.54, showing a CVD reactor for TiC/TiN films, and Figure 4.55, showing a CVD reactor for wafer coating, and complemented by Box 4.4, showing important pyrolysis mechanisms/chemical reaction mechanisms. On the one hand, the reader should know that CVD reactors of the type outlined in Figure 4.54 generate TiC/TiN films following $TiCl_4$ reactions of the type shown in Box 4.4. On the other hand, the reader should know that CVD reactors of the type outlined in Figure 4.55 are widely used for the deposition of films in electronic devices, for example, for the deposition of Si, SiO_2, or Si_3N_4 films on silicon wafers. Of course, in all these cases, it is always worked under reduced pressure.

However, not for each technically necessary film a gaseous counterpart exists. This is one limit of this vapor deposition class. Furthermore, the heat exposure of the substrate is extremly high so that substrates with relatively low melting temperatures and relatively low warping resistance cannot be used. This is another limit of this vapor deposition class. Furthermore, at high temperatures, diffusion processes are excited so that doping alternation and impurity inclusion take place. This is another limit of this vapor deposition class. Therefore, sophisticated variants of CVD allowing us to reduce the working temperature have been developed. In particular, we here quote the plasma-enhanced CVD (PECVD), where a plasma above the substrate is sparked, and the remote plasma-enhanced CVD (RPECVD), where plasma and substrate are spatially separated in order to avoid radiation exposures of the substrate. In addition, we here quote the atmospheric pressure CVD (APCVD), where the process takes place under normal pressure and not under reduced pressure.

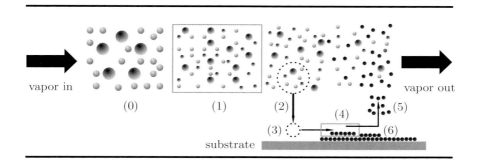

(0) initial compounds in the vapor phase
(1) pyrolysis/chemical reactions in the vapor phase
(2) diffusion of components to the substrate surface
(3) adsorption of components and diffusion on the substrate surface
(4) pyrolysis/chemical reactions on the substrate surface
(5) desorption of components and diffusion into the vapor phase
(6) integration of components in the crystal lattice

Figure 4.53. The principle of CVD.

Box 4.4 (Important data: important reaction equations, film deposition).

Pyrolysis mechanisms:

$$CH_3SiCl_3 \text{ (gaseous)} \rightarrow SiC \text{ (solid)} + 3HCl \text{ (gaseous)}$$
$$(1150°C) .$$

Chemical reaction mechanisms:

$$SiH_4 \text{ (gaseous)} + O_2 \rightarrow SiO_2 \text{ (solid)} + 2H_2$$
$$(450°C) ,$$

$$SiCl_4 \text{ (gaseous)} + CH_4 \text{ (gaseous)} \rightarrow SiC \text{ (solid)} + 4HCl \text{ (gaseous)}$$
$$(1400°C) ,$$

$$TiCl_4 \text{ (gaseous)} + CH_4 \text{ (gaseous)} \rightarrow TiC \text{ (solid)} + 4HCl \text{ (gaseous)}$$
$$(1000°C) ,$$

$$TiCl_4 \text{ (gaseous)} + 1/2N_2 + 2H_2 \rightarrow TiN \text{ (solid)} + 4HCl \text{ (gaseous)}$$
$$(850\text{--}1200°C) .$$

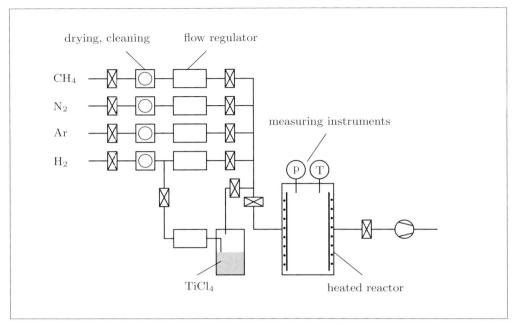

Figure 4.54. CVD reactor for TiC/TiN films.

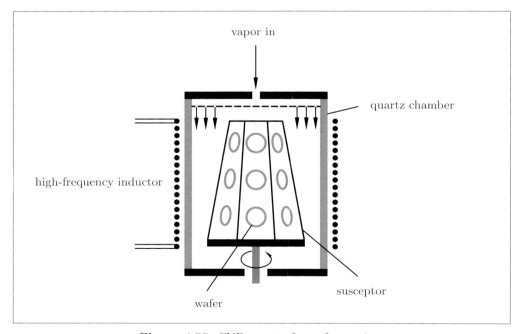

Figure 4.55. CVD reactor for wafer coating.

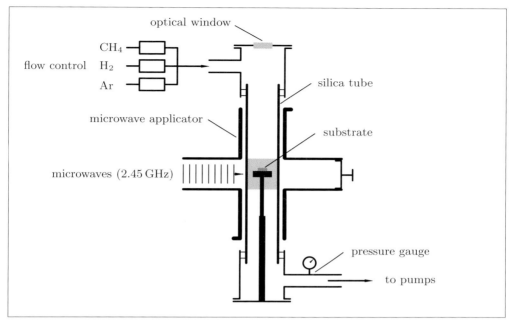

Figure 4.56. Microwave-plasma-assisted CVD (MPACVD).

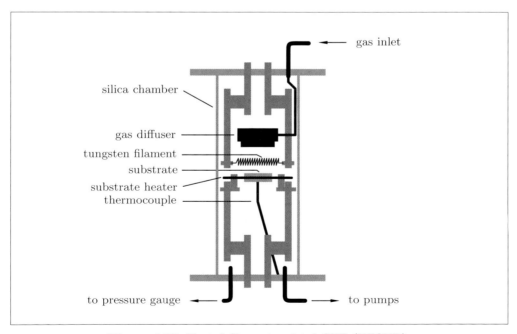

Figure 4.57. Heated-filament-assisted CVD (HFCVD).

A more far-reaching example are vapor-deposited diamond films. Going beyond the high pressure high temperature (HPHT) procedure (which generates artifical diamonds out of graphite by applying high pressures ($\approx 200\,000$ atm), for example, generated by belt presses, and by applying high temperatures ($\approx 3000°C$), for example, generated by heat currents, sometimes utilizing metallic catalyzers generating a metal–graphite solution out of which the artifical diamonds crystallize, see Figure 4.58 and Figure 4.59), thin diamond films are produced as follows. (i) In principle, one creates an environment that allows the C atoms of a gas to settle on a preset substrate. The carbon source, originally a mixture of methane (CH_4) and hydrogen (H_2), is passed over the preset substrate, while an excitation technique, originally a microwave excitation technique, is applied to activate the carbon source, i.e. on the one hand, a particle zoo finally evoking different types of chemical reactions on the substrate is generated, i.e. on the other hand, a particle zoo on a relatively high energetic level is generated. The carbon is deposited on the substrate, on the one hand, forming metastable diamond, and on the other hand, forming stable graphite, however, which is remedied by etchants, originally by atomic hydrogen. (ii) Dealing with vapor-deposited diamond films, two variants of CVD are of particular interest, i.e. microwave-plasma-assisted CVD (MPACVD), we compare with Figure 4.56, and heated-filament-assisted CVD (HFCVD), we compare with Figure 4.57. Beyond it, arc jet CVD (AJCVD) and deposition in a combustion flame sustained by acetylene and oxygen are of practical interest. Exclusively focussing on MPACVD and HFCVD, these remarks should be made. MPACVD means that a microwave field is applied to the carbon source, in this way, generating a plasma, for example, containing carbon, atomic hydrogen, and methyl radicals. HFCVD means that a tungsten filament is heated to $\approx 2000°C$, in this way, generating a plasma, for example, containing carbon, atomic hydrogen, and methyl radicals. More precisely speaking, in the first case, primarily the forces associated with the microwave field generate a mixture of electrons, ions, atoms, and molecules. More precisely speaking, in the second case, primarily the forces associated with the heat collisions generate a mixture of electrons, ions, atoms, and molecules. In this manner, the particle zoo finally evoking different types of chemical reactions on the substrate is generated. (iii) The typical deposition conditions are 0.1–5% methane (CH_4) in an environment mainly consisting of hydrogen (H_2), an activation temperature in the range 1800–2900°C, a substrate temperature in the range 700–1400°C, and a gas pressure in the range 7.5–300 Torr, i.e. a gas pressure less than the atmospheric pressure (1 atm = 760 Torr). The typical deposition thickness are in the range 0.1–10μm. (iv) Experience shows us that an activation, for example, by a microwave field or a tungsten filament, certainly is necessary, however, the special method in no way is critical. Experience shows us that also the special hydrocarbon precursor in no way is critical: crystalline growth of diamond films, for example, has been observed for aliphatic and aromatic hydrocarbons as well as for alcohols and ketones. Experience shows us that also the special substrate in no way is critical: crystalline growth of diamond films, for example, has been observed using the materials Si, Ta, Mo, W, SiC, WC, and diamond. (v) In Box 4.5, a collection of important data is shown. Apart from the effective diamond and etching reactions, it supplies us with two growth mechanisms as well as with a conversion mechanism that transforms graphite structures into diamond structures. [51, 67]

> **Box 4.5** (Important data: important reaction equations, diamond film deposition).
>
> Effective diamond reaction:
> $$CH_4(g) \rightarrow C_{diamond} + 2H_2 \ . \qquad (4.6)$$
>
> Effective etching reaction:
> $$xC_{graphite} + yH^\bullet \rightarrow C_xH_y(g) \ . \qquad (4.7)$$
>
> A growth mechanism of diamond films based upon methyl ($\bullet$ indicates radicals):
> $$\begin{array}{c} C_{diamond}H + H^\bullet \rightarrow C^\bullet_{diamond} + H_2 \\ \Downarrow \\ C^\bullet_{diamond} + CH^\bullet_3(g) \rightarrow C_{diamond}C_{diamond}H_3 \\ \ldots \end{array} \qquad (4.8)$$
>
> A growth mechanism of diamond films based upon acetylene ($\bullet$ indicates radicals):
> $$\begin{array}{c} C_{diamond}H + H^\bullet \rightarrow C^\bullet_{diamond} + H_2 \\ \Downarrow \\ C^\bullet_{diamond} + C_2H_2(g) \rightarrow C_{diamond}C_{diamond}CH^\bullet_2 \\ \ldots \end{array} \qquad (4.9)$$
>
> A conversion mechanism transforming graphite structures into diamond structures (conversion of sp^2 hybrid orbitals in sp^3 hybrid orbitals):
>
> $$\begin{array}{c}
> \text{C}\diagdown\;\;\;\diagup\text{C} \\
> \text{C}=\text{C} \quad + \text{H}^\bullet \quad \longrightarrow \quad \text{H}-\text{C}-\text{C}-\bullet \\
> \text{C}\diagup\;\;\;\diagdown\text{C} \\
> \Downarrow \\
> \text{H}-\text{C}-\text{C}-\bullet \; + \text{H}-\text{H} \quad \longrightarrow \quad \text{H}-\text{C}-\text{C}-\text{H} + \text{H}^\bullet
> \end{array} \qquad (4.10)$$
>
> For instance, according to (4.8) and (4.9), C–H bonds are replaced by C–C bonds, finally building up diamond crystallites.

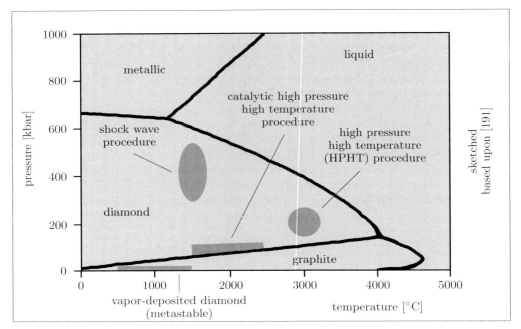

Figure 4.58. *p–T* diagram of diamond.

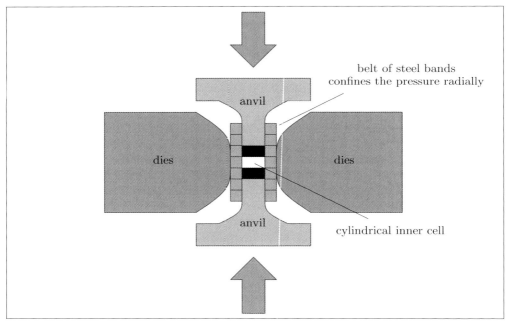

Figure 4.59. Scheme of a special type of belt press.

4.2 Powder-Free Technologies

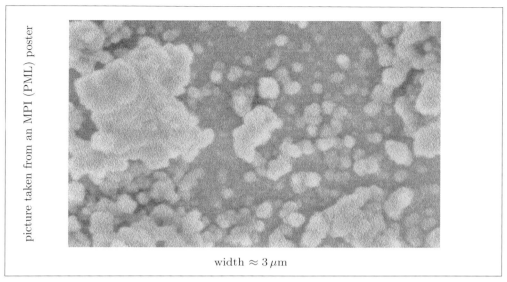

Figure 4.60. c–BN deposition obtained by CVD from $BH_3 + NH_3 + H_2$.

Let us here additionally annotate that in a similar manner, for example, deposits of metastable cubic boron nitride (c–BN) can be produced, we compare with Figure 4.60. In this special example, also the metastable sp^3 configuration is stabilized against the stable (planar) sp^2 configuration of the stable hexagonal boron nitride (h–BN).

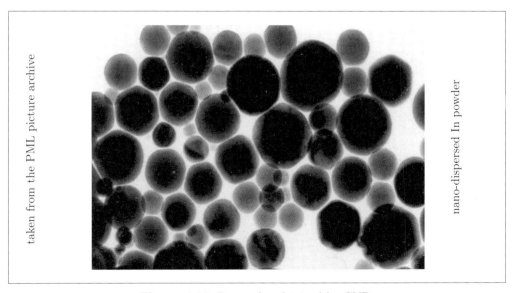

Figure 4.61. In powder obtained by CVD.

While CVD processes make use of relatively hot substrates, PVD processes make use of relatively cold substrates. While CVD processes are deposition processes where chemical reactions take place on the substrate, eventually evoking the layer material, PVD processes are subdivided into "non-reactive processes" and "reactive processes", namely into processes where layer material firstly is brought into the vapor phase, secondly is transported to the substrate, and thirdly is resublimated on the substrate, and processes where material (for example, Al material) firstly is brought into the vapor phase, secondly is mixed with excited gas (for example, excited O_2 gas) evoking the layer material (for example, Al_2O_3 material), thirdly as layer material is transported to the substrate, and fourthly as layer material is resublimated on the substrate, termed "non-reactive processes" and "reactive processes", respectively.

With respect to PVD processes, the reader should appreciate the following further circumstances. (i) The material to be vaporized is brought into the vapor phase by exciting the material in a sagger or crucible applying thermal excitation methods, electric excitation methods, or laser excitation methods, or by sputtering from a target using an electron stream or/and an ion stream. (ii) From a technological point of view, in particular, one distinguishes between thermal evaporation (excitation by heating), arc evaporation (excitation by an arc), pulsed laser deposition/ablation (excitation by a pulsed laser), and electron beam evaporation (excitation by an electron beam). (iii) Furthermore, from a technological point of view, in particular, one distinguishes between diode sputtering (primarily, collisions with electrons cause process gas ions, DC voltage accelerates electrons away from the target and process gas ions towards the target), triode sputtering (primarily, target acts as third electrode, plasma and target are separated by open space), RF sputtering (primarily, AC voltage accelerates electrons, offset voltage accelerates process gas ions), magnetron sputtering (primarily, electric field + magnetic field generate cycloid orbits and thus a wealth of collisions), and diverse types of DC/AC variants of reactive sputtering.

The principle of PVD is illustrated by Figure 4.62. We firstly appreciate that in the case of "non-reactive processes" sources which contain the layer material are applied, allowing the direct access to the layer material. We secondly appreciate that in the case of "reactive processes" a mixture with excited gas has to be produced to evoke the layer material, either using different sources or using a combination of evaporated sources and gases present in the vessel. We thirdly appreciate that in both cases a relatively cold substrate enforces the resublimation. We point out that there are various technical variants of PVD. The magnetron sputtering variant of PVD is illustrated by Figure 4.63. Firstly, it is notable that the process gas (Ar) mainly is ionized by collisions with electrons moving on cycloid orbits generated by electric field + magnetic field. Secondly, it is notable that mainly a stream of process gas ions (Ar^+) comes into being, evoking collision cascades within the cathode (target), leading to the evaporation of cathode (target) material. Thirdly, we point out that in this case and in other cases the target material can reach the substrate and can resublimate only if the vapor pressure is relatively low ("vacuum"). Fifthly, we point out that in this case and in other cases it leads to much better results if the substrate is rotated.

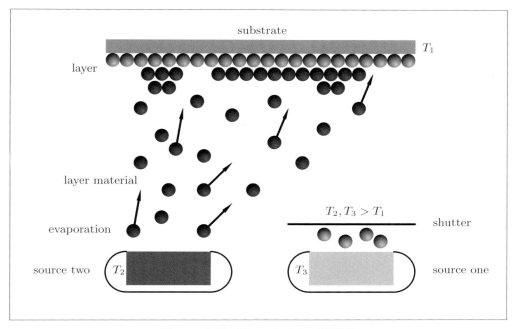

Figure 4.62. The principle of PVD.

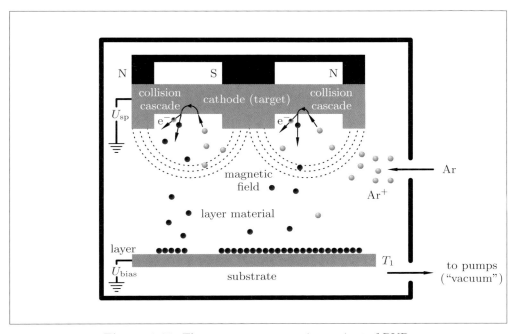

Figure 4.63. The magnetron sputtering variant of PVD.

4.2.3 Bio-Inspired Mineralisation Procedures

The customary ways of producing ceramic films are commonly elaborate and expensive. By contrast, *bio-inspired mineralization procedures* supply us with a much simpler and much cheaper alternative. Such procedures not only allow us to work at temperatures near room temperature, such procedures also allow us to work with simple types of technical setup. (i) The typical technical setup of bio-inspired mineralization procedures is shown in Figure 4.64. We read off from this figure that a temperature-controlled vessel with a holder allowing us to fix the substrate is needed. We read off from this figure, too, that an aequous solution containing the substitutes of the desired ceramic material is needed. (ii) The typical process chain of bio-inspired mineralization procedures is shown in 4.65. We read off from this figure that in the context of this typical process chain one makes use of organic chain molecules, adsorbed from a substrate surface, assembling themselves ("molecular self-organization"), generating an organic template ("self-assembled monolayer (SAM)"). We read off from this figure, too, that in aqueous solution a ceramic film evolves the composition and structure structure of which directly depends on the surface group of the self-assembled monolayer (SAM). We read off from this figure, too, that an SO_3H–SAM template on a silicon wafer is building up a nano-crystalline oxide film, for example, a film of ZrO_2, TiO_2, SnO_2, or V_2O_5, we compare with Figure 4.66, where an example of technical interest is shown, i.e. a film of ZrO_2 formed on a SO_3H–SAM template. We emphasize that the organic template directs the deposition of the inorganic material as it is the case with bio-mineralization (bones, eggshells, snail shells, crustaceans, diatoms, radiolarians etc.).

We take notice that this field of research is rapidly rising, step by step, evoking new kinds of technical setup and process chains. For instance, synthetic polymers such as polyethylene terephthalate (PET), polyethylene imine (PEI), polyacrylamide (PAM), and polyvinylpyrrolidone (PVP) can be incorporated in manifold ways. For instance, control mechanisms are studied allowing us to control the deposition process by utilizing the manifold complexation properties of biological macromolecules such as proteins, peptides, amino acid, and desoxyribonucleic acid (DNA). We also take notice that *chemical liquid deposition (CLD)* is usable in cases where CVD and PVD fail because the substrate exhibits too complex a surface geometry or where CVD fails because the substrate is destroyed if too high a temperature is applied.

Figure 4.67 illustrates microstructures of a test series where titania + vanadia films on SO_3H–SAM templates + silicon wafers were prepared, and Figure 4.68 illustrates microstructures of a test series where titania films on polyethylene terephthalate (PET) were prepared. In the first test series to be considered, Ti/V solutions in HCl obtained from $TiCl_4 \rightarrow Ti(O_2)^{2+} =$ "peroxo complex" and $NH_4VO_3 \rightarrow H_2V_{10}O_{28}^{4-}$ at 353 K were used, and in the second test series to be considered, Ti solutions in HCl obtained from $TiCl_4 \rightarrow Ti(O_2)^{2+} =$ "peroxo complex" at 333 K were used. Without going into details, let us here additionally record that surveys such as Auger depth profile surveys reveal that uniformly distributed nanoparticles such as TiO_2 and V_2O_5 nanoparticles were prepared, we consult Figures 4.69 and 4.70.

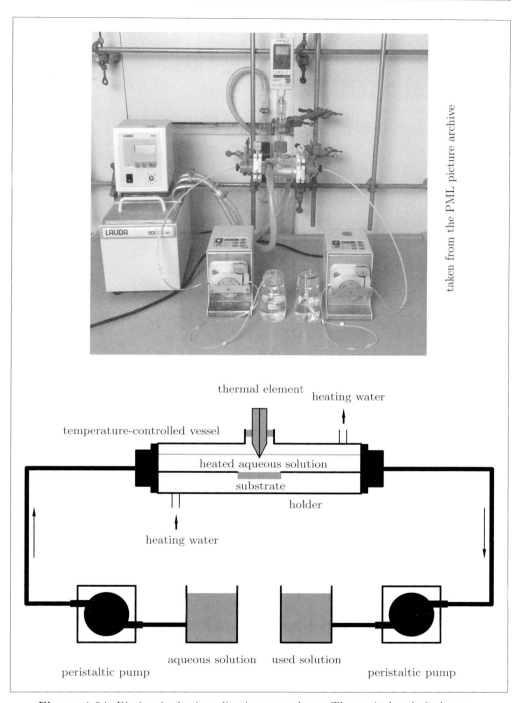

Figure 4.64. Bio-inspired mineralization procedures. The typical technical setup.

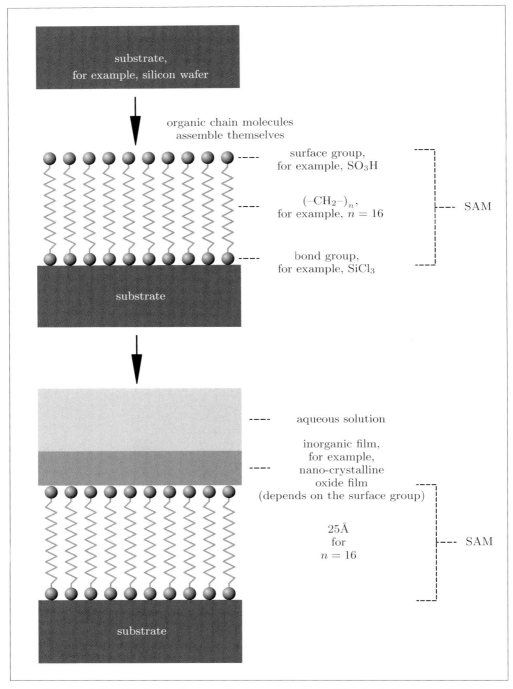

Figure 4.65. Bio-inspired mineralization procedures. The typical process chain.

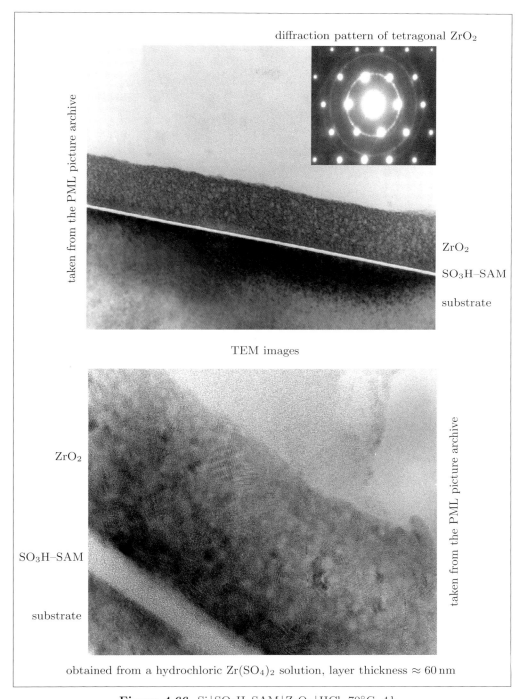

Figure 4.66. Si | SO_3H–SAM | ZrO_2 | HCl, 70°C, 4 h.

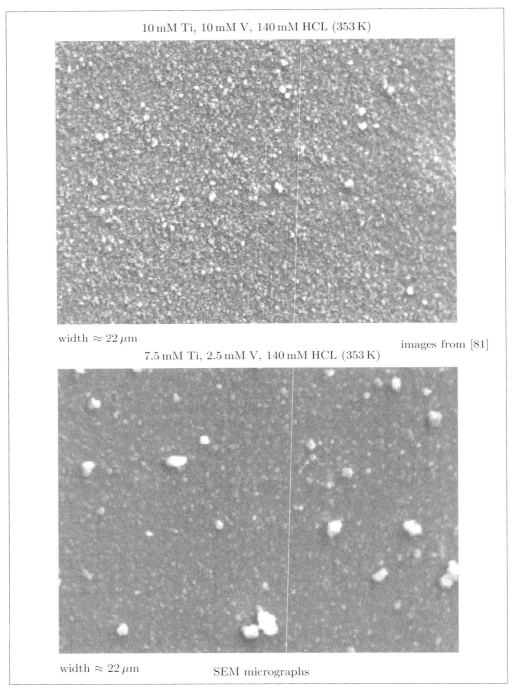

Figure 4.67. Deposition of titania + vanadia films on SO_3H–SAM templates + silicon wafers.

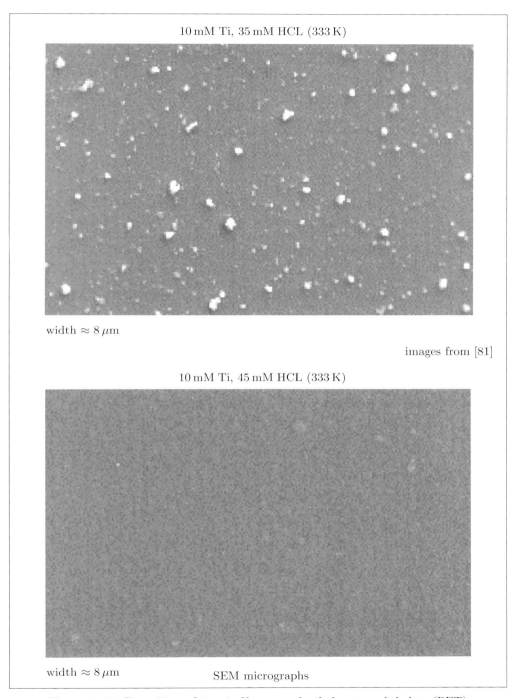

Figure 4.68. Deposition of titania films on polyethylene terephthalate (PET).

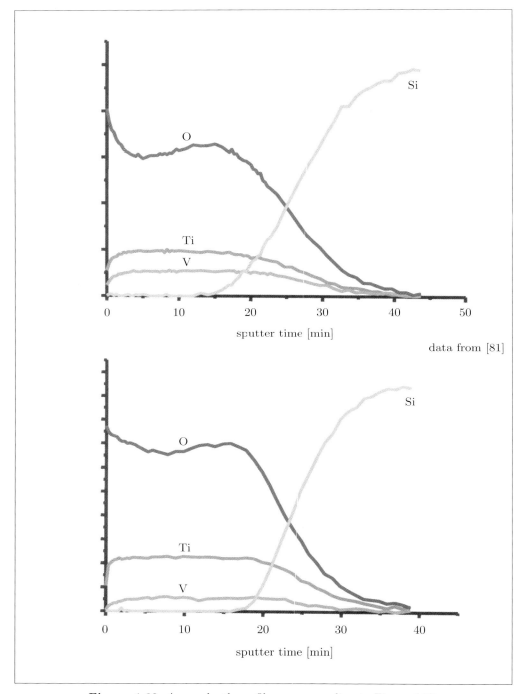

Figure 4.69. Auger depth profiles corresponding to Figure 4.67.

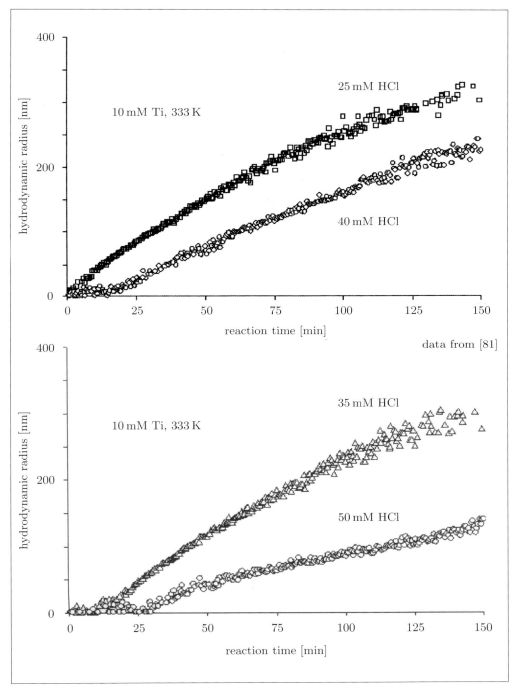

Figure 4.70. Dynamic light scattering. Time dependance of particle size solutions.

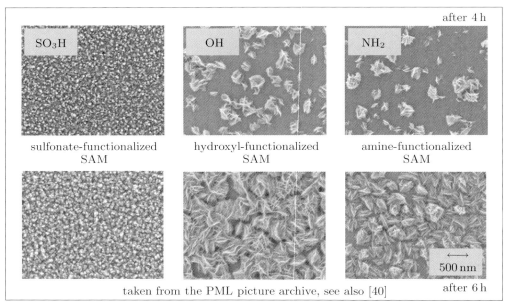

Figure 4.71. SEM images of TiO$_2$ depositions on differently functionalised SAMs.

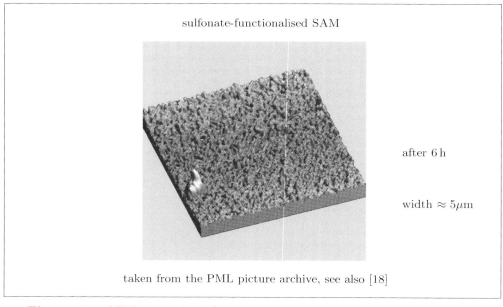

Figure 4.72. AFM image of a TiO$_2$ deposition on a sulfonate-functionalised SAM.

SEM images of a film deposited from an aqueous solution, containing ZnCl$_2$, Zn(CH$_3$COO)$_2$, hexamethylenetetraamine (HMTA), and a copolymer consisting of polymethacrylic acid partially grafted with polyethylenoxide side chains

the surface of the Si substrate was initially modified with γ–mercaptopropyltrimethoxysilane (Si(OCH$_3$)$_3$(CH$_2$)$_3$–SH)

after 6 h

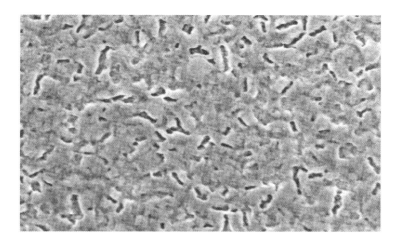

height $\approx 3\mu$m

taken from the PML picture archive, see also [26]

Figure 4.73. SEM images of ZnO depositions.

418   4. Technologies

Let $M_{\text{solution}}(t)$ be the time-dependent ceramic mass in solution and $M_{\text{film}}(t)$ be the time-dependent ceramic mass of the film that comes into being in the course of time. Furthermore, let M_0 be the time-independent total ceramic mass, which concatenates $M_{\text{solution}}(t)$ and $M_{\text{film}}(t)$ according to (4.11). In this notation, a simple potential model for the time-evolution of the ceramic mass of the film and thus of the film thickness is defined by (4.12), where $\phi_{\text{film}}(t)$ is the potential. Provided $c_0 > 0$, $c_1 > 0$, and $c_2 > 0$, (4.12) obviously tells us that $M_{\text{film}}(t)$ is driven into the potential valley which is evoked by $c_0 > 0$, $c_1 > 0$, and $c_2 > 0$ and which indicates the final value $M_{\text{film}}(\infty)$ of $M_{\text{film}}(t)$. Starting from $M_{\text{film}}(t) < M_{\text{film}}(\infty)$ or $M_{\text{film}}(t) > M_{\text{film}}(\infty)$, the trajectories always lead to the final value $M_{\text{film}}(\infty)$, explicitly showing us that this is an equilibrium value. In particular, starting from $M_{\text{film}}(0) = 0$, the c_1 term acts as an acceleration term enforcing the start of the deposition, whereas the c_2 term acts as an retardation term enforcing the stop of the deposition. Figure 4.74 collects measurement results for both a stagnant solution and a flowing solution. Figure 4.75 shows special numerical solutions of (4.12) which are in good agreement with the measurement results. We note in passing that the elimination of $M_{\text{solution}}(t)$ in (4.13) by $M_{\text{solution}}(t) = M_0 - M_{\text{film}}(t)$ leads from (4.13) to (4.12). We note in passing that this shows us that the influence of $M_{\text{solution}}(t)$ is implicitly contained in (4.12). For the reader it might be interesting to know that $c_0 = C_0 - C_s$, $c_1 = 2C_s$, and $c_2 = C_f - C_s$ reflect microscopic system parameters, for example, flow parameters indicating flowing solutions ($\to c_1$) and shielding parameters indicating shielding effects caused by the film that comes into being ($\to c_2$).

Box 4.6 (Important formulae: TiO$_2$ deposition from aqueous solution).

$$M_{\text{solution}}(t) + M_{\text{film}}(t) = M_0 ,$$

$$\frac{d}{dt} M_{\text{solution}}(t) + \frac{d}{dt} M_{\text{film}}(t) = 0 ,$$

(4.11)

$$\frac{d}{dt} M_{\text{film}}(t) = -\frac{\partial}{\partial M_{\text{film}}(t)} \phi_{\text{film}}(t) = c_0 M_0^2 + c_1 M_{\text{film}}(t) - c_2 M_{\text{film}}^2(t) ,$$

(4.12)

$$\phi_{\text{film}}(t) = -c_0 M_0^2 M_{\text{film}}(t) - \frac{c_1}{2} M_{\text{film}}^2(t) + \frac{c_2}{3} M_{\text{film}}^3(t) ,$$

$$\frac{d}{dt} M_{\text{film}}(t) = +C_0 M_0^2 + C_f M_{\text{film}}^2(t) - C_s M_{\text{solution}}^2(t) ,$$

(4.13)

$$\frac{d}{dt} M_{\text{solution}}(t) = -C_0 M_0^2 - C_f M_{\text{film}}^2(t) + C_s M_{\text{solution}}^2(t) .$$

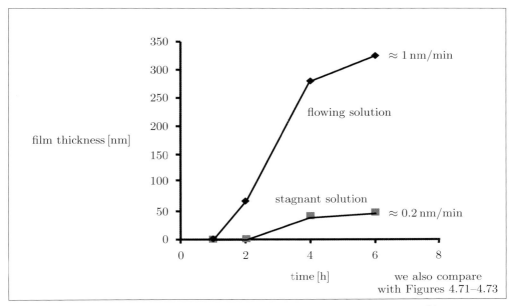

Figure 4.74. TiO$_2$ deposition from aqueous solution. Si | SO$_3$H–SAM | Ti(O$_2$)$^{2+}$ | HCl, 80°C.

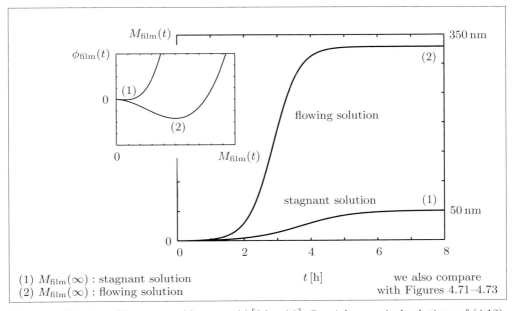

(1) $M_{\text{film}}(\infty)$: stagnant solution
(2) $M_{\text{film}}(\infty)$: flowing solution

Figure 4.75. $M_{\text{film}}(t)$ and $\phi_{\text{film}}(t) = \phi_{\text{film}}(t)\big[M_{\text{film}}(t)\big]$. Special numerical solutions of (4.12).

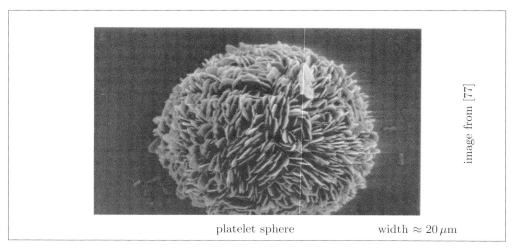

platelet sphere   width ≈ 20 μm

Figure 4.76. Layered zinc salt ($Zn(NO_3)_2$) structure (directed by adenine), part one.

For the convenience of the reader, we show one more case study in Figures 4.76–4.78. The following remarks thereto. (i) This case study demonstrates the structuring effect of biological macromolecules. (ii) The platelet sphere illustrated by Figures 4.76 and 4.77 was obtained by using DNA-derived adenine as additive ($Zn(NO_3)_2$ solution + addenine at 60°C over night). (iii) The rod structure, the plate structure and the sponge structure illustrated by Figure 4.78 were obtained after incorporation of different amino acid combinations ($Zn(NO_3)_2$ solution + amino acid combinations at 60–70°C). (iv) This case study impressively demonstrates that bio-inspired mineralization procedures such as chemical liquid deposition (CLD) facilitate the fabrication of ceramic materials with complicated inherent patterns, in subtle details, adaptable to technical demands.

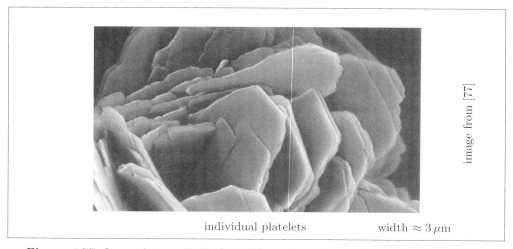

individual platelets   width ≈ 3 μm

Figure 4.77. Layered zinc salt ($Zn(NO_3)_2$) structure (directed by adenine), part two

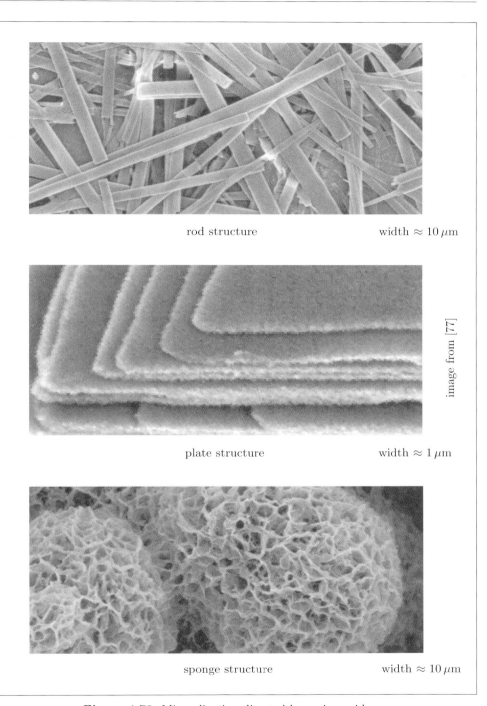

Figure 4.78. Mineralization directed by amino acids.

Sphinx ligustri (privet hawk moth larva)

taken from the VAWsc picture archive

Figure 4.79. A network of biological materials: caterpillars, part one.

Craesus septentrionalis

this is a sawfly larva, no moth larva

taken from the VAWsc picture archive

Orgyia antiqua

Figure 4.80. A network of biological materials: caterpillars, part two.

taken from the VAWsc picture archive

Figure 4.81. Biological materials: cactus spines. Spine array.

taken from the VAWsc picture archive

Figure 4.82. Biological materials: cactus spines. Spine evolution center.

taken from the VAWsc picture archive

Figure 4.83. Romanesco vegetable. A herbal fractal.

Thinking beyond the borders of bio-inspired mineralization procedures such as chemical liquid deposition (CLD), let us finish this chapter with additional remarks concerning a class of future materials which in many ways could overcome the confines of ceramic materials, namely the class of biomimetic materials. In what follows, the reader should compare with Figures 4.79–4.83, supplying us with an intuitve access to the diversity and to the performance of typical biological materials, on the one hand, indicating that biological materials are able to breed highly regular structures, noted by the way, carrying on self-organization principles that could be interesting in the context of technological developments of the future, and on the other hand, indicating that biological materials are able to breed highly irregular, but apart from fluctuations deterministic structures, noted by the way, carrying on structure principles that could also be interesting in the context of technological developments of the future.

A lot of scientific research institutes worldwide have launched scientific research projectes aiming at the development of biomimetic materials, i. e. bio-inspired materials mimeticing biological materials. For example, it would be more than interesting to emulate the feet of the gecko, showing a big amount of microscopic hairs transmitting the van der Waals force, "gluing together" the gecko and the ceiling allowing the gecko to walk upside down at the ceiling. For example, it would be more than interesting to emulate "hightech skills" of nature such as its skill to manage self-similar evolution or fractal evolution in a self-organizing, synergetic manner, i. e. without accomplishing space- and time-dependent actions within the biological system, i. e. simply by changing inherent control parameters of the biological system. Since we are deeply convinced of the performance of ceramic materials, we expect that biomimetic materials will comprise composite materials eventually concatenating ceramic, inorganic units with non-ceramic, organic units – as it is the case for each bug of our blue planet.

Not later than now, nonlinear formalisms allowing us to model biomimetic materials in a self-organizing, synergetic manner are needed. A sophisticated nonlinear formalism that works with biomimetic materials as well as non-biomimetic materials is presented in the chapter that follows, i. e. a self-consistent network of model conceptions, which some of our colleagues may consider as an extension of the Ginzburg–Landau theory of superconductivity [20] or as a semi-classical approach to quantum systems comparable to the Wunner–Main approach to quantum systems [34, 35, 64], however, which could turn out to be the true nonlinear extension of conventional quantum mechanics [63], on all accounts, which works without any known restrictions and is based upon microscopic quantities such as momentum operators, spin operators, and wave functions.

Pointing into the Future of Materials

5. Future Materials

As it is already pointed out in the last segment of the first chapter of the book in hand, going beyond advanced ceramics, in this final Chapter 5, first steps towards new fields of materials research are carried out. The following first introductory remarks already here should be useful. On the one hand, a branch of research is started up dealing with the question "How must materials be structured and how must their technical environment be structured that these can be used as the material basis for an oscillatory system generating a measurable oscillatory gravitational field, so to speak, as the material basis for a gravitational laser?" On the other hand, a branch of research is started up dealing with the question "How must materials be structured and how must their technical environment be structured that these – fundamentally going beyond nuclear techniques and matter–antimatter destruction – can be disaggregated into radiation, which in turn can be used as starting point for new types of energy plants, new types of spacecraft propulsions, and new types of matter conversion techniques?" For this purpose, a self-consistent network of model conceptions is launched, which some colleagues may consider as a direct extension of the Ginzburg–Landau theory of superconductivity [20] or as a semi-classical approach to quantum systems comparable to the Wunner–Main approach to quantum systems [34, 35, 64], however, which could turn out to be the true nonlinear extension of conventional quantum mechanics [63], on all accounts, which works without any known restrictions, and which firstly can supply us with much more information about the principles that govern the constitution, the stabilities, and the instabilities of materials than the methods of thermodynamics and the methods of quantum mechanics can do, and which secondly can supply us with paths to these new fields of materials research. Certainly, it goes far beyond the scope of this book to develop these new fields of research in great detail already here. However, as it is already pointed out in the last segment of the first chapter of the book in hand, the first steps towards this goal indeed are taken here.

Certainly, biomimetic materials are future materials, too, and the nonlinear concept to be launched in the sections that follow works with biomimetic materials as well as non-biomimetic materials. Certainly, we thus refer to biomimetic materials as well as non-biomimetic materials in the sections that follow. Referring to biomimetic materials, the engineers/material scientists among the readers may think of the simulation of the pattern evolution of biomimetic materials. Referring to biomimetic materials, the engineers/material scientists among the readers may think of a quantum synergetics, simplifying the modeling of the pattern evolution of biomimetic materials.

5.1 Advanced Model Conceptions

Let us here systematically develop the self-consistent network of model conceptions mentioned above. Let us begin with an advanced spin model, constituting a central portion of our self-consistent network of model conceptions, supplying us with an access to spin properties that goes far beyond the access to spin properties that is supplied by conventional spin matrix theory, promising us to be best adapted to future materials.

5.1.1 Advanced Modeling of Particles/Spins

In our self-consistent network of model conceptions, exceeding the narrow confines of conventional spin matrix theory, Pauli spin matrices $\hat{\sigma}_i$ and SU(n) spin matrices are no "stand-alone operators", but descendants of differential spin operators, which feed us with much more information about spin properties than the Pauli spin matrices $\hat{\sigma}_i$ and SU(n) spin matrices ever could do. These differential spin operators are part of an advanced spin model. Let us bring this advanced spin model into being as follows.

A mathematical model: an advanced spin model.

In our self-consistent network of model conceptions, the operators (5.2), (5.4), and (5.6) define an advanced spin model provided the unknown vectors $\boldsymbol{A}_{\ni}$ and $\boldsymbol{A}^{\ni}$ as well as the unknown 3×3 matrices $\boldsymbol{\theta}^{\ni}$ and $\boldsymbol{\xi}$ are specified suitably. Together with the operators (5.1), (5.3), and (5.5), these operators can be colorfully established as follows.

$$\hat{T} = -\frac{\hbar^2}{2m_0}(\nabla)\nabla , \tag{5.1}$$

$$\hat{T}' = -\frac{\hbar^2}{2m_0}(\boldsymbol{\theta}^{\ni}\nabla)\nabla , \tag{5.2}$$

$$\hat{V} = -\frac{\hbar^2}{2m_0}(\nabla \times \boldsymbol{A}_{\ni})\boldsymbol{A}^{\ni} \times \nabla , \tag{5.3}$$

$$\hat{V}' = -\frac{\hbar^2}{2m_0}(\nabla \times \boldsymbol{A}_{\ni})\boldsymbol{A}^{\ni} \times \boldsymbol{\theta}^{\ni}\nabla , \tag{5.4}$$

$$\hat{V} \Rightarrow \hat{V}_{\mathrm{L}} = +g_{\mathrm{l}}\frac{e}{2m_{\mathrm{e}}}\boldsymbol{B}\hat{\boldsymbol{l}} , \quad \hat{\boldsymbol{l}} = \boldsymbol{x} \times \hat{\boldsymbol{p}} , \tag{5.5}$$

$$\hat{V}' \Rightarrow \hat{V}_{\mathrm{S}} = +g_{\mathrm{s}}\frac{e}{2m_{\mathrm{e}}}\boldsymbol{B}\hat{\boldsymbol{s}} , \quad \hat{\boldsymbol{s}} = \frac{\hbar}{2}\hat{\boldsymbol{\sigma}} , \tag{5.6}$$

$$\hat{\boldsymbol{p}} = -\mathrm{i}\hbar\nabla , \quad \hat{p}_i = -\mathrm{i}\hbar\frac{\partial}{\partial x^i} ; \quad \hat{\boldsymbol{\sigma}} = -\mathrm{i}\boldsymbol{\xi}\nabla , \quad \hat{\sigma}^j = -\mathrm{i}\sum_{i=1}^{3}\xi^{ji}\frac{\partial}{\partial x^i} . \tag{5.7}$$

5.1 Advanced Conceptions

> **Box 5.1 (A mathematical model: an advanced spin model).**
>
> An advanced spin model, without a magnetic field
> (the transposition sign T here is used in order to point out the matrix character of the
> following quantities, $\tilde{\boldsymbol{p}} = \boldsymbol{p}/\hbar$, $\tilde{\boldsymbol{\omega}} = \boldsymbol{\omega}/c$):
>
> $$T = \frac{1}{2m_0}\left(\boldsymbol{p}\right)^\mathrm{T}\boldsymbol{p} \Rightarrow \tilde{T} = \frac{1}{2}\left(\tilde{\boldsymbol{p}}\right)^\mathrm{T}\tilde{\boldsymbol{p}}\,, \tag{5.8}$$
>
> $$\begin{array}{c}\hat{T} \propto -\dfrac{1}{2}\left(\nabla\right)^\mathrm{T}\nabla\,,\\ \uparrow\quad\uparrow\\ \uparrow\qquad\pm\mathrm{i}\nabla\;\pm\mathrm{i}\nabla\\ \uparrow\quad\uparrow\\ \tilde{T} = +\dfrac{1}{2}\left(\tilde{\boldsymbol{p}}\right)^\mathrm{T}\tilde{\boldsymbol{p}}\,.\end{array} \tag{5.9}$$
>
> $$T = \frac{1}{2}\boldsymbol{L}^\mathrm{T}\boldsymbol{\omega} \Rightarrow \begin{cases}\boldsymbol{L}_\mathrm{L} = \theta\boldsymbol{\omega}\,, & T_\mathrm{L} = \dfrac{1}{2}\theta(\boldsymbol{\omega})^\mathrm{T}\boldsymbol{\omega}\\[6pt] \boldsymbol{L}_\mathrm{S} = \theta\boldsymbol{\omega}\,, & T_\mathrm{S} = \dfrac{1}{2}(\theta\boldsymbol{\omega})^\mathrm{T}\boldsymbol{\omega}\end{cases} \Rightarrow \begin{cases}\tilde{T}_\mathrm{L} = \dfrac{1}{2}(\tilde{\boldsymbol{\omega}})^\mathrm{T}\tilde{\boldsymbol{\omega}}\,,\\[6pt] \tilde{T}_\mathrm{S} = \dfrac{1}{2}(\theta\tilde{\boldsymbol{\omega}})^\mathrm{T}\tilde{\boldsymbol{\omega}}\,,\end{cases} \tag{5.10}$$
>
> $$\begin{array}{cc}\hat{T} \propto -\dfrac{1}{2}(\nabla)^\mathrm{T}\nabla\,, & \hat{T}' \propto -\dfrac{1}{2}(\theta^{\ni}\nabla)^\mathrm{T}\nabla\,,\\ \uparrow\quad\uparrow & \uparrow\quad\uparrow\\ \uparrow\qquad\pm\mathrm{i}\nabla\;\pm\mathrm{i}\nabla & \uparrow\qquad\pm\mathrm{i}\theta^{\ni}\nabla\;\pm\mathrm{i}\nabla\\ \uparrow\quad\uparrow & \uparrow\quad\uparrow\\ \tilde{T}_\mathrm{L} = +\dfrac{1}{2}(\tilde{\boldsymbol{\omega}})^\mathrm{T}\tilde{\boldsymbol{\omega}}\,, & \tilde{T}_\mathrm{S} = +\dfrac{1}{2}(\theta\tilde{\boldsymbol{\omega}})^\mathrm{T}\tilde{\boldsymbol{\omega}}\,,\end{array} \tag{5.11}$$
>
> $$\hat{T} = -\frac{\hbar^2}{2m_0}\left(\nabla\right)^\mathrm{T}\nabla\,,\quad \hat{T}' = -\frac{\hbar^2}{2m_0}\left(\theta^{\ni}\nabla\right)^\mathrm{T}\nabla\,. \tag{5.12}$$
>
> An advanced spin model, with a magnetic field
> (the transposition sign T here is used in order to point out the matrix character of the
> following quantities, $\tilde{\boldsymbol{B}} = (\boldsymbol{B}/|\boldsymbol{B}|)(\omega_0/c)$, $\tilde{\boldsymbol{\omega}}_\mathrm{p} = \boldsymbol{\omega}_\mathrm{p}/\omega_0$, $\tilde{\boldsymbol{\omega}} = \boldsymbol{\omega}/c$, $\omega_0 = e|\boldsymbol{B}|/m_0$):
>
> $$\begin{array}{cc}\hat{V} \propto -\dfrac{1}{2}(\boldsymbol{B}_3)^\mathrm{T}\boldsymbol{A}^{\ni} \times \nabla\,, & \hat{V}' \propto -\dfrac{1}{2}(\boldsymbol{B}_3)^\mathrm{T}\boldsymbol{A}^{\ni} \times \theta^{\ni}\nabla\,,\\ \uparrow\quad\uparrow\quad\uparrow & \uparrow\quad\uparrow\quad\uparrow\\ \uparrow\qquad \boldsymbol{B}_3\;\pm\mathrm{i}\boldsymbol{A}^{\ni}\;\pm\mathrm{i}\nabla & \uparrow\qquad \boldsymbol{B}_3\;\pm\mathrm{i}\boldsymbol{A}^{\ni}\;\pm\mathrm{i}\theta^{\ni}\nabla\\ \uparrow\quad\uparrow\quad\uparrow & \uparrow\quad\uparrow\quad\uparrow\\ \tilde{V}_\mathrm{L} = +\dfrac{1}{2}(\tilde{\boldsymbol{B}})^\mathrm{T}\tilde{\boldsymbol{\omega}}_\mathrm{p} \times \tilde{\boldsymbol{\omega}}\,, & \tilde{V}_\mathrm{S} = +\dfrac{1}{2}(\tilde{\boldsymbol{B}})^\mathrm{T}\tilde{\boldsymbol{\omega}}_\mathrm{p} \times \theta\tilde{\boldsymbol{\omega}}\,,\end{array} \tag{5.13}$$
>
> $$\hat{V} = -\frac{\hbar^2}{2m_0}(\boldsymbol{B}_3)^\mathrm{T}\boldsymbol{A}^{\ni} \times \nabla\,,\quad \hat{V}' = -\frac{\hbar^2}{2m_0}(\boldsymbol{B}_3)^\mathrm{T}\boldsymbol{A}^{\ni} \times \theta^{\ni}\nabla\,. \tag{5.14}$$

Continuation of the mathematical model.

Starting from the expression (5.8), left, for the kinetic energy T, or much better, starting from the expression (5.8), right, for the adjusted kinetic energy $\tilde{T}$ which comes into being by multiplication with $m_0/\hbar^2$, showing the physical dimension of the wave vector square and containing adjusted momentum vectors $\tilde{\boldsymbol{p}} = \boldsymbol{p}/\hbar$ showing the physical dimension of the wave vector, apart from $\hbar^2/m_0$ we gain in a first step the kinetic energy operator $\hat{T}$ according to (5.9) via the replacement rule $\tilde{\boldsymbol{p}} \to \pm i\nabla$, which is the parameter-free version of Jordan's rule. Since the kinetic energy operator is valid, too, for the special case of an orbital particle motion, we conclude in a second step that the replacement rule $\tilde{\boldsymbol{\omega}} \to \pm i\nabla$, which apart from $\hbar^2/m_0$ directly leads from the wave-vector-adjusted orbital rotation energy $\tilde{T}_L$ to the kinetic energy operator $\hat{T}$, is the parameter-free version of Jordan's rule adjusted to the special situation of an orbital rotation. Since the wave-vector-adjusted orbital rotation energy $\tilde{T}_L$ and the wave-vector-adjusted eigenrotation energy $\tilde{T}_S$ differ in a 3×3 matrix $\tilde{\boldsymbol{\theta}}$, we conclude in a third step that the replacement rule $\tilde{\boldsymbol{\theta}}\tilde{\boldsymbol{\omega}} \to \pm i\boldsymbol{\theta}^{\ni}\nabla$ apart from $\hbar^2/m_0$ directly leads from the wave-vector-adjusted eigenrotation energy $\tilde{T}_S$ to a kinetic energy operator $\hat{T}'$ for the spin property without a magnetic field provided the unknown 3×3 matrix $\boldsymbol{\theta}^{\ni}$ is specified suitably. In addition, incorporating $\hbar^2/m_0$, we are led to (5.12), collecting the kinetic energy operator $\hat{T}$ and the kinetic energy operator $\hat{T}'$ for the spin property without a magnetic field provided the unknown 3×3 matrix $\boldsymbol{\theta}^{\ni}$ is specified suitably, i.e. is specified adapted to spin realities.

Picking up the above threads, in addition to the kinetic energy operator $\hat{T}'$ for the spin property without a magnetic field, we here introduce an interaction operator for the spin property with a magnetic field containing differential operators $\hat{\sigma}^i$ superseding the Pauli matrix operators $\hat{\sigma}_i$. Again, we here start from wave-vector-adjusted expressions, namely the adjusted interaction energy $\tilde{V}_L$ which concatenates the magnetic field $\boldsymbol{B}$ with an orbital rotation and the adjusted interaction energy $\tilde{V}_S$ which concatenates the magnetic field $\boldsymbol{B}$ with an eigenrotation. Again, we here realize that a 3×3 matrix $\tilde{\boldsymbol{\theta}}$ makes the difference between an energy that is concatenated with an orbital rotation and an energy that is concatenated with an eigenrotation. Again, we here apply the replacement rules $\tilde{\boldsymbol{\omega}} \to \pm i\nabla$ and $\tilde{\boldsymbol{\theta}}\tilde{\boldsymbol{\omega}} \to \pm i\boldsymbol{\theta}^{\ni}\nabla$. Completing the treatment, in addition, we here apply the replacement rules $\tilde{\boldsymbol{\omega}}_p \to \pm i\boldsymbol{A}^{\ni}$ and $\boldsymbol{B} \to \boldsymbol{B}_{\ni}$, thus substituting the aligned precession frequency vector $\tilde{\boldsymbol{\omega}}_p$ and the aligned magnetic field vector $\boldsymbol{B}$ by the undetermined vectors $\pm i\boldsymbol{A}^{\ni}$ and $\boldsymbol{B}_{\ni}$. Completing the treatment, in addition, incorporating $\hbar^2/m_0$, we are led to (5.14), following the above threads, concluding that $\hat{V}$ is an interaction operator that manages interactions of a magnetic field with an orbital rotation and that $\hat{V}'$ is an interaction operator that manages interactions of a magnetic field with the spin property provided the unknown vectors $\boldsymbol{A}_{\ni}$ and $\boldsymbol{A}^{\ni}$ as well as the unknown 3×3 matrix $\boldsymbol{\theta}^{\ni}$ are specified suitably, i.e. are specified adapted to orbital rotation realities and spin realities, respectively. In the segment that follows, let us work out these circumstances in more detail.

5.1 Advanced Conceptions

Continuation of the mathematical model.

Specifying $\hat{V}$ of (5.14) by setting $\boldsymbol{B}_3 = \nabla \times \boldsymbol{A}_\ni \propto \nabla \times \boldsymbol{A}$ as well as $\boldsymbol{A}^\ni \propto \boldsymbol{x}$, with $\boldsymbol{A}$ being the conventional vector potential and $\boldsymbol{x}$ being the conventional position vector, we immediately realize that $\hat{V}$ of (5.14) includes the interaction operator $\hat{V}_\text{L}$ of an orbital electron motion in a magnetic field as limiting case. (5.5) outlines this relation. Realizing that the 3×3 matrix $\tilde{\boldsymbol{\theta}}$ makes the difference between $\tilde{T}_\text{L}$ and $\tilde{T}_\text{S}$ and makes the difference between $\tilde{V}_\text{L}$ and $\tilde{V}_\text{S}$, in both cases, implementing eigenrotation properties, realizing that the 3×3 matrix $\boldsymbol{\theta}^\ni$ makes the difference between $\hat{T}$ and $\hat{T}'$, in this case, implementing further inner properties such as spin properties, and realizing that this operative difference is reflected by the replacement rules $\tilde{\boldsymbol{\omega}} \to \pm \mathrm{i} \nabla$ and $\tilde{\boldsymbol{\theta}} \tilde{\boldsymbol{\omega}} \to \pm \mathrm{i} \boldsymbol{\theta}^\ni \nabla$, we immediately realize that this difference is also valid for $\hat{V}$ and $\hat{V}'$, in this case, implementing further inner properties such as spin properties. Actually, if we further specify $\hat{V}'$ of (5.14) by setting $\boldsymbol{B}_3 = \nabla \times \boldsymbol{A}_\ni \propto \nabla \times \boldsymbol{A}$ as well as $\boldsymbol{A}^\ni \times \boldsymbol{\theta}^\ni \nabla \propto -\mathrm{i} \boldsymbol{\xi} \nabla$, where the 3×3 matrix $\boldsymbol{\xi}$ is free, we immediately realize that $\hat{V}'$ of (5.14) includes the interaction operator $\hat{V}_\text{S}$ of an electron spin in a magnetic field as limiting case, however, containing differential operators $\hat{\sigma}^i$ superseding the Pauli matrix operators $\hat{\sigma}_i$. (5.6) outlines this relation. $\hat{T}'$ and $\hat{V}'$ together with $\hat{V}_\text{S}$ define our advanced spin model provided the unknown vectors $\boldsymbol{A}_\ni$ and $\boldsymbol{A}^\ni$ as well as the unknown 3×3 matrices $\boldsymbol{\theta}^\ni$ and $\boldsymbol{\xi}$ are specified suitably. The virtue of $\hat{T}'$ is worked out in Box 2.3 by example. The virtue of $\hat{V}_\text{S}$ is worked out in the segments that follow by example.

In Figure 5.1, we consider the energy levels E_n of the electron of the hydrogen atom disturbed by energy shifts $\Delta E_{\uparrow\downarrow}$ caused by the interaction of the electron spin with a magnetic field B_0 applied in z direction. As it is shown in Figure 5.2, there are only two eigenvalues $s_{z,\uparrow}$ and $s_{z,\downarrow}$ of the z component of the electron spin and there is only one eigenvalue s of the length of the electron spin. Let us consider the ground state of the electron of the hydrogen atom, i.e. let us consider the wave function (5.17), where $\psi_{\uparrow\downarrow}$ indicates the spin state of the electron of the hydrogen atom, and where $N'_{1,0}$ indicates a normalization factor. Let us consider the spin matrices $\hat{s}_i$ and the Pauli matrices $\hat{\sigma}_i$ of the conventional spin matrix formalism according to (5.18) as well as the differential operators $\hat{s}^i$ and $\hat{\sigma}^i$ of our advanced spin model according to (5.19). We know that the spin matrices $\hat{s}_i$ and the Pauli matrices $\hat{\sigma}_i$ fulfil the Schrödinger equation (5.15) and the spinor relations (5.24) and (5.33). We show that the differential operators $\hat{s}^i$ and $\hat{\sigma}^i$ likewise allow us to calculate the energy shifts and the spin eigenvalues and likewise fulfil the spin commutator relations of the spin matrices $\hat{s}_i$ and the Pauli matrices $\hat{\sigma}_i$ if the $\xi_{\uparrow\downarrow}^{ij}(\boldsymbol{x})$ are chosen suitably by starting from the Schrödinger equation (5.15) specified by (5.19), i.e. by starting from the Schrödinger equation (5.16). Step by step, we thus derive the special solution (5.22), (5.29), and (5.30). Step by step, we thus prove the applicability of our advanced spin model by example. Last but not least, we realize that the calyciform zero field regions of the function $\xi_{\uparrow\downarrow}^{33}(\boldsymbol{x}) \xi_{\uparrow\downarrow}^{33*}(\boldsymbol{x})$ reflect the precession cones of the classical analogue, giving us a further hint that our advanced spin model supplies us with an access to spin properties that goes far beyond the access to spin properties that is supplied by conventional spin matrix theory.

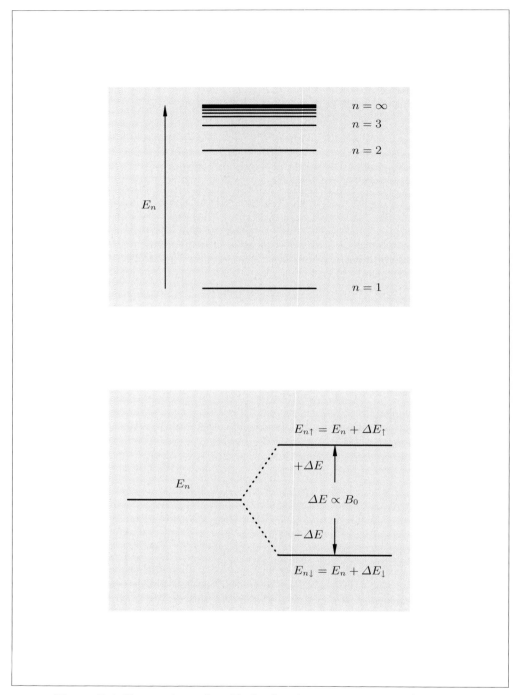

Figure 5.1. Energy eigenvalues E_n (top) and energy shifts $\Delta E_{\uparrow\downarrow}$ (bottom).

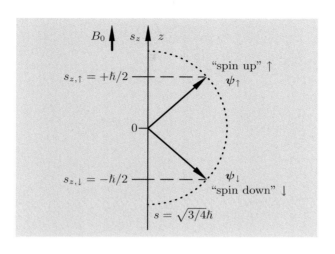
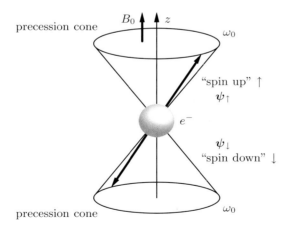

Figure 5.2. Spin eigenvalues (top) and classical analogue (bottom).

Continuation of the mathematical model.

$$\left[+g_{\mathrm{s}}\frac{e}{2m_{\mathrm{e}}}\boldsymbol{B}\hat{\boldsymbol{s}}\right]\psi_{1,0,s,\uparrow\downarrow}(\boldsymbol{x})=\Delta E_{\uparrow\downarrow}\psi_{1,0,s,\uparrow\downarrow}(\boldsymbol{x})\,, \tag{5.15}$$

$$\left[-\mathrm{i}g_{\mathrm{s}}\frac{e}{2m_{\mathrm{e}}}B_0\frac{\hbar}{2}\sum_{i=1}^{3}\xi_{\uparrow\downarrow}^{3i}(\boldsymbol{x})\frac{\partial}{\partial x^i}\right]\psi_{1,0,s,\uparrow\downarrow}(\boldsymbol{x})=\Delta E_{\uparrow\downarrow}\psi_{1,0,s,\uparrow\downarrow}(\boldsymbol{x})\,, \tag{5.16}$$

$$\psi_{1,0,s,\uparrow\downarrow}(\boldsymbol{x})=\psi_{1,0,s}(\boldsymbol{x})\psi_{\uparrow\downarrow}=N'_{1,0}\sqrt{\tfrac{1}{4\pi}}\exp\left[-k\left(E_1\right)r\right]\psi_{\uparrow\downarrow}\,,$$
$$k\left(E_1\right)=\sqrt{-2m_{\mathrm{e}}E_1/\hbar^2}\,,\quad E_1\propto-\frac{m_{\mathrm{e}}e^4}{\hbar^2}\,, \tag{5.17}$$

$$\hat{s}_x=\hat{s}_1=\frac{\hbar}{2}\begin{pmatrix}0 & +1\\+1 & 0\end{pmatrix},\quad \hat{\sigma}_x=\hat{\sigma}_1=\begin{pmatrix}0 & +1\\+1 & 0\end{pmatrix},$$

$$\hat{s}_y=\hat{s}_2=\frac{\hbar}{2}\begin{pmatrix}0 & -\mathrm{i}\\+\mathrm{i} & 0\end{pmatrix},\quad \hat{\sigma}_y=\hat{\sigma}_2=\begin{pmatrix}0 & -\mathrm{i}\\+\mathrm{i} & 0\end{pmatrix}, \tag{5.18}$$

$$\hat{s}_z=\hat{s}_3=\frac{\hbar}{2}\begin{pmatrix}+1 & 0\\0 & -1\end{pmatrix},\quad \hat{\sigma}_z=\hat{\sigma}_3=\begin{pmatrix}+1 & 0\\0 & -1\end{pmatrix},$$

$$\hat{s}^x=\hat{s}^1=-\mathrm{i}\frac{\hbar}{2}\sum_{i=1}^{3}\xi_{\uparrow\downarrow}^{1i}(\boldsymbol{x})\frac{\partial}{\partial x^i}\,,\quad \hat{\sigma}^x=\hat{\sigma}^1=-\mathrm{i}\sum_{i=1}^{3}\xi_{\uparrow\downarrow}^{1i}(\boldsymbol{x})\frac{\partial}{\partial x^i}\,,$$

$$\hat{s}^y=\hat{s}^2=-\mathrm{i}\frac{\hbar}{2}\sum_{i=1}^{3}\xi_{\uparrow\downarrow}^{2i}(\boldsymbol{x})\frac{\partial}{\partial x^i}\,,\quad \hat{\sigma}^y=\hat{\sigma}^2=-\mathrm{i}\sum_{i=1}^{3}\xi_{\uparrow\downarrow}^{2i}(\boldsymbol{x})\frac{\partial}{\partial x^i}\,, \tag{5.19}$$

$$\hat{s}^z=\hat{s}^3=-\mathrm{i}\frac{\hbar}{2}\sum_{i=1}^{3}\xi_{\uparrow\downarrow}^{3i}(\boldsymbol{x})\frac{\partial}{\partial x^i}\,,\quad \hat{\sigma}^z=\hat{\sigma}^3=-\mathrm{i}\sum_{i=1}^{3}\xi_{\uparrow\downarrow}^{3i}(\boldsymbol{x})\frac{\partial}{\partial x^i}\,,$$

$$\Delta E_{\uparrow}=+\Delta E\,,\quad \Delta E_{\downarrow}=-\Delta E\,,\quad \Delta E=\frac{1}{2}\hbar\omega_0\,,\quad \omega_0=g_{\mathrm{s}}\frac{e}{2m_{\mathrm{e}}}B_0\,, \tag{5.20}$$

Continuation of the mathematical model.

$$\mathrm{i}\sum_{i=1}^{3}\xi_{\uparrow}^{3i}(\boldsymbol{x})\frac{\partial\psi_{1,0,s,\uparrow}/\partial x^{i}}{\psi_{1,0,s,\uparrow}}=-1\ ,\quad \mathrm{i}\sum_{i=1}^{3}\xi_{\downarrow}^{3i}(\boldsymbol{x})\frac{\partial\psi_{1,0,s,\downarrow}/\partial x^{i}}{\psi_{1,0,s,\downarrow}}=+1\ , \tag{5.21}$$

$$\xi_{\uparrow}^{33}(\boldsymbol{x})=-\frac{\mathrm{i}}{k\left(E_{1}\right)}\frac{r}{z}\ ,\quad \xi_{\downarrow}^{33}(\boldsymbol{x})=+\frac{\mathrm{i}}{k\left(E_{1}\right)}\frac{r}{z}\ .$$

$$\xi_{\uparrow\downarrow}^{31}(\boldsymbol{x})=\xi_{\uparrow\downarrow}^{32}(\boldsymbol{x})=0\ , \tag{5.22}$$

$$\xi_{\uparrow\downarrow}^{33}(\boldsymbol{x})\xi_{\uparrow\downarrow}^{33\,*}(\boldsymbol{x})=\frac{1}{k^{2}\left(E_{1}\right)}\frac{r^{2}}{z^{2}}\ ,$$

$$\xi_{\uparrow\downarrow}^{31}(\boldsymbol{x})\xi_{\uparrow\downarrow}^{31\,*}(\boldsymbol{x})=\xi_{\uparrow\downarrow}^{32}(\boldsymbol{x})\xi_{\uparrow\downarrow}^{32\,*}(\boldsymbol{x})=0\ , \tag{5.23}$$

$$\hat{s}_{z}\boldsymbol{\psi}_{\uparrow}=\frac{\hbar}{2}\begin{pmatrix}+1&0\\0&-1\end{pmatrix}\begin{pmatrix}1\\0\end{pmatrix}=+\frac{\hbar}{2}\boldsymbol{\psi}_{\uparrow}\ ,\quad \hat{\boldsymbol{\sigma}}_{z}\boldsymbol{\psi}_{\uparrow}=\begin{pmatrix}+1&0\\0&-1\end{pmatrix}\begin{pmatrix}1\\0\end{pmatrix}=+\boldsymbol{\psi}_{\uparrow}\ ,$$

$$\hat{s}_{z}\boldsymbol{\psi}_{\downarrow}=\frac{\hbar}{2}\begin{pmatrix}+1&0\\0&-1\end{pmatrix}\begin{pmatrix}0\\1\end{pmatrix}=-\frac{\hbar}{2}\boldsymbol{\psi}_{\downarrow}\ ,\quad \hat{\boldsymbol{\sigma}}_{z}\boldsymbol{\psi}_{\downarrow}=\begin{pmatrix}+1&0\\0&-1\end{pmatrix}\begin{pmatrix}0\\1\end{pmatrix}=-\boldsymbol{\psi}_{\downarrow}\ , \tag{5.24}$$

$$\hat{s}^{z}\psi_{1,0,s,\uparrow}=-\mathrm{i}\frac{\hbar}{2}\sum_{i=1}^{3}\xi_{\uparrow}^{3i}(\boldsymbol{x})\frac{\partial}{\partial x^{i}}\psi_{1,0,s,\uparrow}=+\frac{\hbar}{2}\psi_{1,0,s,\uparrow}\Rightarrow s_{z,\uparrow}=+\frac{\hbar}{2}\ ,$$

$$\hat{s}^{z}\psi_{1,0,s,\downarrow}=-\mathrm{i}\frac{\hbar}{2}\sum_{i=1}^{3}\xi_{\downarrow}^{3i}(\boldsymbol{x})\frac{\partial}{\partial x^{i}}\psi_{1,0,s,\downarrow}=-\frac{\hbar}{2}\psi_{1,0,s,\downarrow}\Rightarrow s_{z,\downarrow}=-\frac{\hbar}{2}\ , \tag{5.25}$$

$$\hat{\sigma}^{z}\psi_{1,0,s,\uparrow}=-\mathrm{i}\sum_{i=1}^{3}\xi_{\uparrow}^{3i}(\boldsymbol{x})\frac{\partial}{\partial x^{i}}\psi_{1,0,s,\uparrow}=+\psi_{1,0,s,\uparrow}\Rightarrow \sigma_{z,\uparrow}=+1\ ,$$

$$\hat{\sigma}^{z}\psi_{1,0,s,\downarrow}=-\mathrm{i}\sum_{i=1}^{3}\xi_{\downarrow}^{3i}(\boldsymbol{x})\frac{\partial}{\partial x^{i}}\psi_{1,0,s,\downarrow}=-\psi_{1,0,s,\downarrow}\Rightarrow \sigma_{z,\downarrow}=-1\ , \tag{5.26}$$

Continuation of the mathematical model.

$$i \sum_{i=1}^{3} \xi_\uparrow^{1i}(\boldsymbol{x}) \frac{\partial \psi_{1,0,s,\uparrow}/\partial x^i}{\psi_{1,0,s,\downarrow}} = -1 \;,$$

$$i \sum_{i=1}^{3} \xi_\downarrow^{1i}(\boldsymbol{x}) \frac{\partial \psi_{1,0,s,\downarrow}/\partial x^i}{\psi_{1,0,s,\uparrow}} = -1 \;,$$
(5.27)

$$i \sum_{i=1}^{3} \xi_\uparrow^{2i}(\boldsymbol{x}) \frac{\partial \psi_{1,0,s,\uparrow}/\partial x^i}{\psi_{1,0,s,\downarrow}} = -i \;,$$

$$i \sum_{i=1}^{3} \xi_\downarrow^{2i}(\boldsymbol{x}) \frac{\partial \psi_{1,0,s,\downarrow}/\partial x^i}{\psi_{1,0,s,\uparrow}} = +i \;,$$
(5.28)

$$\xi_\uparrow^{11}(\boldsymbol{x}) = -\frac{i}{k(E_1)} \frac{r}{x} \frac{\psi_\downarrow}{\psi_\uparrow} \;, \quad \xi_\downarrow^{11}(\boldsymbol{x}) = -\frac{i}{k(E_1)} \frac{r}{x} \frac{\psi_\uparrow}{\psi_\downarrow} \;,$$

$$\xi_{\uparrow\downarrow}^{12}(\boldsymbol{x}) = \xi_{\uparrow\downarrow}^{13}(\boldsymbol{x}) = 0 \;,$$
(5.29)

$$\xi_\uparrow^{22}(\boldsymbol{x}) = +\frac{1}{k(E_1)} \frac{r}{y} \frac{\psi_\downarrow}{\psi_\uparrow} \;, \quad \xi_\downarrow^{22}(\boldsymbol{x}) = -\frac{1}{k(E_1)} \frac{r}{y} \frac{\psi_\uparrow}{\psi_\downarrow} \;,$$

$$\xi_{\uparrow\downarrow}^{21}(\boldsymbol{x}) = \xi_{\uparrow\downarrow}^{23}(\boldsymbol{x}) = 0 \;,$$
(5.30)

$$\xi_\uparrow^{11}(\boldsymbol{x})\xi_\uparrow^{11*}(\boldsymbol{x}) = \frac{1}{k^2(E_1)} \frac{r^2}{x^2} \;, \quad \xi_\downarrow^{11}(\boldsymbol{x})\xi_\downarrow^{11*}(\boldsymbol{x}) = \frac{1}{k^2(E_1)} \frac{r^2}{x^2} \;,$$

$$\xi_{\uparrow\downarrow}^{12}(\boldsymbol{x})\xi_{\uparrow\downarrow}^{12*}(\boldsymbol{x}) = \xi_{\uparrow\downarrow}^{13}(\boldsymbol{x})\xi_{\uparrow\downarrow}^{13*}(\boldsymbol{x}) = 0 \;,$$
(5.31)

$$\xi_\uparrow^{22}(\boldsymbol{x})\xi_\uparrow^{22*}(\boldsymbol{x}) = \frac{1}{k^2(E_1)} \frac{r^2}{y^2} \;, \quad \xi_\downarrow^{22}(\boldsymbol{x})\xi_\downarrow^{22*}(\boldsymbol{x}) = \frac{1}{k^2(E_1)} \frac{r^2}{y^2} \;,$$

$$\xi_{\uparrow\downarrow}^{21}(\boldsymbol{x})\xi_{\uparrow\downarrow}^{21*}(\boldsymbol{x}) = \xi_{\uparrow\downarrow}^{23}(\boldsymbol{x})\xi_{\uparrow\downarrow}^{23*}(\boldsymbol{x}) = 0 \;,$$
(5.32)

Continuation of the mathematical model.

$$\hat{s}_x \boldsymbol{\psi}_\uparrow = \frac{\hbar}{2}\begin{pmatrix} 0 & +1 \\ +1 & 0 \end{pmatrix}\begin{pmatrix} 1 \\ 0 \end{pmatrix} = +\frac{\hbar}{2}\boldsymbol{\psi}_\downarrow \;,\; \hat{\sigma}_x\boldsymbol{\psi}_\uparrow = \begin{pmatrix} 0 & +1 \\ +1 & 0 \end{pmatrix}\begin{pmatrix} 1 \\ 0 \end{pmatrix} = +\boldsymbol{\psi}_\downarrow \;,$$

$$\hat{s}_x \boldsymbol{\psi}_\downarrow = \frac{\hbar}{2}\begin{pmatrix} 0 & +1 \\ +1 & 0 \end{pmatrix}\begin{pmatrix} 0 \\ 1 \end{pmatrix} = +\frac{\hbar}{2}\boldsymbol{\psi}_\uparrow \;,\; \hat{\sigma}_x\boldsymbol{\psi}_\downarrow = \begin{pmatrix} 0 & +1 \\ +1 & 0 \end{pmatrix}\begin{pmatrix} 0 \\ 1 \end{pmatrix} = +\boldsymbol{\psi}_\uparrow \;,$$

$$\hat{s}_y \boldsymbol{\psi}_\uparrow = \frac{\hbar}{2}\begin{pmatrix} 0 & -\mathrm{i} \\ +\mathrm{i} & 0 \end{pmatrix}\begin{pmatrix} 1 \\ 0 \end{pmatrix} = +\mathrm{i}\frac{\hbar}{2}\boldsymbol{\psi}_\downarrow \;,\; \hat{\sigma}_y\boldsymbol{\psi}_\uparrow = \begin{pmatrix} 0 & -\mathrm{i} \\ +\mathrm{i} & 0 \end{pmatrix}\begin{pmatrix} 1 \\ 0 \end{pmatrix} = +\mathrm{i}\boldsymbol{\psi}_\downarrow \;,$$

$$\hat{s}_y \boldsymbol{\psi}_\downarrow = \frac{\hbar}{2}\begin{pmatrix} 0 & -\mathrm{i} \\ +\mathrm{i} & 0 \end{pmatrix}\begin{pmatrix} 0 \\ 1 \end{pmatrix} = -\mathrm{i}\frac{\hbar}{2}\boldsymbol{\psi}_\uparrow \;,\; \hat{\sigma}_y\boldsymbol{\psi}_\downarrow = \begin{pmatrix} 0 & -\mathrm{i} \\ +\mathrm{i} & 0 \end{pmatrix}\begin{pmatrix} 0 \\ 1 \end{pmatrix} = -\mathrm{i}\boldsymbol{\psi}_\uparrow \;,$$

(5.33)

$$\hat{s}^x \psi_{1,0,s,\uparrow} = -\mathrm{i}\frac{\hbar}{2}\sum_{i=1}^{3}\xi_\uparrow^{1i}(\boldsymbol{x})\frac{\partial}{\partial x^i}\psi_{1,0,s,\uparrow} = +\frac{\hbar}{2}\psi_{1,0,s,\downarrow} \;,$$

$$\hat{s}^x \psi_{1,0,s,\downarrow} = -\mathrm{i}\frac{\hbar}{2}\sum_{i=1}^{3}\xi_\downarrow^{1i}(\boldsymbol{x})\frac{\partial}{\partial x^i}\psi_{1,0,s,\downarrow} = +\frac{\hbar}{2}\psi_{1,0,s,\uparrow} \;,$$

$$\hat{\sigma}^x \psi_{1,0,s,\uparrow} = -\mathrm{i}\sum_{i=1}^{3}\xi_\uparrow^{1i}(\boldsymbol{x})\frac{\partial}{\partial x^i}\psi_{1,0,s,\uparrow} = +\psi_{1,0,s,\downarrow} \;,$$

$$\hat{\sigma}^x \psi_{1,0,s,\downarrow} = -\mathrm{i}\sum_{i=1}^{3}\xi_\downarrow^{1i}(\boldsymbol{x})\frac{\partial}{\partial x^i}\psi_{1,0,s,\downarrow} = +\psi_{1,0,s,\uparrow} \;,$$

(5.34)

$$\hat{s}^y \psi_{1,0,s,\uparrow} = -\mathrm{i}\frac{\hbar}{2}\sum_{i=1}^{3}\xi_\uparrow^{2i}(\boldsymbol{x})\frac{\partial}{\partial x^i}\psi_{1,0,s,\uparrow} = +\mathrm{i}\frac{\hbar}{2}\psi_{1,0,s,\downarrow} \;,$$

$$\hat{s}^y \psi_{1,0,s,\downarrow} = -\mathrm{i}\frac{\hbar}{2}\sum_{i=1}^{3}\xi_\downarrow^{2i}(\boldsymbol{x})\frac{\partial}{\partial x^i}\psi_{1,0,s,\downarrow} = -\mathrm{i}\frac{\hbar}{2}\psi_{1,0,s,\uparrow} \;,$$

$$\hat{\sigma}^y \psi_{1,0,s,\uparrow} = -\mathrm{i}\sum_{i=1}^{3}\xi_\uparrow^{2i}(\boldsymbol{x})\frac{\partial}{\partial x^i}\psi_{1,0,s,\uparrow} = +\mathrm{i}\psi_{1,0,s,\downarrow} \;,$$

$$\hat{\sigma}^y \psi_{1,0,s,\downarrow} = -\mathrm{i}\sum_{i=1}^{3}\xi_\downarrow^{2i}(\boldsymbol{x})\frac{\partial}{\partial x^i}\psi_{1,0,s,\downarrow} = -\mathrm{i}\psi_{1,0,s,\uparrow} \;.$$

(5.35)

Box 5.2 (A mathematical model: an advanced spin–magnetic-field model, spin length).

Spin length (we consider the above spin-precession fields):

$$\hat{s}^*\hat{s}\psi_{1,0,s,\uparrow\downarrow}(\boldsymbol{x}) = s^2\psi_{1,0,s,\uparrow\downarrow}(\boldsymbol{x}) \,,$$

$$\hat{s}^*\hat{s} = \frac{\hbar^2}{4}\hat{\boldsymbol{\sigma}}^*\hat{\boldsymbol{\sigma}}$$

$$\Downarrow$$

$$\hat{s}^*\hat{s}\psi_{1,0,s,\uparrow\downarrow}(\boldsymbol{x}) = s^2\psi_{1,0,s,\uparrow\downarrow}(\boldsymbol{x}) \,,$$

$$\hat{s}^*\hat{s} = \frac{\hbar^2}{4}\left[\sigma^{1*}\sigma^1 + \sigma^{2*}\sigma^2 + \sigma^{3*}\sigma^3\right]$$

$$\Downarrow$$

$$\hat{s}^*\hat{s}\psi_{1,0,s,\uparrow\downarrow}(\boldsymbol{x}) = s^2\psi_{1,0,s,\uparrow\downarrow}(\boldsymbol{x}) \,,$$

$$\hat{s}^*\hat{s} = \frac{\hbar^2}{4}\left[\xi^{11*}_{\uparrow\downarrow}(\boldsymbol{x})\frac{\partial}{\partial x}\xi^{11}_{\uparrow\downarrow}(\boldsymbol{x})\frac{\partial}{\partial x} + \xi^{22*}_{\uparrow\downarrow}(\boldsymbol{x})\frac{\partial}{\partial y}\xi^{22}_{\uparrow\downarrow}(\boldsymbol{x})\frac{\partial}{\partial y} + \xi^{33*}_{\uparrow\downarrow}(\boldsymbol{x})\frac{\partial}{\partial z}\xi^{33}_{\uparrow\downarrow}(\boldsymbol{x})\frac{\partial}{\partial z}\right]$$

$$\Downarrow$$

$$\xi^{11*}_{\uparrow\downarrow}(\boldsymbol{x})\frac{\partial}{\partial x}\left[\mathrm{i}\psi_{1,0,s,\downarrow\uparrow}(\boldsymbol{x})\right] + \xi^{22*}_{\uparrow\downarrow}(\boldsymbol{x})\frac{\partial}{\partial y}\left[\mp \psi_{1,0,s,\downarrow\uparrow}(\boldsymbol{x})\right] +$$
$$+\xi^{33*}_{\uparrow\downarrow}(\boldsymbol{x})\frac{\partial}{\partial z}\left[\pm \mathrm{i}\psi_{1,0,s,\uparrow\downarrow}(\boldsymbol{x})\right] \quad (5.36)$$
$$=$$
$$\frac{4}{\hbar^2}s^2\psi_{1,0,s,\uparrow\downarrow}(\boldsymbol{x})$$

$$\Downarrow$$

$$\psi_{1,0,s,\uparrow\downarrow}(\boldsymbol{x}) + \psi_{1,0,s,\uparrow\downarrow}(\boldsymbol{x}) + \psi_{1,0,s,\uparrow\downarrow}(\boldsymbol{x})$$
$$=$$
$$\frac{4}{\hbar^2}s^2\psi_{1,0,s,\uparrow\downarrow}(\boldsymbol{x})$$

$$\Downarrow$$

$$3\psi_{1,0,s,\uparrow\downarrow}(\boldsymbol{x})$$
$$=$$
$$\frac{4}{\hbar^2}s^2\psi_{1,0,s,\uparrow\downarrow}(\boldsymbol{x})$$

$$\Downarrow$$

$$s = \sqrt{\frac{3}{4}}\hbar \,.$$

Box 5.3 (A mathematical model: an advanced spin–magnetic-field model, commutators).

Commutators (in general):

$$\hat{\sigma}^{y\,*}\hat{\sigma}^z - \hat{\sigma}^{z\,*}\hat{\sigma}^y = 2\mathrm{i}\hat{\sigma}^x \,, \qquad \hat{\sigma}^{2\,*}\hat{\sigma}^3 - \hat{\sigma}^{3\,*}\hat{\sigma}^2 = 2\mathrm{i}\hat{\sigma}^1 \,,$$
$$\hat{\sigma}^{z\,*}\hat{\sigma}^x - \hat{\sigma}^{x\,*}\hat{\sigma}^z = 2\mathrm{i}\hat{\sigma}^y \,, \qquad \hat{\sigma}^{3\,*}\hat{\sigma}^1 - \hat{\sigma}^{1\,*}\hat{\sigma}^3 = 2\mathrm{i}\hat{\sigma}^2 \,, \qquad (5.37)$$
$$\hat{\sigma}^{x\,*}\hat{\sigma}^y - \hat{\sigma}^{y\,*}\hat{\sigma}^x = 2\mathrm{i}\hat{\sigma}^z \,, \qquad \hat{\sigma}^{1\,*}\hat{\sigma}^2 - \hat{\sigma}^{2\,*}\hat{\sigma}^1 = 2\mathrm{i}\hat{\sigma}^3 \,.$$

Commutators (we consider the above spin-precession fields):

$$\xi_{\uparrow\downarrow}^{22\,*}\frac{\partial}{\partial y}\xi_{\uparrow\downarrow}^{33}\frac{\partial}{\partial z} - \xi_{\uparrow\downarrow}^{33\,*}\frac{\partial}{\partial z}\xi_{\uparrow\downarrow}^{22}\frac{\partial}{\partial y} = 2\xi_{\uparrow\downarrow}^{11}\frac{\partial}{\partial x} \,,$$
$$\xi_{\uparrow\downarrow}^{33\,*}\frac{\partial}{\partial z}\xi_{\uparrow\downarrow}^{11}\frac{\partial}{\partial x} - \xi_{\uparrow\downarrow}^{11\,*}\frac{\partial}{\partial x}\xi_{\uparrow\downarrow}^{33}\frac{\partial}{\partial z} = 2\xi_{\uparrow\downarrow}^{22}\frac{\partial}{\partial y} \,, \qquad (5.38)$$
$$\xi_{\uparrow\downarrow}^{11\,*}\frac{\partial}{\partial x}\xi_{\uparrow\downarrow}^{22}\frac{\partial}{\partial y} - \xi_{\uparrow\downarrow}^{22\,*}\frac{\partial}{\partial y}\xi_{\uparrow\downarrow}^{11}\frac{\partial}{\partial x} = 2\xi_{\uparrow\downarrow}^{33}\frac{\partial}{\partial z} \,.$$

Proof (we consider the above spin-precession fields):

$$\left(\xi_\uparrow^{11\,*}\frac{\partial}{\partial x}\xi_\uparrow^{22}\frac{\partial}{\partial y} - \xi_\uparrow^{22\,*}\frac{\partial}{\partial y}\xi_\uparrow^{11}\frac{\partial}{\partial x}\right)\psi_{1,0,s,\uparrow}(\boldsymbol{x}) = 2\xi_\uparrow^{33}\frac{\partial}{\partial z}\psi_{1,0,s,\uparrow}(\boldsymbol{x})$$
$$\Downarrow$$
$$\left(\xi_\uparrow^{11\,*}\frac{\partial}{\partial x}\bigl[-\psi_{1,0,s,\downarrow}(\boldsymbol{x})\bigr] - \xi_\uparrow^{22\,*}\frac{\partial}{\partial y}\bigl[+\mathrm{i}\psi_{1,0,s,\downarrow}(\boldsymbol{x})\bigr]\right) = 2\mathrm{i}\psi_{1,0,s,\uparrow}(\boldsymbol{x}) \qquad (5.39)$$
$$\Downarrow$$
$$\mathrm{i}\psi_{1,0,s,\uparrow}(\boldsymbol{x}) + \mathrm{i}\psi_{1,0,s,\uparrow}(\boldsymbol{x}) = 2\mathrm{i}\psi_{1,0,s,\uparrow}(\boldsymbol{x})$$
$$\Downarrow$$
$$2\mathrm{i}\psi_{1,0,s,\uparrow}(\boldsymbol{x}) = 2\mathrm{i}\psi_{1,0,s,\uparrow}(\boldsymbol{x})$$

with

$$\frac{\psi_\uparrow^*}{\psi_\downarrow^*}\psi_\uparrow = \frac{1}{\psi_\downarrow^*} = \frac{\psi_\downarrow}{\psi_\downarrow^*\psi_\downarrow} = \psi_\downarrow \,, \quad \frac{\psi_\downarrow^*}{\psi_\uparrow^*}\psi_\downarrow = \frac{1}{\psi_\uparrow^*} = \frac{\psi_\uparrow}{\psi_\uparrow^*\psi_\uparrow} = \psi_\uparrow \,, \qquad (5.40)$$

since

$$\psi_\uparrow^*\psi_\uparrow = 1 \,, \quad \psi_\downarrow^*\psi_\downarrow = 1 \qquad (5.41)$$

etc.

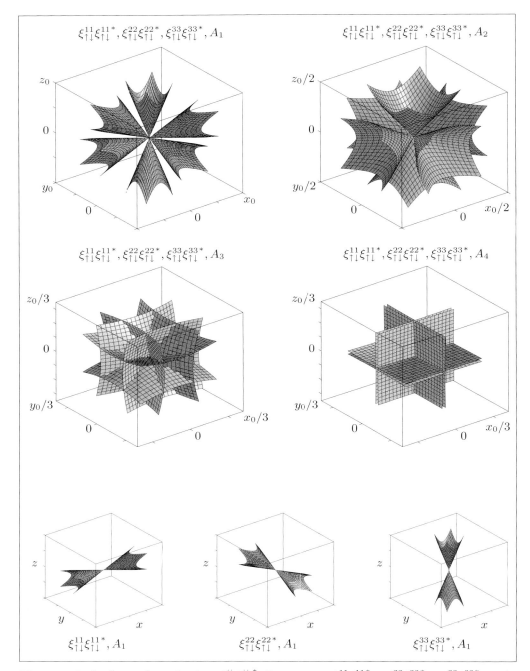

Figure 5.3. Surfaces of equal values $\xi^{ii}_{\uparrow\downarrow}\xi^{ii}_{\uparrow\downarrow}{}^*$. Parameters: $\xi^{11}_{\uparrow\downarrow}\xi^{11}_{\uparrow\downarrow}{}^* = \xi^{22}_{\uparrow\downarrow}\xi^{22}_{\uparrow\downarrow}{}^* = \xi^{33}_{\uparrow\downarrow}\xi^{33}_{\uparrow\downarrow}{}^* := A_i$, $A_1 < A_2 < A_3 \ll A_4$, $x_0 = y_0 = z_0$. The minimal opening angle of the surfaces of equal values $\xi^{ii}_{\uparrow\downarrow}\xi^{ii}_{\uparrow\downarrow}{}^*$ is restricted by the inequality $\xi^{ii}_{\uparrow}\xi^{ii}_{\uparrow}{}^* = \xi^{ii}_{\downarrow}\xi^{ii}_{\downarrow}{}^* \geq 1/k^2$ (E_1).

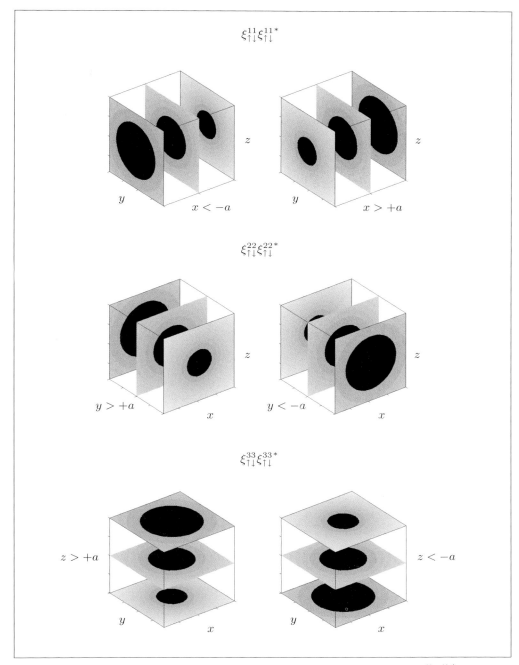

Figure 5.4. The minimal opening angle of the surfaces of equal values $\xi_{\uparrow\downarrow}^{ii}\xi_{\uparrow\downarrow}^{ii\,*}$, which is given by $\xi_{\uparrow}^{ii}\xi_{\uparrow}^{ii\,*} = \xi_{\downarrow}^{ii}\xi_{\downarrow}^{ii\,*} \geq 1/k^2$ (E_1), defines calyciform zero field regions, and these here are indicated by the black areas.

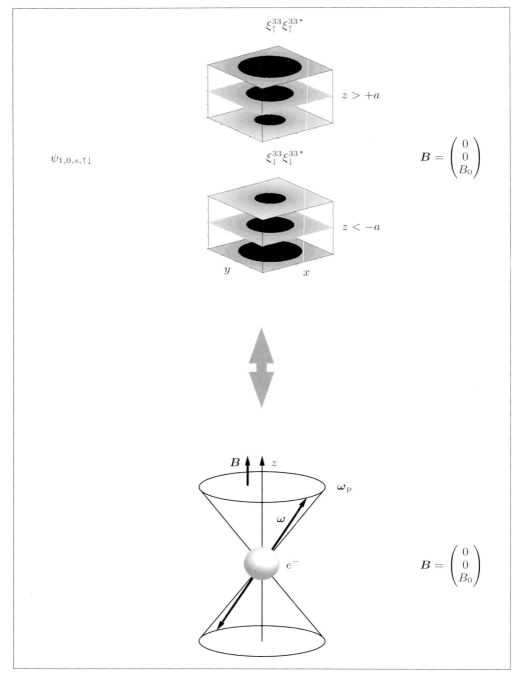

Figure 5.5. The calyciform zero field regions of the function $\xi^{33}_{\uparrow\downarrow}(\boldsymbol{x})\xi^{33\,*}_{\uparrow\downarrow}(\boldsymbol{x})$ (top) reflect the precession cones of the classical analogue (bottom).

5.1.2 Advanced Modeling of Substances/Materials

The method of gaining operative energy terms by starting from classical energy terms which is applied in Section 5.1.1 in close combination with the method of processing operative energy terms within Schrödinger's formalism which is applied in Section 5.1.1 can be continued in order to gain a nonlinear model equation allowing us to model biomimetic materials as well as non-biomimetic materials in a comprehensive manner, in fact, including parts with rest mass (electrons, nucleons, atoms, molecules etc.) and without rest mass (electromagnetic radiation, gravitational radiation etc.), promising us to be best adapted to future materials. Certainly, this here cannot be worked out in detail. However, let us here quote the result, the nonlinear model equation (5.42).

A mathematical model: a nonlinear model equation.

In the spirit of our self-consistent network of model conceptions, in all cases where $\chi_{00}(\boldsymbol{x},t)$ is on a par with an eigenvalue $\chi_{00} = m_0 E/\hbar^2$, the leading energy term (1) stands for the kinetic energy operator of conventional quantum mechanics, whereas the subordinate energy term (12) defines a widely generalized form of the interaction operator of a magnetic field $\boldsymbol{B} = \nabla \times \boldsymbol{A}$ and an orbital particle motion known from conventional quantum mechanics, and whereas the subordinate energy term (13) defines a widely generalized form of the interaction operator of a magnetic field $\boldsymbol{B} = \nabla \times \boldsymbol{A}$ and a particle spin known from conventional quantum mechanics. In the spirit of our self-consistent network of model conceptions, in all cases where $\chi_{00}(\boldsymbol{x},t)$ is not on a par with an eigenvalue $\chi_{00} = m_0 E/\hbar^2$, the energy terms (1), (12), and (13) define further generalizations of energy terms known from conventional quantum mechanics. Going beyond it, in the spirit of our self-consistent network of model conceptions, the energy terms (7) and (9) stand for self-interaction energy terms, i.e. the energy terms (7) and (9) describe the feedback of particle motions onto themselves, in fact, in a modality conventional quantum mechanics does not know. Going beyond it, in the spirit of our self-consistent network of model conceptions, the energy terms (3) and (4) incorporate extra vector fields $\boldsymbol{A}_{\ni}$, whereas the energy terms (5) and (6) incorporate extra tensor fields $\boldsymbol{\theta}_{\ni}$, in fact, in a modality conventional quantum mechanics does not know. Let us here direct the readers' attention to the fact that operators such as the operators of the energy terms (7) and (9), in the spirit of our self-consistent network of model conceptions, eventually build up a nonlinear quantum mechanics! Let us here direct the readers' attention to the fact that operators such as the operators of the energy terms (3) and (4) or (5) and (6), in the spirit of our self-consistent network of model conceptions, eventually build up a bridge to quantum field theory! In the spirit of our self-consistent network of model conceptions, however, much more important is that nonlinear energy terms such as the nonlinear energy terms (3) and (4), (5) and (6), and (7) and (9) lead to a nonlinear model equation exhibiting properties needed to handle material stability, material instability, and phase transitions on microscopic stages, for example, phase transitions separating the primordial soup from macromolecules, macromolecules from biomimetic materials, and normal phases/microstructures from superconducting phases/microstructures. Some details are presented next.

Continuation of the mathematical model.

$$
\begin{aligned}
\mathcal{KT}_{00}^{*} := \chi_{00}(\boldsymbol{x},t)\psi + F_{00}(\boldsymbol{x},t) = & -\frac{1}{2}(\nabla)^{\mathrm{T}}\nabla\psi & (1) \\
& -\frac{1}{2}\left(\boldsymbol{\theta}^{\ni}\nabla\right)^{\mathrm{T}}\nabla\psi & (2) \\
& +\frac{1}{c}\frac{\partial}{\partial t}(\nabla)^{\mathrm{T}}\boldsymbol{A}_{\ni} & (3) \\
& +\frac{1}{c}\frac{\partial}{\partial t}\left(\boldsymbol{\theta}^{\ni}\nabla\right)^{\mathrm{T}}\boldsymbol{A}_{\ni} & (4) \\
& -\frac{1}{2}\frac{1}{c^{2}}\left(1\frac{\partial^{2}}{\partial t^{2}}\right)\bullet(\boldsymbol{\theta}_{\ni}) & (5) \\
& -\frac{1}{2}\frac{1}{c^{2}}\left(\boldsymbol{\theta}^{\ni}\frac{\partial^{2}}{\partial t^{2}}\right)\bullet(\boldsymbol{\theta}_{\ni}) & (6) \\
& -\frac{1}{4}\left[(1-\bar{\psi})\nabla\psi\right]^{\mathrm{T}}\nabla\psi & (7) \\
& -\frac{1}{4}\left[(1-\bar{\psi})\nabla\psi\right]^{\mathrm{T}}\boldsymbol{\theta}^{\ni}\nabla\psi & (8) \\
& -\frac{1}{2}\left[(\boldsymbol{A}^{\ni}\cdot\boldsymbol{A}^{\ni})\nabla\psi\right]^{\mathrm{T}}\nabla\psi & (9) \\
& -\frac{1}{2}\left[(\nabla\cdot\boldsymbol{A}_{\ni})^{\mathrm{T}}\boldsymbol{A}^{\ni}\right]^{\mathrm{T}}\nabla\psi & (10) \\
& -\frac{1}{2}\left[(\nabla\cdot\boldsymbol{A}_{\ni})^{\mathrm{T}}\boldsymbol{A}^{\ni}\right]^{\mathrm{T}}\boldsymbol{\theta}^{\ni}\nabla\psi & (11) \\
& -\frac{1}{2}(\nabla\times\boldsymbol{A}_{\ni})^{\mathrm{T}}\boldsymbol{A}^{\ni}\times\nabla\psi & (12) \\
& -\frac{1}{2}(\nabla\times\boldsymbol{A}_{\ni})^{\mathrm{T}}\boldsymbol{A}^{\ni}\times\boldsymbol{\theta}^{\ni}\nabla\psi & (13) \\
& -\frac{1}{2}\left[(\nabla)^{\mathrm{T}}\boldsymbol{\theta}^{\ni}\right]\nabla\psi & (14) \\
& -\frac{1}{4}\left(\nabla\operatorname{Tr}\boldsymbol{\theta}_{\ni}\right)^{\mathrm{T}}\nabla\psi & (15) \\
& -\frac{1}{4}\left(\nabla\operatorname{Tr}\boldsymbol{\theta}_{\ni}\right)^{\mathrm{T}}\boldsymbol{\theta}^{\ni}\nabla\psi & (16) \\
& -\frac{1}{4}\left[(\boldsymbol{\theta}^{\ni})\bullet(\nabla\circ\boldsymbol{\theta}_{\ni})\right]^{\mathrm{T}}\nabla\psi & (17) \\
& -\frac{1}{4}\left[(\boldsymbol{\theta}^{\ni})\bullet(\nabla\circ\boldsymbol{\theta}_{\ni})\right]^{\mathrm{T}}\boldsymbol{\theta}^{\ni}\nabla\psi & (18)
\end{aligned}
$$

(5.42)

Continuation of the mathematical model.

$$+ \frac{1}{2}\Big[(\boldsymbol{A}^{\ni} \cdot \boldsymbol{A}^{\ni})(\partial \boldsymbol{A}_{\ni}/\partial ct)\Big]^{\mathrm{T}} \nabla \psi \quad (19)$$

$$+ \frac{1}{2}\Big[(\boldsymbol{A}^{\ni} \cdot \boldsymbol{A}^{\ni}) \bullet (\nabla \cdot \boldsymbol{A}_{\ni})\Big] \partial \psi/\partial ct \quad (20)$$

$$- \frac{1}{2}\Big[(\boldsymbol{A}^{\ni} \cdot \boldsymbol{A}^{\ni}) \bullet (\partial \boldsymbol{\theta}_{\ni}/\partial ct)\Big] \partial \psi/\partial ct \quad (21)$$

$$- \frac{1}{4}\Big[\boldsymbol{A}^{\ni\mathrm{T}}(\partial \psi/\partial ct)\Big] \nabla \psi \quad (22)$$

$$+ \frac{1}{2}\Big[\boldsymbol{A}^{\ni\mathrm{T}}(\partial \boldsymbol{A}_{\ni}/\partial ct)\Big] \partial \psi/\partial ct \quad (23)$$

$$+ \frac{1}{2}\Big[(1-\bar{\psi})\partial \boldsymbol{A}_{\ni}/\partial ct\Big]^{\mathrm{T}} \nabla \psi \quad (24)$$

$$+ \frac{1}{2}\Big[(1-\bar{\psi})\partial \boldsymbol{A}_{\ni}/\partial ct\Big]^{\mathrm{T}} \boldsymbol{\theta}^{\ni} \nabla \psi \quad (25)$$

$$- \frac{1}{4}\Big[(1-\bar{\psi})\partial \boldsymbol{\theta}_{\ni}/\partial ct\Big] \bullet (1 \partial \psi/\partial ct) \quad (26)$$

$$- \frac{1}{4}\Big[(1-\bar{\psi})\partial \boldsymbol{\theta}_{\ni}/\partial ct\Big] \bullet (\boldsymbol{\theta}^{\ni} \partial \psi/\partial ct) \quad (27)$$

$$+ \frac{1}{4}\Big[\ (1) \bullet (\nabla \circ \boldsymbol{\theta}_{\ni})\Big]^{\mathrm{T}} \boldsymbol{A}^{\ni} \partial \psi/\partial ct \quad (28)$$

$$+ \frac{1}{4}\Big[(\boldsymbol{\theta}^{\ni}) \bullet (\nabla \circ \boldsymbol{\theta}_{\ni})\Big]^{\mathrm{T}} \boldsymbol{A}^{\ni} \partial \psi/\partial ct \quad (29)$$

$$- \frac{1}{4}\Big[\ (1) \bullet (\partial \boldsymbol{\theta}_{\ni}/\partial ct)\Big] \boldsymbol{A}^{\ni\mathrm{T}} \nabla \psi \quad (30)$$

$$- \frac{1}{4}\Big[\ (\boldsymbol{\theta}^{\ni}) \bullet (\partial \boldsymbol{\theta}_{\ni}/\partial ct)\Big] \boldsymbol{A}^{\ni\mathrm{T}} \nabla \psi \quad (31)$$

$$- \frac{1}{2}\Big[\partial \boldsymbol{A}^{\ni}/\partial ct\Big]^{\mathrm{T}} \nabla \psi \quad (32)$$

$$+ \frac{1}{2}\Big[(\nabla)^{\mathrm{T}} \boldsymbol{A}^{\ni}\Big]^{\mathrm{T}} \partial \psi/\partial ct \quad (33)$$

$$+ \cdots .$$

Continuation of the mathematical model.

The basis of the the nonlinear model equation (5.42) is composed of a sophisticated set of tensors of different rank, namely the scalar functions $\gamma_{00} = \psi$ and $\gamma^{00} = \bar{\psi}$, the vector $\boldsymbol{A}_{\ni}$ with matrix elements γ_{i0} and the vector $\boldsymbol{A}^{\ni}$ with matrix elements γ^{i0}, and the tensor $\boldsymbol{\theta}_{\ni}$ with matrix elements and the tensor γ_{ij}, $\boldsymbol{\theta}^{\ni}$ with matrix elements γ^{ij}, the covariant character of which becomes manifest in the lower indices 0 and $i, j = 1, 2, 3$ and the upper indices 0 and $i, j = 1, 2, 3$. Within the nonlinear model equation (5.42), this sophisticated set of tensors of different rank, on the one hand, is processed via the nabla operator $\nabla = (\partial/\partial x^1, \partial/\partial x^2, \partial/\partial x^3)$, where $x^1 = x, x^2 = y, x^3 = z$, and on the other hand, is processed via the time derivative $\partial/\partial x^0$, where $x^0 = ct$. Within the nonlinear model equation (5.42), applying the mode decomposition (5.43), the nonlinear model equation (5.42) resolves into the mode equations (5.44) specified by the matrix elements (5.45), provided we multiply the resulting nonlinear model equation with $\psi_n^*(\boldsymbol{x})$, subsequently carrying out the integration $\int dV$ with respect to a suitable space volume V.

$$\psi(\boldsymbol{x}, t) = \sum_{k=1}^{\infty} \Xi_k(t)\psi_k(\boldsymbol{x}), \qquad \bar{\psi}(\boldsymbol{x}, t) = \sum_{k=1}^{\infty} \bar{\Xi}_k(t)\bar{\psi}_k(\boldsymbol{x}),$$

$$\gamma_{o0}(\boldsymbol{x}, t) = \sum_{k=1}^{\infty} \Xi_{o0k}(t)\gamma_{o0k}(\boldsymbol{x}), \quad \gamma^{o0}(\boldsymbol{x}, t) = \sum_{k=1}^{\infty} \Xi_k^{o0}(t)\gamma_k^{o0}(\boldsymbol{x}), \qquad (5.43)$$

$$\gamma_{oo'}(\boldsymbol{x}, t) = \sum_{k=1}^{\infty} \Xi_{oo'k}(t)\gamma_{oo'k}(\boldsymbol{x}), \quad \gamma^{oo'}(\boldsymbol{x}, t) = \sum_{k=1}^{\infty} \Xi_k^{oo'}(t)\gamma_k^{oo'}(\boldsymbol{x}).$$

As it is shown in Figure 5.6, speaking of "particle ghosts" or "interaction field ghosts", we assume that the space functions $\psi_k, \bar{\psi}_k, \gamma_{o0k}, \gamma_k^{o0}, \gamma_{oo'k}, \gamma_k^{oo'}$, already indicating those structures (particles) that after the decay of structures (particles) can emerge, specify the notion of sub-structures (sub-particles) of structures (particles). As it is shown in Figure 5.6, we assume that this includes microstructures, crystals, molecules, atoms, nucleons, quarks etc. Naturally, we assume that the interior properties just like the exterior properties of sub-structures (sub-particles) are specified by the space functions $\psi_k, \bar{\psi}_k, \gamma_{o0k}, \gamma_k^{o0}, \gamma_{oo'k}, \gamma_k^{oo'}$, including the translational properties, the spin properties, and the interaction properties of sub-structures (sub-particles). Naturally, we assume that the decay of structures (particles) can be modeled as nonlinear phase transition treatable by synergetic methods like the method of order parameters and slaved modes. Naturally, we assume that the development of structures (particles) and the alteration of structures (particles), for example, the development of macromolecules out of the primordial soup or nonlinear phase transitions leading from one structural modification of a ceramic material to another structural modification of a ceramic material, can be adequately modeled by such mode concepts [63].

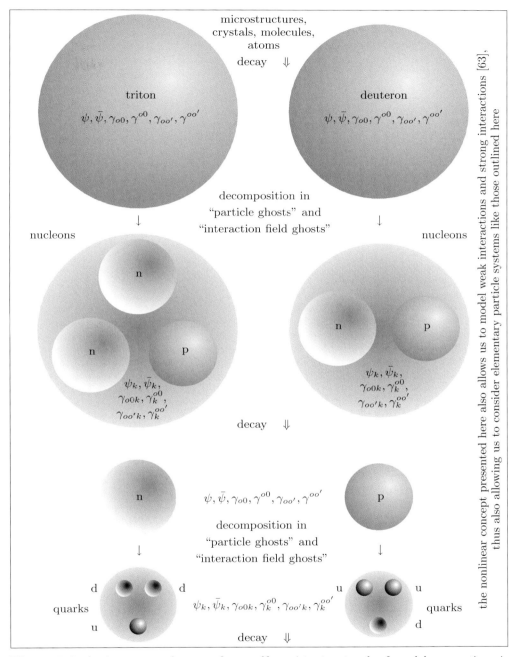

Figure 5.6. An important element of our self-consistent network of model conceptions is that the decomposability of space time functions $\psi, \bar{\psi}, \gamma_{o0}, \gamma^{o0}, \gamma_{oo'}, \gamma^{oo'}$ in space functions $\psi_k, \bar{\psi}_k, \gamma_{o0k}, \gamma_k^{o0}, \gamma_{oo'k}, \gamma_k^{oo'}$ and time-dependent modes $\Xi_k, \bar{\Xi}_k, \Xi_{o0k}, \Xi_k^{o0}, \Xi_{oo'k}, \Xi_k^{oo'}$ reflects the existence of sub-structures (sub-particles) of structures (particles).

Continuation of the mathematical model.

$$\sum_{j=1}^{\infty} \Xi_j \int_V \psi_n^* \chi_{00}(\boldsymbol{x},t) \psi_j \, \mathrm{d}V + \int_V \psi_n^* F_{00}(\boldsymbol{x},t) \, \mathrm{d}V =$$

$$= + \sum_{j=1}^{\infty} \Xi_j T_{nj}^{(1)}$$

$$+ \sum_{j,k=1}^{\infty} \sum_{o,o'=1}^{3} \Xi_j^{oo'} \Xi_k T_{njkoo'}^{(2)}$$

$$+ \sum_{j=1}^{\infty} \sum_{o=1}^{3} \frac{\partial}{\partial t} \Xi_{o0j} F_{njo}^{(3)}$$

$$+ \sum_{j=1}^{\infty} \sum_{o,o'=1}^{3} \frac{\partial^2}{\partial t^2} \Xi_{oo'j} F_{njoo'}^{(5)}$$

$$+ \sum_{j,k,l=1}^{\infty} \sum_{o,o'=1}^{3} \Xi_j^{o0} \Xi_{o'0k} \Xi_l W_{njkloo'}^{(10)} \qquad (5.44)$$

$$+ \sum_{j,k,l,m=1}^{\infty} \sum_{o,o',o''=1}^{3} \Xi_j^{o0} \Xi_k^{o'o''} \Xi_{o'0l} \Xi_m W_{njklmoo'o''}^{(11)}$$

$$+ \sum_{j,k,l=1}^{\infty} \sum_{o,o'=1}^{3} \Xi_j^{o'0} \Xi_{o0k} \Xi_l W_{njkloo'}^{(12,1)}$$

$$+ \sum_{j,k,l=1}^{\infty} \sum_{o,o'=1}^{3} \Xi_j^{o'0} \Xi_{o'0k} \Xi_l W_{njkloo'}^{(12,2)}$$

$$+ \sum_{j,k,l,m=1}^{\infty} \sum_{o,o',o''=1}^{3} \Xi_j^{o'0} \Xi_k^{oo''} \Xi_{o0l} \Xi_m W_{njklmoo'o''}^{(13,1)}$$

$$+ \sum_{j,k,l,m=1}^{\infty} \sum_{o,o',o''=1}^{3} \Xi_j^{o'0} \Xi_k^{oo''} \Xi_{o'0l} \Xi_m W_{njklmoo'o''}^{(13,2)}$$

Continuation of the mathematical model.

$$+ \sum_{j,k=1}^{\infty} \Xi_j \Xi_k S_{njk}^{(7,1)}$$

$$+ \sum_{j,k,l=1}^{\infty} \bar{\Xi}_j \Xi_k \Xi_l S_{njkl}^{(7,2)}$$

$$+ \sum_{j,k,l=1}^{\infty} \sum_{o,o'=1}^{3} \Xi_j^{oo'} \Xi_k \Xi_l S_{njkloo'}^{(8,1)}$$

$$+ \sum_{j,k,l,m=1}^{\infty} \sum_{o,o'=1}^{3} \Xi_j^{oo'} \bar{\Xi}_k \Xi_l \Xi_m S_{njklmoo'}^{(8,2)}$$

$$+ \sum_{j,k,l,m=1}^{\infty} \sum_{o,o'=1}^{3} \Xi_j^{o0} \Xi_k^{o'0} \Xi_l \Xi_m S_{njklmoo'}^{(9)}$$

$$+ \sum_{j,k=1}^{\infty} \sum_{o,o'=1}^{3} \Xi_j^{oo'} \Xi_k \theta_{njkoo'}^{(14)}$$

$$+ \sum_{j,k=1}^{\infty} \sum_{o,o'=1}^{3} \Xi_{o'o'j} \Xi_k \theta_{njkoo'}^{(15)}$$

$$+ \sum_{j,k,l=1}^{\infty} \sum_{o,o',o''=1}^{3} \Xi_j^{oo''} \Xi_{o'o'k} \Xi_l \theta_{njkloo'o''}^{(16)}$$

$$+ \sum_{j,k,l=1}^{\infty} \sum_{o,o',o''=1}^{3} \Xi_j^{oo'} \Xi_{oo'k} \Xi_l \theta_{njkloo'o''}^{(17)}$$

$$+ \sum_{j,k,l,m=1}^{\infty} \sum_{o,o',o'',o'''=1}^{3} \Xi_j^{oo'} \Xi_k^{o''o'''} \Xi_{oo'l} \Xi_m \theta_{njklmoo'o''o'''}^{(18)}$$

$$+ \cdots ,$$

Continuation of the mathematical model.

$$T_{nj}^{(1)} = \int_V \psi_n^* \left(-\frac{1}{2}\triangle\psi_j\right) dV,$$

$$T_{njkoo'}^{(2)} = \int_V \psi_n^* \left(-\frac{1}{2}\gamma_j^{oo'}\frac{\partial}{\partial x^o}\frac{\partial}{\partial x^{o'}}\psi_k\right) dV,$$

$$F_{njo}^{(3)} = \int_V \psi_n^* \left(+\frac{1}{c}\frac{\partial}{\partial x^o}\gamma_{o0j}\right) dV,$$

$$F_{njoo'}^{(5)} = \int_V \psi_n^* \left(-\frac{1}{2}\frac{1}{c^2}\gamma_{oo'j}\right) dV,$$

$$W_{njkloo'}^{(10)} = \int_V \psi_n^* \left(-\frac{1}{2}\gamma_j^{o0}\frac{\partial}{\partial x^o}\gamma_{o'0k}\frac{\partial}{\partial x^{o'}}\psi_l\right) dV,$$

(5.45)

$$W_{njklmoo'o''}^{(11)} = \int_V \psi_n^* \left(-\frac{1}{2}\gamma_j^{o0}\gamma_k^{o'o''}\frac{\partial}{\partial x^o}\gamma_{o'0l}\frac{\partial}{\partial x^{o''}}\psi_m\right) dV,$$

$$W_{njkloo'}^{(12,1)} = \int_V \psi_n^* \left(-\frac{1}{2}\gamma_j^{o0}\frac{\partial}{\partial x^o}\gamma_{o'0k}\frac{\partial}{\partial x^{o'}}\psi_l\right) dV,$$

$$W_{njkloo'}^{(12,2)} = \int_V \psi_n^* \left(+\frac{1}{2}\gamma_j^{o'0}\frac{\partial}{\partial x^o}\gamma_{o'0k}\frac{\partial}{\partial x^o}\psi_l\right) dV,$$

$$W_{njklmoo'o''}^{(13,1)} = \int_V \psi_n^* \left(-\frac{1}{2}\gamma_j^{o'0}\gamma_k^{oo''}\frac{\partial}{\partial x^{o'}}\gamma_{o0l}\frac{\partial}{\partial x^{o''}}\psi_m\right) dV,$$

$$W_{njklmoo'o''}^{(13,2)} = \int_V \psi_n^* \left(+\frac{1}{2}\gamma_j^{o'0}\gamma_k^{oo''}\frac{\partial}{\partial x^o}\gamma_{o'0l}\frac{\partial}{\partial x^{o''}}\psi_m\right) dV,$$

Continuation of the mathematical model.

$$S^{(7,1)}_{njk} = \int_V \psi_n^* \left(-\frac{1}{4} (\nabla \psi_j)^{\mathrm{T}} \nabla \psi_k \right) \mathrm{d}V ,$$

$$S^{(7,2)}_{njkl} = \int_V \psi_n^* \left(+\frac{1}{4} \bar{\psi}_j (\nabla \psi_k)^{\mathrm{T}} \nabla \psi_l \right) \mathrm{d}V ,$$

$$S^{(8,1)}_{njkloo'} = \int_V \psi_n^* \left(-\frac{1}{4} \gamma_j^{oo'} \frac{\partial}{\partial x^o} \psi_k \frac{\partial}{\partial x^{o'}} \psi_l \right) \mathrm{d}V ,$$

$$S^{(8,2)}_{njklmoo'} = \int_V \psi_n^* \left(+\frac{1}{4} \gamma_j^{oo'} \bar{\psi}_k \frac{\partial}{\partial x^o} \psi_l \frac{\partial}{\partial x^{o'}} \psi_m \right) \mathrm{d}V ,$$

$$S^{(9)}_{njklmoo'} = \int_V \psi_n^* \left(-\frac{1}{2} \gamma_j^{o0} \gamma_k^{o'0} \frac{\partial}{\partial x^o} \psi_l \frac{\partial}{\partial x^{o'}} \psi_m \right) \mathrm{d}V ,$$

$$\theta^{(14)}_{njkoo'} = \int_V \psi_n^* \left(-\frac{1}{2} \frac{\partial}{\partial x^o} \gamma_j^{oo'} \frac{\partial}{\partial x^{o'}} \psi_k \right) \mathrm{d}V ,$$

$$\theta^{(15)}_{njkoo'} = \int_V \psi_n^* \left(-\frac{1}{4} \frac{\partial}{\partial x^o} \gamma_{o'o'j} \frac{\partial}{\partial x^o} \psi_k \right) \mathrm{d}V ,$$

$$\theta^{(16)}_{njkloo'o''} = \int_V \psi_n^* \left(-\frac{1}{4} \gamma_j^{oo''} \frac{\partial}{\partial x^o} \gamma_{o'o'k} \frac{\partial}{\partial x^{o''}} \psi_l \right) \mathrm{d}V ,$$

$$\theta^{(17)}_{njkloo'o''} = \int_V \psi_n^* \left(-\frac{1}{4} \gamma_j^{oo'} \frac{\partial}{\partial x^{o''}} \gamma_{oo'k} \frac{\partial}{\partial x^{o''}} \psi_l \right) \mathrm{d}V ,$$

$$\theta^{(18)}_{njklmoo'o''o'''} = \int_V \psi_n^* \left(-\frac{1}{4} \gamma_j^{oo'} \gamma_k^{o''o'''} \frac{\partial}{\partial x^{o''}} \gamma_{oo'l} \frac{\partial}{\partial x^{o'''}} \psi_m \right) \mathrm{d}V ,$$

. . . .

Example 5.1 (A special case of (5.42)).

$$-\frac{1}{2}(\nabla)^{\mathrm{T}}\nabla\psi + \frac{1}{c}\frac{\partial}{\partial t}(\nabla)^{\mathrm{T}}\boldsymbol{A}_{\ni}-$$
$$-\frac{1}{2}\left[(\nabla \cdot \boldsymbol{A}_{\ni})^{\mathrm{T}}\boldsymbol{A}^{\ni}\right]^{\mathrm{T}}\nabla\psi - \frac{1}{2}(\nabla \times \boldsymbol{A}_{\ni})^{\mathrm{T}}\boldsymbol{A}^{\ni} \times \nabla\psi-\quad(5.46)$$
$$-\frac{1}{2}\left[(\boldsymbol{A}^{\ni} \cdot \boldsymbol{A}^{\ni})\nabla\psi\right]^{\mathrm{T}}\nabla\psi = \chi_{00}(\boldsymbol{x},t)\psi + F_{00}(\boldsymbol{x},t)$$

$$\Downarrow$$

$$-\frac{1}{2}\frac{\partial^2}{\partial x^2}\psi + \frac{1}{c}\frac{\partial}{\partial t}\frac{\partial}{\partial x}\gamma_{10}-$$
$$-\frac{1}{2}\gamma^{10}\frac{\partial}{\partial x}\gamma_{10}\frac{\partial}{\partial x}\psi + \frac{1}{2}\gamma^{20}\frac{\partial}{\partial x}\gamma_{20}\frac{\partial}{\partial x}\psi + \frac{1}{2}\gamma^{30}\frac{\partial}{\partial x}\gamma_{30}\frac{\partial}{\partial x}\psi-\quad(5.47)$$
$$-\frac{1}{2}\gamma^{10}\gamma^{10}\frac{\partial}{\partial x}\psi\frac{\partial}{\partial x}\psi = \chi_{00}(x,t)\psi + F_{00}(x,t)$$

$$\Downarrow$$

$$\psi(\boldsymbol{x},t) = \sum_{j=1}^{\infty}\Xi_j(t)\psi_j(\boldsymbol{x}),$$
$$\gamma_{i0}(\boldsymbol{x},t) = \sum_{j=1}^{\infty}\Xi_{i0j}(t)\gamma_{i0j}(\boldsymbol{x}),\ \gamma^{i0}(\boldsymbol{x},t) = \sum_{j=1}^{\infty}\Xi_j^{i0}(t)\gamma_j^{i0}(\boldsymbol{x})\quad(5.48)$$

$$\Downarrow$$

$$-\frac{1}{2}\sum_{j=1}^{\infty}\Xi_j\frac{\partial^2}{\partial x^2}\psi_j + \frac{1}{c}\sum_{j=1}^{\infty}\frac{\partial}{\partial t}\Xi_{10j}\frac{\partial}{\partial x}\gamma_{10j}-$$
$$-\frac{1}{2}\sum_{j,k,l=1}^{\infty}\Xi_j^{10}\Xi_{10k}\Xi_l\,\gamma_j^{10}\frac{\partial}{\partial x}\gamma_{10k}\frac{\partial}{\partial x}\psi_l + \frac{1}{2}\sum_{j,k,l=1}^{\infty}\Xi_j^{20}\Xi_{20k}\Xi_l\,\gamma_j^{20}\frac{\partial}{\partial x}\gamma_{20k}\frac{\partial}{\partial x}\psi_l+$$
$$+\frac{1}{2}\sum_{j,k,l=1}^{\infty}\Xi_j^{30}\Xi_{30k}\Xi_l\,\gamma_j^{30}\frac{\partial}{\partial x}\gamma_{30k}\frac{\partial}{\partial x}\psi_l-\quad(5.49)$$
$$-\frac{1}{2}\sum_{j,k,l,m=1}^{\infty}\Xi_j^{10}\Xi_k^{10}\Xi_l\Xi_m\,\gamma_j^{10}\gamma_k^{10}\frac{\partial}{\partial x}\psi_l\frac{\partial}{\partial x}\psi_m = \chi_{00}(x,t)\sum_{j=1}^{\infty}\Xi_j\psi_j + F_{00}(x,t)$$

Continuation of the example.

$$\Downarrow$$

$$\sum_{j=1}^{\infty} \Xi_j \int_V \psi_n^* \chi_{00}(x,t)\,\psi_j \,\mathrm{d}V + \int_V \psi_n^* F_{00}(x,t)\,\mathrm{d}V =$$

$$= \sum_{j=1}^{\infty} \Xi_j\, T_{nj} + \sum_{j=1}^{\infty} \frac{\partial}{\partial t}\Xi_{10j}\, F_{nj} +$$

$$+ \sum_{j,k,l=1}^{\infty} \Xi_j^{10}\, \Xi_{10k}\, \Xi_l\, M_{njkl}^{(1)} + \sum_{j,k,l=1}^{\infty} \Xi_j^{20}\, \Xi_{20k}\, \Xi_l\, M_{njkl}^{(2)} + \sum_{j,k,l=1}^{\infty} \Xi_j^{30}\, \Xi_{30k}\, \Xi_l\, M_{njkl}^{(3)} +$$

$$+ \sum_{j,k,l,m=1}^{\infty} \Xi_j^{10}\, \Xi_k^{10}\, \Xi_l\, \Xi_m\, S_{njklm}$$

(5.50)

with

$$T_{nj} = \int_V \psi_n^* \left(-\frac{1}{2}\frac{\partial^2}{\partial x^2}\psi_j\right) \mathrm{d}V\,,$$

$$F_{nj} = \int_V \psi_n^* \left(+\frac{1}{c}\frac{\partial}{\partial x}\gamma_{10j}\right) \mathrm{d}V\,,$$

$$M_{njkl}^{(1)} = \int_V \psi_n^* \left(-\frac{1}{2}\gamma_j^{10}\frac{\partial}{\partial x}\gamma_{10k}\frac{\partial}{\partial x}\psi_l\right) \mathrm{d}V\,,$$

(5.51)

$$M_{njkl}^{(2)} = \int_V \psi_n^* \left(+\frac{1}{2}\gamma_j^{20}\frac{\partial}{\partial x}\gamma_{20k}\frac{\partial}{\partial x}\psi_l\right) \mathrm{d}V\,,$$

$$M_{njkl}^{(3)} = \int_V \psi_n^* \left(+\frac{1}{2}\gamma_j^{30}\frac{\partial}{\partial x}\gamma_{30k}\frac{\partial}{\partial x}\psi_l\right) \mathrm{d}V\,,$$

$$S_{njklm} = \int_V \psi_n^* \left(-\frac{1}{2}\gamma_j^{10}\gamma_k^{10}\frac{\partial}{\partial x}\psi_l\frac{\partial}{\partial x}\psi_m\right) \mathrm{d}V\,.$$

456   5. Future Materials

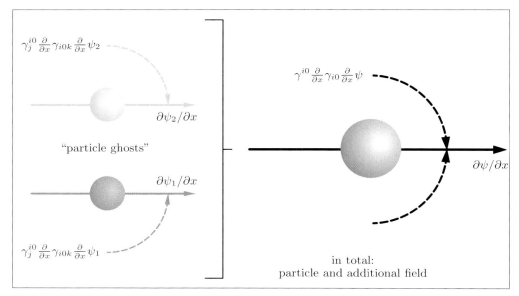

Figure 5.7. A special interaction of the "particle ghosts" $\psi_1(\boldsymbol{x})$ and $\psi_2(\boldsymbol{x})$: interaction with additional fields $\gamma_{iCk}(\boldsymbol{x})$. We compare with the M terms of (5.51).

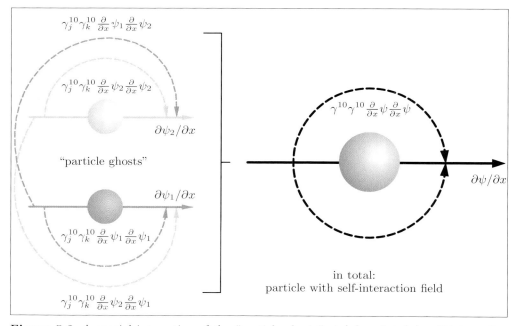

Figure 5.8. A special interaction of the "particle ghosts" $\psi_1(\boldsymbol{x})$ and $\psi_2(\boldsymbol{x})$: self-interaction and cross-interaction. We compare with the S term of (5.51).

5.1 Advanced Conceptions 457

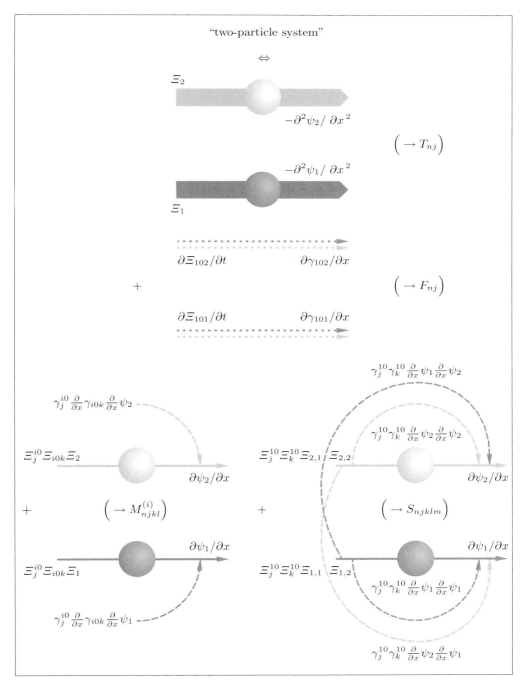

Figure 5.9. The energy aspects that are at the bottom of the "two-particle system": kinetic energy aspects (upper pictograms) and interaction energy aspects (lower pictograms).

Continuation of the mathematical model.

As a simple example, let us consider a special case of (5.42), namely (5.46), in the one-dimensional case reducing to (5.47), after application of (5.48) leading to (5.49), and after carrying out the integration $\int dV$ with respect to a suitable space volume V leading to (5.50). As it is outlined in (5.52), for the sake of simplicity, restricting ourselves to two "particle ghosts" ψ_1 and ψ_2 reflecting two sub-particles 1 and 2, the sum of the two mode equations that remain in the case of the "two-particle system" counts the mean energy (the energy expectation value) of the "two-particle system" within the surrounding of additional fields, while the two summands $\langle\psi_1\rangle_2$ and $\langle\psi_2\rangle_1$ count the mean energies (the energy expectation values) of a "sub-particle 1" within the surrounding of additional fields influenced by "sub-particle 2" and of a "sub-particle 2" within the surrounding of additional fields influenced by "sub-particle 1". Let us here have a brief look at Figures 5.7–5.9, which illustrate the energy aspects that are at the bottom of the "two-particle system". Following Figure 5.9, the first term of (5.51) counts the kinetic energy aspects that are associated with ψ_1 and ψ_2, in fact, in form of parameter-free expectation values T_{nj} the contribution of which is given by the amplitudes Ξ_j. Following Figure 5.9, the second term of (5.51), in close combination with $\partial \Xi_j/\partial t$, counts the kinetic energy aspects that are associated with the γ_{10j}, in fact, in form of parameter-free correlation functions F_{nj}, in close combination with $\partial \Xi_j/\partial t$, evoking wave vectors k_j via $\partial \gamma_{10j}/\partial x$ and wave vectors ω_j/c via $\partial \Xi_{10j}/\partial t$, defining the kinetic energy aspects that are associated with the γ_{10j}, on the one hand, evoking momentum features $p_j = \hbar k_j$, and on the other hand, evoking energy features $E_j = \hbar\omega_j$. Following Figure 5.9, the M terms of (5.51) and the S term of (5.51) count additional interaction energy aspects. On the one hand, the M terms of (5.51) are the result of the interrelation between ψ_1 and additional fields γ_{i0k} and of the interrelation between ψ_2 and additional fields γ_{i0k}. On the other hand, the S term of (5.51) is the result of self-interactions ψ_1–ψ_1 and ψ_2–ψ_2 and of cross-interactions ψ_1–ψ_2. We here point out that in the M case the carriers of interaction are the γ_j^{i0} fields, while in the S case the carriers of interaction are the γ_j^{10} fields. We here point out that in the M case the amplitudes Ξ_j^{i0}, Ξ_{i0k}, and Ξ_l weight the diverse field components, while in the S case the amplitudes Ξ_j^{10} and Ξ_l weight the diverse field components.

$$\int_V \psi^* \chi_{00}(x,t)\psi \, dV + \int_V \psi^* F_{00}(x,t) \, dV = \Big\langle \psi_1 \Big\rangle_2 + \Big\langle \psi_2 \Big\rangle_1 , \qquad (5.52)$$

$$\begin{aligned}\Big\langle \psi_1 \Big\rangle_2 &= \Xi_1 \Big\langle \psi_1 \Big| \chi_{00}(x,t) \Big| \psi_1 \Big\rangle + \Xi_2 \Big\langle \psi_1 \Big| \chi_{00}(x,t) \Big| \psi_2 \Big\rangle + \Big\langle \psi_1 \Big| F_{00}(x,t) \Big\rangle , \\ \Big\langle \psi_2 \Big\rangle_1 &= \Xi_1 \Big\langle \psi_2 \Big| \chi_{00}(x,t) \Big| \psi_1 \Big\rangle + \Xi_2 \Big\langle \psi_2 \Big| \chi_{00}(x,t) \Big| \psi_2 \Big\rangle + \Big\langle \psi_2 \Big| F_{00}(x,t) \Big\rangle ,\end{aligned} \qquad (5.53)$$

$$\begin{aligned}\Big\langle \psi_n \Big| \chi_{00}(x,t) \Big| \psi_j \Big\rangle &= \int_V \psi_n^* \chi_{00}(x,t)\psi_j \, dV , \\ \Big\langle \psi_n \Big| F_{00}(x,t) \Big\rangle &= \int_V \psi_n^* F_{00}(x,t) \, dV .\end{aligned} \qquad (5.54)$$

5.1.3 Advanced Modeling of Threshold Ranges

The partial differential equation (5.42) can be gained as non-relativistic limiting case of the energy contribution of generally relativistic energy momentum balance equations, in the sections that follow, denoted as *generalized Einstein field equations* [63].

A mathematical model: the generalized Einstein field equations.

In the following discussion, the nonlinear model equations defined by (5.55) are termed *generalized Einstein field equations*. We note that R is the scalar of curvature, $R_{\mu\nu}$ is the Ricci tensor, and the dots outline contributions that expand its buildup to infinity. We note that $\mathcal{K}$ can be any constant and $\mathcal{T}_{\mu\nu}$ can be any energy momentum tensor. We note that $R_{\mu\nu}$ is a function of the metric tensor $g_{\mu\nu}$, which in turn is a function of the generalized space–time coordinates q^μ ($\mu = 1, 2, 3, 0$). It should be clear that setting $\mathcal{K}$ equal to Einstein's constant of gravitation K, restricting ourselves to the domain of mass-related energy momentum tensors $\mathcal{T}_{\mu\nu}$, and restricting ourselves to the domain of the Ricci tensor $R_{\mu\nu}$, Einstein's field equations of gravitation result. It should be clear that other constants $\mathcal{K}$ and other energy momentum tensors $\mathcal{T}_{\mu\nu}$ lead to field equations that are not identical to Einstein's field equations of gravitation. It should be clear that such extensions take for granted the validity of the generalized metric field model that is launched at the end of this section and that interweaves the notions *mass*, *charge*, and *metric* to a unity [63].

$$-R_{\mu\nu} + \cdots = \mathcal{K}\mathcal{T}_{\mu\nu} - \frac{R}{2}g_{\mu\nu}. \tag{5.55}$$

Focussing on the energy contribution of the generalized Einstein field equations, in the general case we derive the first equation of (5.56) and in the stationary limiting case we derive the second equation of (5.56). Introducing basic tensors, basic operators, and further quantities according to Boxes 5.4–5.8, the individual terms $\mathcal{W}_{00}^{(a)}$ and $\mathcal{S}_{00}^{(a)}$, respectively, are readily expressed as (5.90)–(5.128) and (5.79)–(5.88), respectively. It should be pointed out that the metric tensor decomposition $g_{\mu\nu} = \eta_{\mu\nu} + \gamma_{\mu\nu}$ that leads from (5.55) to (5.56) consists of the pseudo-Euclidean metric tensor $\eta_{\mu\nu}$ with $\eta_{11} = \eta_{22} = \eta_{33} = 1, \eta_{00} = -1$, and $\eta_{\mu\neq\nu} = 0$ and of the deviation tensor $\gamma_{\mu\nu}$ which is not restricted in any way since we do not prohibit to set $\gamma_{\mu\nu} = -\eta_{\mu\nu} + g_{\mu\nu}$. Therefore, this is not an "Ansatz" in the perturbation-theoretical sense. In fact, this is an "Ansatz" that via the term $\eta_{\mu\nu}$ paves the way to non-curved reference frames and via the term $\gamma_{\mu\nu}$ paves the way to potentials, finally paving the way from Einstein's curvilinear world to Newton's and Schrödinger's rectilinear world [63].

$$+\mathcal{W}_{00} + \cdots = \mathcal{K}\mathcal{T}_{00} + \frac{R}{2} - \frac{R}{2}\gamma_{00}, \quad \mathcal{W}_{00} = \sum_{a=1}^{10} \mathcal{W}_{00}^{(a)},$$
$$+\mathcal{S}_{00} + \cdots = \mathcal{K}\mathcal{T}_{00} + \frac{R}{2} - \frac{R}{2}\gamma_{00}, \quad \mathcal{S}_{00} = \sum_{a=1}^{10} \mathcal{S}_{00}^{(a)}. \tag{5.56}$$

Continuation of the mathematical model.

Starting from the first equation of (5.56), the partial differential equation (5.42) is obtained as follows. On the one hand, we consider relatively small space–time domains, finally allowing us to replace generalized space–time coordinates q^μ ($\mu = 1, 2, 3, 0$) by Cartesian space–time coordinates x^μ ($\mu = 1, 2, 3, 0$) with $x^1 = x, x^2 = y, x^3 = z, x^0 = ct$, meeting the demands of Newton's and Schrödinger's rectilinear world. Observing that the generalized Nabla operator ∇_3, which is based upon the q^i ($i = 1, 2, 3$), then passes into the conventional Nabla operator ∇, which is based upon the x^i ($i = 1, 2, 3$), we especially realize that (5.91) leads to (1) + (2) of the partial differential equation (5.42) provided we set $\gamma_{00} = \psi$ and we especially realize that (5.112) leads to (12) + (13) of the partial differential equation (5.42) provided we set $\gamma_{00} = \psi$. Proceeding in this way, we rediscover the individual terms (k) of the partial differential equation (5.42), provided we additionally set $\gamma^{00} = \bar\psi$. On the other hand, we consider the relation (5.57), which according to the first part $\chi_{00}(\boldsymbol{x}, t)\psi$ implements the notion of microscopic systems that are characterized by wave properties ($\to \psi$) *and* particle properties ($\to \chi_{00}(\boldsymbol{x}, t)$) and according to the second part $F_{00}(\boldsymbol{x}, t)$ implements the notion of additional fields as, for example, given by gravitational fields or electromagnetic fields. Proceeding in this way, we rediscover the individual terms (k) of the partial differential equation (5.42), provided we additionally set $\gamma^{00} = \bar\psi$, and we rediscover the individual terms $\chi_{00}(\boldsymbol{x}, t)\psi$ and $F_{00}(\boldsymbol{x}, t)$ defining the entity of the partial differential equation (5.42).

$$\mathcal{K}\mathcal{T}_{00} + \frac{R}{2} - \frac{R}{2}\gamma_{00} := \chi_{00}(\boldsymbol{x}, t)\gamma_{00} + F_{00}(\boldsymbol{x}, t) := \chi_{00}(\boldsymbol{x}, t)\psi + F_{00}(\boldsymbol{x}, t) \ . \qquad (5.57)$$

Beyond that, adjusting the constant $\mathcal{K}$ and the energy momentum tensor $\mathcal{T}_{\mu\nu}$ covariantly to microscopic systems exhibiting wave properties and particle properties coexistently and inseparably, starting from (5.55), model equations can be established, allowing us to bring a covariant formalism on its way that covers threshold ranges of particles and materials on microscopic stages such as the behavior of materials at velocities near the light velocity on microscopic stages and the behavior of materials in the presence of high energies on microscopic stages. In this context, we especially annotate that product terms such as $\chi_{\mu\nu}(\boldsymbol{x}, t)\gamma_{\mu\nu}$ terms certainly are a good choice if we want to study non-relativistic specifications of the generalized Einstein field equations on microscopic stages, but more general terms are a good choice if we want to study relativistic or even covariant, i. e. (regarding any frame transformations) forminvariant specifications of the generalized Einstein field equations on microscopic stages. In this context, we especially annotate that this general approach generates complete sets of energy momentum balance equations, configuring momentum balance equations that complete the energy balance equation (5.42), supplying us with enough equations for the determination of the "fundamental tensors" γ_{00} (ψ), γ^{00} ($\bar\psi$), $\boldsymbol{A}_\ni$, $\boldsymbol{A}^\ni$, $\boldsymbol{\theta}_\ni$, $\boldsymbol{\theta}^\ni$. In this context, we especially annotate that completely different specifications generate energy momentum balance equations working on mesoscopic or macroscopic stages. At all events, however, we may start from generalized Einstein field equations that resort to the $g_{\mu\nu}$ or to the $\eta_{\mu\nu}$ and $\gamma_{\mu\nu}$.

Box 5.4 (A mathematical model: generalized Einstein field equations, basic tensors).

$$\gamma_{00}, \quad \gamma^{00}, \tag{5.58}$$

$$\boldsymbol{A}_{\ni} = \begin{pmatrix} \gamma_{10} \\ \gamma_{20} \\ \gamma_{30} \end{pmatrix} = \begin{pmatrix} \gamma_{01} \\ \gamma_{02} \\ \gamma_{03} \end{pmatrix}, \quad \boldsymbol{A}_{\ni}^{\mathrm{T}} = (\gamma_{01}, \gamma_{02}, \gamma_{03}) = (\gamma_{10}, \gamma_{20}, \gamma_{30}),$$

$$\boldsymbol{A}^{\ni} = \begin{pmatrix} \gamma^{10} \\ \gamma^{20} \\ \gamma^{30} \end{pmatrix} = \begin{pmatrix} \gamma^{01} \\ \gamma^{02} \\ \gamma^{03} \end{pmatrix}, \quad \boldsymbol{A}^{\ni\,\mathrm{T}} = (\gamma^{01}, \gamma^{02}, \gamma^{03}) = (\gamma^{10}, \gamma^{20}, \gamma^{30}), \tag{5.59}$$

$$\boldsymbol{\theta}_{\ni} = \begin{pmatrix} \gamma_{11} & \gamma_{12} & \gamma_{13} \\ \gamma_{21} & \gamma_{22} & \gamma_{23} \\ \gamma_{31} & \gamma_{32} & \gamma_{33} \end{pmatrix}, \quad \boldsymbol{\theta}_{\ni}^{\mathrm{T}} = \begin{pmatrix} \gamma_{11} & \gamma_{21} & \gamma_{31} \\ \gamma_{12} & \gamma_{22} & \gamma_{32} \\ \gamma_{13} & \gamma_{23} & \gamma_{33} \end{pmatrix},$$

$$\boldsymbol{\theta}^{\ni} = \begin{pmatrix} \gamma^{11} & \gamma^{12} & \gamma^{13} \\ \gamma^{21} & \gamma^{22} & \gamma^{23} \\ \gamma^{31} & \gamma^{32} & \gamma^{33} \end{pmatrix}, \quad \boldsymbol{\theta}^{\ni\,\mathrm{T}} = \begin{pmatrix} \gamma^{11} & \gamma^{21} & \gamma^{31} \\ \gamma^{12} & \gamma^{22} & \gamma^{32} \\ \gamma^{13} & \gamma^{23} & \gamma^{33} \end{pmatrix}. \tag{5.60}$$

Box 5.5 (A mathematical model: generalized Einstein field equations, basic operators).

$$\nabla_3 = \begin{pmatrix} \partial/\partial q^1 \\ \partial/\partial q^2 \\ \partial/\partial q^3 \end{pmatrix}, \quad \nabla_0 = \frac{\partial}{\partial q^0}, \tag{5.61}$$

$$\nabla_3 Q = \begin{pmatrix} \partial Q/\partial q^1 \\ \partial Q/\partial q^2 \\ \partial Q/\partial q^3 \end{pmatrix}, \quad \nabla_3 \cdot \boldsymbol{Q} = \begin{pmatrix} \partial Q_1/\partial q^1 & \partial Q_2/\partial q^1 & \partial Q_3/\partial q^1 \\ \partial Q_1/\partial q^2 & \partial Q_2/\partial q^2 & \partial Q_3/\partial q^2 \\ \partial Q_1/\partial q^3 & \partial Q_2/\partial q^3 & \partial Q_3/\partial q^3 \end{pmatrix},$$

$$\nabla_3 \circ \mathsf{Q} = \begin{pmatrix} \partial \mathsf{Q}/\partial q^1 \\ \partial \mathsf{Q}/\partial q^2 \\ \partial \mathsf{Q}/\partial q^3 \end{pmatrix}, \quad \nabla_3 \times \boldsymbol{Q} = \begin{pmatrix} \partial Q_3/\partial q^2 - \partial Q_2/\partial q^3 \\ \partial Q_1/\partial q^3 - \partial Q_3/\partial q^1 \\ \partial Q_2/\partial q^1 - \partial Q_1/\partial q^2 \end{pmatrix}, \tag{5.62}$$

$$(\nabla_3)^{\mathsf{T}} \boldsymbol{Q} = \partial Q_1/\partial q^1 + \partial Q_2/\partial q^2 + \partial Q_3/\partial q^3,$$

$$\mathsf{Q} \bullet \mathsf{Q}' = \sum_{i,j=1}^{n} Q_{ij} Q'_{ij}. \tag{5.63}$$

Box 5.6 (A mathematical model: Laplace-type operators).

$$\triangle_{L,3,3} = (\nabla_3)^T \nabla_3 = \triangle_3 \,,$$

$$\triangle_{S,3,3} = (\theta^\ni \nabla_3)^T \nabla_3 = \triangle_3' \,,$$

$$\overline{\triangle}_{L,3,0} = (\nabla_3)^T \nabla_0 \,,$$

$$\overline{\triangle}_{S,3,0} = (\theta^\ni \nabla_3)^T \nabla_0 \,, \qquad (5.64)$$

$$\triangle_{L,0,0} = \nabla_0^2 \,,$$

$$\overline{\overline{\triangle}}_{L,0,0} = \mathbf{1} \nabla_0^2 \,,$$

$$\overline{\overline{\triangle}}_{S,0,0} = \theta^\ni \nabla_0^2 \,,$$

$$\triangle_{L,3,3} = \sum_{i=1}^{3} \frac{\partial^2}{\partial q^{i\,2}} \,,$$

$$\triangle_{S,3,3} = \sum_{i,j=1}^{3} \gamma^{ji} \frac{\partial}{\partial q^i} \frac{\partial}{\partial q^j} \,, \qquad (5.65)$$

$$\overline{\triangle}_{L,3,0} = \left(\frac{\partial}{\partial q^1} \frac{\partial}{\partial q^0}, \frac{\partial}{\partial q^2} \frac{\partial}{\partial q^0}, \frac{\partial}{\partial q^3} \frac{\partial}{\partial q^0} \right) \,,$$

$$\overline{\triangle}_{S,3,0} = \left(\sum_{i=1}^{3} \gamma^{1i} \frac{\partial}{\partial q^i} \frac{\partial}{\partial q^0}, \sum_{i=1}^{3} \gamma^{2i} \frac{\partial}{\partial q^i} \frac{\partial}{\partial q^0}, \sum_{i=1}^{3} \gamma^{3i} \frac{\partial}{\partial q^i} \frac{\partial}{\partial q^0} \right) \,, \qquad (5.66)$$

$$\triangle_{L,0,0} = \frac{\partial^2}{\partial q^{0\,2}} \,,$$

$$\overline{\overline{\triangle}}_{L,0,0} = \begin{pmatrix} 1 & 0 & 0 \\ 0 & 1 & 0 \\ 0 & 0 & 1 \end{pmatrix} \frac{\partial^2}{\partial q^{0\,2}} \,, \qquad (5.67)$$

$$\overline{\overline{\triangle}}_{S,0,0} = \begin{pmatrix} \gamma^{11} & \gamma^{12} & \gamma^{13} \\ \gamma^{21} & \gamma^{22} & \gamma^{23} \\ \gamma^{31} & \gamma^{32} & \gamma^{33} \end{pmatrix} \frac{\partial^2}{\partial q^{0\,2}} \,.$$

Box 5.7 (A mathematical model: momentum-type operators).

$$\hat{\mu}_{L,0} = \frac{1}{2}\nabla_0 \ ,$$

$$\hat{\boldsymbol{\mu}}_{L,0} = \frac{1}{2}\mathbf{1}\nabla_0 \ , \quad \hat{\boldsymbol{\mu}}_{S,0} = \frac{1}{2}\boldsymbol{\theta}^{\ni}\nabla_0 \ ,$$

$$\hat{\boldsymbol{\mu}}_{L,3} = \frac{1}{2}\nabla_3 \ , \quad \hat{\boldsymbol{\mu}}_{S,3} = \frac{1}{2}\boldsymbol{\theta}^{\ni}\nabla_3 \ , \qquad (5.68)$$

$$\hat{\boldsymbol{\mu}}'_{L,3} = \frac{1}{2}\left(\boldsymbol{A}^{\ni} \times \nabla_3\right) \ , \quad \hat{\boldsymbol{\mu}}'_{S,3} = \frac{1}{2}\left(\boldsymbol{A}^{\ni} \times \boldsymbol{\theta}^{\ni}\nabla_3\right) \ ,$$

$$\hat{\mu}_{L,0} = \frac{1}{2}\frac{\partial}{\partial q^0} \ ,$$

$$\hat{\boldsymbol{\mu}}_{L,0} = \frac{1}{2}\begin{pmatrix} 1 & 0 & 0 \\ 0 & 1 & 0 \\ 0 & 0 & 1 \end{pmatrix}\frac{\partial}{\partial q^0} \ , \quad \hat{\boldsymbol{\mu}}_{S,0} = \frac{1}{2}\begin{pmatrix} \gamma^{11} & \gamma^{12} & \gamma^{13} \\ \gamma^{21} & \gamma^{22} & \gamma^{23} \\ \gamma^{31} & \gamma^{32} & \gamma^{33} \end{pmatrix}\frac{\partial}{\partial q^0} \ ,$$

$$\hat{\boldsymbol{\mu}}_{L,3} = \begin{pmatrix} \dfrac{1}{2}\dfrac{\partial}{\partial q^1} \\[4pt] \dfrac{1}{2}\dfrac{\partial}{\partial q^2} \\[4pt] \dfrac{1}{2}\dfrac{\partial}{\partial q^3} \end{pmatrix} \ , \quad \hat{\boldsymbol{\mu}}_{S,3} = \begin{pmatrix} \dfrac{\gamma^{11}}{2}\dfrac{\partial}{\partial q^1} + \dfrac{\gamma^{12}}{2}\dfrac{\partial}{\partial q^2} + \dfrac{\gamma^{13}}{2}\dfrac{\partial}{\partial q^3} \\[4pt] \dfrac{\gamma^{21}}{2}\dfrac{\partial}{\partial q^1} + \dfrac{\gamma^{22}}{2}\dfrac{\partial}{\partial q^2} + \dfrac{\gamma^{23}}{2}\dfrac{\partial}{\partial q^3} \\[4pt] \dfrac{\gamma^{31}}{2}\dfrac{\partial}{\partial q^1} + \dfrac{\gamma^{32}}{2}\dfrac{\partial}{\partial q^2} + \dfrac{\gamma^{33}}{2}\dfrac{\partial}{\partial q^3} \end{pmatrix} \ , \qquad (5.69)$$

$$\hat{\boldsymbol{\mu}}'_{L,3} = \begin{pmatrix} \dfrac{\gamma^{20}}{2}\dfrac{\partial}{\partial q^3} - \dfrac{\gamma^{03}}{2}\dfrac{\partial}{\partial q^2} \\[4pt] \dfrac{\gamma^{30}}{2}\dfrac{\partial}{\partial q^1} - \dfrac{\gamma^{01}}{2}\dfrac{\partial}{\partial q^3} \\[4pt] \dfrac{\gamma^{10}}{2}\dfrac{\partial}{\partial q^2} - \dfrac{\gamma^{02}}{2}\dfrac{\partial}{\partial q^1} \end{pmatrix} \ , \quad \hat{\boldsymbol{\mu}}'_{S,3} = \begin{pmatrix} \dfrac{\gamma^{20}}{2}\sum_{i=1}^{3}\gamma^{3i}\dfrac{\partial}{\partial q^i} - \dfrac{\gamma^{03}}{2}\sum_{i=1}^{3}\gamma^{2i}\dfrac{\partial}{\partial q^i} \\[4pt] \dfrac{\gamma^{30}}{2}\sum_{i=1}^{3}\gamma^{1i}\dfrac{\partial}{\partial q^i} - \dfrac{\gamma^{01}}{2}\sum_{i=1}^{3}\gamma^{3i}\dfrac{\partial}{\partial q^i} \\[4pt] \dfrac{\gamma^{10}}{2}\sum_{i=1}^{3}\gamma^{2i}\dfrac{\partial}{\partial q^i} - \dfrac{\gamma^{02}}{2}\sum_{i=1}^{3}\gamma^{1i}\dfrac{\partial}{\partial q^i} \end{pmatrix} \ ,$$

$$\hat{\boldsymbol{\mu}}'_{L,0}\boldsymbol{A}_{\ni} = \frac{1}{2}\boldsymbol{A}^{\ni} \times \mathbf{1}\nabla_0\boldsymbol{A}_{\ni} \ ,$$

$$\hat{\boldsymbol{\mu}}'_{S,0}\boldsymbol{A}_{\ni} = \frac{1}{2}\boldsymbol{A}^{\ni} \times \boldsymbol{\theta}^{\ni}\nabla_0\boldsymbol{A}_{\ni} \ . \qquad (5.70)$$

Box 5.8 (A mathematical model: field definitions).

$$\boldsymbol{w}^{(1)} = \boldsymbol{w}^{(1,1)} + \boldsymbol{w}^{(1,2)},$$

$$\boldsymbol{w}^{(1,1)} = -\frac{1}{2}\begin{pmatrix} \frac{\partial \gamma_{00}}{\partial q^1} \\ \frac{\partial \gamma_{00}}{\partial q^2} \\ \frac{\partial \gamma_{00}}{\partial q^3} \end{pmatrix}, \quad \boldsymbol{w}^{(1,2)} = \frac{1}{2}\gamma^{00}\begin{pmatrix} \frac{\partial \gamma_{00}}{\partial q^1} \\ \frac{\partial \gamma_{00}}{\partial q^2} \\ \frac{\partial \gamma_{00}}{\partial q^3} \end{pmatrix}, \quad (5.71)$$

$$\boldsymbol{w}^{(2)} = \mathsf{A}^{\ni}\,\nabla_3\gamma_{00}\,,$$

$$\mathsf{A}^{\ni} = \boldsymbol{A}^{\ni}\cdot\boldsymbol{A}^{\ni} = \begin{pmatrix} \gamma^{10}\gamma^{10} & \gamma^{10}\gamma^{20} & \gamma^{10}\gamma^{30} \\ \gamma^{20}\gamma^{10} & \gamma^{20}\gamma^{20} & \gamma^{20}\gamma^{30} \\ \gamma^{30}\gamma^{10} & \gamma^{30}\gamma^{20} & \gamma^{30}\gamma^{30} \end{pmatrix}, \quad (5.72)$$

$$\boldsymbol{w}^{(3)} = \mathsf{A}_3^{\mathrm{T}}\boldsymbol{A}^{\ni}\,,$$

$$\mathsf{A}_3 = \nabla_3 \cdot \boldsymbol{A}_{\ni} = \begin{pmatrix} \frac{\partial \gamma_{10}}{\partial q^1} & \frac{\partial \gamma_{20}}{\partial q^1} & \frac{\partial \gamma_{30}}{\partial q^1} \\ \frac{\partial \gamma_{10}}{\partial q^2} & \frac{\partial \gamma_{20}}{\partial q^2} & \frac{\partial \gamma_{30}}{\partial q^2} \\ \frac{\partial \gamma_{10}}{\partial q^3} & \frac{\partial \gamma_{20}}{\partial q^3} & \frac{\partial \gamma_{30}}{\partial q^3} \end{pmatrix}, \quad (5.73)$$

$$\boldsymbol{B}_3 = \nabla_3 \times \boldsymbol{A}_{\ni} = \begin{pmatrix} \frac{\partial \gamma_{30}}{\partial q^2} - \frac{\partial \gamma_{02}}{\partial q^3} \\ \frac{\partial \gamma_{10}}{\partial q^3} - \frac{\partial \gamma_{03}}{\partial q^1} \\ \frac{\partial \gamma_{20}}{\partial q^1} - \frac{\partial \gamma_{01}}{\partial q^2} \end{pmatrix}. \quad (5.74)$$

> Continuation of Box.

$$
\boldsymbol{X}^{(1)} = \begin{pmatrix} \dfrac{\partial \gamma^{11}}{\partial q^1} + \dfrac{\partial \gamma^{21}}{\partial q^2} + \dfrac{\partial \gamma^{31}}{\partial q^3} \\ \dfrac{\partial \gamma^{12}}{\partial q^1} + \dfrac{\partial \gamma^{22}}{\partial q^2} + \dfrac{\partial \gamma^{32}}{\partial q^3} \\ \dfrac{\partial \gamma^{13}}{\partial q^1} + \dfrac{\partial \gamma^{23}}{\partial q^2} + \dfrac{\partial \gamma^{33}}{\partial q^3} \end{pmatrix},
$$

$$
\boldsymbol{X}^{(2)} = \begin{pmatrix} \dfrac{1}{2} \dfrac{\partial}{\partial q^1} \sum_i \gamma_{ii} \\ \dfrac{1}{2} \dfrac{\partial}{\partial q^2} \sum_i \gamma_{ii} \\ \dfrac{1}{2} \dfrac{\partial}{\partial q^3} \sum_i \gamma_{ii} \end{pmatrix}, \qquad (5.75)
$$

$$
\boldsymbol{X}^{(3)} = \begin{pmatrix} \sum_{i,j} \dfrac{\gamma^{ij}}{2} \dfrac{\partial \gamma_{ij}}{\partial q^1} \\ \sum_{i,j} \dfrac{\gamma^{ij}}{2} \dfrac{\partial \gamma_{ij}}{\partial q^2} \\ \sum_{i,j} \dfrac{\gamma^{ij}}{2} \dfrac{\partial \gamma_{ij}}{\partial q^3} \end{pmatrix},
$$

$$
\mathsf{X}^{(1)} = \begin{pmatrix} \gamma^{22} + \gamma^{33} & -\gamma^{12} & -\gamma^{13} \\ -\gamma^{21} & \gamma^{11} + \gamma^{33} & -\gamma^{23} \\ -\gamma^{31} & -\gamma^{32} & \gamma^{11} + \gamma^{22} \end{pmatrix},
$$

$$
\mathsf{X}^{(2)} = \begin{pmatrix} \gamma^{22}\gamma^{33} - \gamma^{23}\gamma^{32} & \gamma^{23}\gamma^{31} - \gamma^{33}\gamma^{12} & \gamma^{32}\gamma^{21} - \gamma^{22}\gamma^{13} \\ \gamma^{23}\gamma^{31} - \gamma^{33}\gamma^{12} & \gamma^{11}\gamma^{33} - \gamma^{13}\gamma^{31} & \gamma^{21}\gamma^{13} - \gamma^{11}\gamma^{32} \\ \gamma^{32}\gamma^{21} - \gamma^{22}\gamma^{13} & \gamma^{21}\gamma^{13} - \gamma^{11}\gamma^{32} & \gamma^{11}\gamma^{22} - \gamma^{12}\gamma^{21} \end{pmatrix}, \qquad (5.76)
$$

$$
\mathsf{X}^{(3)} = \begin{pmatrix} \dfrac{\partial \gamma_{11}}{\partial q^0} & \dfrac{\partial \gamma_{12}}{\partial q^0} & \dfrac{\partial \gamma_{13}}{\partial q^0} \\ \dfrac{\partial \gamma_{21}}{\partial q^0} & \dfrac{\partial \gamma_{22}}{\partial q^0} & \dfrac{\partial \gamma_{23}}{\partial q^0} \\ \dfrac{\partial \gamma_{31}}{\partial q^0} & \dfrac{\partial \gamma_{32}}{\partial q^0} & \dfrac{\partial \gamma_{33}}{\partial q^0} \end{pmatrix}, \qquad \mathsf{X}^{(4)} = \begin{pmatrix} \dfrac{\partial \gamma^{11}}{\partial q^0} & \dfrac{\partial \gamma^{12}}{\partial q^0} & \dfrac{\partial \gamma^{13}}{\partial q^0} \\ \dfrac{\partial \gamma^{21}}{\partial q^0} & \dfrac{\partial \gamma^{22}}{\partial q^0} & \dfrac{\partial \gamma^{23}}{\partial q^0} \\ \dfrac{\partial \gamma^{31}}{\partial q^0} & \dfrac{\partial \gamma^{32}}{\partial q^0} & \dfrac{\partial \gamma^{33}}{\partial q^0} \end{pmatrix}. \qquad (5.77)
$$

Continuation of the mathematical model.

$$\mathcal{S}_{00} = \sum_{a=1}^{10} \mathcal{S}_{00}^{(a)} , \tag{5.78}$$

$$\mathcal{S}_{00}^{(1)} = \mathcal{S}_{00}^{(1,1)} + \mathcal{S}_{00}^{(1,2)} ,$$

$$\mathcal{S}_{00}^{(1,1)} = -\frac{1}{2}(\nabla_3)^{\mathrm{T}} \nabla_3 \gamma_{00} = -\frac{1}{2}\triangle_{\mathrm{L},3,3}\gamma_{00} = -\frac{1}{2}\triangle_3 \gamma_{00} , \tag{5.79}$$

$$\mathcal{S}_{00}^{(1,2)} = -\frac{1}{2}(\boldsymbol{\theta}^{\ni}\nabla_3)^{\mathrm{T}} \nabla_3 \gamma_{00} = -\frac{1}{2}\triangle_{\mathrm{S},3,3}\gamma_{00} = -\frac{1}{2}\triangle'_3 \gamma_{00} ,$$

$$\mathcal{S}_{00}^{(2)} = \mathcal{S}_{00}^{(2,1)} + \mathcal{S}_{00}^{(2,2)} ,$$

$$\begin{aligned}\mathcal{S}_{00}^{(2,1)} &= -\frac{1}{4}(\nabla_3 \gamma_{00})^{\mathrm{T}} \nabla_3 \gamma_{00} \\ &= -(\hat{\boldsymbol{\mu}}_{\mathrm{L},3}\gamma_{00})^{\mathrm{T}} \hat{\boldsymbol{\mu}}_{\mathrm{L},3}\gamma_{00} = +\boldsymbol{w}^{(1,1)\mathrm{T}} \hat{\boldsymbol{\mu}}_{\mathrm{L},3}\gamma_{00} ,\end{aligned} \tag{5.80}$$

$$\begin{aligned}\mathcal{S}_{00}^{(2,2)} &= -\frac{1}{4}(\nabla_3 \gamma_{00})^{\mathrm{T}} \boldsymbol{\theta}^{\ni} \nabla_3 \gamma_{00} \\ &= -(\hat{\boldsymbol{\mu}}_{\mathrm{L},3}\gamma_{00})^{\mathrm{T}} \hat{\boldsymbol{\mu}}_{\mathrm{S},3}\gamma_{00} = +\boldsymbol{w}^{(1,1)\mathrm{T}} \hat{\boldsymbol{\mu}}_{\mathrm{S},3}\gamma_{00} ,\end{aligned}$$

$$\mathcal{S}_{00}^{(3)} = \mathcal{S}_{00}^{(3,1)} + \mathcal{S}_{00}^{(3,2)} ,$$

$$\begin{aligned}\mathcal{S}_{00}^{(3,1)} &= +\frac{1}{4}(\gamma^{00}\nabla_3 \gamma_{00})^{\mathrm{T}} \nabla_3 \gamma_{00} \\ &= +(\gamma^{00}\hat{\boldsymbol{\mu}}_{\mathrm{L},3}\gamma_{00})^{\mathrm{T}} \hat{\boldsymbol{\mu}}_{\mathrm{L},3}\gamma_{00} = +\boldsymbol{w}^{(1,2)\mathrm{T}} \hat{\boldsymbol{\mu}}_{\mathrm{L},3}\gamma_{00} ,\end{aligned} \tag{5.81}$$

$$\begin{aligned}\mathcal{S}_{00}^{(3,2)} &= +\frac{1}{4}(\gamma^{00}\nabla_3 \gamma_{00})^{\mathrm{T}} \boldsymbol{\theta}^{\ni} \nabla_3 \gamma_{00} \\ &= +(\gamma^{00}\hat{\boldsymbol{\mu}}_{\mathrm{L},3}\gamma_{00})^{\mathrm{T}} \hat{\boldsymbol{\mu}}_{\mathrm{S},3}\gamma_{00} = +\boldsymbol{w}^{(1,2)\mathrm{T}} \hat{\boldsymbol{\mu}}_{\mathrm{S},3}\gamma_{00} ,\end{aligned}$$

Continuation of the mathematical model.

$$\mathcal{S}_{00}^{(4)} = \mathcal{S}_{00}^{(4,1)} = -\frac{1}{2}\bigl[\bigl(\boldsymbol{A}^{\ni}\cdot\boldsymbol{A}^{\ni}\bigr)\bigl(\nabla_3\gamma_{00}\bigr)\bigr]^{\mathrm{T}}\nabla_3\gamma_{00} = -\boldsymbol{w}^{(2)\mathrm{T}}\hat{\boldsymbol{\mu}}_{\mathrm{L},3}\gamma_{00} \ , \tag{5.82}$$

$$\begin{aligned}
\mathcal{S}_{00}^{(5)} &= \mathcal{S}_{00}^{(5,1)} + \mathcal{S}_{00}^{(5,2)} \ , \\
\mathcal{S}_{00}^{(5,1)} &= -\frac{1}{2}\bigl[\bigl(\nabla_3\cdot\boldsymbol{A}_{\ni}\bigr)^{\mathrm{T}}\boldsymbol{A}^{\ni}\bigr]^{\mathrm{T}}\nabla_3\gamma_{00} = -\boldsymbol{w}^{(3)\mathrm{T}}\hat{\boldsymbol{\mu}}_{\mathrm{L},3}\gamma_{00} \ , \\
\mathcal{S}_{00}^{(5,2)} &= -\frac{1}{2}\bigl[\bigl(\nabla_3\cdot\boldsymbol{A}_{\ni}\bigr)^{\mathrm{T}}\boldsymbol{A}^{\ni}\bigr]^{\mathrm{T}}\boldsymbol{\theta}^{\ni}\nabla_3\gamma_{00} = -\boldsymbol{w}^{(3)\mathrm{T}}\hat{\boldsymbol{\mu}}_{\mathrm{S},3}\gamma_{00} \ ,
\end{aligned} \tag{5.83}$$

$$\begin{aligned}
\mathcal{S}_{00}^{(6)} &= \mathcal{S}_{00}^{(6,1)} + \mathcal{S}_{00}^{(6,2)} \ , \\
\mathcal{S}_{00}^{(6,1)} &= -\frac{1}{2}\bigl(\nabla_3\times\boldsymbol{A}_{\ni}\bigr)^{\mathrm{T}}\boldsymbol{A}^{\ni}\times\nabla_3\gamma_{00} = -\boldsymbol{B}_3^{\mathrm{T}}\hat{\boldsymbol{\mu}}_{\mathrm{L},3}'\gamma_{00} \ , \\
\mathcal{S}_{00}^{(6,2)} &= -\frac{1}{2}\bigl(\nabla_3\times\boldsymbol{A}_{\ni}\bigr)^{\mathrm{T}}\boldsymbol{A}^{\ni}\times\boldsymbol{\theta}^{\ni}\nabla_3\gamma_{00} = -\boldsymbol{B}_3^{\mathrm{T}}\hat{\boldsymbol{\mu}}_{\mathrm{S},3}'\gamma_{00} \ ,
\end{aligned} \tag{5.84}$$

$$\mathcal{S}_{00}^{(7)} = \mathcal{S}_{00}^{(7,1)} = -\boldsymbol{X}^{(1)\mathrm{T}}\hat{\boldsymbol{\mu}}_{\mathrm{L},3}\gamma_{00} \ , \tag{5.85}$$

$$\begin{aligned}
\mathcal{S}_{00}^{(8)} &= \mathcal{S}_{00}^{(8,1)} + \mathcal{S}_{00}^{(8,2)} \ , \\
\mathcal{S}_{00}^{(8,1)} &= -\boldsymbol{X}^{(2)\mathrm{T}}\hat{\boldsymbol{\mu}}_{\mathrm{L},3}\gamma_{00} \ , \quad \mathcal{S}_{00}^{(8,2)} = -\boldsymbol{X}^{(2)\mathrm{T}}\hat{\boldsymbol{\mu}}_{\mathrm{S},3}\gamma_{00} \ ,
\end{aligned} \tag{5.86}$$

$$\begin{aligned}
\mathcal{S}_{00}^{(9)} &= \mathcal{S}_{00}^{(9,1)} + \mathcal{S}_{00}^{(9,2)} \ , \\
\mathcal{S}_{00}^{(9,1)} &= -\boldsymbol{X}^{(3)\mathrm{T}}\hat{\boldsymbol{\mu}}_{\mathrm{L},3}\gamma_{00} \ , \quad \mathcal{S}_{00}^{(9,2)} = -\boldsymbol{X}^{(3)\mathrm{T}}\hat{\boldsymbol{\mu}}_{\mathrm{S},3}\gamma_{00} \ ,
\end{aligned} \tag{5.87}$$

$$\begin{aligned}
\mathcal{S}_{00}^{(10)} &= \mathcal{S}_{00}^{(10,1)} + \mathcal{S}_{00}^{(10,2)} + \mathcal{S}_{00}^{(10,3)} \ , \\
\mathcal{S}_{00}^{(10,1)} &= +\frac{1}{2}\boldsymbol{B}_3^{\mathrm{T}}\boldsymbol{B}_3 \ , \\
\mathcal{S}_{00}^{(10,2)} &= +\frac{1}{2}\boldsymbol{B}_3^{\mathrm{T}}\mathsf{X}^{(1)}\boldsymbol{B}_3 \ , \\
\mathcal{S}_{00}^{(10,3)} &= +\frac{1}{2}\boldsymbol{B}_3^{\mathrm{T}}\mathsf{X}^{(2)}\boldsymbol{B}_3 \ .
\end{aligned} \tag{5.88}$$

Continuation of the mathematical model.

$$\mathcal{W}_{00} = \sum_{a=1}^{10} \mathcal{W}_{00}^{(a)} \,, \tag{5.89}$$

$$\mathcal{W}_{00}^{(1)} = \sum_{b=1}^{6} \mathcal{W}_{00}^{(1,b)} \,, \tag{5.90}$$

$$\begin{aligned}
\mathcal{W}_{00}^{(1,1)} &= -\frac{1}{2} \left(\nabla_3 \right)^{\mathrm{T}} \nabla_3 \gamma_{00} \\
&= -\frac{1}{2} \triangle_{\mathrm{L},3,3} \gamma_{00} = -\frac{1}{2} \triangle_3 \gamma_{00} \,, \\
\mathcal{W}_{00}^{(1,2)} &= -\frac{1}{2} \left(\boldsymbol{\theta}^{\ni} \nabla_3 \right)^{\mathrm{T}} \nabla_3 \gamma_{00} \\
&= -\frac{1}{2} \triangle_{\mathrm{S},3,3} \gamma_{00} = -\frac{1}{2} \triangle_3' \gamma_{00} \,,
\end{aligned} \tag{5.91}$$

$$\begin{aligned}
\mathcal{W}_{00}^{(1,3)} &= + \left(\nabla_3 \right)^{\mathrm{T}} \nabla_0 \boldsymbol{A}_{\ni} \\
&= + \overline{\triangle}_{\mathrm{L},3,0} \boldsymbol{A}_{\ni} \,, \\
\mathcal{W}_{00}^{(1,4)} &= + \left(\boldsymbol{\theta}^{\ni} \nabla_3 \right)^{\mathrm{T}} \nabla_0 \boldsymbol{A}_{\ni} \\
&= + \overline{\triangle}_{\mathrm{S},3,0} \boldsymbol{A}_{\ni} \,,
\end{aligned} \tag{5.92}$$

$$\begin{aligned}
\mathcal{W}_{00}^{(1,5)} &= -\frac{1}{2} \left(1 \nabla_0^2 \right) \bullet \left(\boldsymbol{\theta}_{\ni} \right) \\
&= -\frac{1}{2} \overline{\overline{\triangle}}_{\mathrm{L},0,0} \bullet \left(\boldsymbol{\theta}_{\ni} \right) \,, \\
\mathcal{W}_{00}^{(1,6)} &= -\frac{1}{2} \left(\boldsymbol{\theta}^{\ni} \nabla_0^2 \right) \bullet \left(\boldsymbol{\theta}_{\ni} \right) \\
&= -\frac{1}{2} \overline{\overline{\triangle}}_{\mathrm{S},0,0} \bullet \left(\boldsymbol{\theta}_{\ni} \right) \,,
\end{aligned} \tag{5.93}$$

Continuation of the mathematical model.

$$\mathcal{W}_{00}^{(2)} = \sum_{b=1}^{6} \mathcal{W}_{00}^{(2,b)}, \tag{5.94}$$

$$\begin{aligned}
\mathcal{W}_{00}^{(2,1)} &= -\frac{1}{4}\left(\nabla_3 \gamma_{00}\right)^{\mathrm{T}} \nabla_3 \gamma_{00} \\
&= -\left(\hat{\boldsymbol{\mu}}_{\mathrm{L},3} \gamma_{00}\right)^{\mathrm{T}} \hat{\boldsymbol{\mu}}_{\mathrm{L},3} \gamma_{00} \\
&= +\boldsymbol{w}^{(1,1)\,\mathrm{T}} \hat{\boldsymbol{\mu}}_{\mathrm{L},3} \gamma_{00}, \\
\mathcal{W}_{00}^{(2,2)} &= -\frac{1}{4}\left(\nabla_3 \gamma_{00}\right)^{\mathrm{T}} \boldsymbol{\theta}^{\ni} \nabla_3 \gamma_{00} \\
&= -\left(\hat{\boldsymbol{\mu}}_{\mathrm{L},3} \gamma_{00}\right)^{\mathrm{T}} \hat{\boldsymbol{\mu}}_{\mathrm{S},3} \gamma_{00} \\
&= +\boldsymbol{w}^{(1,1)\,\mathrm{T}} \hat{\boldsymbol{\mu}}_{\mathrm{S},3} \gamma_{00},
\end{aligned} \tag{5.95}$$

$$\begin{aligned}
\mathcal{W}_{00}^{(2,3)} &= +\frac{1}{2}\left(\nabla_0 \boldsymbol{A}_{\ni}\right)^{\mathrm{T}} \nabla_3 \gamma_{00} \\
&= +2\left(\hat{\mu}_{\mathrm{L},0} \boldsymbol{A}_{\ni}\right)^{\mathrm{T}} \hat{\boldsymbol{\mu}}_{\mathrm{L},3} \gamma_{00}, \\
\mathcal{W}_{00}^{(2,4)} &= +\frac{1}{2}\left(\nabla_0 \boldsymbol{A}_{\ni}\right)^{\mathrm{T}} \boldsymbol{\theta}^{\ni} \nabla_3 \gamma_{00} \\
&= +2\left(\hat{\mu}_{\mathrm{L},0} \boldsymbol{A}_{\ni}\right)^{\mathrm{T}} \hat{\boldsymbol{\mu}}_{\mathrm{S},3} \gamma_{00},
\end{aligned} \tag{5.96}$$

$$\begin{aligned}
\mathcal{W}_{00}^{(2,5)} &= -\frac{1}{4}\left(\nabla_0 \boldsymbol{\theta}_{\ni}\right) \bullet \left(1 \nabla_0 \gamma_{00}\right) \\
&= -\left(\hat{\mu}_{\mathrm{L},0} \boldsymbol{\theta}_{\ni}\right) \bullet \left(\hat{\mu}_{\mathrm{L},0} \gamma_{00}\right), \\
\mathcal{W}_{00}^{(2,6)} &= -\frac{1}{4}\left(\nabla_0 \boldsymbol{\theta}_{\ni}\right) \bullet \left(\boldsymbol{\theta}^{\ni} \nabla_0 \gamma_{00}\right) \\
&= -\left(\hat{\mu}_{\mathrm{L},0} \boldsymbol{\theta}_{\ni}\right) \bullet \left(\hat{\mu}_{\mathrm{S},0} \gamma_{00}\right),
\end{aligned} \tag{5.97}$$

Continuation of the mathematical model.

$$\mathcal{W}_{00}^{(3)} = \sum_{b=1}^{6} \mathcal{W}_{00}^{(3,b)}, \tag{5.98}$$

$$\begin{aligned}
\mathcal{W}_{00}^{(3,1)} &= +\frac{1}{4}\left(\gamma^{00}\nabla_3\gamma_{00}\right)^{\mathrm{T}}\nabla_3\gamma_{00} \\
&= +\left(\gamma^{00}\hat{\boldsymbol{\mu}}_{\mathrm{L},3}\gamma_{00}\right)^{\mathrm{T}}\hat{\boldsymbol{\mu}}_{\mathrm{L},3}\gamma_{00} \\
&= +\boldsymbol{w}^{(1,2)\mathrm{T}}\hat{\boldsymbol{\mu}}_{\mathrm{L},3}\gamma_{00}, \\
\mathcal{W}_{00}^{(3,2)} &= +\frac{1}{4}\left(\gamma^{00}\nabla_3\gamma_{00}\right)^{\mathrm{T}}\boldsymbol{\theta}^{\ni}\nabla_3\gamma_{00} \\
&= +\left(\gamma^{00}\hat{\boldsymbol{\mu}}_{\mathrm{L},3}\gamma_{00}\right)^{\mathrm{T}}\hat{\boldsymbol{\mu}}_{\mathrm{S},3}\gamma_{00} \\
&= +\boldsymbol{w}^{(1,2)\mathrm{T}}\hat{\boldsymbol{\mu}}_{\mathrm{S},3}\gamma_{00},
\end{aligned} \tag{5.99}$$

$$\begin{aligned}
\mathcal{W}_{00}^{(3,3)} &= -\frac{1}{2}\left(\gamma^{00}\nabla_0\boldsymbol{A}_{\ni}\right)^{\mathrm{T}}\nabla_3\gamma_{00} \\
&= -2\left(\gamma^{00}\hat{\mu}_{\mathrm{L},0}\boldsymbol{A}_{\ni}\right)^{\mathrm{T}}\hat{\boldsymbol{\mu}}_{\mathrm{L},3}\gamma_{00}, \\
\mathcal{W}_{00}^{(3,4)} &= -\frac{1}{2}\left(\gamma^{00}\nabla_0\boldsymbol{A}_{\ni}\right)^{\mathrm{T}}\boldsymbol{\theta}^{\ni}\nabla_3\gamma_{00} \\
&= -2\left(\gamma^{00}\hat{\mu}_{\mathrm{L},0}\boldsymbol{A}_{\ni}\right)^{\mathrm{T}}\hat{\boldsymbol{\mu}}_{\mathrm{S},3}\gamma_{00},
\end{aligned} \tag{5.100}$$

$$\begin{aligned}
\mathcal{W}_{00}^{(3,5)} &= +\frac{1}{4}\left(\gamma^{00}\nabla_0\boldsymbol{\theta}_{\ni}\right)\bullet\left(\mathbf{1}\nabla_0\gamma_{00}\right) \\
&= +\left(\gamma^{00}\hat{\mu}_{\mathrm{L},0}\boldsymbol{\theta}_{\ni}\right)\bullet\left(\hat{\mu}_{\mathrm{L},0}\gamma_{00}\right), \\
\mathcal{W}_{00}^{(3,6)} &= +\frac{1}{4}\left(\gamma^{00}\nabla_0\boldsymbol{\theta}_{\ni}\right)\bullet\left(\boldsymbol{\theta}^{\ni}\nabla_0\gamma_{00}\right) \\
&= +\left(\gamma^{00}\hat{\mu}_{\mathrm{L},0}\boldsymbol{\theta}_{\ni}\right)\bullet\left(\hat{\mu}_{\mathrm{S},0}\gamma_{00}\right),
\end{aligned} \tag{5.101}$$

Continuation of the mathematical model.

$$\mathcal{W}_{00}^{(4)} = \sum_{b=1}^{6} \mathcal{W}_{00}^{(4,b)} , \tag{5.102}$$

$$\begin{aligned}
\mathcal{W}_{00}^{(4,1)} &= -\frac{1}{2}\Big[\big(\boldsymbol{A}^{\ni} \cdot \boldsymbol{A}^{\ni}\big)\big(\nabla_{3}\gamma_{00}\big)\Big]^{\mathrm{T}} \nabla_{3}\gamma_{00} \\
&= -2\Big[\big(\boldsymbol{A}^{\ni} \cdot \boldsymbol{A}^{\ni}\big)\big(\hat{\boldsymbol{\mu}}_{\mathrm{L},3}\gamma_{00}\big)\Big]^{\mathrm{T}} \hat{\boldsymbol{\mu}}_{\mathrm{L},3}\gamma_{00} \\
&= -\boldsymbol{w}^{(2)\,\mathrm{T}} \hat{\boldsymbol{\mu}}_{\mathrm{L},3}\gamma_{00} ,
\end{aligned} \tag{5.103}$$

$$\begin{aligned}
\mathcal{W}_{00}^{(4,2)} &= +\frac{1}{2}\Big[\big(\boldsymbol{A}^{\ni} \cdot \boldsymbol{A}^{\ni}\big)\big(\nabla_{0}\boldsymbol{A}_{\ni}\big)\Big]^{\mathrm{T}} \nabla_{3}\gamma_{00} \\
&= +2\Big[\big(\boldsymbol{A}^{\ni} \cdot \boldsymbol{A}^{\ni}\big)\big(\hat{\mu}_{\mathrm{L},0}\boldsymbol{A}_{\ni}\big)\Big]^{\mathrm{T}} \hat{\boldsymbol{\mu}}_{\mathrm{L},3}\gamma_{00} ,
\end{aligned}$$

$$\begin{aligned}
\mathcal{W}_{00}^{(4,3)} &= +\frac{1}{2}\Big[\big(\boldsymbol{A}^{\ni} \cdot \boldsymbol{A}^{\ni}\big) \bullet \big(\nabla_{3} \cdot \boldsymbol{A}_{\ni}\big)\Big] \nabla_{0}\gamma_{00} \\
&= +2\Big[\big(\boldsymbol{A}^{\ni} \cdot \boldsymbol{A}^{\ni}\big) \bullet \big(\hat{\boldsymbol{\mu}}_{\mathrm{L},3} \cdot \boldsymbol{A}_{\ni}\big)\Big] \hat{\mu}_{\mathrm{L},0}\gamma_{00} , \\
\mathcal{W}_{00}^{(4,4)} &= -\frac{1}{2}\Big[\big(\boldsymbol{A}^{\ni} \cdot \boldsymbol{A}^{\ni}\big) \bullet \big(\nabla_{0}\boldsymbol{\theta}_{\ni}\big)\Big] \nabla_{0}\gamma_{00} \\
&= -2\Big[\big(\boldsymbol{A}^{\ni} \cdot \boldsymbol{A}^{\ni}\big) \bullet \big(\hat{\mu}_{\mathrm{L},0}\boldsymbol{\theta}_{\ni}\big)\Big] \hat{\mu}_{\mathrm{L},0}\gamma_{00} ,
\end{aligned} \tag{5.104}$$

$$\begin{aligned}
\mathcal{W}_{00}^{(4,5)} &= -\frac{1}{4}\Big[\boldsymbol{A}^{\ni\,\mathrm{T}}\big(\nabla_{0}\gamma_{00}\big)\Big] \nabla_{3}\gamma_{00} \\
&= -1\Big[\boldsymbol{A}^{\ni\,\mathrm{T}}\big(\hat{\mu}_{\mathrm{L},0}\gamma_{00}\big)\Big] \hat{\boldsymbol{\mu}}_{\mathrm{L},3}\gamma_{00} , \\
\mathcal{W}_{00}^{(4,6)} &= +\frac{1}{2}\Big[\boldsymbol{A}^{\ni\,\mathrm{T}}\big(\nabla_{0}\boldsymbol{A}_{\ni}\big)\Big] \nabla_{0}\gamma_{00} \\
&= +2\Big[\boldsymbol{A}^{\ni\,\mathrm{T}}\big(\hat{\mu}_{\mathrm{L},0}\boldsymbol{A}_{\ni}\big)\Big] \hat{\mu}_{\mathrm{L},0}\gamma_{00} ,
\end{aligned} \tag{5.105}$$

Continuation of the mathematical model.

$$\mathcal{W}_{00}^{(5)} = \sum_{b=1}^{8} \mathcal{W}_{00}^{(5,b)}, \tag{5.106}$$

$$\mathcal{W}_{00}^{(5,1)} = -\frac{1}{2} \Big[\big(\nabla_3 \cdot \boldsymbol{A}_{\ni}\big)^{\mathrm{T}} \boldsymbol{A}^{\ni} \Big]^{\mathrm{T}} \nabla_3 \gamma_{00}$$

$$= -\boldsymbol{w}^{(3)\mathrm{T}} \hat{\boldsymbol{\mu}}_{\mathrm{L},3} \gamma_{00}, \tag{5.107}$$

$$\mathcal{W}_{00}^{(5,2)} = -\frac{1}{2} \Big[\big(\nabla_3 \cdot \boldsymbol{A}_{\ni}\big)^{\mathrm{T}} \boldsymbol{A}^{\ni} \Big]^{\mathrm{T}} \boldsymbol{\theta}^{\ni} \nabla_3 \gamma_{00}$$

$$= -\boldsymbol{w}^{(3)\mathrm{T}} \hat{\boldsymbol{\mu}}_{\mathrm{S},3} \gamma_{00},$$

$$\mathcal{W}_{00}^{(5,3)} = +\frac{1}{2} \Big[\big(\nabla_3 \cdot \boldsymbol{A}_{\ni}\big)^{\mathrm{T}} \boldsymbol{A}^{\ni} \Big]^{\mathrm{T}} \mathbf{1} \nabla_0 \boldsymbol{A}_{\ni}$$

$$= +\boldsymbol{w}^{(3)\mathrm{T}} \hat{\boldsymbol{\mu}}_{\mathrm{L},0} \boldsymbol{A}_{\ni}, \tag{5.108}$$

$$\mathcal{W}_{00}^{(5,4)} = +\frac{1}{2} \Big[\big(\nabla_3 \cdot \boldsymbol{A}_{\ni}\big)^{\mathrm{T}} \boldsymbol{A}^{\ni} \Big]^{\mathrm{T}} \boldsymbol{\theta}^{\ni} \nabla_0 \boldsymbol{A}_{\ni}$$

$$= +\boldsymbol{w}^{(3)\mathrm{T}} \hat{\boldsymbol{\mu}}_{\mathrm{S},0} \boldsymbol{A}_{\ni},$$

$$\mathcal{W}_{00}^{(5,5)} = +\frac{1}{2} \Big[\big(\nabla_3 \cdot \boldsymbol{A}_{\ni}\big) \boldsymbol{A}^{\ni} \Big]^{\mathrm{T}} \mathbf{1} \nabla_0 \boldsymbol{A}_{\ni}, \tag{5.109}$$

$$\mathcal{W}_{00}^{(5,6)} = +\frac{1}{2} \Big[\big(\nabla_3 \cdot \boldsymbol{A}_{\ni}\big) \boldsymbol{A}^{\ni} \Big]^{\mathrm{T}} \boldsymbol{\theta}^{\ni} \nabla_0 \boldsymbol{A}_{\ni},$$

$$\mathcal{W}_{00}^{(5,7)} = -\Big[\big(\nabla_0 \boldsymbol{\theta}_{\ni}\big) \boldsymbol{A}^{\ni} \Big]^{\mathrm{T}} \mathbf{1} \nabla_0 \boldsymbol{A}_{\ni}, \tag{5.110}$$

$$\mathcal{W}_{00}^{(5,8)} = -\Big[\big(\nabla_0 \boldsymbol{\theta}_{\ni}\big) \boldsymbol{A}^{\ni} \Big]^{\mathrm{T}} \boldsymbol{\theta}^{\ni} \nabla_0 \boldsymbol{A}_{\ni},$$

Continuation of the mathematical model.

$$\mathcal{W}_{00}^{(6)} = \sum_{b=1}^{10} \mathcal{W}_{00}^{(6,b)}, \tag{5.111}$$

$$\begin{aligned}
\mathcal{W}_{00}^{(6,1)} &= -\frac{1}{2}\big(\nabla_3 \times \boldsymbol{A}_\ni\big)^{\mathrm{T}} \boldsymbol{A}^\ni \times \nabla_3 \gamma_{00} \\
&= -\boldsymbol{B}_3^{\mathrm{T}} \hat{\boldsymbol{\mu}}'_{\mathrm{L},3} \gamma_{00}, \\
\mathcal{W}_{00}^{(6,2)} &= -\frac{1}{2}\big(\nabla_3 \times \boldsymbol{A}_\ni\big)^{\mathrm{T}} \boldsymbol{A}^\ni \times \boldsymbol{\theta}^\ni \nabla_3 \gamma_{00} \\
&= -\boldsymbol{B}_3^{\mathrm{T}} \hat{\boldsymbol{\mu}}'_{\mathrm{S},3} \gamma_{00},
\end{aligned} \tag{5.112}$$

$$\begin{aligned}
\mathcal{W}_{00}^{(6,3)} &= +\frac{1}{2}\big(\nabla_3 \times \boldsymbol{A}_\ni\big)^{\mathrm{T}} \boldsymbol{A}^\ni \times \mathbf{1} \nabla_0 \boldsymbol{A}_\ni \\
&= +\boldsymbol{B}_3^{\mathrm{T}} \hat{\boldsymbol{\mu}}'_{\mathrm{L},0} \boldsymbol{A}_\ni, \\
\mathcal{W}_{00}^{(6,4)} &= +\frac{1}{2}\big(\nabla_3 \times \boldsymbol{A}_\ni\big)^{\mathrm{T}} \boldsymbol{A}^\ni \times \boldsymbol{\theta}^\ni \nabla_0 \boldsymbol{A}_\ni \\
&= +\boldsymbol{B}_3^{\mathrm{T}} \hat{\boldsymbol{\mu}}'_{\mathrm{S},0} \boldsymbol{A}_\ni,
\end{aligned} \tag{5.113}$$

$$\begin{aligned}
\mathcal{W}_{00}^{(6,5)} &= +\frac{1}{4}\Big[\;(1) \bullet \big(\nabla_3 \circ \boldsymbol{\theta}_\ni\big)\Big]^{\mathrm{T}} \boldsymbol{A}^\ni \nabla_0 \gamma_{00} \\
&= +\boldsymbol{X}^{(2)\mathrm{T}} \boldsymbol{A}^\ni \hat{\mu}_{\mathrm{L},0} \gamma_{00}, \\
\mathcal{W}_{00}^{(6,6)} &= +\frac{1}{4}\Big[\big(\boldsymbol{\theta}^\ni\big) \bullet \big(\nabla_3 \circ \boldsymbol{\theta}_\ni\big)\Big]^{\mathrm{T}} \boldsymbol{A}^\ni \nabla_0 \gamma_{00} \\
&= +\boldsymbol{X}^{(3)\mathrm{T}} \boldsymbol{A}^\ni \hat{\mu}_{\mathrm{L},0} \gamma_{00},
\end{aligned} \tag{5.114}$$

$$\begin{aligned}
\mathcal{W}_{00}^{(6,7)} &= +\frac{1}{2}\Big[\;(1) \bullet \big(\nabla_0 \boldsymbol{\theta}_\ni\big)\Big] \boldsymbol{A}^{\ni\mathrm{T}} \nabla_0 \boldsymbol{A}_\ni, \\
\mathcal{W}_{00}^{(6,8)} &= +\frac{1}{2}\Big[\big(\boldsymbol{\theta}^\ni\big) \bullet \big(\nabla_0 \boldsymbol{\theta}_\ni\big)\Big] \boldsymbol{A}^{\ni\mathrm{T}} \nabla_0 \boldsymbol{A}_\ni,
\end{aligned} \tag{5.115}$$

$$\begin{aligned}
\mathcal{W}_{00}^{(6,9)} &= -\frac{1}{4}\Big[\;(1) \bullet \big(\nabla_0 \boldsymbol{\theta}_\ni\big)\Big] \boldsymbol{A}^{\ni\mathrm{T}} \nabla_3 \gamma_{00}, \\
\mathcal{W}_{00}^{(6,10)} &= -\frac{1}{4}\Big[\big(\boldsymbol{\theta}^\ni\big) \bullet \big(\nabla_0 \boldsymbol{\theta}_\ni\big)\Big] \boldsymbol{A}^{\ni\mathrm{T}} \nabla_3 \gamma_{00},
\end{aligned} \tag{5.116}$$

Continuation of the mathematical model.

$$\mathcal{W}_{00}^{(7)} = \sum_{b=1}^{4} \mathcal{W}_{00}^{(7,b)}, \tag{5.117}$$

$$\begin{aligned}\mathcal{W}_{00}^{(7,1)} &= -\frac{1}{2}\bigl[(\nabla_3)^{\mathrm{T}}\boldsymbol{\theta}^{\ni}\bigr]\nabla_3\gamma_{00} \\ &= -1\boldsymbol{X}^{(1)\mathrm{T}}\hat{\boldsymbol{\mu}}_{\mathrm{L},3}\gamma_{00}, \\ \mathcal{W}_{00}^{(7,4)} &= +1\bigl[(\nabla_3)^{\mathrm{T}}\boldsymbol{\theta}^{\ni}\bigr]\nabla_0\boldsymbol{A}_{\ni} \\ &= +2\boldsymbol{X}^{(1)\mathrm{T}}\hat{\boldsymbol{\mu}}_{\mathrm{L},0}\boldsymbol{A}_{\ni},\end{aligned} \tag{5.118}$$

$$\begin{aligned}\mathcal{W}_{00}^{(7,2)} &= -\frac{1}{2}\bigl[(\nabla_0)^{\mathrm{T}}\boldsymbol{A}^{\ni}\bigr]^{\mathrm{T}}\nabla_3\gamma_{00}, \\ \mathcal{W}_{00}^{(7,3)} &= +\frac{1}{2}\bigl[(\nabla_3)^{\mathrm{T}}\boldsymbol{A}^{\ni}\bigr]^{\mathrm{T}}\nabla_0\gamma_{00},\end{aligned} \tag{5.119}$$

$$\mathcal{W}_{00}^{(8)} = \sum_{b=1}^{4} \mathcal{W}_{00}^{(8,b)}, \tag{5.120}$$

$$\begin{aligned}\mathcal{W}_{00}^{(8,1)} &= -\frac{1}{4}\bigl(\nabla_3\operatorname{Tr}\boldsymbol{\theta}_{\ni}\bigr)^{\mathrm{T}}\nabla_3\gamma_{00} \\ &= -\boldsymbol{X}^{(2)\mathrm{T}}\hat{\boldsymbol{\mu}}_{\mathrm{L},3}\gamma_{00}, \\ \mathcal{W}_{00}^{(8,2)} &= -\frac{1}{4}\bigl(\nabla_3\operatorname{Tr}\boldsymbol{\theta}_{\ni}\bigr)^{\mathrm{T}}\boldsymbol{\theta}^{\ni}\nabla_3\gamma_{00} \\ &= -\boldsymbol{X}^{(2)\mathrm{T}}\hat{\boldsymbol{\mu}}_{\mathrm{S},3}\gamma_{00},\end{aligned} \tag{5.121}$$

$$\begin{aligned}\mathcal{W}_{00}^{(8,3)} &= +\frac{1}{2}\bigl(\nabla_3\operatorname{Tr}\boldsymbol{\theta}_{\ni}\bigr)^{\mathrm{T}}\mathbf{1}\nabla_0\boldsymbol{A}_{\ni} \\ &= +2\boldsymbol{X}^{(2)\mathrm{T}}\hat{\boldsymbol{\mu}}_{\mathrm{L},0}\boldsymbol{A}_{\ni}, \\ \mathcal{W}_{00}^{(8,4)} &= +\frac{1}{2}\bigl(\nabla_3\operatorname{Tr}\boldsymbol{\theta}_{\ni}\bigr)^{\mathrm{T}}\boldsymbol{\theta}^{\ni}\nabla_0\boldsymbol{A}_{\ni} \\ &= +2\boldsymbol{X}^{(2)\mathrm{T}}\hat{\boldsymbol{\mu}}_{\mathrm{S},0}\boldsymbol{A}_{\ni},\end{aligned} \tag{5.122}$$

> **Continuation of the mathematical model.**

$$\mathcal{W}_{00}^{(9)} = \sum_{b=1}^{4} \mathcal{W}_{00}^{(9,b)}, \tag{5.123}$$

$$\mathcal{W}_{00}^{(9,1)} = -\frac{1}{4}\big[(\boldsymbol{\theta}^{\ni}) \bullet (\nabla_3 \circ \boldsymbol{\theta}_{\ni})\big]^{\mathrm{T}} \nabla_3 \gamma_{00} \quad = -\boldsymbol{X}^{(3)\,\mathrm{T}} \hat{\boldsymbol{\mu}}_{\mathrm{L},3} \gamma_{00}\,,$$

$$\mathcal{W}_{00}^{(9,2)} = -\frac{1}{4}\big[(\boldsymbol{\theta}^{\ni}) \bullet (\nabla_3 \circ \boldsymbol{\theta}_{\ni})\big]^{\mathrm{T}} \boldsymbol{\theta}^{\ni} \nabla_3 \gamma_{00} = -\boldsymbol{X}^{(3)\,\mathrm{T}} \hat{\boldsymbol{\mu}}_{\mathrm{S},3} \gamma_{00}\,, \tag{5.124}$$

$$\mathcal{W}_{00}^{(9,3)} = +\frac{1}{2}\big[(\boldsymbol{\theta}^{\ni}) \bullet (\nabla_3 \circ \boldsymbol{\theta}_{\ni})\big]^{\mathrm{T}} 1 \nabla_0 \boldsymbol{A}_{\ni} \quad = +2\boldsymbol{X}^{(3)\,\mathrm{T}} \hat{\boldsymbol{\mu}}_{\mathrm{L},0} \boldsymbol{A}_{\ni}\,,$$

$$\mathcal{W}_{00}^{(9,4)} = +\frac{1}{2}\big[(\boldsymbol{\theta}^{\ni}) \bullet (\nabla_3 \circ \boldsymbol{\theta}_{\ni})\big]^{\mathrm{T}} \boldsymbol{\theta}^{\ni} \nabla_0 \boldsymbol{A}_{\ni} = +2\boldsymbol{X}^{(3)\,\mathrm{T}} \hat{\boldsymbol{\mu}}_{\mathrm{S},0} \boldsymbol{A}_{\ni}\,, \tag{5.125}$$

$$\mathcal{W}_{00}^{(10)} = \sum_{b=1}^{7} \mathcal{W}_{00}^{(10,b)}, \tag{5.126}$$

$$\mathcal{W}_{00}^{(10,1)} = +\frac{1}{2}\big(\nabla_3 \times \boldsymbol{A}_{\ni}\big)^{\mathrm{T}}\big(\nabla_3 \times \boldsymbol{A}_{\ni}\big) \quad = +\frac{1}{2} \boldsymbol{B}_3 \boldsymbol{B}_3\,,$$

$$\mathcal{W}_{00}^{(10,2)} = +\frac{1}{2}\big(\nabla_3 \times \boldsymbol{A}_{\ni}\big)^{\mathrm{T}} \mathsf{X}^{(1)}\big(\nabla_3 \times \boldsymbol{A}_{\ni}\big) = +\frac{1}{2} \boldsymbol{B}_3 \mathsf{X}^{(1)} \boldsymbol{B}_3\,, \tag{5.127}$$

$$\mathcal{W}_{00}^{(10,3)} = +\frac{1}{2}\big(\nabla_3 \times \boldsymbol{A}_{\ni}\big)^{\mathrm{T}} \mathsf{X}^{(2)}\big(\nabla_3 \times \boldsymbol{A}_{\ni}\big) = +\frac{1}{2} \boldsymbol{B}_3 \mathsf{X}^{(2)} \boldsymbol{B}_3\,,$$

$$\mathcal{W}_{00}^{(10,4)} = -\frac{1}{2}\big(\nabla_0 \boldsymbol{\theta}^{\ni}\big) \bullet \big(\nabla_0 \boldsymbol{\theta}_{\ni}\big) \quad = -\frac{1}{2} \mathsf{X}^{(4)} \bullet \mathsf{X}^{(3)}\,,$$

$$\mathcal{W}_{00}^{(10,5)} = -\frac{1}{4}\big(\nabla_0 \boldsymbol{\theta}_{\ni}\big) \bullet \big(\nabla_0 \boldsymbol{\theta}_{\ni}\big) \quad = -\frac{1}{4} \mathsf{X}^{(3)} \bullet \mathsf{X}^{(3)}\,, \tag{5.128}$$

$$\mathcal{W}_{00}^{(10,6)} = -\frac{1}{2}\big(\nabla_0 \boldsymbol{\theta}_{\ni}\big) \bullet \big(\boldsymbol{\theta}^{\ni} \nabla_0 \boldsymbol{\theta}_{\ni}\big) \quad = -\frac{1}{2} \mathsf{X}^{(3)} \bullet \boldsymbol{\theta}^{\ni} \mathsf{X}^{(3)}\,,$$

$$\mathcal{W}_{00}^{(10,7)} = -\frac{1}{4}\big(\boldsymbol{\theta}^{\ni} \nabla_0 \boldsymbol{\theta}_{\ni}\big) \bullet \big(\boldsymbol{\theta}^{\ni} \nabla_0 \boldsymbol{\theta}_{\ni}\big) = -\frac{1}{4} \boldsymbol{\theta}^{\ni} \mathsf{X}^{(3)} \bullet \boldsymbol{\theta}^{\ni} \mathsf{X}^{(3)}\,.$$

476   5. Future Materials

Let us consider the equation of geodetic lines (5.129) ($\mu, \nu, \epsilon = 0, 1, 2, 3$) split in an equation of motion for the generalized time coordinate q^0 and in an equation of motion for the generalized space coordinates q^i as given by (5.130) and (5.131) ($i = 1, 2, 3$), where the Christoffel symbols without metric tensor splitting are given by (5.132) and with metric tensor splitting are given by (5.133). As it is shown in [63], restricting ourselves to relatively small space time domains, in the non-relativistic limiting case, the Newtonian equation of motion (5.134) specified by the Newtonian forces collected in Box 5.9 and Box 5.10 results. In the context of the Einstein field theory, which is defined by the Einstein field equations and the equation of geodetic lines, these are considered as gravitational forces, in particular, defining gravitational analogues of electromagnetic forces. In the context of the generalized Einstein field theory (GEFT), which in the first instance is defined by the generalized Einstein field equations that are presented above, these are considered as general forces, in particular, including gravitational forces and electromagnetic forces.

$$\frac{\mathrm{d}^2 q^\epsilon}{\mathrm{d}s^2} + \sum_{\mu,\nu} \Gamma^\epsilon_{\mu\nu} \frac{\mathrm{d}q^\mu}{\mathrm{d}s} \frac{\mathrm{d}q^\nu}{\mathrm{d}s} = 0 \;, \tag{5.129}$$

$$\frac{\mathrm{d}^2 q^0}{\mathrm{d}s^2} + \sum_{\mu,\nu} \Gamma^0_{\mu\nu} \frac{\mathrm{d}q^\mu}{\mathrm{d}s} \frac{\mathrm{d}q^\nu}{\mathrm{d}s} = 0 \;, \tag{5.130}$$

$$\frac{\mathrm{d}^2 q^i}{\mathrm{d}s^2} + \sum_{\mu,\nu} \Gamma^i_{\mu\nu} \frac{\mathrm{d}q^\mu}{\mathrm{d}s} \frac{\mathrm{d}q^\nu}{\mathrm{d}s} = 0 \;, \tag{5.131}$$

$$\Gamma^\epsilon_{\mu\nu} = \frac{1}{2} \sum_\lambda g^{\epsilon\lambda} \left(\frac{\partial g_{\lambda\mu}}{\partial q^\nu} - \frac{\partial g_{\mu\nu}}{\partial q^\lambda} + \frac{\partial g_{\nu\lambda}}{\partial q^\mu} \right), \tag{5.132}$$

$$\begin{aligned} \Gamma^\epsilon_{\mu\nu} = &+ \frac{1}{2} \sum_\lambda \eta^{\epsilon\lambda} \left(\frac{\partial \gamma_{\lambda\mu}}{\partial q^\nu} - \frac{\partial \gamma_{\mu\nu}}{\partial q^\lambda} + \frac{\partial \gamma_{\nu\lambda}}{\partial q^\mu} \right) \\ &+ \frac{1}{2} \sum_\lambda \gamma^{\epsilon\lambda} \left(\frac{\partial \gamma_{\lambda\mu}}{\partial q^\nu} - \frac{\partial \gamma_{\mu\nu}}{\partial q^\lambda} + \frac{\partial \gamma_{\nu\lambda}}{\partial q^\mu} \right), \end{aligned} \tag{5.133}$$

$$m_0 \frac{\mathrm{d}^2 \boldsymbol{x}}{\mathrm{d}t^2} = \sum_{a=1}^{14} \boldsymbol{F}_a \;. \tag{5.134}$$

A mathematical model: the generalized metric field model

In the spirit of the generalized Einstein field theory (GEFT), which in the first instance is defined by the generalized Einstein field equations that are presented above, this is managed by the mass–charge exchange relation (5.136). Sure enough, specifying $\boldsymbol{A}_{\ni}$ of $\boldsymbol{F}_3$ according to (5.138), where c is the light velocity, $\boldsymbol{A} = \boldsymbol{A}_\mathrm{C}$ is the vector potential, and U is a yet undefined voltage, via the Newtonian equation of motion (5.134), we arrive at the electric force (5.139) if we apply the mass–charge exchange relation (5.136), identifying U as the voltage that leads to the same acceleration of the mass m_0 as the charge q. By contrast, (5.140) and (5.141) here define the gravitational analogue. Sure enough, specifying $\boldsymbol{B}_3$ of $\boldsymbol{F}_8$ according to (5.142), where $\boldsymbol{B} = \boldsymbol{B}_\mathrm{C}$ is the vector of the magnetic field, in the same manner, we arrive at the Lorentz force (5.143) if we again apply the mass–charge exchange relation (5.136), again identifying U as the voltage that leads to the same acceleration of the mass m_0 as the charge q. By contrast, (5.144) and (5.145) here define the gravitational analogue. Sure enough, combining the vector potential $\boldsymbol{A} = \boldsymbol{A}_\mathrm{C}$ and its gravitational analogue $\boldsymbol{A}_\mathrm{g}$, combining the vector of the magnetic field $\boldsymbol{B} = \boldsymbol{B}_\mathrm{C}$ and its gravitational analogue $\boldsymbol{B}_\mathrm{g}$, and last but not least, combining the Coulomb potential ϕ_C and the gravitational potential ϕ_g, according to (5.147), we arrive at the Newtonian equation of motion (5.146) if we again apply the mass–charge exchange relation (5.136), and the Newtonian equation of motion (5.146) covers both mass scenarios and charge scenarios.

The mass–charge exchange relation (5.136) defines the first central step towards the total unification of the notions *mass*, *charge*, and *metric* in our generalized type of Einstein field theory. The combined mass–charge density (5.151) defines the second central step towards the total unification of the notions *mass*, *charge*, and *metric* in our generalized type of Einstein field theory. According to (5.148), concatenating the combined mass–charge potential (5.150) and the combined mass–charge density (5.151), the potential equations for mass densities and charge densities are readily obtained according to (5.153) and (5.154), where K is Einstein's constant of gravitation and G is Newton's constant of gravitation. According to (5.156), implementing the combined mass–charge density (5.151) within the energy momentum tensor, the field equation (5.155) with $\Phi = -\gamma_{00}/2 = \phi_{\mathrm{g,C}}$ results in (5.148). Certainly, the field equation (5.155) is a limiting case of the Einstein field equations. However, since the Einstein field theory certainly does know the alliance of masses and metric, but does not know the alliance of charges and metric, it is better to consider the field equation (5.155) as limiting case of the generalized Einstein field equations. At all events, eventually establishing a generalized metric field model, this interweaves the notions *mass*, *charge*, and *metric* to a unity. At all events, this supplies us with a justification for the actual situation that we apply the generalized Einstein field equations and their limiting cases to both mass scenarios and charge scenarios. In particular, specifying such field equations by mass-specific energy impulse tensors, we apply such field equations to cosmic systems, but specifying these by wave-specific *and* particle-specific energy impulse tensors, we apply these to wave–particle systems ("quantum systems"). Figures 5.10–5.14 should supply the reader of an impression of this central idea.

Box 5.9 (Newtonian forces, $\Phi = -\gamma_{00}/2$, $v = dx/dt$, $B_3 = \nabla \times A_\ni$).

$$F_1 = -m_0 c^2 \nabla \Phi \qquad \Leftrightarrow \qquad F_2 = -m_0 c^2 \theta^\ni \nabla \Phi ,$$

$$F_3 = -m_0 c \frac{\partial A_\ni}{\partial t} \qquad \Leftrightarrow \qquad F_4 = -m_0 c \theta^\ni \frac{\partial A_\ni}{\partial t} ,$$

$$F_6 = -m_0 \frac{\partial \theta_\ni}{\partial t} v \qquad \Leftrightarrow \qquad F_7 = -m_0 \theta^\ni \frac{\partial \theta_\ni}{\partial t} v , \qquad (5.135)$$

$$F_8 = +m_0 c (v \times B_3) \qquad \Leftrightarrow \qquad F_9 = +m_0 c \theta^\ni (v \times B_3) ,$$

$$F_{13} = -\frac{1}{2} m_0 \left[(\nabla \circ \theta_\ni)^T v \right] v \qquad \Leftrightarrow \qquad F_{14} = -\frac{1}{2} m_0 \theta^\ni \left[(\nabla \circ \theta_\ni)^T v \right] v .$$

Continuation of the mathematical model.

$$\begin{aligned} qU &= m_0 c^2 \\ &\Leftrightarrow \\ m_0 &= qU/c^2 \\ &\Leftrightarrow \\ q &= m_0 c^2/U \\ &\Leftrightarrow \\ U &= m_0 c^2/q . \end{aligned} \qquad (5.136)$$

Box 5.10 (Newtonian forces, $\Phi = -\gamma_{00}/2$, $v = dx/dt$).

$$F_5 = +2m_0 c A^\ni \frac{\partial \Phi}{\partial t} ,$$

$$F_{10} = +2m_0 c A^\ni \left[\nabla \Phi \right]^T v ,$$

$$F_{11} = -m_0 A^\ni \left[\left(\nabla \cdot A_\ni \right)^T v \right]^T v , \qquad (5.137)$$

$$F_{12} = +\frac{1}{2c} m_0 A^\ni \left[\left(\frac{\partial \theta_\ni}{\partial t} \right)^T v \right]^T v .$$

Example 5.2 ($\boldsymbol{F}_3$: electric force, gravitational analogue).

$$\boldsymbol{A}_\ni := \frac{c}{U}\boldsymbol{A} := \frac{c}{U}\boldsymbol{A}_\mathrm{C}\,,\quad \mathrm{Dim}\,[\boldsymbol{A}_\ni] = 1 = \mathrm{Dim}\left[\frac{c}{U}\boldsymbol{A}_\mathrm{C}\right] = \frac{\mathrm{m}}{\mathrm{s}}\frac{1}{\mathrm{V}}\frac{\mathrm{V\,s}}{\mathrm{m}}\,, \quad (5.138)$$

$$m_0\frac{\mathrm{d}^2\boldsymbol{x}}{\mathrm{d}t^2} = -\frac{m_0 c^2}{U}\frac{\partial \boldsymbol{A}_\mathrm{C}}{\partial t} = -q\frac{\partial \boldsymbol{A}_\mathrm{C}}{\partial t}\,, \quad (5.139)$$

$$\boldsymbol{A}_\ni := \frac{1}{c}\boldsymbol{A}_\mathrm{g}\,,\quad \mathrm{Dim}\,[\boldsymbol{A}_\ni] = 1 = \mathrm{Dim}\left[\frac{1}{c}\boldsymbol{A}_\mathrm{g}\right] = \frac{\mathrm{s}}{\mathrm{m}}\frac{\mathrm{m}}{\mathrm{s}}\,, \quad (5.140)$$

$$m_0\frac{\mathrm{d}^2\boldsymbol{x}}{\mathrm{d}t^2} = -m_0\frac{\partial \boldsymbol{A}_\mathrm{g}}{\partial t}\,. \quad (5.141)$$

Example 5.3 ($\boldsymbol{F}_8$: Lorentz force, gravitational analogue).

$$\boldsymbol{B}_3 := \frac{c}{U}\boldsymbol{B} := \frac{c}{U}\boldsymbol{B}_\mathrm{C}\,,\quad \mathrm{Dim}\,[\boldsymbol{B}_3] = \frac{1}{\mathrm{m}} = \mathrm{Dim}\left[\frac{c}{U}\boldsymbol{B}_\mathrm{C}\right] = \frac{\mathrm{m}}{\mathrm{s}}\frac{1}{\mathrm{V}}\frac{\mathrm{V\,s}}{\mathrm{m}^2}\,, \quad (5.142)$$

$$m_0\frac{\mathrm{d}^2\boldsymbol{x}}{\mathrm{d}t^2} = \frac{m_0 c^2}{U}(\boldsymbol{v}\times\boldsymbol{B}_\mathrm{C}) = q(\boldsymbol{v}\times\boldsymbol{B}_\mathrm{C})\,, \quad (5.143)$$

$$\boldsymbol{B}_3 := \frac{1}{c}\boldsymbol{B}_\mathrm{g}\,,\quad \mathrm{Dim}\,[\boldsymbol{B}_3] = \frac{1}{\mathrm{m}} = \mathrm{Dim}\left[\frac{1}{c}\boldsymbol{B}_\mathrm{g}\right] = \frac{\mathrm{s}}{\mathrm{m}}\frac{1}{\mathrm{s}}\,, \quad (5.144)$$

$$m_0\frac{\mathrm{d}^2\boldsymbol{x}}{\mathrm{d}t^2} = m_0(\boldsymbol{v}\times\boldsymbol{B}_\mathrm{g})\,. \quad (5.145)$$

Example 5.4 ($\boldsymbol{F}_1 + \boldsymbol{F}_3 + \boldsymbol{F}_8$, $\Phi = -\gamma_{00}/2 = \phi_{\mathrm{g,C}}$).

$$m_0\frac{\mathrm{d}^2\boldsymbol{x}}{\mathrm{d}t^2} = -m_0 c^2 \nabla_{\boldsymbol{x}}\phi_{\mathrm{g,C}} - m_0 c^2 \frac{\partial \boldsymbol{A}_{\mathrm{g,C}}}{\partial t} + m_0 c^2(\boldsymbol{v}\times\boldsymbol{B}_{\mathrm{g,C}})$$

$$= -qU\nabla_{\boldsymbol{x}}\phi_{\mathrm{g,C}} - qU\frac{\partial \boldsymbol{A}_{\mathrm{g,C}}}{\partial t} + qU(\boldsymbol{v}\times\boldsymbol{B}_{\mathrm{g,C}})\,, \quad (5.146)$$

$$\phi_{\mathrm{g,C}} = \frac{\phi_\mathrm{g}}{c^2} + \frac{\phi_\mathrm{C}}{U}\,,\quad \boldsymbol{A}_{\mathrm{g,C}} = \frac{\boldsymbol{A}_\mathrm{g}}{c^2} + \frac{\boldsymbol{A}_\mathrm{C}}{U} = \frac{\boldsymbol{A}_\mathrm{g}}{c^2} + \frac{\boldsymbol{A}}{U}\,,$$

$$\boldsymbol{B}_{\mathrm{g,C}} = \frac{\boldsymbol{B}_\mathrm{g}}{c^2} + \frac{\boldsymbol{B}_\mathrm{C}}{U} = \frac{\boldsymbol{B}_\mathrm{g}}{c^2} + \frac{\boldsymbol{B}}{U}\,. \quad (5.147)$$

Continuation of the mathematical model.

$$-\frac{1}{2}\triangle \phi_{g,C} = +\frac{K}{4}\rho_{g,C}(\boldsymbol{x})c^2 \;, \tag{5.148}$$

$$\phi_{g,C} = -\frac{Kc^2}{8\pi}\int_{V'}\frac{\rho_{g,C}(\boldsymbol{x}')}{|\boldsymbol{x}-\boldsymbol{x}'|}\,\mathrm{d}V' \;, \tag{5.149}$$

$$\phi_{g,C} = \frac{\phi_g}{c^2} + \frac{\phi_C}{U} \;, \tag{5.150}$$

$$\rho_{g,C}(\boldsymbol{x}) = \rho_g(\boldsymbol{x}) + \lambda_C \rho_C(\boldsymbol{x}) \;, \tag{5.151}$$

$$\lambda_C = -\frac{2}{Kc^2\epsilon_0 U} \;, \tag{5.152}$$

$$\begin{aligned}-\frac{1}{2}\triangle \phi_g &= +\frac{Kc^2}{4}\rho_g(\boldsymbol{x})c^2 \\ &\Downarrow \\ -\frac{1}{2}\triangle \phi_g &= +2\pi G \rho_g(\boldsymbol{x}) \;,\end{aligned} \tag{5.153}$$

$$\begin{aligned}-\frac{1}{2}\triangle \phi_C &= +\frac{Kc^2}{4}\lambda_C \rho_C(\boldsymbol{x})U \\ &\Downarrow \\ -\frac{1}{2}\triangle \phi_C &= -\frac{1}{2\epsilon_0}\rho_C(\boldsymbol{x}) \;,\end{aligned} \tag{5.154}$$

$$\triangle_3 \gamma_{00} = -2KT^{*\,\prime}_{00} \;,\quad T^{*\,\prime}_{00} = T_{00} - \frac{1}{2}T^{00} \;, \tag{5.155}$$

$$T_{\mu\nu} = \begin{pmatrix}-\rho_g c^2 - \lambda_C \rho_C c^2 & 0 & 0 & 0 \\ 0 & 0 & 0 & 0 \\ 0 & 0 & 0 & 0 \\ 0 & 0 & 0 & 0\end{pmatrix} = T^{\mu\nu} \;. \tag{5.156}$$

We here point at a further aspect of our self-consistent network of model conceptions. In the first instance, (3.156) and (3.157) of Box 3.22 reveal us by example that there is a path that leads from thermodynamic functions specified by microsystem functions to microsystem equations defining (meta-)stable microsystem states. In second instance, (3.156) and (3.157) of Box 3.22 reveal us by example that the microsystem equations defining (meta-)stable microsystem states are covered by our self-consistent network of model conceptions. Testing cases that go beyond this special example, we easily realize that not only microsystem equations defining (meta-)stable microsystem states of normal conducting materials and superconducting materials can be set up, but also microsystem equations defining (meta-)stable microsystem states of atoms, molecules, non-conducting non-biomimetic materials, non-conducting biomimetic materials, etc. can be set up. Testing cases that go beyond this special example, we additionally realize that the variation, in practice, the differentiation of a thermodynamic function F specified by microsystem functions X with respect to the microsystem functions X at minimum points formally manages the transition from the first level of state to the second level of state. Testing cases that go beyond this special example, we additionally realize that absolute minima are associated with stable microsystem states, whereas relative minima are associated with meta-stable microsystem states. Finally, we here want to note that some research groups go ahead and do research work that exactly points into this direction [27]. Certainly, these research groups tread only similar paths, but apart from that these research groups indeed explore the interrelation of energy landscapes spanned by thermodynamic functions (such as the free enthalpy G) and stable/meta-stable states of molecules (such as $YBa_2Cu_3O_{7-\delta}$).

We here point at a further aspect of the generalized Einstein field theory, which includes our self-consistent network of model conceptions, which exclusively deals with microscopic systems, as limiting case. Firstly, directly starting from our self-consistent network of model conceptions, reversing the above argumentations, it is possible to develop a non-covariant thermodynamics based upon microscopic functions. Secondly, directly starting from the generalized Einstein field theory, reversing and generalizing the above argumentations, it is possible to develop a covariant thermodynamics, in the first instance, based upon microscopic functions, and in second instance, based upon mesoscopic, macroscopic, or cosmologic functions. Picking up the above threads, we also annotate that such covariant interrelations are a prerequisite for the theroetical treatment of the behavior of materials at velocities near the light velocity and the behavior of materials in the presence of high energies, for example, in the vicinity of extremly energy-rich objects such as extremly mass-rich planets and stars. Upgrading the above threads, we also annotate that the covariance of the starting equations, the generalized Einstein field equations, implies the many possibilities of specification of the starting equations, the generalized Einstein field equations. Certainly, for the engineer of the present, who develops materials for ignition plugs, engines, turbines etc., this knowledge is not important. However, for the engineer of the future, who will develop much more advanced materials for much more advanced needs such as advanced rockets, this knowledge will be important.

482   5. Future Materials

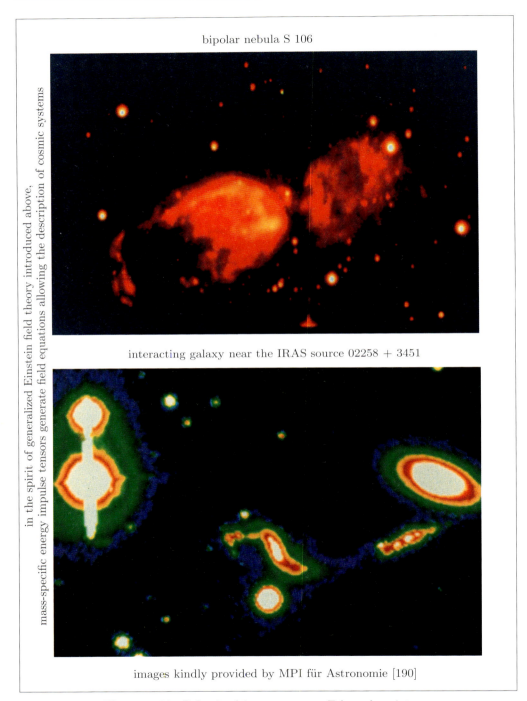

Figure 5.10. Galactic objects, part one. False color pictures.

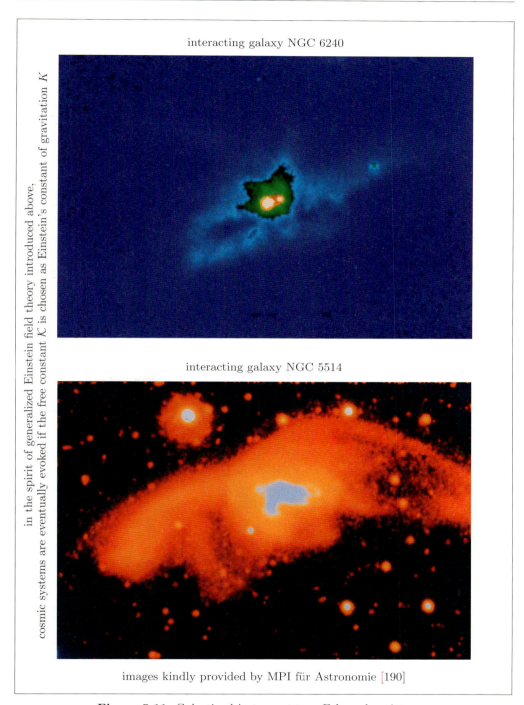

Figure 5.11. Galactic objects, part two. False color pictures.

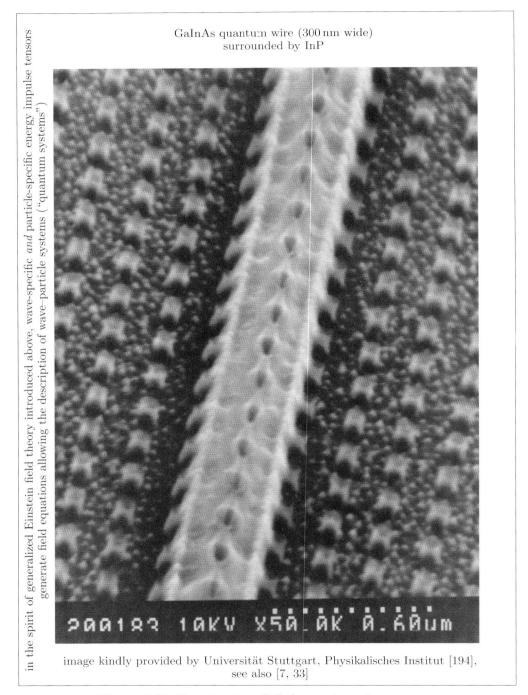

Figure 5.12. Nanostructure. GaInAs quantum wire, part one.

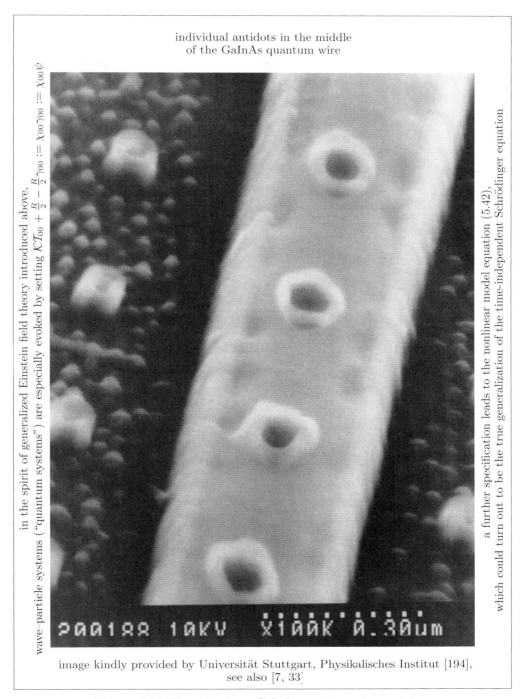

Figure 5.13. Nanostructure. GaInAs quantum wire, part two.

486   5. Future Materials

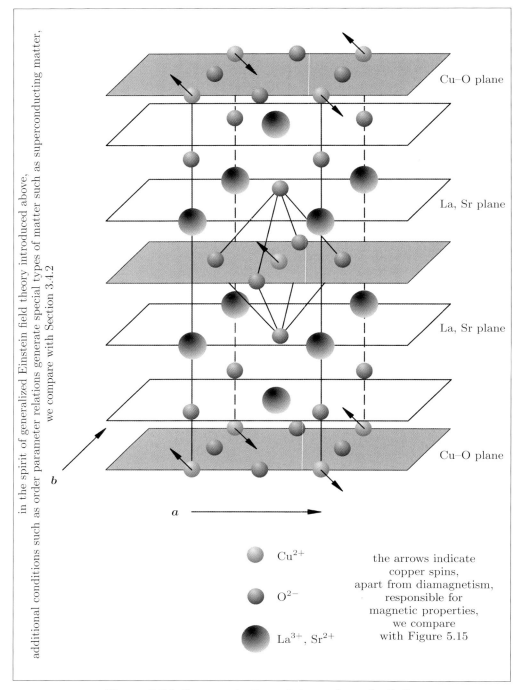

Figure 5.14. Superconducting substance. $La_{2-x}Sr_xCuO_4$.

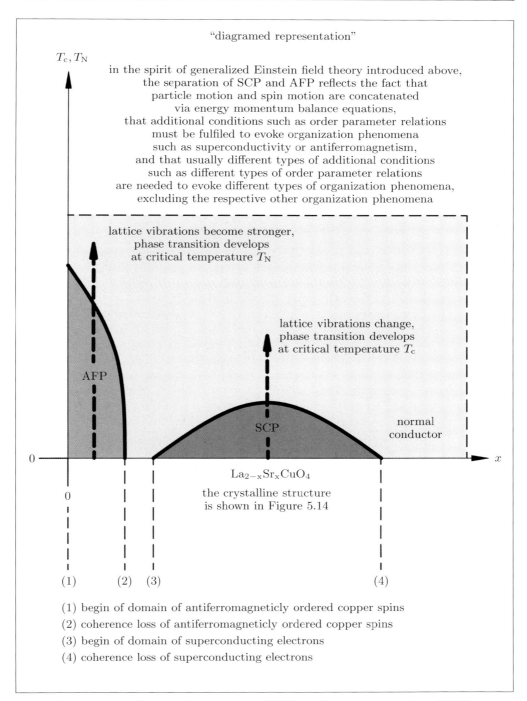

Figure 5.15. Superconducting phase (SCP), antiferromagnetic phase (AFP).

5.2 Advanced Applications

One of the most worrying problems of today is the lack of technically available energy, i. e. solar energy is not enough to satisfy the earth's needs, nuclear energy is inseparably combined with highly dangerous radioactive waste, and the combustion of coal and oil has a bad influence on the earth's climate! It was a dream of ancient scholars to convert lead into gold, and it is a vision of modern science fiction authors to convert people into radiation beams, to send these radiation beams to distant places, and finally to re-convert these radiation beams into people! There is no doubt, the human race is on the way to conquer the outer space, but this way will be more than painful provided we do this without having technologies allowing us to generate artifical gravitation! But how can we produce enough technically available energy without obtaining unwanted remains? But how can we convert one form of matter into another form of matter? But how can we manipulate, or better, how can we generate artifical gravitation?

Certainly, we here cannot answer these questions in great detail. However, reflecting a little bit about the generalized Einstein field theory, which in the first instance is defined by the generalized Einstein field equations presented above, we finally derive the following ideas. Consequently applying the nonlinear concept presented above in elementary particle physics, atom physics, molecule physics, and crystal physics, in many ways going beyond conventional quantum mechanics and quantum field theory, we expect to obtain the internal control parameters, the external control parameters, and the conditions that control the behavior of elementary particles, atoms, molecules, and crystals including the diverse kinds of transitions, i. e. we expect to obtain the "on/off switches" and "controllers" that manage the behavior of elementary particles, atoms, molecules, and crystals including the diverse kinds of transitions. Consequently analyzing the "on/off switches" and "controllers", it should be possible to construct machines that make use of the "on/off switches" and the "controllers", in a first step, gaining a comprehensive access to materials of any kind, and in a second step, designing future technologies such as the future technologies that are mentioned above. Indeed, the application of this nonlinear concept to superconducting states, we compare with Section 3.4.2, illustrates the power of this nonlinear concept plainly.

Certainly, the engineers/material scientists among the readers will ask themselves what they at once can do to widen the gateway to these new fields of materials research. Well, on the one hand, we should launch research projects aiming at the development of mass oscillators, including suitable mass systems and suitable excitation mechanisms, and on the other hand, we should launch research projects aiming at the simulation of mass oscillators, preparing the ground for artifical gravitation technologies. Well, in a first step, we should specify the superconductivity model presented in Section 3.4.2 for selected crystal structures and should learn more about intrinsic transition conditions, and in a second step, we should extend the superconductivity model presented in Section 3.4.2 for electrons/protons/ions under the influence of electromagnetic modes instead of vibrational modes and should search for decay conditions, in particular, for decays conditions "particle–radiation", preparing the ground for energy production technologies that go beyond chemical and nuclear reactions.

5.2.1 Artifical Gravitation Technologies

Regarding the conception of artifical gravitation technologies, let us make the following more precise proposal. Let us think about a functional scenario of the following kind. Let us explore the oscillatory capability of different classes of materials, in particular, let us capitalize on the structuring effect of biological macromolecules in the framework of bio-inspired mineralization procedures, let us generate ceramic classes of materials characterized by different classes of structural patterns (rods, plates, sponges etc.), and let us explore the oscillatory capability of the ceramic classes of materials. Let us try out different classes of excitation mechanisms and different classes of vibrational modes (transversal, longitudinal, circular, torsional etc.). However, since most ceramic classes of materials must be characterized as brittle, it could be necessary to design composites that concatenate ceramic parts with metallic parts and/or polymeric parts. Our task then should be to search for mass systems and excitation mechanisms enabling extreme oscillatory motion, preparing the ground for artifical gravitation technologies.

5.2.2 Energy Production Technologies

Regarding the conception of energy production technologies that go beyond chemical and nuclear reactions, let us make the following more precise proposal. Let us think about a functional scenario of the following kind. Let us collide two electron beams (negative charge!) and two proton/ion beams (positive charge!) starlike and centrally in one point, preparing a mass ratio and a charge ratio comparable to the mass ratio and the charge ratio of an electron–positron (matter–antimatter) system. Let us try out different classes of constraints, for example, different materials that are placed at the point of collision implementing secondary conditions. Our task then should be to search for decay conditions "particle–radiation", preparing the ground for energy production technologies that go beyond chemical and nuclear reactions.

Certainly, these ideas are more than visionary. Certainly, it would be much work to develop the materials and the technological environments allowing the implementation of all this. However, we – the authors of this book – think that it is high time to pluck up courage and to follow Wernher von Braun, the famous rocket engineer, who lived according to the principle that everything of which man can develop a vision is feasible. Having said this, we want to finish our survey of ceramic materials as well as our (hopefully motivating) reflections about the future of materials.

Bibliography

[A] *References: Scientific Publications*

1. Aldinger F.: Ultrafine Structures and Superplasticity of Ceramics, *in* Focus on Materials 2 (Max-Planck-Institut für Metallforschung 2006)
2. Arlt G., Hennings D., de With G.: J. Appl. Phys. 58, 1619 (1985)
3. Bardeen J., Cooper L. N., Schrieffer J. R.: Phys. Rev. 108 (5), 1175–1205 (1957)
4. Barsoum M. W.: Fundamentals of Ceramics (McGraw-Hill 1997)
5. Batchelder D. N., Simmons R. O.: Lattice Constants and Thermal Expansivities of Silicon and of Calcium Fluoride between 6 and 322 K, JCPSA6, 41, 2324–2329 (1964)
6. Bednorz J. G., Müller K. A.: Z. Phys. B 64, 189 (1986)
7. Bergmann R., Menschig A., Lichtenstein N., Hommel J., Härle V., Scholz F., Schweizer H., Grützmacher D.: Investigation of Boundary Scattering in Dry Etched Quantum Wires, Microelect. Eng. 23, 429–432 (Elsevier Science, Amsterdam 1994)
8. Bloch C.: Dominisis C, Nucl. Phys. 7, 459 (1958)
9. Bogoljubov N. N.: Zh. Eksp. Teor. Fiz. 34, 58 (1958)
10. Bonnig K. et al.: Quantitative analysis of the corrosion rates of palladium alloy, Dtsch. Azhnarztl. A 45, 508–510 (1990)
11. Brill H.: Mikrobielle Materialzerstörung und Materialschutz – Schädigungsmechanismen und Schutzmaßnahmen (Fischer Verlag, Stuttgart 1995)
12. Deng S., Simon A., Köhler J.: International Conference on Dynamic Inhomogenities in Complex Oxides, June 14–20, Bled, Slovenia (2003)
13. Deng S., Simon A., Köhler J.: J. Supercond. 17, 227 (2004)
14. Downs J. W., Gibbs G. V.: An exploratory examination of the electron density and electrostatic potenial of phenakite, AMMIAY, 72, 769–777 (1987)
15. Fabrichnaya O., Lakiza S., Wang C., Zinkevich M., Levi C. G., Aldinger F.: Thermodynamic database for the ZrO_2-$YO_{3/2}$-$GdO_{3/2}$-$AlO_{3/2}$ system: application for thermal barrier coatings, Journal of Phase Equilibria and Diffusion, 27, 4, 343–352 (2006)
16. Finger, L. W., Hazen, R. M.: Crystal structure and compression of ruby to 46 kbar JAPIAU,49,5823-5826 (1978)
17. Fuchs G.: Allgemeine Mikrobiologie, 8. Auflage. (Thieme, Stuttgart 2006)
18. Fuchs T., Hoffmann R., Niesen T., Tew H., Bill J., Aldinger F.: J. Mater. Chem. 12, 1597–1601 (2002)
19. Gallardo-López A., Muñoz A., Martínez-Fernández J., Domínguez-Rodríguez A.: Acta Mater 47, 2185 (1999)
20. Ginzburg V. L., Landau L. D.: Zh. Eksp. Teor. Fiz. 20, 1064 (1950)
21. Gonschorek W.: X-Ray charge density study of rutile (TiO_2), ZEKRDZ, 160, 187–203 (1982)
22. Gruner J. W.: The Crystal Structures of Talc and Pyrophyllite, ZEKGAX, 88, 412–419 (1934)
23. Hanke K.: Beitraege zu Kristallstrukturen vom Olivin-Typ, BMUPA4, 11, 535–558 (1965)

24. Hassan I., Grundy H. D.: The Crystal Structures of Sodalite-Group Minerals, ASBSDK, 40, 6–13 (1984)
25. Hassel O.: Ueber die Kristallstruktur des Graphits, ZEPYAA, 25, 317–337 (1924)
26. Hoffmann R., Fuchs T., Niesen T., Bill J., Aldinger F.: Surf Interf. Anal. 34, 708–711(2002)
27. Jansen M.: The Deductive Approach to Chemistry, a Paradigm Shift, in: Turning points in Solid-State, Materials and Surface Science, eds.: K. M. Harris and P. Edwards (RSC Publishing, Cambridge 2008)
28. Kingery W. D.: Introduction to Ceramics, 2nd edition (Wiley-VCH 1976)
29. Kresin V., Wolf S.: Physica C 169, 476 (1990)
30. London F., London H.: Z. Phys. 96, 359–364 (1935)
31. Loskalenko V.: Fiz Metal Metallov 8, 503 (1959)
32. Maher P., K., Hunter F. D., Scherzer J.: Crystal Structures of Ultrastable Faujasites, ADCSA, 101, 266–278 (1971)
33. Mahler G., Weberruß V. A.: Quantum Networks. Dynamics of Open Nanostructures, 2nd edition (Springer, New York Berlin Heidelberg 1998), ISBN: 3-540-58850-7
34. Main J., Wunner G.: Hydrogen Atom in a Magnetic Field: Ghost Orbits, Catastrophes, and Uniform Semiclassical Approximations (Physical Review A, 55, 3, 1997)
35. Main J., Wunner G.: Uniform Semiclassical Approximations for Umbilic Bifurcation Catastrophes (Physical Review E, 57, 6, 1998)
36. Mehring M., Weberruß V. A.: Object-Oriented Magnetic Resonance. Classes and Objects, Calculations and Computations (Academic Press, London 2001), ISBN: 0-12-740620-4
37. Micnas R., Robaszkiewicz S., Bussmann-Holder A.: Two-Component Scenarios for Non-Conventional Superconductors, Struct. Bond. 114, 13 (2005)
38. Morosin B.: Structure and thermal expansion of beryl, ACBCAR, 28, 1899–1903 (1972)
39. Mursic Z., Vogt T., Boysen H., Frey F.: Single-crystal neutron diffraction study of metamict zircon up to 2000 K, JACGAR, 25, 519–523 (1992)
40. Niesen T. P., Bill J., Aldinger F.: Deposition of titania thin films by a peroxide route on different functionalized self-assembled monolayers, J. Mater. Chem. 13, 1552–1559 (2001)
41. Northrup P. A., Leinenweber K., Parise J. B.: The location of H in the high-pressure synthetic Al2 Si O4 (O H)2 topaz analogue, AMMIAY, 79, 401–404 (1994)
42. Pauling L.: The principles determining the structure of complex ionic crystals, J. Am. Chem. Soc., 51 (4), 1010–1026 (1929)
43. Pechini M.: Method of Preparing Lead and Alkaline Earth Titanates and Niobates and Coating Method Using the Same to Form a Capacitor, U.S. Pat. No. 3330697 (1967)
44. Pillai C. K. et al.:Interaction of palladium with DNA, Biochem Biophys Acta, 474, 11–16 (1977)
45. Pluth J. J., Smith J. V., Faber J.: Crystal structure of low cristobalite at 10, 293, and 473 K: Variation of framework geometry with temperature, JAPIAU, 57, 1045–1049 (1985)
46. Sasaki S., Prewitt C. T., Bass J. D.: Orthorhombic Perovskite $CaTiO_3$ and $CdTiO_3$: Structure and Space Group, ACSCEE, 43, 1668–1674 (1987)
47. Schedle A. et al.: Response of fibroblasts to various metal cation, J. Dent. Res., 74, 1513–1520 (1995)
48. Shultz M. D. et al.: Palladium – a new inhibitor of cellulase enzyme activit, Biochem. Biophys. Res. Commun., 209, 1046–1052 (1995)
49. Shannon C. E.: Prediction and entropy of printed English, The Bell System Technical Journal, 30, 50-64 (1950)
50. Sommerfeld A.: Vorlesungen über theoretische Physik, Band V, Thermodynamik und Statistik (Harri Deutsch 1977)
51. Spear K. E.: Diamond – Ceramic Coating of the Future, J. Am. Ceram. Soc., 72, 171–191 (1989)

52. Stern O.: Z. Electrochem., 30, 508 (1924)
53. Straumanis M. E., Aka E. Z.: Precision Determination of Lattice Parameter, Coefficient of Thermal Expansion and Atomic Weight of Carbon in Diamond JACSAT, 73, 5643–5646 (1951)
54. Suhl H., Matthias B., Walker L.: Phys. Rev. Lett. 3, 552 (1959)
55. Takeda S. et al.: Corrosion behavior of Ag-Pd alloys and its cytotoxicity, Shika Zairyo Kikai, 9, 825–830 (1990)
56. Tas C. A.: Preparation of Lead Zirconate Titanate by Homogeneous Precipitation and Calcination, J. Am. Ceram. Soc. 82, 1582–1584 (1999)
57. Tas C. A., Majewski P. J., Aldinger F.: Chemical Preparation of Pure and Sr- and/or Mg-Doped Lanthanum Gallate Powders, J. Am. Ceram. Soc. 83, 2954–2960 (2000)
58. Tas C. A., Majewski P. J., Aldinger F.: Preparation of Sr- and Zn-Doped $LaGaO_3$ via Precipitation in the Presence of Urea and/or Enzyme Urease, J. Am. Ceram. Soc. 85, 1414–1420 (2002)
59. Tas C. A., Majewski P. J., Aldinger F.: Synthesis of Gallium Oxide Hydroxide Crystals in Aqueous Solutions with or without urea and their Calcination Behavior, J. Am. Ceram. Soc. 85, 1421–1429 (2002)
60. Ulrich F., Zachariasen W.: Ueber die Kristallstruktur des Cd S, sowie des Wurtzits, ZEKGAX, 62, 260–273 (1925)
61. Vezzalini G., Quartieri S., Passaglia E.: Crystal structure of a K-rich natural gmelinite and comparison with the other refined gmelinite samples, NJMMAW, 1990, 504–516 (1990)
62. Weberruß V. A.: Quantenphysik im Überblick (Oldenbourg, München Wien 1998), ISBN: 3-486-24418-3
63. Weberruß V. A.: Nichtlineare Quantenphysik (Shaker-Verlag, Aachen 2008), ISBN: 978-3-8322-7100-8
64. Weibert K., Main J., Wunner G.: Periodic Orbit Quantization of Chaotic Maps by Harmonic Inversion (Physics Letters A, 289, 329–332, 2001)
65. Wiederhorn S. M., Hockey B. J., French J. D.: J. Eur. Ceram. Soc. 19, 2273 (1999)
66. Yamanaka T., Takeuchi Y.: Order-disorder transition in Mg Al2 O4 spinel at high temperatures up to 1700 degrees C, ZEKRDZ, 165, 65–78 (1983)
67. Yarbrough W. A.: Vapor-Phase-Deposited Diamond – Problems and Potential, J. Am. Ceram. Soc., 75, 3179–3200 (1992)
68. Yeh C., Lu Z. W., Froyen S., Zunger A.: Zinc-blende–Wurtzite polytypism in semiconductors, PRBMDO, 46, 10086–10097 (1992)
69. Zachariasen W. H., Plettinger H. A.: Extinction in quartz, ACCRA9, 18, 710–714 (1965)
70. Zern A., Mayer J., Janakiraman N., Weinmann M., Bill J., Rühle M.: Quantitative EFTEM study of precursor-derived Si–B–C–N ceramics, J. Europ. Ceram. Soc. 22, 1621–1629 (2002)

[B] *References: Industry Material*

71. Material kindly made available by CeramTec, Plochingen (CeramTec Platz 1–9, 73207 Plochingen) and Marktredwitz (CeramTec-Weg 1, 95615 Marktredwitz), Germany.
72. Material kindly made available by SGL Brakes GmbH, Werner-von-Siemens-Str. 18, 86405 Meitingen, Germany

[C] *References: Scientific Material*

73. Danzer R.: Vorlesungen in Struktur- und Funktionskeramik (Montanuniversität Leoben, Austria, and Max-Planck-Institut für Metallforschung, Germany)
74. Simon A.: Lectures on Solid State Chemistry (Cornell University, USA, and Max-Planck-Institut für Festkörperforschung, Germany)
75. Weberruß V. A., Simon A., Köhler J.: Superconductivity, Advanced Modeling. First studies. (Max-Planck-Institut für Festkörperforschung, Stuttgart 2009)

[D] *References: MPI (PML) Posters*

76. Biswas K., Rixecker G., Aldinger F.: Mechanical Properties of SiC-Ceramics Sintered with Rare Earth Oxide Additives, Max-Planck-Institut für Metallforschung, PML, Stuttgart
77. Burghard Z., Durupthy O., Gerstel P., Hoffmann R., Jost M., Lipowsky P., Marhofer A., Pitta Bauermann L., Qiu Y., Röhberg J., Al-Shakran M., Tucic A., Bill J.: Bio-inspired Materials Synthesis, Max-Planck-Institut für Metallforschung, PML, Stuttgart
78. Carle V., Schmid E.: The Metallography of Palladium Dental Alloys, Max-Planck-Institut für Metallforschung, PML, Stuttgart
79. Carle V., Schmid E.: Primary Solidified γ-Phase Be_2Fe with Magnetic Domains, Max-Planck-Institut für Metallforschung, PML, Stuttgart
80. Fabrichnaya O., Lakiza S., Wang C., Zinkevitch M., Aldinger F.: Calculations of phase relations, Max-Planck-Institut für Metallforschung, PML, Stuttgart
81. Hoffmann R. C., Bill J.: Bio-inspired Materials Synthesis, Max-Planck-Institut für Metallforschung, PML, Stuttgart
82. Ittlinger E.: Hardening of Dental Amalgams, Max-Planck-Institut für Metallforschung, PML, Stuttgart
83. Peng J., Seifert H.-J., Aldinger F.: Precursor-Derived Si–(B–)C–N Ceramics – Phase Equilibria, Phase Reactions and Thermal Stability, Max-Planck-Institut für Metallforschung, PML, Stuttgart
84. Rozumek M., Datta P., Liu N., Majewski P., Maldener T., Aldinger F.: LSGM Materials for SOFCs, Max-Planck-Institut für Metallforschung, PML, Stuttgart
85. Schäfer U., Alan M., Jaeger H.: Nucleation and Growth Kinetics of the Tetragonal/Orthorhombic Transformation in $YBa_2Cu_3O_{7-x}$, Max-Planck-Institut für Metallforschung, PML, Stuttgart
86. Schäfer U., Carle V.: Siliciumcarbid-Hochleistungskeramik, Max-Planck-Institut für Metallforschung, PML, Stuttgart
87. Schäfer U., Schubert H.: TZP- and PSZ-Keramik, Max-Planck-Institut für Metallforschung, PML, Stuttgart
88. Su H.-L., Majewski P., Aldinger F.: Der Zusammenhang zwischen dem Sauerstoffgehalt und der Kationenstöchiometrie der hochtemperatursupraleitenden Phase $Bi_{2+x}(Sr, Ca)_3Cu_2O_{8+d}$ (2212-Phase), Max-Planck-Institut für Metallforschung, PML, Stuttgart
89. Täffner U., Hofmann H.: Metallographic Investigation of the Fe–Ni–W system, Max-Planck-Institut für Metallforschung, PML, Stuttgart
90. Täffner U., Rostek A.: Transmission Optical Micrograph of a Thin Section of Barium Titanate, Max-Planck-Institut für Metallforschung, PML, Stuttgart
91. Täffner U., Rostek A.: Domain Switching in $BaTiO_3$ during crack growth, Max-Planck-Institut für Metallforschung, PML, Stuttgart
92. Tas C. A., Majewski P. J., Aldinger F.: Preparation of Sr-, Mg-, or Zn-Doped $LaGaO_3$ Solid Oxide Fuel Cell Ceramics by using Chemical Synthesis Routes, Max-Planck-Institut für Metallforschung, PML, Stuttgart
93. Weinmann M., Müller A., Gerstel P., Kamphowe T. W., Nast S., Bill J., Aldinger F.: Synthese metallorganischer Polymere als molekulare Vorstufen für hochtemperaturstabile Keramiken, Max-Planck-Institut für Metallforschung, PML, Stuttgart
94. Wetzel K., Rixecker G., Aldinger F.: Herstellung dichter nanokristalliner Siliciumcarbid-Keramiken durch Sinterschmieden, Max-Planck-Institut für Metallforschung, PML, Stuttgart
95. Wildhack S., Aldinger F.: Gefriergießen von Aluminiumnitrid, Max-Planck-Institut für Metallforschung, PML, Stuttgart
96. Zinkevich M., Cancarevic M., Djurovic D., Fabrichnaya O., Golczewski J., Geupel S., Manga V. R., Solak N., Wang C., Wu B.: Computational and Experimental Phase Studies, Max-Planck-Institut für Metallforschung, PML, Stuttgart

[E] *Further Literature: MPI (PML) Works*

97. Barsoum M. W., Golczewski J., Seifert H. J., Aldinger F.: Fabrication and electrical and thermal properties of Ti2InC, Hf2InC and (Ti,Hf)2InC, Journal of Alloys and Compounds, 340, 1-2, 173–179 (2002)
98. Barsoum M. W., Salama I., El-Raghy T., Golczewski J., Porter W. D., Wang H., Seifert H. J., Aldinger F.: Thermal and electrical properties of Nb2AlC, (Ti, Nb)(2)AlC and Ti2AlC, Metallurgical and Materials Transactions A-Physical Metallurgy and Materials Science, 33, 9, 2775–2779 (2002)
99. Bill J., Aldinger F., Petzow G., Sloma M., Maier J., Riedel R.: Electrical conductivity of amine-borane-derived boron carbide nitride, Journal of Materials Science Letters, 18, 1513–1516 (1999)
100. Biswas K., Rixecker G., Aldinger F.: Creep and visco-elastic behaviour of LPS-SiC sintered with Lu2O23-AlN additive, Materials Chemistry and Physics, 104, 1, 10–17 (2007)
101. Bunjes N., Müller A., Sigle W., Aldinger F.: Crystallization of polymer-derived SiC/BN/C composites investigated by TEM, Journal of Non-Crystalline Solids, 353, 1567–1576 (2007)
102. Burghard Z., Tucic A., Jeurgens L. P. H., Hoffmann R., Bill J., Aldinger F.: Nanomechanical properties of bioinspired organic-inorganic composite films, Advanced Materials, 19, 970–974 (2007)
103. Cancarevic M.: Thermodynamic optimization of the PbO-ZrO2-TiO2 (PZT) system and its application to the processing of composites of PZT ceramics and copper, Universität Stuttgart, PhD-Thesis (2007)
104. Cancarevic M., Zinkevich M., Aldinger F.: Thermodynamic assessment of the PZT system, Journal of the Ceramic Society of Japan, 114, 11, 937–949 (2006)
105. Cancarevic M., Zinkevich M., Aldinger, F.: Thermodynamic description of the Ti-O system using the associate model for the liquid phase, CALPHAD, 31, 330–342 (2007)
106. Coelho G. C., Golczewski J. A., Fischmeister H. F.: Thermodynamic calculations for Nb-containing high-speed steels and white-cast-iron alloys, Metallurgical and Materials Transactions A, 34, 9, 1749–1758 (2003)
107. Datta P., Majewski P., Aldinger F.: Synthesis and microstructural characterization of Sr- and Mg-substituted LaGaO3 solid electrolyte, Materials Chemistry and Physics, 102, 240–244 (2007)
108. Datta P., Majewski P., Aldinger F.: Synthesis and characterization of strontium and magnesium substituted lanthanum gallate-nickel cermet anode for solid oxide fuel cells, Materials Chemistry and Physics, 102, 125–131 (2007)
109. Datta P., Majewski P., Aldinger F.: Synthesis and characterization of gadolinia-doped ceria-silver cermet cathode material for solid oxide fuel cells, Materials Chemistry and Physics, 107, 370–376 (2008)
110. Djurovic D., Zinkevich M., Aldinger F.: Thermodynamic modeling of the yttrium-oxygen system, Calphad, 31, 4, 560–566 (2007)
111. Djurovic D., Zinkevich M., Boskovic S., Srot V., Aldinger F.: Densification behaviour of nano-sized CeO2, Materials Science Forum, 555, 189–194 (2007)
112. Durupthy O., Bill J., Aldinger F.: Bioinspired synthesis of crystalline TiO2: effect of amino acids on nanoparticles structure and shape, Crystal Growth & Design, 7, 12, 2696–2704 (2007)
113. Fabrichnaya O., Lakiza S., Wang C., Zinkevich M., Aldinger F.: Assessment of thermodynamic functions in the ZrO2-La2O3-Al2O3 system, Journal of Alloys and Compounds, 453, 271–281 (2008)
114. Fabrichnaya O., Zinkevich M., Aldinger F.: Thermodynamic modeling in the ZrO2-La2O3-Y2O3-Al2O3 system, International Journal of Materials Research, 98, 9, 838–846 (2007)

115. Fischer A., Jentoft F. C., Weinberg G., Schlögl R., Niesen T., Bill J., Aldinger F., De Guire M., Ruehle M.: Characterization of thin films containing zirconium, oxygen, and sulfur by scanning electron and atomic force microscopy, Journal of Materials Research, 14, 9, 3725–3733 (1999)
116. Gautam D.: Characterization of the conduction properties of alkali metal conducting solid electrolytes using thermoelectric measurements, Universität Stuttgart, PhD-Thesis (2006)
117. Gerstel P., Hoffmann R. C., Lipowsky P., Jeurgens L. P. H., Bill J., Aldinger F.: Mineralization from aqueous solutions of zinc salts directed by amino acids and peptides, Chemistry of Materials, 18, 179–186 (2006)
118. Gerstel P., Lipowsky P., Durupthy O., Hoffmann R. C., Bellina P., Bill J., Aldinger F.: Deposition of zinc oxide and layered basic zinc salts from aqueous solutions containing amino acids and dipeptides, Journal of the Ceramic Society of Japan, 114, 11, 911–917 (2006)
119. Golczewski J. A.: Thermodynamic analysis of structural transformations induced by annealing of amorphous Si-C-N ceramics derived from polymer precursors, International Journal of Materials Research, 97, 6, 729–736 (2006)
120. Golczewski J. A.: Thermodynamic analysis of isothermal crystallization of amorphous Si-C-N ceramics derived from polymer precursors, Journal of the Ceramic Society of Japan, 114, 11, 950–957 (2006)
121. Golczewski J. A., Aldinger F.: Phase separation in Si-(B)-C-N polymer-derived ceramcis, Zeitschrift für Metallkunde, 97, 114–118 (2006)
122. Grieb B., Henig E. T., Reinsch B., Petzow G., Buschow K. H. J., de Mooij D. B., Stadelmaier H. H.: The ternary system Fe-Nd-C, Zeitschrift für Metallkunde, 92, 2, 172–178 (2001)
123. Hoffmann R. C., Jia S., Jeurgens L. P. H., Bill J., Aldinger F.: Influence of polyvinyl pyrrolidone on the formation and properties of ZnO thin films in chemical bath deposition, Materials Science and Engineering C, 26, 41–45 (2006)
124. Ishihara S., Bill J., Aldinger F., Shinoda Y., Wakai F., Nishimura T., Tanaka, H.: High-temperature deformation of Si-C-N monoliths containing residual amorphous phase derived from polyvinylsilazane, Journal of the Ceramic Society of Japan, 114, 1330, 575–579 (2006)
125. Jung H.-J., Ahn K. Y., Choi S. C., Lee S. H., Aldinger F.: Low pressure sintering of sialon using different sintering additives, Journal of the Ceramic Society of Japan, 116, 1, 130–136 (2008)
126. Kamphowe T. W., Bill J., Aldinger F., Sindermann K., Oberacker R., Hofmann M. J.: Herstellung faserverstärkter Verbundwerkstoffe aus siliciumorganischen Precursoren und deren Charakterisierung, in Forschungsergebnisse, Keramikverbund Karlsruhe-Stuttgart, 01.01.1996 – 31.12.1998 (Deutsche Keramische Gesellschaft 2000)
127. Katsuda Y., Gerstel P., Narayanan J., Bill J., Aldinger F.: Reinforcement of precursor-derived Si-C-N ceramics with carbon nanotubes, Journal of the European Ceramic Society, 26, 3399–3405 (2006)
128. Kumar R., Phillipp F., Aldinger F.: Oxidation induced effects on the creep properties of nano-crystalline porous Si-B-C-N ceramics, Materials Science and Engineering A, 445-446, 251–258 (2007)
129. Kumar R., Rixecker G., Aldinger F.: Anelasticity of precursor derived Si-B-C-N ceramics, Journal of the European Ceramic Society, 27, 2-3, 1475–1480 (2007)
130. Lakiza S., Fabrichnya O., Wang C., Zinkevich M., Aldinger F.: Phase diagram of the ZrO2-Gd2O3-Al2O3 system, Journal of the European Ceramic Society, 26, 3, 233–246 (2006)
131. Lakiza S., Fabrichnaya O., Zinkevich M., Aldinger F.: On the phase relations in the ZrO2-YO1.5-AlO1.5 system, Journal of Alloys and Compounds, 420, 1-2, 237–245 (2006)

132. Lee S.-H., Kaiser G., Rixecker G., Aldinger, F., Park, J.-Y., Auh K.-H., Choi S.-C.: Hydrothermal treatment of Si3N4 for the improvement of oxidation resistance at 1400° C, Journal of the American Ceramic Society, 91, 2, 679–682 (2008)
133. Lee S.-H., Weinmann M., Aldinger F.: Fabrication of fiber-reinforced ceramic composites by the modified slurry infiltration technique, Journal of the American Ceramic Society, 90, 8, 2657–2660 (2007)
134. Lee S.-H., Weinmann M., Aldinger F.: Processing and properties of C/Si-B-C-N fiber-reinforced ceramic matrix composites prepared by precursor impregnation and pyrolysis, Acta Materialia, 56, 1529–1538 (2008)
135. Lee S.-H., Weinmann M., Gerstel P., Rixecker G., Choi S.C., Aldinger F.: Novel precursor-derived Al-C-N-(O)-based ceramic additive for the low-temperature pressure-less sintering of silicon nitride, Materials Research, 23, 6, 1713–1721 (2008)
136. Lee S. S.: Mullite coatings on liquid-phase sintered silicon carbide by chemical vapour deposition, Universität Stuttgart, PhD-Thesis (2007)
137. Lipowsky P., Hedin N., Bill J., Hoffmann R. C., Ahniyaz A., Aldinger F., Bergström L.: Controlling the assembly of nanocrystalline ZnO films by a transient amorphous phase in solution, Journal of Physical Chemistry C, 112, 5373–5383 (2008)
138. Lipowsky P., Hoffmann R. C., Welzel U., Bill J., Aldinger F.: Site-selective deposition of nanostructured ZnO thin films from solutions containing polyvinylpyrrolidone, Advanced Functional Materials, 17, 2151–2159 (2007)
139. Liu N., Shi M., Wang C., Yuan Y. P., Majewski P., Aldinger F.: Microstructure and ionic conductivity of Sr- and Mg-doped LaGaO3, Journal of Materials Science, 41, 4205–4213 (2006)
140. Liu N., Shi M., Yuan Y. P., Chao S., Feng J. P., Majewski P., Aldinger F.: Thermal shock and thermal fatigue study of Sr- and Mg-doped lanthanum gallate, International Journal of Fatigue, 28, 3. 237–242 (2006)
141. Ludwig T.: Thermoanalytische und konstitutionelle Charakterisierung des Systems Si3N4-Y2O3-Al2O3-SiO2. Universitätt Stuttgart, PhD-Thesis (2008)
142. Maier N., Nickel K. G., Rixecker G.: Formation and stability of Gd, Y, Yb and Lu disilicates and their solid solutions, Journal of Solid State Chemistry, 179, 1630–1635 (2006)
143. Maier H.-P., Maile K., Lauf S., Schmauder S., Henke T., Krenkel W., Greiner A., Bill J., Aldinger F.: Entwicklung von keramischen Bauteilen für die Anwendung im Hochtemperaturbereich von kohlebefeuerten Kraftwerken, in Forschungsergebnisse, Keramikverbund Karlsruhe-Stuttgart, 01.01.1996 – 31.12.1998 (Deutsche Keramische Gesellschaft 2000)
144. Manga R. V.: Effect of H2S on the thermodynamic stability and electrochemical performance of Ni cermet-type of anodes for solid oxide fuel cells, Universität Stuttgart, PhD-Thesis (2006)
145. Matovic B., Rixecker G., Boskovic S., Aldinger F.: Sintering of Si3N4 with Li-exchanged zeolite additive, International Journal of Materials Research, 97, 9, 1264–1267 (2006)
146. Matovic B., Rixecker G., Boskovic S., Aldinger F.: Effect of LiYO2 addition on sintering behavior and indentation properties of silicon nitride ceramics, International Journal of Materials Research, 97, 9, 1268–1272 (2006)
147. Matovic B., Rixecker G., Golczewski J., Aldinger F.: Thermal conductivity of pressureless sintered silicon nitride materials with LiYO2 additive, Science of Sintering, 36, 1, 3–9 (2004)
148. Näfe, H.: Voltage of a solid electrolyte galvanic cell in terms of the activity of the mobile species of the electrolyte, Electrochimica Acta, 52, 7409–7411 (2007)
149. Näfe, H.: Thermodynamics of solid and molten alkali carbonates, Electrochemical and Solid State Letters, 10, 3, 9–10 (2007)
150. Näfe H., Aldinger F.: Potentialverteilung in einem sauerstoffionenleitenden Zweischicht-Festelektrolyten bei Belastung, CFI - Fortschrittsberichte der Deutschen Keramischen Gesellschaft, 15, 285–304 (2000)

151. Näfe H., Amin R., Aldinger F.: Thermodynamic characterization of the eutectic phase mixture NaNbO3/Na3NbO4. II: Solid-state electrochemical investigation, Journal of the American Ceramic Society, 90, 10, 3227–3232 (2007)
152. Näfe H., Gollhofer S., Aldinger F.: Determination of the hole conductivity of sodium-beta-alumina by potentiometric measurements, in Solid State Ionic Devices II, Ceramic Sensors, 339–349 (The Electrochemical Society, Pennington 2000)
153. Näfe H., Karpukhina N.: Na-modified cubic zirconia-link between sodium zirconate and zirconia in the Na2O-ZrO2 phase diagram, Journal of the American Ceramic Society, 90, 5, 1597–1602 (2007)
154. Näfe H., Meyer F., Aldinger F.: The equilibrium between Na-β- and Na-β''-alumina as a function of the phase composition, Electrochimica Acta, 45, 10, 1631–1638 (2000)
155. Näfe H., Subasri R.: Indication of bivariance in the phase system sodium zirconate/zirconia, The Journal of Chemical Thermodynamics, 39, 6, 972–977 (2007)
156. Niesen T., De Guire M., Bill J., Aldinger F., Rühle M., Fischer A., Jentoft F. C., Schlögl R.: Atomic force microscopy studies of oxide thin films on organic self-assembled monolayers, Journal of Materials Research, 14, 6, 2464–2475 (1999)
157. Olevsky E. A., Wang X., Bruce E., Stern M. B., Wildhack S., Aldinger F.: Synthesis of gold micro- and nano-wires by infiltration and thermolysis, Scripta Materialia, 56, 867–869 (2007)
158. Peng H., Salamon D., Bill J., Rixecker G., Burghard Z., Aldinger F., Shen Z. J.: Consolidating and deforming SiC nanoceramics via dynamic grain sliding, Advanced Engineering Materials, 9, 4, 303–306 (2007)
159. Petzow G., Herrmann M.: Silicon Nitride Ceramics, in High Performance Non-Oxide Ceramics II, 47–167 (Springer, New York Berlin Heidelberg 2002)
160. Pitta Bauermann L., Bill J., Aldinger F.: Bio-friendly synthesis of ZnO nanoparticles in aqueous solution at near-neutral pH and low temperature, Journal of Physical Chemistry B, 110, 5182–5185 (2006)
161. Pitta Bauermann L., Bill J., Aldinger F.: Bio-inspired syntheses of ZnO-protein composites, International Journal of Materials Research, 98, 9, 879–883 (2007)
162. Qian P., Gu H., Aldinger F.: Initial stage of bi-modal microstructure in TiO2-doped α-Al2O3 ceramics induced by SiO2 impurity, International Journal of Materials Research, 3, 240–244 (2008)
163. Qiu Y., Bellina P., Jeurgens L. P. H., Leineweber A., Welzel U., Gerstel P., Jiang L., van Aken P., Bill J., Aldinger F.: Aqueous deposition of ultraviolet luminescent columnar tin-doped indium hydroxide films, Advanced Functional Materials, 18, 2572–2583 (2008)
164. Rager T., Golczewski J.: Solar-thermal zinc oxide reduction assisted by a second redox pair, Zeitschrift für Physikalische Chemie, 219, 2, 235–246 (2005)
165. Saltykov P., Fabrichnaya O., Golczewski J., Aldinger F.: Thermodynamic modeling of oxidation of Al-Cr-Ni alloys, Journal of Alloys and Compounds, 381, 99–113 (2004)
166. Seifert H. J., Lukas H. L., Aldinger F.: Development of Si-B-C-N Ceramics Supported by Constitution and Thermochemistry, Ber. Bunsenges. Phys. Chem. 102, 1309–1313 (1998)
167. Seitz J., Clauß B., Schuhmacher J., Müller K., Canel J., Thurn G., Bill J., Aldinger F.: Struktur- und Funktionskeramiken aus neuartigen Precursoren im System Si/B/C/N, in Forschungsergebnisse, Keramikverbund Karlsruhe-Stuttgart, 01.01.1996 – 31.12.1998 (Deutsche Keramische Gesellschaft 2000)
168. Shi M., Liu N., Xu Y. D., Yuan Y. P., Majewski P., Aldinger F.: Synthesis and characterization of Sr- and Mg-doped LaGaO3 by using glycine-nitrate combustion method, Journal of Alloys and Compounds, 425, 1-2, 348–352 (2006)
169. Sinha A., Näfe H., Sharma, B. P., Gopalan P.: Study on ionic and electronic transport properties of calcium-doped GdAlO3, Journal of the Electrochemical Society, 155, B309–B314 (2008)
170. Stadelmaier H. H., Petzow G.: Metastable alloys at moderate cooling rates, Zeitschrift für Metallkunde, 93, 10, 1019–1023 (2002)

171. Subasri R., Näfe H.: Phase evolution on heat treatment of sodium silicate water glass, Journal of Non-Crystalline Solids, 354, 896–900 (2008)
172. Turchi P. E. A., Abrikosov I., Burton B., Fries S., Grimvall G., Kaufman L., Korzhavyi P., Manga V. R., Ohno M., Pisch A., Scott A., Zhang W.: Interface between quantum-mechanical-based approaches, experiments, and CALPHAD methodology, Calphad, 31, 1, 4–27 (2007)
173. Wang C., Fabrichnaya O., Zinkevich M., Du Y., Aldinger F.: Experimental study and thermodynamic modelling of the ZrO2-LaO1.5 system, Calphad, 32, 1, 111–120 (2008)
174. Wang Z. C., Kamphowe T. W., Katz S., Peng J. Q., Seifert H. J., Bill J., Aldinger F.: Effects of polymer thermolysis on composition, structure and high-temperature stability of amorphous silicoboron carbonitride ceramics, Journal of Materials Science Letters, 19, 19, 1701–1704 (2000)
175. Wang C., Zinkevich M., Aldinger F.: Experimental study and thermodynamic assessment of the ZrO2-DyO1.5 system, International Journal of Materials Research, 98, 2, 91–98 (2007)
176. Wang C., Zinkevich M., Aldinger F.: Experimental investigation and thermodynamic modeling of the ZrO2-SmO1.5 system, Journal of the American Ceramic Society, 90, 7, 2210–2219 (2007)
177. Wang L. W., Yanez J., Sigmund W. M., Aldinger F.: Rapid prototype of ceramic parts via stereolithography molds, Industrial Ceramics, 20, 2, 93–94 (2000)
178. Weinmann M., Schuhmacher J., Kummer H., Prinz S., Peng J. Q., Seifert H. J., Christ M., Müller K., Bill J., Aldinger F.: Design of polymeric Si-B-C-N ceramic precursors for application in fiber-reinforced composite materials, Chemistry of Materials, 12, 3, 623–632 (2000)
179. Wetzel K.: Nanostrukturiertes flüssigphasengesintertes Siliziumcarbid, Universität Stuttgart, PhD-Thesis (2007)
180. Witusiewicz V., Arpshofen I., Seifert H. J., Sommer F., Aldinger F.: Enthalpy of mixing of liquid Cu-Ni-Si alloys, Zeitschrift für Metallkunde, 91, 128–142 (2000)
181. Witusiewicz V., Arpshofen I., Seifert H.-J., Sommer F., Aldinger F.: Thermodynamics of liquid and undercooled liquid Al-Ni-Si alloys, Journal of Alloys and Compounds, 305, 157–171 (2000)
182. Wu B., Zinkevich M., Aldinger F., Chu M., Shen J.: Prediction of the ordering behaviours of the orthorhombic phase based on Ti2AlNb alloys by combining thermodynamic model with ab initio calculation, Intermetallics, 16, 1, 42–51 (2008)
183. Wu B., Zinkevich M., Aldinger F., Wen D., Chen L.: Ab initio study on structure and phase transition of A-type and B-type rare earth sesquioxides Ln2O3 (Ln = La-Lu, Y, and Sc) based on density function theory, Journal of Solid State Chemistry, 180, 3280–3287 (2007)
184. Wu B., Zinkevich M., Aldinger F., Zhang W.: Ab initio structural and energetic study of LaMO3 (M = Al, Ga) perovskites, Journal of Physics and Chemistry of Solids, 68, 570–575 (2007)
185. Zhou L., Hoffmann R. C., Zhao Z., Bill J., Aldinger F.: Chemical bath deposition of thin TiO2-anatase films for dielectric applications, Thin Solid Films, 516, 7661–7666 (2008)
186. Zhou L., Rixecker G., Aldinger F.: Electric fatigue in ferroelectric lead zirconate stannate titanate ceramics prepared by spark plasma sintering, Journal of the American Ceramic Society, 89, 12, 3868–3870 (2006)
187. Zhou L., Rixecker G., Aldinger F.: Bipolar electric fatigue in ferroelectric Nb-doped PZST ceramics, Key Engineering Materials, 336–338, 359–362 (2007)
188. Zinkevich M., Aldinger F.: Structures of ceramic materials: thermodynamics and constitution, Ceramics Science and Technology. Vol. 1, Structures, 183–229 (Wiley-VCH 2008)
189. Zinkevich M., Solak N., Nitsche H., Ahrens M., Aldinger F.: Stability and thermodynamic functions of lanthanum nickelates, Journal of Alloys and Compounds, 438, 92–99 (2007)

[F]   *Archived Material*

190. Archived material from the Max-Planck-Institut für Astronomie (Max-Planck-Institut für Astronomie, Königstuhl 17, D-69117 Heidelberg, Germany)
191. Archived material from the Pulvermetallurgischen Laboratorium (Max-Planck-Institut für Metallforschung, PML, Heisenbergstr. 3, D-70569 Stuttgart, Germany). Contributors: FHG/IWM (Fraunhofer-Gesellschaft/Institut für Werkstoffmechanik); Murata Manufacturing Co., Yasu 520–2393, Japan; Pennsylvania State University, University Park, PA 16802–4801, USA; TDK Corporation, 570-2 Matsugashita, Chiba-Prefecture 286-8588, Japan et al.
192. Archived material from the Stuttgart Center for Electron Microscopy (Max-Planck-Institut für Metallforschung, StEM, Heisenbergstr. 3, D-70569 Stuttgart, Germany). Contributors: van Aken P., Phillipp F., Küstner V., Bunjes N., Sigle W., Sung Bo Lee et al.
193. Archived material from the Andersen group of the Max-Planck-Institut für Festkörperforschung (Max-Planck-Institut für Festkörperforschung, Abteilung Andersen, Heisenbergstr. 1, D-70569 Stuttgart, Germany). Contributors: Andersen O. K., Saha-Dasgupta T., Zurek E., Boeri L. et al.
194. Archived material from the Physikalischen Institut, Universität Stuttgart (Universität Stuttgart, Physikalisches Institut, Pfaffenwaldring 57, D-70569 Stuttgart, Germany). Contributors: Schweizer H., Bergmann R. et al.

Index

Acheson process, 347
advanced spin model, 38, 430, 431, 433, 440, 441
aliphatic hydrocarbons, 402
alkoxide process, 380
allomorphism, 104
aluminum nitride, 341
aluminum oxide, 215, 341
anion vacancy model, 122
antiferromagnetism, 230, 234
aromatic hydrocarbons, 402
Ashby–Verall mechanism, 272
atmospheric pressure CVD (APCVD), 398

bacterial anaerobic corrosion, 199
band gap, 53, 280
band structure, 54
Bayer process, 341, 347
BCS model Hamiltonian, 316
BCS theory, 280, 313
beryl, 98
binder, 350
bio-inspired mineralization, 151, 380, 408
biological materials, 427
biomimetic materials, 427
Bloch functions, 56
Bloch theorem, 54, 58
Boltzmann distribution, 127
Boltzmann entropy, 127, 138
bond, 16
– bridge, 16
– covalent, 16
– ionic, 16
– metallic, 16
– van der Waals, 16
bond sensitivity, 326
bond stiffness, 245
bond tetrahedron, 16
bond type, 59
Born–Oppenheimer approximation, 33, 37
Born–Oppenheimer separation, 33, 37, 326
Bragg condition, 54, 55, 58

Bragg diffraction, 55
brain, 218
Bravais lattices, 65
Bravais–Miller indices, 64
Brillouin zone, 54
brittle fracture, 260
Burgers vector, 111, 251
burnout, 357

calcination, 347
canonical ensemble, 127
capacitance, 218
carbon-thermal nitridation, 341
cation–anion radius, 20, 25
cation interstitial model, 122
ceramic magnet, 232
ceramics, 1
– definition, 1
chemical bond, 16, 326
chemical liquid deposition (CLD), 380
chemical potential, 128, 130
chemical properties, 171
chemical vapor deposition (CVD), 341, 380, 398
clock generator, 220
closed pores, 143
closed system, 126
Coble creep, 256
coefficient of compressibility, 211
coherence length, 280
composite materials, 156
compression stress, 242, 248
conductance, 219
conduction band, 49–51
conductivity, 219
conductor, 49
configurational entropy, 127
Cooper pair, 280, 284
coordination geometry, 18, 21
coordination number, 18, 21
cordierite automotive catalytic converters, 150

corrosion, 199
- acids, 206
- carbon, 207
- fused salts, 206
- halogens, 207
- hydrogen, 207
- water vapor, 203
corrosion pit, 204
corundum, 78
Coulomb force, 22
Coulomb model, 24
covalent bond, 29
crack growth, 260
crack resistance, 270
creep, 150, 256
criterion of von Mises, 248
crystal orbital overlap population (COOP), 54
crystal systems, 65
cuprates, 272
Curie temperature, 220, 230
Curie–Weiss law, 231
Curie's law, 231
cyclosilicates, 89

Debye interaction, 47
defect spinel, 78
density of states (DOS), 54
Desulfovibrio, 199
diamagnetic polarization, 230, 231
diamagnetism, 230
diamond, 74, 402, 404
- artifical, 402
- vapor-deposited, 402
diamond films, 402
dielectric constant, 33
dielectrics, 218
diimide process, 341
diode sputtering, 406
direct nitridation, 341
dislocation creep, 256
dislocation line, 111
dry bag press, 354
dry pressing, 354
duplex microstructure, 143

edge dislocation, 111
Einstein field equations, 476, 477
Einstein field theory, 476, 477
Einstein's constant of gravitation, 459, 483
elastic constants, 241
elasticity, 238
electric conductivity, 49
electric flux density, 219

electric potential, 220
electrical conductivity, 328
electrical resistance, 280, 281
electromagnetic properties, 218
electron beam evaporation, 406
electron gas, 55
electron pair, 29, 32
electron scattering, 55
electronegativity, 16
elementary charge, 33
energy bands, 49, 327
energy release rate, 265
enthalpy, 126, 128, 130
entropy, 122, 126, 128, 130
epitaxy, 398
equation of geodetic lines, 476
extrinsic defects, 118
extruding, 355

failure condition, 264, 265
- Griffith, 264, 265
- Irwin, 264, 265
failure probability, 264, 265
faujastite, 103
Fermi energy, 50
Fermi gas, 55
ferrimagnetism, 230
ferroelectricity, 220
ferromagnetism, 230, 234
fiber optic cable, 236, 335
fluorite, 78
Fokker–Planck equation, 15
formation enthalpy, 122, 133, 196, 197
formation entropy, 122, 133
forming, 352
forsterite, 95
fracture, 260
fracture strength, 265
fracture toughness, 150, 264
free enthalpy, 128, 305
freeze casting, 156, 352–354
Frenkel defect, 111, 118
fuel cell, 15, 358, 376
fullerenes, 74
fully stabilized zirconia (FSZ), 104
future materials, 429

gahnite, 215
gallium arsenid (GaAs), 18, 49, 51, 78
GAM, 179
GAP, 179
gas pressure sintering, 151
gas sensor, 225

generalized Einstein field equations, 459–461, 476, 477
generalized Einstein field theory (GEFT), 476, 477, 488
Gibbs entropy, 138
Gibbs free energy, 122
Ginzburg–Landau parameter, 307
Ginzburg–Landau theory, 280
glass phase, 364
gliding planes, 250
gmelinite, 102
goniometer, 55
grain, 143
grain boundaries, 64, 111, 143, 215
grain boundary layer, 149
grain boundary sliding, 272
grain deformations, 248
grain shape, 150
grain size, 150
grand canonical ensemble, 127
graphite, 47, 74, 327
– energy bands, 327
gravitational laser system, 12, 429
green body, 341
gridline, 64
grinding, 365
Grüneisen model, 210, 211
Grüneisen parameter, 211

hard ferrite, 232
hard machining, 365
hardness, 248
heat, 128
heat capacity, 126, 130, 280
heat flow density, 210, 211
heated-filament-assisted CVD (HFCVD), 402
Helmholtz free energy, 128
heteroepitaxy, 398
hexagonal close-packed lattice, 65
high pressure high temperature (HPHT) procedure, 402, 404
high voltage insulator, 225
hip joint endoprosthesis, 1, 8
homoepitaxy, 398
hot gas corrosion, 203, 204
hot isostatic pressing (HIP), 359
hydrocarbon precursor, 402
hydrothermal oxidation, 341

ideal gas law, 15, 208
ignition plug, 225
information, 138
information entropy, 138

infrared radiation, 210
injection molding, 355, 358
insulator, 49, 218
interfacial (surface) tension, 145, 149
internal energy, 128
intrinsic defects, 118
ion beam aided deposition (IBAD), 398
ionic bond, 22
– bond energy, 22
– lattice enthalpy, 27
irreversible deformations, 248
irreversible process, 126
isomorphism, 104
isostatic pressing, 354, 358
isostatic pressure, 243, 248
isotropic solid, 241

jump temperature, 274, 278, 330

Keesom interaction, 47
ketone, 402
Kubaschewski Ansatz, 131, 194

λ sensor, 224
Lamé's parameters, 240
Landau potential, 128
Lanxide process, 372
lapping, 365
lasering, 365
lattice constants, 64
lattice defect, 215
lattice enthalpy, 22, 27
lattice indices, 64
lattice parameter, 55
lattice plane, 64
lattice vector, 64
lattice vibrations, 210, 280, 284
laws of thermodynamics, 128
lead titanate zirconate (PZT), 220, 225
Lenz's rule, 230
line defect, 111
linear fracture mechanics, 264
liquid crystals, 15
liquid-phase epitaxy (LPE), 398
liquid-phase sintering, 364
lithium aluminosilicate (LAS), 216
loading mode, 264
London dispersion interaction, 47
London equations, 15, 280, 306
London penetration depth, 272, 280, 307
lonsdalite, 74, 78
LSGM ceramics, 372, 376
LSGZ ceramics, 376

macrocrack, 116
macropore, 116
Madelung constant, 27
magnesium aluminate spinel, 215
magnesium aluminosilicate ("cordierite"), 216
magnesium silicate, 95
magnetite, 232
magnetization, 231
magnetostriction, 232
martensitic transformation, 157
mass, 341, 347
mass fabrication, 347
Maxwell equations, 15, 305
Maxwell relations, 129
mean free phonon path, 210
mechanical properties, 238
Meißner–Ochsenfeld effect, 272, 280
Meißner phase, 272, 280
melting and casting, 380
metal matrix composite (MMC) structures, 156, 359
metal oxide varistor (MOV), 224
metallic alloy, 156
metallic bond, 49, 55
metalloid, 49
metals, 49
microcanonical ensemble, 127
microcrack, 104, 116, 270
micropore, 116
microporosity, 359
microstructure, 143, 150
microwave-plasma-assisted CVD (MPACVD), 402
Miller indices, 64
mixed dislocation, 111
molar heat capacity, 130
molecular beam epitaxy (MBE), 398
molecular orbital, 29, 40
monocrystal, 64
monotropic transformation, 109
mullite, 215

Nabarro–Herring creep, 256
Navier–Stokes equations, 15
near-net-shape ceramic parts, 365
Néel temperature, 230
Nernst sensors, 224
nesosilicates, 89
neural network, 218
neutron scattering, 55
non-stoichiometric defects, 118
notch sensitivity, 264

Ohm's law, 305
open pores, 143
open system, 126
order parameter relations, 284, 293
oxidation, 199, 200
– oxygen environment, 200

paraelectricity, 220
paramagnetism, 230
partly stabilized zirconia (PSZ), 104
Pauli principle, 280
Pauling rules, 20, 59
Pauling system, 16
Pechini method, 376
perovskite, 78
phase boundaries, 178
phase diagram, 172
– binary, 173
phase equilibria, 171
phase transformation, 64
phase transition, 108
– foreign ions, 108
– irreversibilities, 108
phenakite, 96
phonon, 210, 215, 280
phyllosilicates, 89
physical vapor deposition (PVD), 380, 406
π antibonding orbital, 327
π bonding orbital, 327
piezoelectricity, 220
piezomagnetic effect, 232
planar defect, 111
Planck's constant, 33
plasma-enhanced CVD (PECVD), 398
plasticity, 248
plastification mass, 350
Poinsot ellipsoid, 39
point defect, 111
Poisson's ratio, 240, 241, 245
polarization, 218
– dielectric, 219
polishing, 365
polycrystal, 64
polymorphism, 104
polytypism, 109
pores, 215
porosity, 150
porous printing, 372
post-HIP, 359
potassium chloride, 24
powder, 341
powder fabrication, 341
powder preparation, 352
powder rolling, 358

powder suspension, 350
precursor ceramics, 209
precursor thermolysis, 381
preparation, 157
pressure casting, 356
pressure sintering, 363
principal quantum number, 40
protective coatings, 199–202
– siliceous, 200
pulsed laser deposition, 406
pyrochlor-type phase, 179, 184
pyroelectricity, 220

quantum mechanics, 32
quartz, 60
quartz glass, 60

radius ratio, 20, 22, 59
reaction bonding, 372
reaction enthalpy, 122, 132
reaction entropy, 122, 142
reaction sintering, 372
reciprocal lattice, 54
refraction law, 237
refractive index, 237
remote plasma-enhanced CVD (RPECVD), 398
resin transfer molding (RTM), 156
resistance, 219
resistivity, 219
reversible deformations, 238
reversible process, 126
rigid unit modes (RUMs), 216
roll compaction, 355
rutile, 78

sandwich structure, 89
scanning electron microscope (SEM), 157
Schottky defect, 111, 118
Schrödinger's theory, 54
Schrödinger equation, 33, 40, 54
screw dislocation, 111
secondary phase, 215
seed crystals, 64
self-propagating high-temperature synthesis, 373
self-healing effect, 104, 270
self-evolutional nanostructures, 151
semi-conductor, 49–51, 218, 225
semi-metals, 49
shear modulus, 240
shear stress, 238, 243, 248
silica gel process, 380
silicate, 89

– two-layer, 89
silicate glass, 61
silicon nitride, 341
sinter forging, 359
sintering, 151, 357
sintering equation, 131
slip, 356
slip casting, 352–354
sodalite, 101
sodium discharge lamps, 150
soft ferrite, 232
sol, 380
sol–gel processes, 380
solid electrolyte, 224, 225
solid solution, 122
solid-state sintering, 364
soliton, 111
sorosilicates, 89
specific heat capacity, 128, 130
spin, 38, 51, 280
– antiparallely oriented, 280
spin exchange interaction, 234, 235
spinel, 78
sputtering, 406
stamping, 365
Stern layer, 351
stiffness tensor, 240
stoichiometric defects, 118
stoichiometry, 59
strain tensor, 128, 240
strains, 240
strength, 260
stress, 240
stress intensity factor, 260, 265
stress tensor, 128, 240
structure prototype, 74
substitutionally mixed crystal, 118
substrate, 225
superconducting phase transition, 307
superconductivity, 272, 280
– critical temperature, 272
superconductor, 272
– type I, 272
– type II, 272
superparticles, 284, 294
superplasticity, 272
surface energy, 265

talc, 89, 100
tape casting, 352, 354, 358
tectosilicates, 89
tensile stress, 238, 242, 248, 272
tensor of inertia, 39
tetragonal zirconia polycrystal (TZP), 104

thermal conductivity, 210, 211
– dependencies, 215, 216
thermal entropy, 127
thermal evaporation, 406
thermal expansion, 211
thermal properties, 210
thermal shock, 260
thermistor, 224, 225
thermochemical topologies, 171
thermodynamic potential, 128
topaz, 97
transmission electron microscope (TEM), 157
transverse vibrational modes, 216
triode sputtering, 406
triple junction, 143
twin, 114, 143

uniaxial hot pressing, 359
unit cell, 64

valence band, 49–51
valency, 20, 59
van der Waals bond, 47
van der Waals equation, 15
van der Waals interaction, 47
van der Waals radius, 47
vapor-deposited diamond films, 402
varistor, 224, 225
vector potential, 305
Vegard's slope, 179
Vickers hardness, 248
Vickers indentations, 248

voltage dependent resistor (VDR), 224
volume defects, 111
von Neumann entropy, 138
vortices, 272

wafer coating, 398
wave–particle systems, 29, 33, 111
wavefunction, 29, 53
Weibull distribution, 264, 265
Weibull modulus, 264
Weiss domain, 220, 230
wet bag press, 354
wetting angle of contact, 145
work, 128
wuestite, 124
wurtzite, 78, 104

X-ray diffraction, 55

YAG, 179
YAM, 179
YAP, 179
YBCO, 274, 275
Young's modulus, 240, 265
yttrium iron garnet, 232

zinc blende, 78, 104
zircon, 94
zirconium oxide, 179, 341, 380
zirconium oxide (ZrO_2) polymorphs, 104
zirconium silicate, 94
zone scheme, 53
– reduced, 53